Cube

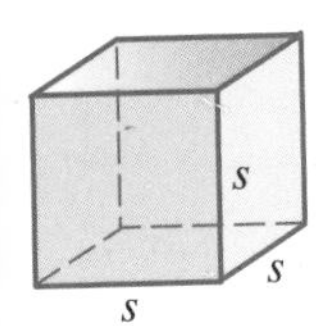

$V = s^3$ $\quad S = 6s^2$

V	volume
S	total surface area
s	length of a side

Rectangular Prism

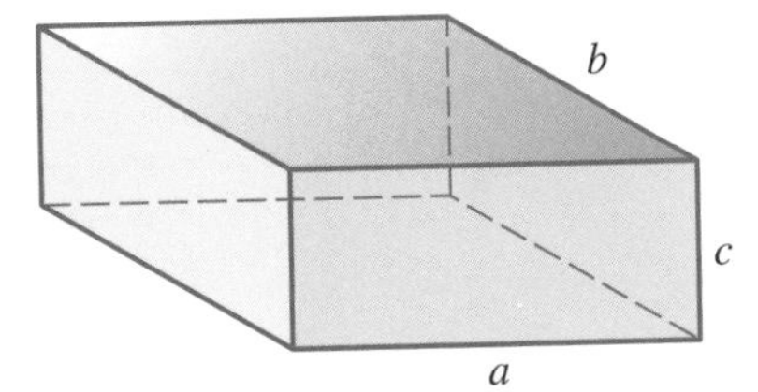

$V = abc$ $\quad S = 2ac + 2ab + 2bc$

V	volume
S	total surface area
a	width
b	length
c	height

Right Circular Cylinder

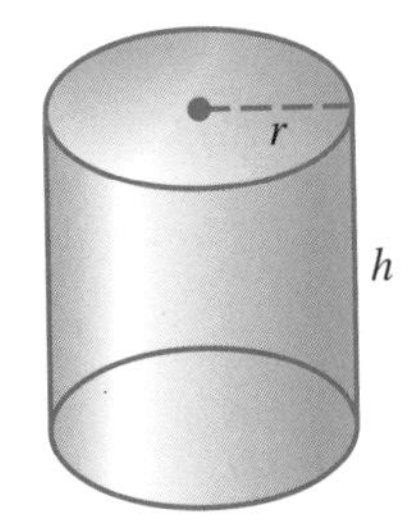

$V = \pi r^2 h$ $\quad S = 2\pi r^2 + 2\pi rh$

V	volume
S	total surface area
r	radius
h	altitude (height)

Sphere

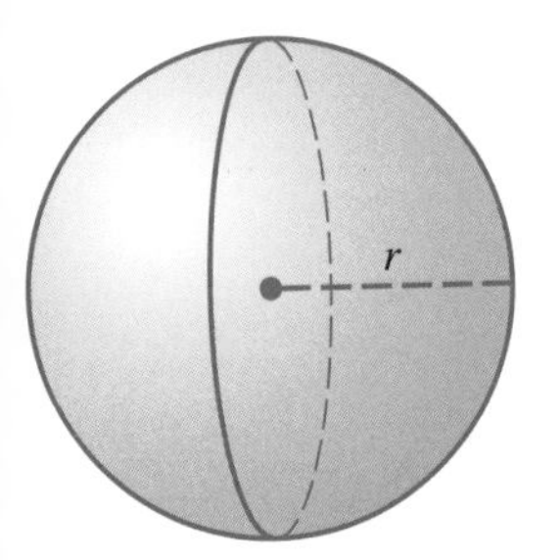

$V = \frac{4}{3}\pi r^3$ $\quad S = 4\pi r^2$

V	volume
S	total surface area
r	radius

Pyramid

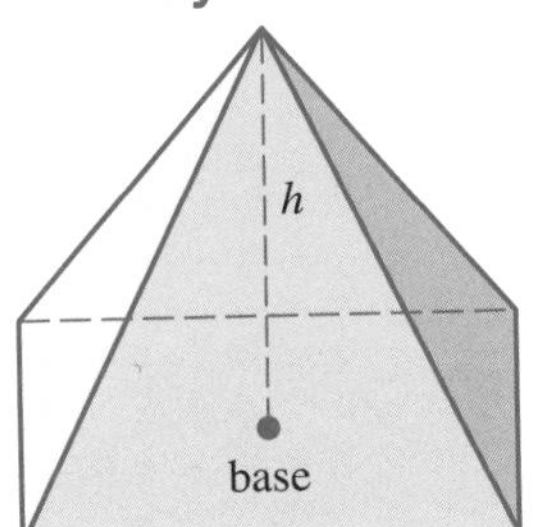

$V = \frac{1}{3}Bh$

V	volume
B	area of base
h	altitude (height)

Right Circular Cone

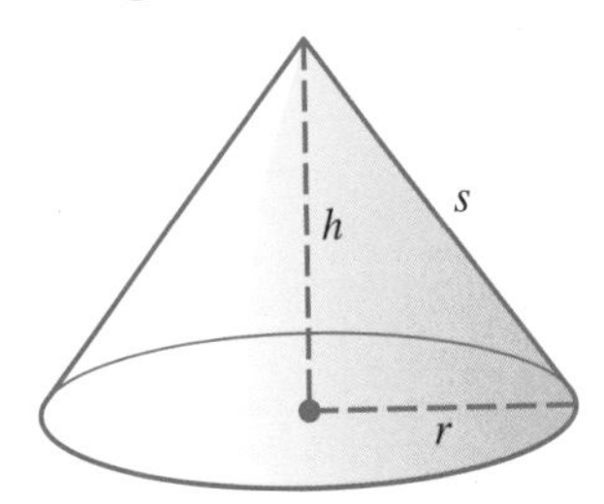

$V = \frac{1}{3}\pi r^2 h$ $\quad S = \pi r^2 + \pi rs$

V	volume
S	total surface area
r	radius
h	altitude (height)
s	slant height

Elementary Algebra

EIGHTH EDITION

Elementary Algebra

Jerome E. Kaufmann

Karen L. Schwitters

Seminole Community College

Australia • Brazil • Japan • Korea • Mexico • Singapore • Spain • United Kingdom • United States

Elementary Algebra, Eighth Edition
Jerome E. Kaufmann, Karen L. Schwitters

Editor: Gary Whalen

Assistant Editor: Rebecca Subity

Editorial Assistant: Katherine Cook and Dianne Muhammad

Technology Project Manager: Sarah Woicicki

Marketing Manager: Greta Kleinert

Marketing Assistant: Brian R. Smith

Marketing Communications Manager: Darlene Amidon-Brent

Project Manager, Editorial Production: Harold P. Humphrey

Art Director: Vernon T. Boes

Print Buyer: Barbara Britton

Permissions Editor: Bob Kauser

Production Service: Susan Graham

Text Designer: John Edeen

Art Editor: Susan Graham

Photo Researcher: Sarah Evertson

Copy Editor: Susan Graham

Illustrator: Network Graphics and G&S Typesetters

Cover Designer: Lisa Henry

Cover Image: Doug Smock/Getty Images

Compositor: G&S Typesetters, Inc.

For product information and technology assistance, contact us at
Cengage Learning Academic Resource Center, 1-800-354-9706
For permission to use material from this text or product, submit all requests online at **www.cengage.com/permissions**
Further permissions questions can be emailed to
permissionrequest@cengage.com

Library of Congress Control Number: 2005938484

ISBN-13: 978-1-4390-4586-2
ISBN-10: 1-4390-4586-0

Brooks/Cole
10 Davis Drive
Belmont, CA 94002-3098
USA

Cengage Learning products are represented in Canada by Nelson Education, Ltd.

To learn more about Brooks/Cole, visit **www.cengage.com/brookscole**

Purchase any of our products at your local college store or at our preferred online store **www.ichapters.com**

Printed in China
5 6 7 8 11 10 09

Contents

Chapter 7 Algebraic Fractions 274

Chapter 8 Coordinate Geometry and Linear Systems 321

Chapter 9 Roots and Radicals 385

Chapter 10 Quadratic Equations 423

Preface

When preparing *Elementary Algebra, Eighth Edition,* we attempted to preserve the features that made the previous editions successful; at the same time we incorporated several improvements suggested by reviewers.

This text was written for those students who have never had an elementary algebra course, and for those who need a review before taking additional mathematics courses. The basic concepts of elementary algebra are presented in a simple, straightforward manner. Concepts are developed through examples, continuously reinforced through additional examples, and then applied in problem-solving situations.

Algebraic ideas are developed in a logical sequence, and in an easy-to-read manner, without excessive vocabulary and formalism. Whenever possible, the algebraic concepts are allowed to develop from their arithmetic counterparts. The following are two specific examples of this development.

- Manipulation with simple algebraic fractions begins early (Sections 2.1 and 2.2) when we review operations with rational numbers.
- Multiplying monomials, without any of the formal vocabulary, is introduced in Section 2.4 when we work with exponents.

A common thread runs throughout the book: namely, **learn a skill;** next, **use the skill to help solve equations;** and then **use the equations to solve application problems.** This thread influenced some of the decisions we made in preparing this text.

- Approximately 550 word problems are scattered throughout this text. (Appendix B contains another 150 word problems.) Every effort was made to start with easy problems, in order to build students' confidence in solving word problems. We offer numerous problem-solving suggestions with special discussions in several sections. We feel that the key to solving word problems is to work with various problem-solving techniques rather than to be overly concerned about whether all the traditional types of problems are being covered.
- Newly acquired skills are used as soon as possible to solve equations and applications. Therefore, the work with solving equations is introduced early—in Chapter 3—and is developed throughout the text. This concept continues through the sections on solving equations in two variables (in Chapter 8).
- Chapter 6 ties together the concepts of factoring, solving equations, and solving applications.

In approximately 700 worked-out examples, we demonstrate a large variety of situations, but we leave some things for students to think about in the problem sets. We also use examples to guide students in organizing their work and to help them decide when they may try a shortcut. The progression from showing all steps to demonstrating a suggested shortcut format is gradual.

As recommended by the American Mathematical Association of Two-Year Colleges, we integrate some geometric concepts into a problem-solving setting, which show the connections among algebra, geometry, and the real world. Approximately 25 examples and 180 problems are designed to review basic geometry ideas. The following sections contain the bulk of the geometry material:

Section 2.5: Linear measurement concepts
Section 3.3: Complementary and supplementary angles; the sum of the measures of the three angles of a triangle equals 180°
Section 4.3: Area and volume formulas
Section 6.3: The Pythagorean theorem
Section 10.1: More on the Pythagorean theorem, including work with isosceles right triangles and 30°–60° right triangles

New in This Edition

- Sections 8.1 and 8.2 have been reorganized so that only linear equations in two variables are graphed in Section 8.1. Added to Section 8.1 are examples and exercises that just cover plotting points on a coordinate system. Graphing higher-order equations has been moved to the further investigations section of the problem set for Section 8.1.
- A focal point of every revision is the problem sets. Some users of the previous editions have suggested that the "very good" problems sets could be made even better by adding a few problems in different places. Based on these suggestions, problems have been added for plotting points in a coordinate system; more equations with solutions of null set or all real numbers; and more factoring problems that contain a common factor.
- For the annotated instructor's edition, problems in the problem set that are in the *iLrn* test question bank are identified by an underline. The *iLrn* test question bank includes 20 problems from each problem set.

Other Special Features

- **Thoughts into Words** are included in every problem set except the review exercises. These problems are designed to give students an opportunity to express in written form their thoughts about various mathematical ideas. For examples of Thoughts into Words, see Problem Sets 3.2, 3.4, 4.1, and 6.2.
- **Further Investigations** are included in many problem sets. These problems are "extras" but they allow some students to pursue more complicated ideas. Many of these investigations lend themselves to small group work. For examples of Further Investigations, see Problem Sets 4.3, 5.3, 6.2, 6.3, and 8.1.

- A **Chapter Test** appears at the end of each chapter. Along with Chapter Review Problem Sets, these practice tests provide the students with ample opportunity to prepare for "real" tests.
- **Cumulative Review Problem Sets** are included at the end of Chapters 2, 4, 5, 7, and 9. Working these problems should help the students review some basic skills introduced earlier in the text and tie together various mathematical concepts.
- **Interval notation** is introduced in Section 3.5, and it is used when appropriate after that point. Solution sets for inequalities are given in both set-builder notation and in interval notation. Instructors have the option of selecting their preferred form of the answer. In this edition we are also using the interval notation symbolism (parentheses and brackets) on the number line graphs for inequalities.
- **Problem sets** have been constructed on an even-odd basis; that is, all variations of skill-development exercises are contained in both the even- and the odd-numbered problems. It is important to note that problem sets are a focal point of every revision. We constantly add, subtract, and reword problems based on the suggestions of users of the previous editions.
- All **answers** for Chapter Review Problem Sets, Chapter Tests, Cumulative Review Problem Sets, and Appendix B Problems appear in the back of the text, along with answers to the odd-numbered problems.
- Please note the very pleasing **design features,** including the functional use of color. The open format makes for a continuous and easy reading flow of material instead of working through a maze of flags, caution symbols, reminder symbols, etc.

Additional Comments About Some of the Chapters

- Chapter 3 presents an early introduction to an important part of elementary algebra. Problem solving and the solving of equations and inequalities are introduced early so they can be used as unifying themes throughout the text.
- Chapter 4 builds upon Chapter 3 by expanding on both solving equations and problem solving. Many geometric formulas and relationships are reviewed in a problem solving setting. Consumer oriented problems are a focal point of this chapter.
- Chapter 6 clearly illustrates the theme (learn a skill → use the skill to solve equations → use equations to solve problems) mentioned earlier in the preface. In this chapter we develop some factoring techniques and skills that can be used to solve equations. Then the equations are used to expand our problem solving capabilities.
- Chapter 8 introduces some basic concepts of coordinate geometry. Some graphing ideas are presented with an emphasis on graphing linear equations and inequalities of two variables. The last three sections are devoted to solving systems of two linear equations in two variables.
- Chapter 11 remains an "extra" chapter, in which most of the topics are a continuation of topics studied earlier in the text. For example, Section 11.1 (Equations and Inequalities Involving Absolute Value) could follow, as it does in our *Intermediate Algebra* text, after Section 3.6 (Inequalities, Compound Inequalities,

and Problem Solving). Certainly this chapter could be very beneficial for students who plan to take additional mathematics courses.

Ancillaries

For the Instructor

Annotated Instructor's Edition. This special version of the complete student text contains a Resource Integration Guide and answers printed next to all respective exercises. Graphs, tables, and other answers appear in a special answer section at the back of the text.

Test Bank. The *Test Bank* includes eight tests per chapter as well as three final exams. The tests are made up of a combination of multiple-choice, free-response, true/false, and fill-in-the-blank questions.

Complete Solutions Manual. The *Complete Solutions Manual* provides worked-out solutions to all of the problems in the text.

iLrn™ Instructor Version. Providing instructors and students with unsurpassed control, variety, and all-in-one utility, *iLrn™* is a powerful and fully integrated teaching and learning system. *iLrn* ties together five fundamental learning activities: diagnostics, tutorials, homework, quizzing, and testing. Easy to use, *iLrn* offers instructors complete control when creating assessments in which they can draw from the wealth of exercises provided or create their own questions. *iLrn* features a great variety of problem types—allowing instructors to assess the way they teach. A real timesaver for instructors, *iLrn* offers automatic grading of homework, quizzes, and tests, with results flowing directly into the gradebook. The auto-enrollment feature also saves time with course setup as students self-enroll into the course gradebook. *iLrn* provides seamless integration with Blackboard™ and WebCT™.

Text-Specific Videotapes. These text-specific videotape sets, available at no charge to qualified adopters of the text, feature 10- to 20-minute problem-solving lessons that cover each section of every chapter.

For the Student

Student Solutions Manual. The *Student Solutions Manual* provides worked-out solutions to the odd-numbered problems, and all chapter review, chapter test, and cumulative review problems in the text.

Website (www.cengage.com/math). Instructors and students have access to a variety of teaching and learning resources. This website features everything from book-specific resources to newsgroups.

iLrn™ Tutorial Student Version. Featuring a variety of approaches that connect with all types of learners, *iLrn™ Tutorial* offers text-specific tutorials that require no setup by instructors. Students can begin exploring active examples from the text by using the access code packaged free with a new book. *iLrn Tutorial* supports students with explanations from the text, examples, step-by-step problem-solving help, unlimited practice, and chapter-by-chapter video lessons. With this self-paced system, students can even check their comprehension along the way by taking quizzes and receiving feedback. If they still are having trouble, students can easily access *vMentor*™ for online help from a live math instructor. Students can ask any question and get personalized help through the interactive whiteboard and by using their computer microphones to speak with the instructor. While designed for self-study, instructors can also assign the individual tutorial exercises.

Interactive Video Skillbuilder CD-ROM. Think of it as portable instructor office hours. The *Interactive Video Skillbuilder CD-ROM* contains video instruction covering each chapter of the text. The problems worked during each video lesson are shown first so that students can try working them before watching the solution. To help students evaluate their progress, each section contains a 10-question web quiz (the results of which can be e-mailed to the instructor) and each chapter contains a chapter test, with the answer to each problem on each test. A new learning tool on this CD-ROM is a graphing calculator tutorial for precalculus and college algebra, featuring examples, exercises, and video tutorials. Also new, English/Spanish closed caption translations can be selected to display along with the video instruction. This CD-ROM also features *MathCue* tutorial and testing software. Keyed to the text, *MathCue* offers these components:

- *MathCue Skill Builder*—Presents problems to solve, evaluates answers, and tutors students by displaying complete solutions with step-by-step explanations.
- *MathCue Quiz*—Allows students to generate large numbers of quiz problems keyed to problem types from each section of the book.
- *MathCue Chapter Test*—Also provides large numbers of problems keyed to problem types from each chapter.
- *MathCue Solution Finder*—This unique tool allows students to enter their own basic problems and receive step-by-step help as if they were working with a tutor.
- Score reports for any *MathCue* session can be printed and handed in for credit or extra credit.
- Print or e-mail score reports—Score reports for any *MathCue* session can be printed or sent to instructors via *MathCue's* secure e-mail score system.

vMentor™ *Live, Online Tutoring.* Packaged free with every text. Accessed seamlessly through *iLrn Tutorial, vMentor* provides tutorial help that can substantially improve student performance, increase test scores, and enhance technical aptitude. Students have access, via the web, to highly qualified tutors with thorough knowledge of our textbooks. When students get stuck on a particular problem or concept, they need only log on to *vMentor*, where they can talk (using their own computer microphones) to *vMentor* tutors who will skillfully guide them through the problem using the interactive whiteboard for illustration. Brooks/Cole also offers *Elluminate Live!*, an online virtual classroom environment that is customizable and easy to use. *Elluminate Live!* keeps students engaged with full two-way audio, instant messaging, and an interactive whiteboard—all in one, intuitive, graphical interface. For information about obtaining an *Elluminate Live!* site license, instructors may contact their Cengage Learning representative. *For proprietary, college, and university adopters only. For additional information, instructors may consult their Cengage Learning representative.*

Explorations in Beginning and Intermediate Algebra Using the TI-82/83/83-Plus/85/86 Graphing Calculator, Third Edition (0-534-40644-0)
Deborah J. Cochener and Bonnie M. Hodge, both of Austin Peay State University
This user-friendly workbook improves students' understanding and their retention of algebra concepts through a series of activities and guided explorations using the graphing calculator. An ideal supplement for any beginning or intermediate algebra course, *Explorations in Beginning and Intermediate Algebra, Third Edition* is an ideal tool for integrating technology without sacrificing course content. By clearly and succinctly teaching keystrokes, class time is devoted to investigations instead of how to use a graphing calculator.

The Math Student's Guide to the TI-83 Graphing Calculator (0-534-37802-1)
The Math Student's Guide to the TI-86 Graphing Calculator (0-534-37801-3)
The Math Student's Guide to the TI-83 Plus Graphing Calculator (0-534-42021-4)
The Math Student's Guide to the TI-89 Graphing Calculator (0-534-42022-2)
Trish Cabral of Butte College
These videos are designed for students who are new to the graphing calculator or for those who would like to brush up on their skills. Each instructional graphing calculator videotape covers basic calculations, the custom menu, graphing, advanced graphing, matrix operations, trigonometry, parametric equations, polar coordinates, calculus, Statistics I and one-variable data, and Statistics II with linear regression. These wonderful tools are each 105 minutes in length and cover all of the important functions of a graphing calculator.

Mastering Mathematics: How to Be a Great Math Student, Third Edition (0-534-34947-1)
Richard Manning Smith, Bryant College
Providing solid tips for every stage of study, *Mastering Mathematics* stresses the importance of a positive attitude and gives students the tools to succeed in their math course.

Activities for Beginning and Intermediate Algebra, Second Edition
Instructor Edition (0-534-99874-7); Student Edition (0-534-99873-9)
Debbie Garrison, Judy Jones, and Jolene Rhodes, all of Valencia Community College
Designed as a stand-alone supplement for any beginning or intermediate algebra text, *Activities in Beginning and Intermediate Algebra* is a collection of activities written to incorporate the recommendations from the NCTM and from AMATYC's Crossroads. Activities can be used during class or in a laboratory setting to introduce, teach, or reinforce a topic.

Conquering Math Anxiety: A Self-Help Workbook, Second Edition (0-534-38634-2)
Cynthia Arem, Pima Community College
A comprehensive workbook that provides a variety of exercises and worksheets along with detailed explanations of methods to help "math-anxious" students deal with and overcome math fears. This edition now comes with a free relaxation CD-ROM and a detailed list of Internet resources.

Active Arithmetic and Algebra: Activities for Prealgebra and Beginning Algebra (0-534-36771-2)
Judy Jones, Valencia Community College
This activities manual includes a variety of approaches to learning mathematical concepts. Sixteen activities, including puzzles, games, data collection, graphing, and writing activities are included.

Math Facts: Survival Guide to Basic Mathematics, Second Edition (0-534-94734-4)
Algebra Facts: Survival Guide to Basic Algebra (0-534-19986-0)
Theodore John Szymanski, Tompkins-Cortland Community College
This booklet gives easy access to the most crucial concepts and formulas in algebra. Although it is bound, this booklet is structured to work like flash cards.

Acknowledgments

We would like to take this opportunity to thank the following people who served as reviewers for the new editions of this series of texts:

Yusuf Abdl
Rutgers University

Lynda Fish
St. Louis Community College at Forest Park

Cindy Fleck
Wright State University

James Hodge
Mountain State University

Barbara Laubenthal
University of North Alabama

Karolyn Morgan
University of Montevallo

Jayne Prude
University of North Alabama

Renee Quick
Wallace State Community College

We would like to express our sincere gratitude to the staff of Brooks/Cole, especially Gary Whalen, for his continuous cooperation and assistance throughout this project; and to Susan Graham and Hal Humphrey, who carry out the many details of production. Finally, very special thanks are due to Arlene Kaufmann, who spends numerous hours reading page proofs.

Jerome E. Kaufmann
Karen L. Schwitters

Some Basic Concepts of Arithmetic and Algebra

Golfers are familiar with positive and negative integers.

Karla started 2006 with \$500 in her savings account, and she planned to save an additional \$15 per month for all of 2006. Without considering any accumulated interest, the numerical expression $500 + 12(15)$ represents the amount in her savings account at the end of 2006.

The numbers $+2, -1, -3, +1$, and -4 represent Woody's scores relative to par for five rounds of golf. The numerical expression $2 + (-1) + (-3) + 1 + (-4)$ can be used to determine how Woody stands relative to par at the end of the five rounds.

The temperature at 4 A.M. was -14°F. By noon the temperature had increased by 23°F. The numerical expression $-14 + 23$ can be used to determine the temperature at noon.

In the first two chapters of this text the concept of a **numerical expression** is used as a basis for reviewing addition, subtraction, multiplication, and division of various kinds of numbers. Then the concept of a **variable** allows us to move from numerical expressions to **algebraic expressions**; that is, to start the transition from arithmetic to algebra. Keep in mind that algebra is simply a generalized approach to arithmetic. Many algebraic concepts are extensions of arithmetic ideas; your knowledge of arithmetic will help you with your study of algebra.

1.1 Numerical and Algebraic Expressions

In arithmetic, we use symbols such as 4, 8, 17, and π to represent numbers. We indicate the basic operations of addition, subtraction, multiplication, and division by the symbols $+$, $-$, $\cdot$, and $\div$, respectively. Thus we can formulate specific **numerical expressions**. For example, we can write the indicated sum of eight and four as $8 + 4$.

In algebra, **variables** allow us to generalize. By letting x and y represent *any* number, we can use the expression $x + y$ to represent the indicated sum of *any two* numbers. The x and y in such an expression are called variables and the phrase $x + y$ is called an **algebraic expression**. We commonly use letters of the alphabet such as x, y, z, and w as variables; the key idea is that they represent numbers. Our review of various operations and properties pertaining to numbers establishes the foundation for our study of algebra.

Many of the notational agreements made in arithmetic are extended to algebra with a few slight modifications. The following chart summarizes these notational agreements pertaining to the four basic operations. Notice the variety of ways to write a product by including parentheses to indicate multiplication. Actually the ab form is the simplest and probably the most used form; expressions such as abc, $6x$, and $7xyz$ all indicate multiplication. Also note the various forms for indicating division; the fractional form, $\frac{c}{d}$, is usually used in algebra although the other forms do serve a purpose at times.

Operation	Arithmetic	Algebra	Vocabulary
Addition	$4 + 6$	$x + y$	The sum of x and y
Subtraction	$7 - 2$	$w - z$	The difference of w and z
Multiplication	$9 \cdot 8$	$a \cdot b$, $a(b)$, $(a)b$, $(a)(b)$, or ab	The product of a and b
Division	$8 \div 2$, $\frac{8}{2}$, $2\overline{)8}$	$c \div d$, $\frac{c}{d}$, or $d\overline{)c}$	The quotient of c and d

As we review arithmetic ideas and introduce algebraic concepts, it is important to include some of the basic vocabulary and symbolism associated with sets. A **set** is a collection of objects, and the objects are called **elements** or **members** of the set. In arithmetic and algebra the elements of a set are often numbers. To communicate about sets, we use set braces, { }, to enclose the elements (or a description of the elements) and we use capital letters to name sets. For example, we can represent a set A, which consists of the vowels of the alphabet, as

$A = \{\text{Vowels of the alphabet}\}$ Word description, or

$A = \{a, e, i, o, u\}$ List or roster description

We can modify the listing approach if the number of elements is large. For example, all of the letters of the alphabet can be listed as

$\{a, b, c, \ldots, z\}$

We begin by simply writing enough elements to establish a pattern, then the three dots indicate that the set continues in that pattern. The final entry indicates the last element of the pattern. If we write

$\{1, 2, 3, \ldots\}$

the set begins with the counting numbers 1, 2, and 3. The three dots indicate that it continues in a like manner forever; there is no last element. A set that consists of no elements is called the **null set** (written $\varnothing$).

Two sets are said to be *equal* if they contain exactly the same elements. For example,

$\{1, 2, 3\} = \{2, 1, 3\}$

because both sets contain the same elements; the order in which the elements are written doesn't matter. The slash mark through the equality symbol denotes *not equal to*. Thus if $A = \{1, 2, 3\}$ and $B = \{1, 2, 3, 4\}$, we can write $A \neq B$, which we read as "set A is not equal to set B."

Simplifying Numerical Expressions

Now let's simplify some numerical expressions that involve the set of **whole numbers**, that is, the set $\{0, 1, 2, 3, \ldots\}$.

EXAMPLE 1

Simplify $8 + 7 - 4 + 12 - 7 + 14$.

Solution

The additions and subtractions should be performed from left to right in the order that they appear. Thus, $8 + 7 - 4 + 12 - 7 + 14$ simplifies to 30. ■

EXAMPLE 2

Simplify $7(9 + 5)$.

Solution

The parentheses indicate the product of 7 and the quantity $9 + 5$. Perform the addition inside the parentheses first and then multiply; $7(9 + 5)$ thus simplifies to $7(14)$, which becomes 98. ■

EXAMPLE 3

Simplify $(7 + 8) \div (4 - 1)$.

Solution

First, we perform the operations inside the parentheses; $(7 + 8) \div (4 - 1)$ thus becomes $15 \div 3$, which is 5. ■

We frequently express a problem like Example 3 in the form $\frac{7+8}{4-1}$. We don't need parentheses in this case because the fraction bar indicates that the sum of 7 and 8 is to be divided by the difference, $4 - 1$. A problem may, however, contain parentheses and fraction bars, as the next example illustrates.

EXAMPLE 4

Simplify $\frac{(4+2)(7-1)}{9} + \frac{4}{7-3}$.

Solution

First, simplify above and below the fraction bars, and then proceed to evaluate as follows.

$$\frac{(4+2)(7-1)}{9} + \frac{4}{7-3} = \frac{(6)(6)}{9} + \frac{4}{4}$$

$$= \frac{36}{9} + 1 = 4 + 1 = 5$$

■

EXAMPLE 5

Simplify $7 \cdot 9 + 5$.

Solution

If there are no parentheses to indicate otherwise, multiplication takes precedence over addition. First perform the multiplication, and then do the addition; $7 \cdot 9 + 5$ therefore simplifies to $63 + 5$, which is 68. ■

Remark: Compare Example 2 to Example 5, and note the difference in meaning.

EXAMPLE 6

Simplify $8 + 4 \cdot 3 - 14 \div 2$.

Solution

The multiplication and division should be done first in the order that they appear, from left to right. Thus $8 + 4 \cdot 3 - 14 \div 2$ simplifies to $8 + 12 - 7$. We perform the addition and subtraction in the order that they appear, which simplifies $8 + 12 - 7$ to 13. ■

EXAMPLE 7

Simplify $8 \cdot 5 \div 4 + 7 \cdot 3 - 32 \div 8 + 9 \div 3 \cdot 2$.

Solution

When we perform the multiplications and divisions first in the order that they appear and then do the additions and subtractions, our work takes on the following format.

$$8 \cdot 5 \div 4 + 7 \cdot 3 - 32 \div 8 + 9 \div 3 \cdot 2 = 10 + 21 - 4 + 6 = 33$$

■

EXAMPLE 8

Simplify $5 + 6[2(3 + 9)]$.

Solution

We use brackets for the same purpose as parentheses. In such a problem we need to simplify *from the inside out*; perform the operations inside the innermost parentheses first.

$$\begin{aligned} 5 + 6[2(3 + 9)] &= 5 + 6[2(12)] \\ &= 5 + 6[24] \\ &= 5 + 144 \\ &= 149 \end{aligned}$$

■

Let's now summarize the ideas presented in the previous examples regarding **simplifying numerical expressions**. When simplifying a numerical expression, use the following order of operations.

Order of Operations

1. Perform the operations inside the symbols of inclusion (parentheses and brackets) and above and below each fraction bar. Start with the innermost inclusion symbol.
2. Perform all multiplications and divisions in the order that they appear from left to right.
3. Perform all additions and subtractions in the order that they appear from left to right.

■ Evaluating Algebraic Expressions

We can use the concept of a variable to generalize from numerical expressions to algebraic expressions. Each of the following is an example of an algebraic expression.

$$3x + 2y, \qquad 5a - 2b + c, \qquad 7(w + z)$$

$$\frac{5d + 3e}{2c - d}, \qquad 2xy + 5yz, \qquad (x + y)(x - y)$$

An algebraic expression takes on a numerical value whenever each variable in the expression is replaced by a specific number. For example, if x is replaced by 9 and z by 4, the algebraic expression $x - z$ becomes the numerical expression $9 - 4$, which simplifies to 5. We say that $x - z$ **has a value of 5** when x equals 9 and z equals 4. The value of $x - z$, when x equals 25 and z equals 12, is 13. The general algebraic expression $x - z$ has a specific value each time x and z are replaced by numbers.

Consider the following examples, which illustrate the process of finding a value of an algebraic expression. We call this process **evaluating algebraic expressions**.

EXAMPLE 9

Find the value of $3x + 2y$ when x is replaced by 5 and y by 17.

Solution

The following format is convenient for such problems.

$$\begin{aligned} 3x + 2y &= 3(5) + 2(17) \quad \text{when } x = 5 \text{ and } y = 17 \\ &= 15 + 34 \\ &= 49 \end{aligned}$$

■

Note that in Example 9, for the algebraic expression, $3x + 2y$, the multiplications "3 times x" and "2 times y" are implied without the use of parentheses. Substituting the numbers switches the algebraic expression to a numerical expression, and then parentheses are used to indicate the multiplication.

EXAMPLE 10

Find the value of $12a - 3b$ when $a = 5$ and $b = 9$.

Solution

$$\begin{aligned} 12a - 3b &= 12(5) - 3(9) \quad \text{when } a = 5 \text{ and } b = 9 \\ &= 60 - 27 \\ &= 33 \end{aligned}$$

■

EXAMPLE 11

Evaluate $4xy + 2xz - 3yz$ when $x = 8$, $y = 6$, and $z = 2$.

Solution

$$\begin{aligned} 4xy + 2xz - 3yz &= 4(8)(6) + 2(8)(2) - 3(6)(2) \quad \text{when } x = 8, y = 6, \text{ and } z = 2 \\ &= 192 + 32 - 36 \\ &= 188 \end{aligned}$$

■

EXAMPLE 12 Evaluate $\frac{5c + d}{3c - d}$ for $c = 12$ and $d = 4$.

Solution

$$\frac{5c + d}{3c - d} = \frac{5(12) + 4}{3(12) - 4} \quad \text{for } c = 12 \text{ and } d = 4$$

$$= \frac{60 + 4}{36 - 4} = \frac{64}{32} = 2$$

■

EXAMPLE 13 Evaluate $(2x + 5y)(3x - 2y)$ when $x = 6$ and $y = 3$.

Solution

$$\begin{aligned}(2x + 5y)(3x - 2y) &= (2 \cdot 6 + 5 \cdot 3)(3 \cdot 6 - 2 \cdot 3) \quad \text{when } x = 6 \text{ and } y = 3 \\ &= (12 + 15)(18 - 6) \\ &= (27)(12) \\ &= 324\end{aligned}$$

■

Problem Set 1.1

For Problems 1–34, simplify each numerical expression.

1. $9 + 14 - 7$

2. $32 - 14 + 6$

3. $7(14 - 9)$

4. $8(6 + 12)$

5. $16 + 5 \cdot 7$

6. $18 - 3(5)$

7. $4(12 + 9) - 3(8 - 4)$

8. $7(13 - 4) - 2(19 - 11)$

9. $4(7) + 6(9)$

10. $8(7) - 4(8)$

11. $6 \cdot 7 + 5 \cdot 8 - 3 \cdot 9$

12. $8(13) - 4(9) + 2(7)$

13. $(6 + 9)(8 - 4)$

14. $(15 - 6)(13 - 4)$

15. $6 + 4[3(9 - 4)]$

16. $92 - 3[2(5 - 2)]$

17. $16 \div 8 \cdot 4 + 36 \div 4 \cdot 2$

18. $7 \cdot 8 \div 4 - 72 \div 12$

19. $\frac{8 + 12}{4} - \frac{9 + 15}{8}$

20. $\frac{19 - 7}{6} + \frac{38 - 14}{3}$

21. $56 - [3(9 - 6)]$

22. $17 + 2[3(4 - 2)]$

23. $7 \cdot 4 \cdot 2 \div 8 + 14$

24. $14 \div 7 \cdot 8 - 35 \div 7 \cdot 2$

25. $32 \div 8 \cdot 2 + 24 \div 6 - 1$

26. $48 \div 12 + 7 \cdot 2 \div 2 - 1$

27. $4 \cdot 9 \div 12 + 18 \div 2 + 3$

28. $5 \cdot 8 \div 4 - 8 \div 4 \cdot 3 + 6$

29. $\frac{6(8 - 3)}{3} + \frac{12(7 - 4)}{9}$

30. $\frac{3(17 - 9)}{4} + \frac{9(16 - 7)}{3}$

31. $83 - \frac{4(12 - 7)}{5}$

32. $78 - \frac{6(21 - 9)}{4}$

33. $\frac{4 \cdot 6 + 5 \cdot 3}{7 + 2 \cdot 3} + \frac{7 \cdot 9 + 6 \cdot 5}{3 \cdot 5 + 8 \cdot 2}$

34. $\frac{7 \cdot 8 + 4}{5 \cdot 8 - 10} + \frac{9 \cdot 6 - 4}{6 \cdot 5 - 20}$

For Problems 35–54, evaluate each algebraic expression for the given values of the variables.

35. $7x + 4y$ for $x = 6$ and $y = 8$

36. $8x + 6y$ for $x = 9$ and $y = 5$

37. $16a - 9b$ for $a = 3$ and $b = 4$

38. $14a - 5b$ for $a = 7$ and $b = 9$

39. $4x + 7y + 3xy$ for $x = 4$ and $y = 9$

40. $x + 8y + 5xy$ for $x = 12$ and $y = 3$

41. $14xz + 6xy - 4yz$ for $x = 8$, $y = 5$, and $z = 7$

42. $9xy - 4xz + 3yz$ for $x = 7$, $y = 3$, and $z = 2$

43. $\frac{54}{n} + \frac{n}{3}$ for $n = 9$

44. $\frac{n}{4} + \frac{60}{n} - \frac{n}{6}$ for $n = 12$

45. $\frac{y + 16}{6} + \frac{50 - y}{3}$ for $y = 8$

46. $\frac{w + 57}{9} + \frac{90 - w}{7}$ for $w = 6$

47. $(x + y)(x - y)$ for $x = 8$ and $y = 3$

48. $(x + 2y)(2x - y)$ for $x = 7$ and $y = 4$

49. $(5x - 2y)(3x + 4y)$ for $x = 3$ and $y = 6$

50. $(3a + b)(7a - 2b)$ for $a = 5$ and $b = 7$

51. $6 + 3[2(x + 4)]$ for $x = 7$

52. $9 + 4[3(x + 3)]$ for $x = 6$

53. $81 - 2[5(n + 4)]$ for $n = 3$

54. $78 - 3[4(n - 2)]$ for $n = 4$

For Problems 55–60, find the value of $\frac{bh}{2}$ for each set of values for the variables b and h.

55. $b = 8$ and $h = 12$

56. $b = 6$ and $h = 14$

57. $b = 7$ and $h = 6$

58. $b = 9$ and $h = 4$

59. $b = 16$ and $h = 5$

60. $b = 18$ and $h = 13$

For Problems 61–66, find the value of $\frac{h(b_1 + b_2)}{2}$ for each set of values for the variables h, b_1, and b_2. (Subscripts are used to indicate that b_1 and b_2 are different variables.)

61. $h = 17$, $b_1 = 14$, and $b_2 = 6$

62. $h = 9$, $b_1 = 12$, and $b_2 = 16$

63. $h = 8$, $b_1 = 17$, and $b_2 = 24$

64. $h = 12$, $b_1 = 14$, and $b_2 = 5$

65. $h = 18$, $b_1 = 6$, and $b_2 = 11$

66. $h = 14$, $b_1 = 9$, and $b_2 = 7$

67. You should be able to do calculations like those in Problems 1–34 *with* and *without* a calculator. Be sure that you can do Problems 1–34 *with* your calculator, and make use of the parentheses key when appropriate.

THOUGHTS INTO WORDS

68. Explain the difference between a numerical expression and an algebraic expression.

69. Your friend keeps getting an answer of 45 when simplifying $3 + 2(9)$. What mistake is he making and how would you help him?

FURTHER INVESTIGATIONS

Grouping symbols can affect the order in which the arithmetic operations are performed. For the following problems, insert parentheses so that the expression is equal to the given value.

70. Insert parentheses so that $36 + 12 \div 3 + 3 + 6 \cdot 2$ is equal to 20.

71. Insert parentheses so that $36 + 12 \div 3 + 3 + 6 \cdot 2$ is equal to 50.

72. Insert parentheses so that $36 + 12 \div 3 + 3 + 6 \cdot 2$ is equal to 38.

73. Insert parentheses so that $36 + 12 \div 3 + 3 + 6 \cdot 2$ is equal to 55.

1.2 Prime and Composite Numbers

Occasionally, terms in mathematics are given a special meaning in the discussion of a particular topic. Such is the case with the term *divides* as it is used in this section. We say that 6 *divides* 18 because 6 times the whole number 3 produces 18; but 6 *does not divide* 19 because there is no whole number such that 6 times the number produces 19. Likewise, 5 *divides* 35 because 5 times the whole number 7 produces 35; 5 *does not divide* 42 because there is no whole number such that 5 times the number produces 42. We present the following general definition.

Definition 1.1

Given that a and b are whole numbers, with a not equal to zero, *a divides b* if and only if there exists a whole number k such that $a \cdot k = b$.

Remark: Notice the use of variables, a, b, and k, in the statement of a *general* definition. Also note that the definition merely generalizes the concept of *divides*, which was introduced in the specific examples prior to the definition.

The following statements further clarify Definition 1.1. Pay special attention to the italicized words because they indicate some of the terminology used for this topic.

1. 8 *divides* 56 because $8 \cdot 7 = 56$.
2. 7 *does not divide* 38 because there is no whole number, k, such that $7 \cdot k = 38$.
3. 3 is a *factor* of 27 because $3 \cdot 9 = 27$.
4. 4 is *not a factor* of 38 because there is no whole number, k, such that $4 \cdot k = 38$.
5. 35 is a *multiple* of 5 because $5 \cdot 7 = 35$.
6. 29 is *not a multiple* of 7 because there is no whole number, k, such that $7 \cdot k = 29$.

We use the *factor* terminology extensively. We say that 7 and 8 are factors of 56 because $7 \cdot 8 = 56$; 4 and 14 are also factors of 56 because $4 \cdot 14 = 56$. The factors of a number are also divisors of the number.

Now consider two special kinds of whole numbers called **prime numbers** and **composite numbers** according to the following definition.

Definition 1.2

A **prime number** is a whole number, greater than 1, that has no factors (divisors) other than itself and 1. Whole numbers, greater than 1, that are not prime numbers are called **composite numbers**.

The prime numbers less than 50 are 2, 3, 5, 7, 11, 13, 17, 19, 23, 29, 31, 37, 41, 43, and 47. Notice that each of these has no factors other than itself and 1. The set of prime numbers is an infinite set; that is, the prime numbers go on forever, and there is no *largest* prime number.

We can express every composite number as the indicated product of prime numbers. Consider the following examples.

$$4 = 2 \cdot 2, \qquad 6 = 2 \cdot 3, \qquad 8 = 2 \cdot 2 \cdot 2, \qquad 10 = 2 \cdot 5, \qquad 12 = 2 \cdot 2 \cdot 3$$

In each case we expressed a composite number as the indicated product of prime numbers. The indicated product form is sometimes called the **prime factored form** of the number.

There are various procedures to find the prime factors of a given composite number. For our purposes, the simplest technique is to factor the given composite number into any two easily recognized factors and then to continue to factor each of these until we obtain only prime factors. Consider these examples.

$$18 = 2 \cdot 9 = 2 \cdot 3 \cdot 3 \qquad 27 = 3 \cdot 9 = 3 \cdot 3 \cdot 3$$

$$24 = 4 \cdot 6 = 2 \cdot 2 \cdot 2 \cdot 3 \qquad 150 = 10 \cdot 15 = 2 \cdot 5 \cdot 3 \cdot 5$$

It does not matter which two factors we choose first. For example, we might start by expressing 18 as $3 \cdot 6$ and then factor 6 into $2 \cdot 3$, which produces a final result of $18 = 3 \cdot 2 \cdot 3$. Either way, 18 contains two prime factors of 3 and one prime factor of 2. The order in which we write the prime factors is not important.

Greatest Common Factor

We can use the prime factorization form of two composite numbers to conveniently find their **greatest common factor**. Consider the following example.

$$42 = 2 \cdot 3 \cdot 7$$

$$70 = 2 \cdot 5 \cdot 7$$

Notice that 2 is a factor of both, as is 7. Therefore, 14 (the product of 2 and 7) is the greatest common factor of 42 and 70. In other words, 14 is the largest whole number that divides both 42 and 70. The following examples should further clarify the process of finding the greatest common factor of two or more numbers.

EXAMPLE 1

Find the greatest common factor of 48 and 60.

Solution

$$48 = 2 \cdot 2 \cdot 2 \cdot 2 \cdot 3$$

$$60 = 2 \cdot 2 \cdot 3 \cdot 5$$

Since two 2s and one 3 are common to both, the greatest common factor of 48 and 60 is $2 \cdot 2 \cdot 3 =$ **12**. ■

EXAMPLE 2

Find the greatest common factor of 21 and 75.

Solution

$$21 = 3 \cdot 7$$

$$75 = 3 \cdot 5 \cdot 5$$

Since only one 3 is common to both, the greatest common factor is **3**. ■

EXAMPLE 3

Find the greatest common factor of 24 and 35.

Solution

$$24 = 2 \cdot 2 \cdot 2 \cdot 3$$

$$35 = 5 \cdot 7$$

Since there are no common prime factors, the greatest common factor is **1**. ■

The concept of greatest common factor can be extended to more than two numbers, as the next example demonstrates.

EXAMPLE 4

Find the greatest common factor of 24, 56, and 120.

Solution

$$24 = 2 \cdot 2 \cdot 2 \cdot 3$$

$$56 = 2 \cdot 2 \cdot 2 \cdot 7$$

$$120 = 2 \cdot 2 \cdot 2 \cdot 3 \cdot 5$$

Since three 2s are common to the numbers, the greatest common factor of 24, 56, and 120 is $2 \cdot 2 \cdot 2 =$ **8**. ■

Least Common Multiple

We stated earlier in this section that 35 is a *multiple of* 5 because $5 \cdot 7 = 35$. The set of all whole numbers that are multiples of 5 consists of 0, 5, 10, 15, 20, 25, and so on. In other words, 5 times each successive whole number ($5 \cdot 0 = \mathbf{0}$, $5 \cdot 1 = \mathbf{5}$, $5 \cdot 2 = \mathbf{10}$, $5 \cdot 3 = \mathbf{15}$, and so on) produces the multiples of 5. In a like manner, the set of multiples of 4 consists of 0, 4, 8, 12, 16, and so on.

It is sometimes necessary to determine the smallest common *nonzero* multiple of two or more whole numbers. We use the phrase **least common multiple** to designate this nonzero number. For example, the least common multiple of 3 and 4 is 12, which means that 12 is the smallest nonzero multiple of both 3 and 4. Stated another way, 12 is the smallest nonzero whole number that is divisible by both 3 and 4. Likewise, we say that the least common multiple of 6 and 8 is 24.

If we cannot determine the least common multiple by inspection, then the prime factorization form of composite numbers is helpful. Study the solutions to the following examples very carefully so that we can develop a systematic technique for finding the least common multiple of two or more numbers.

EXAMPLE 5

Find the least common multiple of 24 and 36.

Solution

Let's first express each number as a product of prime factors.

$$24 = 2 \cdot 2 \cdot 2 \cdot 3$$

$$36 = 2 \cdot 2 \cdot 3 \cdot 3$$

Since 24 contains three 2s, the least common multiple must have three 2s. Also, since 36 contains two 3s, we need to put two 3s in the least common multiple. The least common multiple of 24 and 36 is therefore $2 \cdot 2 \cdot 2 \cdot 3 \cdot 3 = \mathbf{72}$. ■

If the least common multiple is not obvious by inspection, then we can proceed as follows.

Step 1 Express each number as a product of prime factors.

Step 2 The least common multiple contains each different prime factor as many times as the *most* times it appears in any one of the factorizations from step 1.

EXAMPLE 6

Find the least common multiple of 48 and 84.

Solution

$$48 = 2 \cdot 2 \cdot 2 \cdot 2 \cdot 3$$

$$84 = 2 \cdot 2 \cdot 3 \cdot 7$$

We need four 2s in the least common multiple because of the four 2s in 48. We need one 3 because of the 3 in each of the numbers, and one 7 is needed because of the 7 in 84. The least common multiple of 48 and 84 is $2 \cdot 2 \cdot 2 \cdot 2 \cdot 3 \cdot 7 = \mathbf{336}$. ■

EXAMPLE 7 Find the least common multiple of 12, 18, and 28.

Solution

$12 = 2 \cdot 2 \cdot 3$

$18 = 2 \cdot 3 \cdot 3$

$28 = 2 \cdot 2 \cdot 7$

The least common multiple is $2 \cdot 2 \cdot 3 \cdot 3 \cdot 7 = \mathbf{252}$. ■

EXAMPLE 8 Find the least common multiple of 8 and 9.

Solution

$8 = 2 \cdot 2 \cdot 2$

$9 = 3 \cdot 3$

The least common multiple is $2 \cdot 2 \cdot 2 \cdot 3 \cdot 3 = \mathbf{72}$. ■

Problem Set 1.2

For Problems 1–20, classify each statement as true or false.

1. 8 divides 56

2. 9 divides 54

3. 6 does not divide 54

4. 7 does not divide 42

5. 96 is a multiple of 8

6. 78 is a multiple of 6

7. 54 is not a multiple of 4

8. 64 is not a multiple of 6

9. 144 is divisible by 4

10. 261 is divisible by 9

11. 173 is divisible by 3

12. 149 is divisible by 7

13. 11 is a factor of 143

14. 11 is a factor of 187

15. 9 is a factor of 119

16. 8 is a factor of 98

17. 3 is a prime factor of 57

18. 7 is a prime factor of 91

19. 4 is a prime factor of 48

20. 6 is a prime factor of 72

For Problems 21–30, fill in the blanks with a pair of numbers that has the indicated product and the indicated sum. For example, $\underline{8} \cdot \underline{5} = 40$ and $\underline{8} + \underline{5} = 13$.

21. ___ · ___ = 24 and ___ + ___ = 11

22. ___ · ___ = 12 and ___ + ___ = 7

23. ___ · ___ = 24 and ___ + ___ = 14

24. ___ · ___ = 25 and ___ + ___ = 26

25. ___ · ___ = 36 and ___ + ___ = 13

26. ___ · ___ = 18 and ___ + ___ = 11

27. ___ · ___ = 50 and ___ + ___ = 15

28. ___ · ___ = 50 and ___ + ___ = 27

29. ___ · ___ = 9 and ___ + ___ = 10

30. ___ · ___ = 48 and ___ + ___ = 16

For Problems 31–40, classify each number as prime or composite.

31. 53 **32.** 57

33. 59 **34.** 61

35. 91 **36.** 81

37. 89 **38.** 97

39. 111 **40.** 101

For Problems 41–50, familiarity with a few basic divisibility rules will be helpful for determining the prime factors. The divisibility rules for 2, 3, 5, and 9 are as follows.

Rule for 2

A whole number is divisible by 2 if and only if the units digit of its base-ten numeral is divisible by 2. (In other words, the units digit must be 0, 2, 4, 6, or 8.)

EXAMPLE 68 is divisible by 2 because 8 is divisible by 2.

EXAMPLE 57 is not divisible by 2 because 7 is not divisible by 2.

Rule for 3

A whole number is divisible by 3 if and only if the sum of the digits of its base-ten numeral is divisible by 3.

EXAMPLES 51 is divisible by 3 because $5 + 1 = 6$, and 6 is divisible by 3.
144 is divisible by 3 because $1 + 4 + 4 = 9$, and 9 is divisible by 3.
133 is not divisible by 3 because $1 + 3 + 3 = 7$, and 7 is not divisible by 3.

Rule for 5

A whole number is divisible by 5 if and only if the units digit of its base-ten numeral is divisible by 5. (In other words, the units digit must be 0 or 5.)

EXAMPLES 115 is divisible by 5 because 5 is divisible by 5.
172 is not divisible by 5 because 2 is not divisible by 5.

Rule for 9

A whole number is divisible by 9 if and only if the sum of the digits of its base-ten numeral is divisible by 9.

EXAMPLES 765 is divisible by 9 because $7 + 6 + 5 = 18$, and 18 is divisible by 9.
147 is not divisible by 9 because $1 + 4 + 7 = 12$, and 12 is not divisible by 9.

Use these divisibility rules to help determine the prime factorization of the following numbers.

41. 118 **42.** 76

43. 201 **44.** 123

45. 85 **46.** 115

47. 117 **48.** 441

49. 129 **50.** 153

For Problems 51–62, factor each composite number into a product of prime numbers. For example, $18 = 2 \cdot 3 \cdot 3$.

51. 26 **52.** 16

53. 36 **54.** 80

55. 49 **56.** 92

57. 56 **58.** 144

59. 120 **60.** 84

61. 135 **62.** 98

For Problems 63–74, find the greatest common factor of the given numbers.

63. 12 and 16 **64.** 30 and 36

65. 56 and 64 **66.** 72 and 96

67. 63 and 81 **68.** 60 and 72

69. 84 and 96 **70.** 48 and 52

71. 36, 72, and 90 **72.** 27, 54, and 63

73. 48, 60, and 84 **74.** 32, 80, and 96

For Problems 75–86, find the least common multiple of the given numbers.

75. 6 and 8 **76.** 8 and 12

77. 12 and 16 **78.** 9 and 12

79. 28 and 35

80. 42 and 66

81. 49 and 56

82. 18 and 24

83. 8, 12, and 28

84. 6, 10, and 12

85. 9, 15, and 18

86. 8, 14, and 24

THOUGHTS INTO WORDS

87. How would you explain the concepts of "greatest common factor" and "least common multiple" to a friend who missed class during that discussion?

88. Is it always true that the greatest common factor of two numbers is less than the least common multiple of those same two numbers? Explain your answer.

FURTHER INVESTIGATIONS

89. The numbers 0, 2, 4, 6, 8, and so on are multiples of 2. They are also called *even* numbers. Why is 2 the only even prime number?

90. Find the smallest nonzero whole number that is divisible by 2, 3, 4, 5, 6, 7, and 8.

91. Find the smallest whole number, greater than 1, that produces a remainder of 1 when divided by 2, 3, 4, 5, or 6.

92. What is the greatest common factor of x and y if x and y are both prime numbers, and x does not equal y? Explain your answer.

93. What is the greatest common factor of x and y if x and y are nonzero whole numbers, and y is a multiple of x? Explain your answer.

94. What is the least common multiple of x and y if they are both prime numbers, and x does not equal y? Explain your answer.

95. What is the least common multiple of x and y if the greatest common factor of x and y is 1? Explain your answer.

1.3 Integers: Addition and Subtraction

"A record temperature of 35° *below* zero was recorded on this date in 1904." "The PO stock closed *down* 3 points yesterday." "On a first-down sweep around the left end, Moser *lost* 7 yards." "The Widget Manufacturing Company reported *assets* of 50 million dollars and *liabilities* of 53 million dollars for 1981." These examples illustrate our need for negative numbers.

The number line is a helpful visual device for our work at this time. We can associate the set of whole numbers with evenly spaced points on a line as indicated in Figure 1.1. For each nonzero whole number we can associate its *negative*

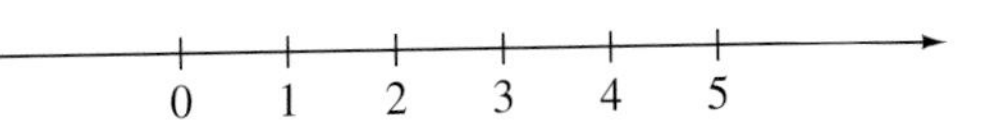

Figure 1.1

to the left of zero; with 1 we associate -1, with 2 we associate -2, and so on, as indicated in Figure 1.2. The set of whole numbers along with $-1, -2, -3$, and so on, is called the set of **integers**.

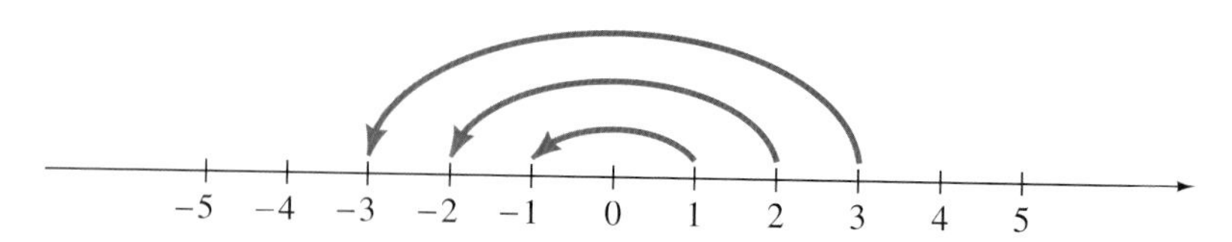

Figure 1.2

The following terminology is used with reference to the integers.

$\{\ldots, -3, -2, -1, 0, 1, 2, 3, \ldots\}$	Integers
$\{1, 2, 3, 4, \ldots\}$	Positive integers
$\{0, 1, 2, 3, 4, \ldots\}$	Nonnegative integers
$\{\ldots, -3, -2, -1\}$	Negative integers
$\{\ldots, -3, -2, -1, 0\}$	Nonpositive integers

The symbol -1 can be read as "negative one," "opposite of one," or "additive inverse of one." The *opposite-of* and *additive-inverse-of* terminology is very helpful when working with variables. The symbol $-x$, read as "opposite of x" or "additive inverse of x," emphasizes an important issue. Since x can be any integer, $-x$ (the opposite of x) can be zero, positive, or negative. If x is a positive integer, then $-x$ is negative. If x is a negative integer, then $-x$ is positive. If x is zero, then $-x$ is zero. These statements are written as follows and illustrated on the number lines in Figure 1.3.

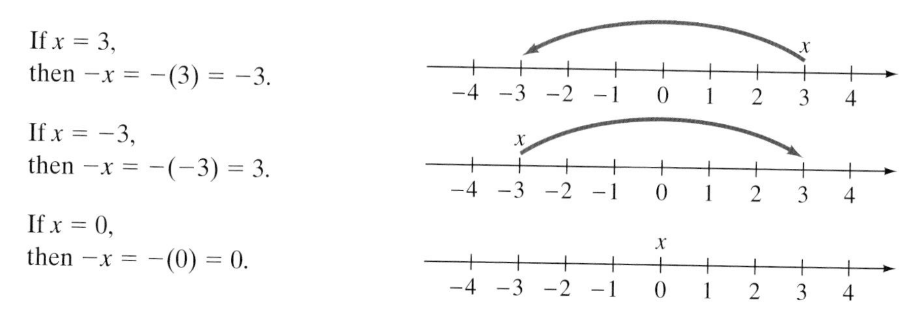

If $x = 3$,
then $-x = -(3) = -3$.
If $x = -3$,
then $-x = -(-3) = 3$.
If $x = 0$,
then $-x = -(0) = 0$.
x
−4 −3 −2 −1 0 1 2 3 4

Figure 1.3

From this discussion we also need to recognize the following general property.

Property 1.1

If a is any integer, then

$$-(-a) = a$$

(The opposite of the opposite of any integer is the integer itself.)

Addition of Integers

The number line is also a convenient visual aid for interpreting **addition of integers**. In Figure 1.4 we see number line interpretations for the following examples.

Problem	Number line interpretation	Sum
$3 + 2$	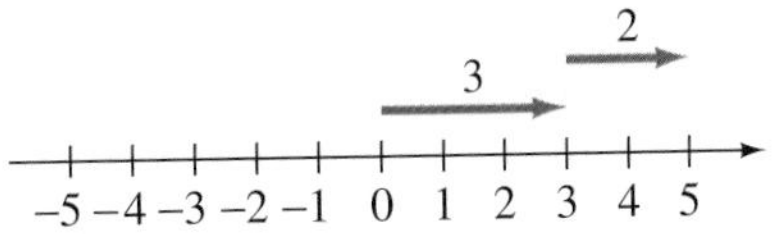 3, 2 −5 −4 −3 −2 −1 0 1 2 3 4 5	$3 + 2 = 5$
$3 + (-2)$	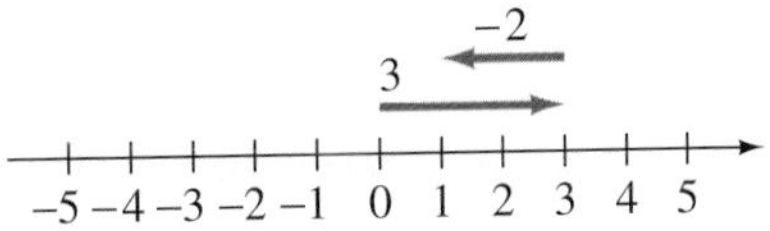 3, −2 −5 −4 −3 −2 −1 0 1 2 3 4 5	$3 + (-2) = 1$
$-3 + 2$	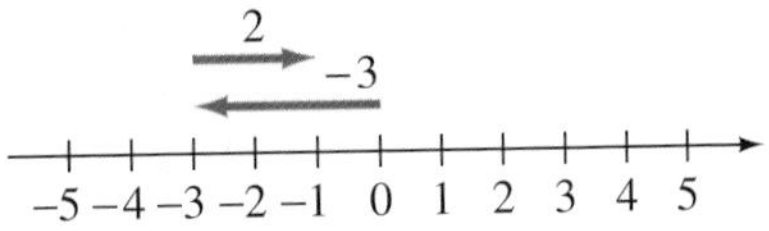 −3, 2 −5 −4 −3 −2 −1 0 1 2 3 4 5	$-3 + 2 = -1$
$-3 + (-2)$	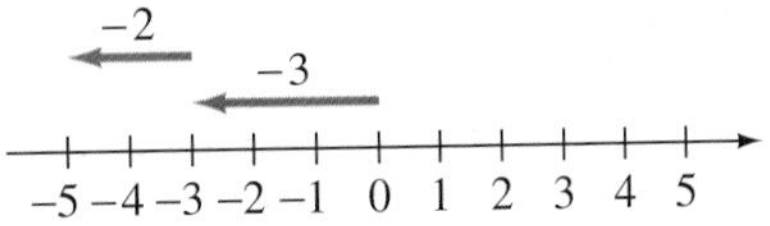 −3, −2 −5 −4 −3 −2 −1 0 1 2 3 4 5	$-3 + (-2) = -5$

Figure 1.4

Once you acquire a feeling of movement on the number line, a mental image of this movement is sufficient. Consider the following addition problems, and mentally picture the number line interpretation. Be sure that you agree with all of our answers.

$$5 + (-2) = 3, \qquad -6 + 4 = -2, \qquad -8 + 11 = 3,$$
$$-7 + (-4) = -11, \qquad -5 + 9 = 4, \qquad 9 + (-2) = 7,$$
$$14 + (-17) = -3, \qquad 0 + (-4) = -4, \qquad 6 + (-6) = 0$$

The last example illustrates a general property that should be noted: **Any integer plus its opposite equals zero.**

Remark: Profits and losses pertaining to investments also provide a good physical model for interpreting addition of integers. A loss of \$25 on one investment along with a profit of \$60 on a second investment produces an overall profit of \$35. This can be expressed as $-25 + 60 = 35$. Perhaps it would be helpful for you to check the previous examples using a profit and loss interpretation.

Even though all problems involving the addition of integers could be done by using the number line interpretation, it is sometimes convenient to give a more precise description of the addition process. For this purpose we need to briefly consider the concept of absolute value. The **absolute value** of a number is the distance between the number and 0 on the number line. For example, the absolute value of 6 is 6. The absolute value of -6 is also 6. The absolute value of 0 is 0. Symbolically, absolute value is denoted with vertical bars. Thus, we write

$$|6| = 6, \qquad |-6| = 6, \qquad |0| = 0$$

Notice that the absolute value of a positive number is the number itself, but the absolute value of a negative number is its opposite. Thus, the absolute value of any number except 0 is positive, and the absolute value of 0 is 0.

We can describe precisely the process of **adding integers** by using the concept of absolute value as follows.

Two Positive Integers

The sum of two positive integers is the sum of their absolute values. (The sum of two positive integers is a positive integer.)

$$43 + 54 = |43| + |54| = 43 + 54 = 97$$

Two Negative Integers

The sum of two negative integers is the opposite of the sum of their absolute values. (The sum of two negative integers is a negative integer.)

$$\begin{aligned}(-67) + (-93) &= -(|-67| + |-93|)\\ &= -(67 + 93)\\ &= -160\end{aligned}$$

One Positive and One Negative Integer

We can find the sum of a positive and a negative integer by subtracting the smaller absolute value from the larger absolute value and giving the result the sign of the original number that has the larger absolute value. If the integers have the same absolute value, then their sum is 0.

$$\begin{aligned}82 + (-40) &= |82| - |-40|\\ &= 82 - 40\\ &= 42\end{aligned}$$

$$74 + (-90) = -(|-90| - |74|)$$
$$= -(90 - 74)$$
$$= -16$$
$$(-17) + 17 = |-17| - |17|$$
$$= 17 - 17$$
$$= 0$$

Zero and Another Integer

The sum of 0 and any integer is the integer itself.

$$0 + (-46) = -46$$
$$72 + 0 = 72$$

The following examples further demonstrate how to add integers. Be sure that you agree with each of the results.

$$-18 + (-56) = -(|-18| + |-56|) = -(18 + 56) = -74,$$
$$-71 + (-32) = -(|-71| + |-32|) = -(71 + 32) = -103,$$
$$64 + (-49) = |64| - |-49| = 64 - 49 = 15,$$
$$-56 + 93 = |93| - |-56| = 93 - 56 = 37,$$
$$-114 + 48 = -(|-114| - |48|) = -(114 - 48) = -66,$$
$$45 + (-73) = -(|-73| - |45|) = -(73 - 45) = -28,$$
$$46 + (-46) = 0, \qquad -48 + 0 = -48,$$
$$(-73) + 73 = 0, \qquad 0 + (-81) = -81$$

It is true that this absolute value approach does precisely describe the process of adding integers, but don't forget about the number line interpretation. Included in the next problem set are other physical models for interpreting the addition of integers. You may find these models helpful.

Subtraction of Integers

The following examples illustrate a relationship between addition and subtraction of whole numbers.

$7 - 2 = 5$ because $2 + 5 = 7$

$9 - 6 = 3$ because $6 + 3 = 9$

$5 - 1 = 4$ because $1 + 4 = 5$

This same relationship between addition and subtraction holds for *all integers*.

$$5 - 6 = -1 \quad \text{because } 6 + (-1) = 5$$
$$-4 - 9 = -13 \quad \text{because } 9 + (-13) = -4$$
$$-3 - (-7) = 4 \quad \text{because } -7 + 4 = -3$$
$$8 - (-3) = 11 \quad \text{because } -3 + 11 = 8$$

Now consider a further observation:

$$5 - 6 = -1 \quad \text{and} \quad 5 + (-6) = -1$$
$$-4 - 9 = -13 \quad \text{and} \quad -4 + (-9) = -13$$
$$-3 - (-7) = 4 \quad \text{and} \quad -3 + 7 = 4$$
$$8 - (-3) = 11 \quad \text{and} \quad 8 + 3 = 11$$

The previous examples help us realize that we can state the subtraction of integers in terms of the addition of integers. More precisely, a general description for the subtraction of integers follows.

Subtraction of Integers

If a and b are integers, then $a - b = a + (-b)$.

It may be helpful for you to read $a - b = a + (-b)$ as "a minus b is equal to a plus the opposite of b." Every subtraction problem can be changed to an equivalent addition problem as illustrated by the following examples.

$$6 - 13 = 6 + (-13) = -7,$$
$$9 - (-12) = 9 + 12 = 21,$$
$$-8 - 13 = -8 + (-13) = -21,$$
$$-7 - (-8) = -7 + 8 = 1$$

It should be apparent that the addition of integers is a key operation. The ability to effectively add integers is a necessary skill for further algebraic work.

Evaluating Algebraic Expressions

Let's conclude this section by evaluating some algebraic expressions using negative and positive integers.

EXAMPLE 1

Evaluate each algebraic expression for the given values of the variables.

(a) $x - y$ for $x = -12$ and $y = 20$

(b) $-a + b$ for $a = -8$ and $b = -6$

(c) $-x - y$ for $x = 14$ and $y = -7$

Solution

(a) $x - y = -12 - 20$ when $x = -12$ and $y = 20$

$= -12 + (-20)$ Change to addition

$= -32$

(b) $-a + b = -(-8) + (-6)$ when $a = -8$ and $b = -6$

$= 8 + (-6)$ Note the use of parentheses when substituting the values

$= 2$

(c) $-x - y = -(14) - (-7)$ when $x = 14$ and $y = -7$

$= -14 + 7$ Change to addition

$= -7$

Problem Set 1.3

For Problems 1–10, use the number line interpretation to find each sum.

1. $5 + (-3)$ **2.** $7 + (-4)$

3. $-6 + 2$ **4.** $-9 + 4$

5. $-3 + (-4)$ **6.** $-5 + (-6)$

7. $8 + (-2)$ **8.** $12 + (-7)$

9. $5 + (-11)$ **10.** $4 + (-13)$

For Problems 11–30, find each sum.

11. $17 + (-9)$ **12.** $16 + (-5)$

13. $8 + (-19)$ **14.** $9 + (-14)$

15. $-7 + (-8)$ **16.** $-6 + (-9)$

17. $-15 + 8$ **18.** $-22 + 14$

19. $-13 + (-18)$ **20.** $-15 + (-19)$

21. $-27 + 8$ **22.** $-29 + 12$

23. $32 + (-23)$ **24.** $27 + (-14)$

25. $-25 + (-36)$ **26.** $-34 + (-49)$

27. $54 + (-72)$ **28.** $48 + (-76)$

29. $-34 + (-58)$ **30.** $-27 + (-36)$

For Problems 31–50, subtract as indicated.

31. $3 - 8$ **32.** $5 - 11$

33. $-4 - 9$ **34.** $-7 - 8$

35. $5 - (-7)$ **36.** $9 - (-4)$

37. $-6 - (-12)$ **38.** $-7 - (-15)$

39. $-11 - (-10)$ **40.** $-14 - (-19)$

41. $-18 - 27$ **42.** $-16 - 25$

43. $34 - 63$ **44.** $25 - 58$

45. $45 - 18$ **46.** $52 - 38$

47. $-21 - 44$ **48.** $-26 - 54$

49. $-53 - (-24)$ **50.** $-76 - (-39)$

For Problems 51–66, add or subtract as indicated.

51. $6 - 8 - 9$ **52.** $5 - 9 - 4$

53. $-4 - (-6) + 5 - 8$ **54.** $-3 - 8 + 9 - (-6)$

55. $5 + 7 - 8 - 12$

56. $-7 + 9 - 4 - 12$

57. $-6 - 4 - (-2) + (-5)$

58. $-8 - 11 - (-6) + (-4)$

59. $-6 - 5 - 9 - 8 - 7$

60. $-4 - 3 - 7 - 8 - 6$

61. $7 - 12 + 14 - 15 - 9$

62. $8 - 13 + 17 - 15 - 19$

63. $-11 - (-14) + (-17) - 18$

64. $-15 + 20 - 14 - 18 + 9$

65. $16 - 21 + (-15) - (-22)$

66. $17 - 23 - 14 - (-18)$

The horizontal format is used extensively in algebra, but occasionally the vertical format shows up. Some exposure to the vertical format is therefore needed. Find the following sums for Problems 67–78.

67. $\begin{array}{r} 5 \\ -9 \\ \hline \end{array}$

68. $\begin{array}{r} 8 \\ -13 \\ \hline \end{array}$

69. $\begin{array}{r} -13 \\ -18 \\ \hline \end{array}$

70. $\begin{array}{r} -14 \\ -28 \\ \hline \end{array}$

71. $\begin{array}{r} -18 \\ 9 \\ \hline \end{array}$

72. $\begin{array}{r} -17 \\ 9 \\ \hline \end{array}$

73. $\begin{array}{r} -21 \\ 39 \\ \hline \end{array}$

74. $\begin{array}{r} -15 \\ 32 \\ \hline \end{array}$

75. $\begin{array}{r} 27 \\ -19 \\ \hline \end{array}$

76. $\begin{array}{r} 31 \\ -18 \\ \hline \end{array}$

77. $\begin{array}{r} -53 \\ 24 \\ \hline \end{array}$

78. $\begin{array}{r} 47 \\ -28 \\ \hline \end{array}$

For Problems 79–90, do the subtraction problems in vertical format.

79. $\begin{array}{r} 5 \\ 12 \\ \hline \end{array}$

80. $\begin{array}{r} 8 \\ 19 \\ \hline \end{array}$

81. $\begin{array}{r} 6 \\ -9 \\ \hline \end{array}$

82. $\begin{array}{r} 13 \\ -7 \\ \hline \end{array}$

83. $\begin{array}{r} -7 \\ -8 \\ \hline \end{array}$

84. $\begin{array}{r} -6 \\ -5 \\ \hline \end{array}$

85. $\begin{array}{r} 17 \\ -19 \\ \hline \end{array}$

86. $\begin{array}{r} 18 \\ -14 \\ \hline \end{array}$

87. $\begin{array}{r} -23 \\ 16 \\ \hline \end{array}$

88. $\begin{array}{r} -27 \\ 15 \\ \hline \end{array}$

89. $\begin{array}{r} -12 \\ 12 \\ \hline \end{array}$

90. $\begin{array}{r} -13 \\ -13 \\ \hline \end{array}$

For Problems 91–100, evaluate each algebraic expression for the given values of the variables.

91. $x - y$ for $x = -6$ and $y = -13$

92. $-x - y$ for $x = -7$ and $y = -9$

93. $-x + y - z$ for $x = 3$, $y = -4$, and $z = -6$

94. $x - y + z$ for $x = 5$, $y = 6$, and $z = -9$

95. $-x - y - z$ for $x = -2$, $y = 3$, and $z = -11$

96. $-x - y + z$ for $x = -8$, $y = -7$, and $z = -14$

97. $-x + y + z$ for $x = -11$, $y = 7$, and $z = -9$

98. $-x - y - z$ for $x = 12$, $y = -6$, and $z = -14$

99. $x - y - z$ for $x = -15$, $y = 12$, and $z = -10$

100. $x + y - z$ for $x = -18$, $y = 13$, and $z = 8$

A game such as football can also be used to interpret addition of integers. A gain of 3 yards on one play followed by a loss of 5 yards on the next play places the ball 2 yards behind the initial line of scrimmage; this could be expressed as $3 + (-5) = -2$. Use this football interpretation to find the following sums (Problems 101–110).

101. $4 + (-7)$

102. $3 + (-5)$

103. $-4 + (-6)$

104. $-2 + (-5)$

105. $-5 + 2$

106. $-10 + 6$

107. $-4 + 15$

108. $-3 + 22$

109. $-12 + 17$

110. $-9 + 21$

For Problems 111–120, refer to the Remark on page 17 and use the profit and loss interpretation for the addition of integers.

111. $60 + (-125)$

112. $50 + (-85)$

113. $-55 + (-45)$

114. $-120 + (-220)$

115. $-70 + 45$

116. $-125 + 45$

117. $-120 + 250$

118. $-75 + 165$

119. $145 + (-65)$

120. $275 + (-195)$

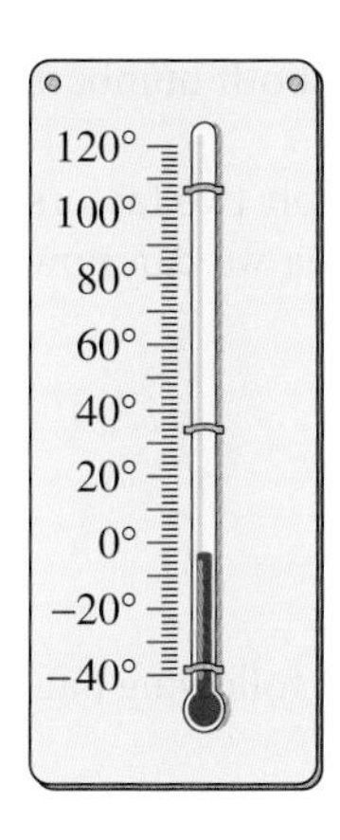

Figure 1.5

121. The temperature at 5 A.M. was $-17°F$. By noon the temperature had increased by 14°F. Use the addition of integers to describe this situation and to determine the temperature at noon (see Figure 1.5).

122. The temperature at 6 P.M. was $-6°F$, and by 11 P.M. the temperature had dropped 5°F. Use the subtraction of integers to describe this situation and to determine the temperature at 11 P.M. (see Figure 1.5).

123. Megan shot rounds of 3 over par, 2 under par, 3 under par, and 5 under par for a four-day golf tournament. Use the addition of integers to describe this situation and to determine how much over or under par she was for the tournament.

124. The annual report of a company contained the following figures: a loss of \$615,000 for 2002, a loss of \$275,000 for 2003, a loss of \$70,000 for 2004, and a profit of \$115,000 for 2005. Use the addition of integers to describe this situation and to determine the company's total loss or profit for the four-year period.

125. Suppose that during a five-day period a share of Dell's stock recorded the following gains and losses:

Monday	**Tuesday**	**Wednesday**
lost \$2	gained \$1	gained \$3
Thursday	**Friday**	
gained \$1	lost \$2	

Use the addition of integers to describe this situation and to determine the amount of gain or loss for the five-day period.

126. The Dead Sea is approximately thirteen hundred ten feet below sea level. Suppose that you are standing eight hundred five feet above the Dead Sea. Use the addition of integers to describe this situation and to determine your elevation.

127. Use your calculator to check your answers for Problems 51–66.

THOUGHTS INTO WORDS

128. The statement $-6 - (-2) = -6 + 2 = -4$ can be read as "negative six minus negative two equals negative six plus two, which equals negative four." Express in words each of the following.

(a) $8 + (-10) = -2$

(b) $-7 - 4 = -7 + (-4) = 11$

(c) $9 - (-12) = 9 + 12 = 21$

(d) $-5 + (-6) = -11$

129. The algebraic expression $-x - y$ can be read as "the opposite of x minus y." Express in words each of the following.

(a) $-x + y$

(b) $x - y$

(c) $-x - y + z$

1.4 Integers: Multiplication and Division

Multiplication of whole numbers may be interpreted as repeated addition. For example, $3 \cdot 4$ means the sum of three 4s; thus, $3 \cdot 4 = 4 + 4 + 4 = 12$. Consider the following examples that use the repeated addition idea to find the product of a positive integer and a negative integer:

$$3(-2) = -2 + (-2) + (-2) = -6,$$

$$2(-4) = -4 + (-4) = -8,$$

$$4(-1) = -1 + (-1) + (-1) + (-1) = -4$$

Note the use of parentheses to indicate multiplication. Sometimes both numbers are enclosed in parentheses so that we have $(3)(-2)$.

When multiplying whole numbers, the order in which we multiply two factors does not change the product: $2(3) = 6$ and $3(2) = 6$. Using this idea we can now handle a negative number times a positive integer as follows:

$$(-2)(3) = (3)(-2) = (-2) + (-2) + (-2) = -6$$

$$(-3)(2) = (2)(-3) = (-3) + (-3) = -6$$

$$(-4)(3) = (3)(-4) = (-4) + (-4) + (-4) = -12$$

Finally, let's consider the product of two negative integers. The following pattern helps us with the reasoning for this situation:

$$4(-3) = -12$$

$$3(-3) = -9$$

$$2(-3) = -6$$

$$1(-3) = -3$$

$$0(-3) = 0$$ The product of 0 and any integer is 0

$$(-1)(-3) = ?$$

Certainly, to continue this pattern, the product of -1 and -3 has to be 3. In general, this type of reasoning helps us to realize that the product of any two negative integers is a positive integer.

Using the concept of absolute value, these three facts precisely describe the **multiplication of integers**:

1. The product of two positive integers or two negative integers is the product of their absolute values.
2. The product of a positive and a negative integer (either order) is the opposite of the product of their absolute values.
3. The product of zero and any integer is zero.

The following are examples of the multiplication of integers:

$$(-5)(-2) = |-5| \cdot |-2| = 5 \cdot 2 = 10$$

$$(7)(-6) = -(|7| \cdot |-6|) = -(7 \cdot 6) = -42$$

$$(-8)(9) = -(|-8| \cdot |9|) = -(8 \cdot 9) = -72$$

$$(-14)(0) = 0$$

$$(0)(-28) = 0$$

These examples show a step-by-step process for multiplying integers. In reality, however, the key issue is to remember whether the product is positive or negative. In other words, we need to remember that **the product of two positive integers or two negative integers is a positive integer; and the product of a positive integer and a negative integer (in either order) is a negative integer**. Then we can avoid the step-by-step analysis and simply write the results as follows:

$$(7)(-9) = -63$$
$$(8)(7) = 56$$
$$(-5)(-6) = 30$$
$$(-4)(12) = -48$$

Division of Integers

By looking back at our knowledge of whole numbers, we can get some guidance for our work with integers. We know, for example, that $\frac{8}{2} = 4$ because $2 \cdot 4 = 8$. In other words, we can find the quotient of two whole numbers by looking at a related multiplication problem. In the following examples we use this same link between multiplication and division to determine the quotients.

$$\frac{8}{-2} = -4 \quad \text{because } (-2)(-4) = 8$$

$$\frac{-10}{5} = -2 \quad \text{because } (5)(-2) = -10$$

$$\frac{-12}{-4} = 3 \quad \text{because } (-4)(3) = -12$$

$$\frac{0}{-6} = 0 \quad \text{because } (-6)(0) = 0$$

$\frac{-9}{0}$ is undefined because no number times 0 produces -9

$\frac{0}{0}$ is undefined because any number times 0 equals 0. **Remember that division by zero is undefined!**

The following three statements precisely describe the division of integers:

1. The quotient of two positive or two negative integers is the quotient of their absolute values.
2. The quotient of a positive integer and a negative integer (or a negative and a positive) is the opposite of the quotient of their absolute values.
3. The quotient of zero and any nonzero number (zero divided by any nonzero number) is zero.

The following are examples of the division of integers:

$$\frac{-8}{-4} = \frac{|-8|}{|-4|} = \frac{8}{4} = 2, \qquad \frac{-14}{2} = -\left(\frac{|-14|}{|2|}\right) = -\left(\frac{14}{2}\right) = -7,$$

$$\frac{0}{-4} = 0, \qquad \frac{15}{-3} = -\left(\frac{|15|}{|-3|}\right) = -\left(\frac{15}{3}\right) = -5$$

For practical purposes, the key is to remember whether the quotient is positive or negative. We need to remember that **the quotient of two positive integers or two negative integers is positive; and the quotient of a positive integer and a negative integer or a negative integer and a positive integer is negative**. We can then simply write the quotients as follows without showing all of the steps:

$$\frac{-18}{-6} = 3, \qquad \frac{-24}{12} = -2, \qquad \frac{36}{-9} = -4$$

Remark: Occasionally, people use the phrase "two negatives make a positive." We hope they realize that the reference is to multiplication and division only; in addition the sum of two negative integers is still a negative integer. It is probably best to avoid such imprecise statements.

Simplifying Numerical Expressions

Now we can simplify numerical expressions involving any or all of the four basic operations with integers. Keep in mind the order of operations given in Section 1.1.

EXAMPLE 1

Simplify $-4(-3) - 7(-8) + 3(-9)$.

Solution

$$\begin{aligned} -4(-3) - 7(-8) + 3(-9) &= 12 - (-56) + (-27) \\ &= 12 + 56 + (-27) \\ &= 41 \end{aligned}$$

■

EXAMPLE 2

Simplify $\dfrac{-8 - 4(5)}{-4}$.

Solution

$$\begin{aligned} \frac{-8 - 4(5)}{-4} &= \frac{-8 - 20}{-4} \\ &= \frac{-28}{-4} \\ &= 7 \end{aligned}$$

■

Evaluating Algebraic Expressions

Evaluating algebraic expressions will often involve the use of two or more operations with integers. We use the final examples of this section to represent such situations.

EXAMPLE 3

Find the value of $3x + 2y$ when $x = 5$ and $y = -9$.

Solution

$$3x + 2y = 3(5) + 2(-9) \quad \text{when } x = 5 \text{ and } y = -9$$
$$= 15 + (-18)$$
$$= -3$$

■

EXAMPLE 4

Evaluate $-2a + 9b$ for $a = 4$ and $b = -3$.

Solution

$$-2a + 9b = -2(4) + 9(-3) \quad \text{when } a = 4 \text{ and } b = -3$$
$$= -8 + (-27)$$
$$= -35$$

■

EXAMPLE 5

Find the value of $\frac{x - 2y}{4}$ when $x = -6$ and $y = 5$.

Solution

$$\frac{x - 2y}{4} = \frac{-6 - 2(5)}{4} \quad \text{when } x = -6 \text{ and } y = 5$$
$$= \frac{-6 - 10}{4}$$
$$= \frac{-16}{4}$$
$$= -4$$

■

Problem Set 1.4

For Problems 1–40, find the product or quotient (multiply or divide) as indicated.

1. $5(-6)$

2. $7(-9)$

3. $\frac{-27}{3}$

4. $\frac{-35}{5}$

5. $\frac{-42}{-6}$

6. $\frac{-72}{-8}$

7. $(-7)(8)$

8. $(-6)(9)$

9. $(-5)(-12)$

10. $(-7)(-14)$

11. $\frac{96}{-8}$

12. $\frac{-91}{7}$

13. $14(-9)$

14. $17(-7)$

15. $(-11)(-14)$

16. $(-13)(-17)$

17. $\frac{135}{-15}$

18. $\frac{-144}{12}$

19. $\frac{-121}{-11}$

20. $\frac{-169}{-13}$

21. $(-15)(-15)$

22. $(-18)(-18)$

23. $\frac{112}{-8}$

24. $\frac{112}{-7}$

25. $\frac{0}{-8}$

26. $\frac{-8}{0}$

27. $\frac{-138}{-6}$

28. $\frac{-105}{-5}$

29. $\frac{76}{-4}$

30. $\frac{-114}{6}$

31. $(-6)(-15)$

32. $\frac{0}{-14}$

33. $(-56) \div (-4)$

34. $(-78) \div (-6)$

35. $(-19) \div 0$

36. $(-90) \div 15$

37. $(-72) \div 18$

38. $(-70) \div 5$

39. $(-36)(27)$

40. $(42)(-29)$

For Problems 41–60, simplify each numerical expression.

41. $3(-4) + 5(-7)$

42. $6(-3) + 5(-9)$

43. $7(-2) - 4(-8)$

44. $9(-3) - 8(-6)$

45. $(-3)(-8) + (-9)(-5)$

46. $(-7)(-6) + (-4)(-3)$

47. $5(-6) - 4(-7) + 3(2)$

48. $7(-4) - 8(-7) + 5(-8)$

49. $\frac{13 + (-25)}{-3}$

50. $\frac{15 + (-36)}{-7}$

51. $\frac{12 - 48}{6}$

52. $\frac{16 - 40}{8}$

53. $\frac{-7(10) + 6(-9)}{-4}$

54. $\frac{-6(8) + 4(-14)}{-8}$

55. $\frac{4(-7) - 8(-9)}{11}$

56. $\frac{5(-9) - 6(-7)}{3}$

57. $-2(3) - 3(-4) + 4(-5) - 6(-7)$

58. $2(-4) + 4(-5) - 7(-6) - 3(9)$

59. $-1(-6) - 4 + 6(-2) - 7(-3) - 18$

60. $-9(-2) + 16 - 4(-7) - 12 + 3(-8)$

For Problems 61–76, evaluate each algebraic expression for the given values of the variables.

61. $7x + 5y$ for $x = -5$ and $y = 9$

62. $4a + 6b$ for $a = -6$ and $b = -8$

63. $9a - 2b$ for $a = -5$ and $b = 7$

64. $8a - 3b$ for $a = -7$ and $b = 9$

65. $-6x - 7y$ for $x = -4$ and $y = -6$

66. $-5x - 12y$ for $x = -5$ and $y = -7$

67. $\frac{5x - 3y}{-6}$ for $x = -6$ and $y = 4$

68. $\frac{-7x + 4y}{-8}$ for $x = 8$ and $y = 6$

69. $3(2a - 5b)$ for $a = -1$ and $b = -5$

70. $4(3a - 7b)$ for $a = -2$ and $b = -4$

71. $-2x + 6y - xy$ for $x = 7$ and $y = -7$

72. $-3x + 7y - 2xy$ for $x = -6$ and $y = 4$

73. $-4ab - b$ for $a = 2$ and $b = -14$

74. $-5ab + b$ for $a = -1$ and $b = -13$

75. $(ab + c)(b - c)$ for $a = -2$, $b = -3$, and $c = 4$

76. $(ab - c)(a + c)$ for $a = -3$, $b = 2$, and $c = 5$

For Problems 77–82, find the value of $\frac{5(F - 32)}{9}$ for each of the given values for F.

77. $F = 59$

78. $F = 68$

79. $F = 14$

80. $F = -4$

81. $F = -13$

82. $F = -22$

For Problems 83–88, find the value of $\frac{9C}{5} + 32$ for each of the given values for C.

83. $C = 25$

84. $C = 35$

85. $C = 40$

86. $C = 0$

87. $C = -10$

88. $C = -30$

89. On Monday morning, Thad bought 800 shares of a stock at $19 per share. During that workweek, the stock went up $2 per share on one day and dropped $1 per share on each of the other four days. Use multiplication and addition of integers to describe this situation and to determine the value of the 800 shares by closing time on Friday.

90. In one workweek a small company showed a profit of $475 for one day and a loss of $65 for each of the other four days. Use multiplication and addition of integers to describe this situation and to determine the company's profit or loss for the week.

91. At 6 P.M. the temperature was 5°F. For the next four hours the temperature dropped 3° per hour. Use multiplication and addition of integers to describe this situation and to find the temperature at 10 P.M.

92. For each of the first three days of a golf tournament, Jason shot 2 strokes under par. Then for each of the last two days of the tournament he shot 4 strokes over par. Use multiplication and addition of integers to describe this situation and to determine how Jason shot relative to par for the five-day tournament.

93. Use a calculator to check your answers for Problems 41–60.

THOUGHTS INTO WORDS

94. Your friend keeps getting an answer of -7 when simplifying the expression $-6 + (-8) \div 2$. What mistake is she making and how would you help her?

95. Make up a problem that could be solved using $6(-4) = -24$.

96. Make up a problem that could be solved using $(-4)(-3) = 12$.

97. Explain why $\frac{0}{4} = 0$ but $\frac{4}{0}$ is undefined.

1.5 Use of Properties

We will begin this section by listing and briefly commenting on some of the basic properties of integers. We will then show how these properties facilitate manipulation with integers and also serve as a basis for some algebraic computation.

Commutative Property of Addition

If a and b are integers, then

$$a + b = b + a$$

Commutative Property of Multiplication

If a and b are integers, then

$$ab = ba$$

Addition and multiplication are said to be commutative operations. This means that the order in which you add or multiply two integers does not affect the result. For example, $3 + 5 = 5 + 3$ and $7(8) = 8(7)$. It is also important to realize that subtraction and division *are not* commutative operations; order does make a difference. For example, $8 - 7 \neq 7 - 8$ and $16 \div 4 \neq 4 \div 16$.

Associative Property of Addition

If a, b, and c are integers, then

$$(a + b) + c = a + (b + c)$$

Associative Property of Multiplication

If a, b, and c are integers, then

$$(ab)c = a(bc)$$

Our arithmetic operations are binary operations. We only operate (add, subtract, multiply, or divide) on two numbers at a time. Therefore, when we need to operate on three or more numbers, the numbers must be grouped. The associative properties can be thought of as grouping properties. For a sum of three numbers, changing the grouping of the numbers does not affect the final result. For example, $(-8 + 3) + 9 = -8 + (3 + 9)$. This is also true for multiplication as $[(-6)(5)](-4) = (-6)[(5)(-4)]$ illustrates. Addition and multiplication are associative operations. Subtraction and division *are not* associative operations. For example, $(8 - 4) - 7 = -3$, whereas $8 - (4 - 7) = 11$ shows that subtraction is not an associative operation, Also, $(8 \div 4) \div 2 = 1$, whereas $8 \div (4 \div 2) = 4$ shows that division is not associative.

Identity Property of Addition

If a is an integer, then

$$a + 0 = 0 + a = a$$

We refer to zero as the identity element for addition. This simply means that the sum of any integer and zero is exactly the same integer. For example, $-197 + 0 = 0 + (-197) = -197$.

Identity Property of Multiplication

If a is an integer, then

$$a(1) = 1(a) = a$$

We call one the identity element for multiplication. The product of any integer and one is exactly the same integer. For example, $(-573)(1) = (1)(-573) = -573$.

Additive Inverse Property

For every integer a, there exists an integer $-a$ such that

$$a + (-a) = (-a) + a = 0$$

The integer $-a$ is called the additive inverse of a or the opposite of a. Thus 6 and -6 are additive inverses, and their sum is 0. The additive inverse of 0 is 0.

Multiplication Property of Zero

If a is an integer, then

$$(a)(0) = (0)(a) = 0$$

The product of zero and any integer is zero. For example, $(-873)(0) = (0)(-873) = 0$.

Multiplicative Property of Negative One

If a is an integer, then

$$(a)(-1) = (-1)(a) = -a$$

The product of any integer and -1 is the opposite of the integer. For example, $(-1)(48) = (48)(-1) = -48$.

Distributive Property

If a, b, and c are integers, then

$$a(b + c) = ab + ac$$

The distributive property involves both addition and multiplication. We say that **multiplication distributes over addition**. For example, $3(4 + 7) = 3(4) + 3(7)$. Since $b - c = b + (-c)$, it follows that **multiplication also distributes over subtraction**. This could be stated as $a(b - c) = ab - ac$. For example, $7(8 - 2) = 7(8) - 7(2)$.

Let's now consider some examples that use the properties to help with certain types of manipulations.

EXAMPLE 1

Find the sum $(43 + (-24)) + 24$.

Solution

In this problem it is much more advantageous to group -24 and 24. Thus,

$$\begin{aligned}(43 + (-24)) + 24 &= 43 + ((-24) + 24) && \text{Associative property for addition}\\ &= 43 + 0\\ &= 43\end{aligned}$$

■

EXAMPLE 2

Find the product $[(-17)(25)](4)$.

Solution

In this problem it is easier to group 25 and 4. Thus,

$$\begin{aligned}[(-17)(25)](4) &= (-17)[(25)(4)] && \text{Associative property for multiplication}\\ &= (-17)(100)\\ &= -1700\end{aligned}$$

■

EXAMPLE 3

Find the sum $17 + (-24) + (-31) + 19 + (-14) + 29 + 43$.

Solution

Certainly we could add in the order that the numbers appear. However, since addition is *commutative* and *associative* we could change the order and group any convenient way. For example, we could add all of the positive integers and add all of the negative integers, and then add these two results. In that case it is convenient to use the vertical format as follows.

$$\begin{array}{r} 17\\ 19\\ 29\\ \underline{43}\\ 108 \end{array} \qquad \begin{array}{r} \\ -24\\ -31\\ \underline{-14}\\ -69 \end{array} \qquad \begin{array}{r} \\ \\ 108\\ \underline{-69}\\ 39 \end{array}$$

■

For a problem such as Example 3 it might be advisable to first add in the order that the numbers appear, and then use the rearranging and regrouping idea as a check. Don't forget the link between addition and subtraction. A problem such as $18 - 43 + 52 - 17 - 23$ can be changed to $18 + (-43) + 52 + (-17) + (-23)$.

EXAMPLE 4

Simplify $(-75)(-4 + 100)$.

Solution

For such a problem, it is convenient to apply the *distributive property* and then to simplify.

$$\begin{aligned}(-75)(-4 + 100) &= (-75)(-4) + (-75)(100)\\ &= 300 + (-7500)\\ &= -7200\end{aligned}$$

■

EXAMPLE 5

Simplify $19(-26 + 25)$.

Solution

For this problem we are better off *not* applying the distributive property, but simply adding the numbers inside the parentheses first and then finding the indicated product. Thus,

$$19(-26 + 25) = 19(-1) = -19$$

■

EXAMPLE 6

Simplify $27(104) + 27(-4)$.

Solution

Keep in mind that the distributive property allows us to change from the form $a(b + c)$ to $ab + ac$ or from $ab + ac$ to $a(b + c)$. In this problem we want to use the latter change. Thus,

$$\begin{aligned}27(104) + 27(-4) &= 27(104 + (-4))\\ &= 27(100) = 2700\end{aligned}$$

■

Examples 4, 5, and 6 demonstrate an important issue. Sometimes the form $a(b + c)$ is the most convenient, but at other times the form $ab + ac$ is better. A suggestion in regard to this issue — as well as to the use of the other properties — is to think first, and then decide whether or not the properties can be used to make the manipulations easier.

Combining Similar Terms

Algebraic expressions such as

$$3x, \quad 5y, \quad 7xy, \quad -4abc, \quad \text{and} \quad z$$

are called **terms**. A term is an indicated product, and it may have any number of factors. We call the variables in a term **literal factors**, and we call the numerical factor the **numerical coefficient**. Thus, in $7xy$, the x and y are literal factors, and 7 is the numerical coefficient. The numerical coefficient of the term $-4abc$ is -4. Since $z = 1(z)$, the numerical coefficient of the term z is 1. Terms that have the same literal factors are called **like terms** or **similar terms**. Some examples of similar terms are

$$3x \text{ and } 9x, \qquad 14abc \text{ and } 29abc,$$

$$7xy \text{ and } -15xy, \qquad 4z,\ 9z, \text{ and } -14z$$

We can simplify algebraic expressions that contain similar terms by using a form of the distributive property. Consider the following examples:

$$3x + 5x = (3 + 5)x$$
$$= 8x$$
$$-9xy + 7xy = (-9 + 7)xy$$
$$= -2xy$$
$$18abc - 27abc = (18 - 27)abc$$
$$= (18 + (-27))abc$$
$$= -9abc$$
$$4x + x = (4 + 1)x \quad \text{Don't forget that } x = 1(x)$$
$$= 5x$$

More complicated expressions might first require some rearranging of terms by using the commutative property.

$$7x + 3y + 9x + 5y = 7x + 9x + 3y + 5y$$
$$= (7 + 9)x + (3 + 5)y$$
$$= 16x + 8y$$
$$9a - 4 - 13a + 6 = 9a + (-4) + (-13a) + 6$$
$$= 9a + (-13a) + (-4) + 6$$
$$= (9 + (-13))a + 2$$
$$= -4a + 2$$

As you become more adept at handling the various simplifying steps, you may want to do the steps mentally and thereby go directly from the given expression to the simplified form as follows.

$$19x - 14y + 12x + 16y = 31x + 2y,$$
$$17ab + 13c - 19ab - 30c = -2ab - 17c,$$
$$9x + 5 - 11x + 4 + x - 6 = -x + 3$$

Simplifying some algebraic expressions requires repeated applications of the distributive property as the next examples demonstrate.

$$5(x - 2) + 3(x + 4) = 5(x) - 5(2) + 3(x) + 3(4)$$
$$= 5x - 10 + 3x + 12$$
$$= 5x + 3x - 10 + 12$$
$$= 8x + 2$$

$$
\begin{aligned}
-7(y+1)-4(y-3) &= -7(y)-7(1)-4(y)-4(-3) \\
&= -7y-7-4y+12 \quad \text{Be careful with this sign} \\
&= -7y-4y-7+12 \\
&= -11y+5
\end{aligned}
$$

$$
\begin{aligned}
5(x+2)-(x+3) &= 5(x+2)-1(x+3) \quad \text{Remember } -a = -1a \\
&= 5(x)+5(2)-1(x)-1(3) \\
&= 5x+10-x-3 \\
&= 5x-x+10-3 \\
&= 4x+7
\end{aligned}
$$

After you are sure of each step, you can use a more simplified format.

$$
\begin{aligned}
5(a+4)-7(a-2) &= 5a+20-7a+14 \\
&= -2a+34
\end{aligned}
$$

$$
\begin{aligned}
9(z-7)+11(z+6) &= 9z-63+11z+66 \\
&= 20z+3
\end{aligned}
$$

$$
\begin{aligned}
-(x-2)+(x+6) &= -x+2+x+6 \\
&= 8
\end{aligned}
$$

Back to Evaluating Algebraic Expressions

To simplify by combining similar terms aids in the process of evaluating some algebraic expressions. The last examples of this section illustrate this idea.

EXAMPLE 7

Evaluate $8x - 2y + 3x + 5y$ for $x = 3$ and $y = -4$.

Solution

Let's first simplify the given expression.

$$8x - 2y + 3x + 5y = 11x + 3y$$

Now we can evaluate for $x = 3$ and $y = -4$.

$$
\begin{aligned}
11x + 3y &= 11(3) + 3(-4) \\
&= 33 + (-12) = 21
\end{aligned}
$$

■

EXAMPLE 8 Evaluate $2ab + 5c - 6ab + 12c$ for $a = 2$, $b = -3$, and $c = 7$.

Solution

$$\begin{aligned} 2ab + 5c - 6ab + 12c &= -4ab + 17c \\ &= -4(2)(-3) + 17(7) \quad \text{when } a = 2, b = -3, \text{ and } c = 7 \\ &= 24 + 119 = 143 \end{aligned}$$

■

EXAMPLE 9 Evaluate $8(x - 4) + 7(x + 3)$ for $x = 6$.

Solution

$$\begin{aligned} 8(x - 4) + 7(x + 3) &= 8x - 32 + 7x + 21 \\ &= 15x - 11 \\ &= 15(6) - 11 \quad \text{when } x = 6 \\ &= 79 \end{aligned}$$

■

Problem Set 1.5

For Problems 1–12, state the property that justifies each statement. For example, $3 + (-4) = (-4) + 3$ because of the commutative property for addition.

1. $3(7 + 8) = 3(7) + 3(8)$

2. $(-9)(17) = 17(-9)$

3. $-2 + (5 + 7) = (-2 + 5) + 7$

4. $-19 + 0 = -19$

5. $143(-7) = -7(143)$

6. $5(9 + (-4)) = 5(9) + 5(-4)$

7. $-119 + 119 = 0$

8. $-4 + (6 + 9) = (-4 + 6) + 9$

9. $-56 + 0 = -56$

10. $5 + (-12) = -12 + 5$

11. $[5(-8)]4 = 5[-8(4)]$

12. $[6(-4)]8 = 6[-4(8)]$

For Problems 13–30, simplify each numerical expression. Don't forget to take advantage of the properties if they can be used to simplify the computation.

13. $(-18 + 56) + 18$

14. $-72 + [72 + (-14)]$

15. $36 - 48 - 22 + 41$

16. $-24 + 18 + 19 - 30$

17. $(25)(-18)(-4)$

18. $(2)(-71)(50)$

19. $(4)(-16)(-9)(-25)$

20. $(-2)(18)(-12)(-5)$

21. $37(-42 - 58)$

22. $-46(-73 - 27)$

23. $59(36) + 59(64)$

24. $-49(72) - 49(28)$

25. $15(-14) + 16(-8)$

26. $-9(14) - 7(-16)$

27. $17 + (-18) - 19 - 14 + 13 - 17$

28. $-16 - 14 + 18 + 21 + 14 - 17$

29. $-21 + 22 - 23 + 27 + 21 - 19$

30. $24 - 26 - 29 + 26 + 18 + 29 - 17 - 10$

For Problems 31–62, simplify each algebraic expression by combining similar terms.

31. $9x - 14x$

32. $12x - 14x + x$

33. $4m + m - 8m$

34. $-6m - m + 17m$

35. $-9y + 5y - 7y$

36. $14y - 17y - 19y$

37. $4x - 3y - 7x + y$

38. $9x + 5y - 4x - 8y$

39. $-7a - 7b - 9a + 3b$

40. $-12a + 14b - 3a - 9b$

41. $6xy - x - 13xy + 4x$

42. $-7xy - 2x - xy + x$

43. $5x - 4 + 7x - 2x + 9$

44. $8x + 9 + 14x - 3x - 14$

45. $-2xy + 12 + 8xy - 16$

46. $14xy - 7 - 19xy - 6$

47. $-2a + 3b - 7b - b + 5a - 9a$

48. $-9a - a + 6b - 3a - 4b - b + a$

49. $13ab + 2a - 7a - 9ab + ab - 6a$

50. $-ab - a + 4ab + 7ab - 3a - 11ab$

51. $3(x + 2) + 5(x + 6)$

52. $7(x + 8) + 9(x + 1)$

53. $5(x - 4) + 6(x + 8)$

54. $-3(x + 2) - 4(x - 10)$

55. $9(x + 4) - (x - 8)$

56. $-(x - 6) + 5(x - 9)$

57. $3(a - 1) - 2(a - 6) + 4(a + 5)$

58. $-4(a + 2) + 6(a + 8) - 3(a - 6)$

59. $-2(m + 3) - 3(m - 1) + 8(m + 4)$

60. $5(m - 10) + 6(m - 11) - 9(m - 12)$

61. $(y + 3) - (y - 2) - (y + 6) - 7(y - 1)$

62. $-(y - 2) - (y + 4) - (y + 7) - 2(y + 3)$

For Problems 63–80, simplify each algebraic expression and then evaluate the resulting expression for the given values of the variables.

63. $3x + 5y + 4x - 2y$ for $x = -2$ and $y = 3$

64. $5x - 7y - 9x - 3y$ for $x = -1$ and $y = -4$

65. $5(x - 2) + 8(x + 6)$ for $x = -6$

66. $4(x - 6) + 9(x + 2)$ for $x = 7$

67. $8(x + 4) - 10(x - 3)$ for $x = -5$

68. $-(n + 2) - 3(n - 6)$ for $n = 10$

69. $(x - 6) - (x + 12)$ for $x = -3$

70. $(x + 12) - (x - 14)$ for $x = -11$

71. $2(x + y) - 3(x - y)$ for $x = -2$ and $y = 7$

72. $5(x - y) - 9(x + y)$ for $x = 4$ and $y = -4$

73. $2xy + 6 + 7xy - 8$ for $x = 2$ and $y = -4$

74. $4xy - 5 - 8xy + 9$ for $x = -3$ and $y = -3$

75. $5x - 9xy + 3x + 2xy$ for $x = 12$ and $y = -1$

76. $-9x + xy - 4xy - x$ for $x = 10$ and $y = -11$

77. $(a - b) - (a + b)$ for $a = 19$ and $b = -17$

78. $(a + b) - (a - b)$ for $a = -16$ and $b = 14$

79. $-3x + 7x + 4x - 2x - x$ for $x = -13$

80. $5x - 6x + x - 7x - x - 2x$ for $x = -15$

81. Use a calculator to check your answers for Problems 13–30.

THOUGHTS INTO WORDS

82. State in your own words the associative property for addition of integers.

83. State in your own words the distributive property for multiplication over addition.

84. Is $2 \cdot 3 \cdot 5 \cdot 7 \cdot 11 + 7$ a prime or composite number? Defend your answer.

FURTHER INVESTIGATIONS

For Problems 85–90, state whether the expressions in each problem are equivalent and explain why or why not.

85. $15a(x + y)$ and $5a(3x + 3y)$

86. $(-6a + 7b) + 11c$ and $7b + (11c - 6a)$

87. $2x - 3y + 4z$ and $2x - 4z + 3y$

88. $a + 5(x + y)$ and $(a + 5)x + y$

89. $7x + 6(y - 2z)$ and $6(2z - y) + 7x$

90. $9m + 8(3p - q)$ and $8(3p - q) + 9m$

Chapter 1 Summary

(1.1) To **simplify a numerical expression**, perform the operations in the following order.

1. Perform the operations inside the symbols of inclusion (parentheses and brackets) and above and below each fraction bar. Start with the innermost inclusion symbol.
2. Perform all multiplications and divisions in the order that they appear from left to right.
3. Perform all additions and subtractions in the order that they appear from left to right.

To **evaluate an algebraic expression**, substitute the given values for the variables into the algebraic expression and simplify the resulting numerical expression.

(1.2) A **prime number** is a whole number greater than 1 that has no factors (divisors) other than itself and 1. Whole numbers greater than 1 that are not prime numbers are called **composite numbers**. Every composite number has one and only one prime factorization.

The **greatest common factor** of 6 and 8 is 2, which means that 2 is the largest whole number divisor of both 6 and 8.

The **least common multiple** of 6 and 8 is 24, which means that 24 is the smallest nonzero multiple of both 6 and 8.

(1.3) The number line is a convenient visual aid for interpreting **addition of integers**.

Subtraction of integers is defined in terms of addition: *$a - b$ means $a + (-b)$.*

(1.4) To **multiply integers** we must remember that the product of two positives or two negatives is positive, and the product of a positive and a negative (either order) is negative.

To **divide integers** we must remember that the quotient of two positives or two negatives is positive, and the quotient of a positive and a negative (or a negative and a positive) is negative.

(1.5) The following basic properties help with numerical manipulations and serve as a basis for algebraic computations.

Commutative Properties $a + b = b + a$
$ab = ba$

Associative Properties $(a + b) + c = a + (b + c)$
$(ab)c = a(bc)$

Identity Properties $a + 0 = 0 + a = a$
$a(1) = 1(a) = a$

Additive Inverse Property
$a + (-a) = (-a) + a = 0$

Multiplication Property of Zero $a(0) = 0(a) = 0$

Multiplication Property of Negative One
$-1(a) = a(-1) = -a$

Distributive Properties $a(b + c) = ab + ac$
$a(b - c) = ab - ac$

Chapter 1 Review Problem Set

In Problems 1–10, perform the indicated operations.

1. $7 + (-10)$

2. $(-12) + (-13)$

3. $8 - 13$

4. $-6 - 9$

5. $-12 - (-11)$

6. $-17 - (-19)$

7. $(13)(-12)$

8. $(-14)(-18)$

9. $(-72) \div (-12)$

10. $117 \div (-9)$

In Problems 11–15, classify each of the numbers as *prime* or *composite*.

11. 73

12. 87

13. 63

14. 81

15. 91

In Problems 16–20, express each of the numbers as the product of prime factors.

16. 24

17. 63

18. 57

19. 64

20. 84

21. Find the greatest common factor of 36 and 54.

22. Find the greatest common factor of 48, 60, and 84.

23. Find the least common multiple of 18 and 20.

24. Find the least common multiple of 15, 27, and 35.

For Problems 25–38, simplify each of the numerical expressions.

25. $(19 + 56) + (-9)$

26. $43 - 62 + 12$

27. $8 + (-9) + (-16) + (-14) + 17 + 12$

28. $19 - 23 - 14 + 21 + 14 - 13$

29. $3(-4) - 6$

30. $(-5)(-4) - 8$

31. $(5)(-2) + (6)(-4)$

32. $(-6)(8) + (-7)(-3)$

33. $(-6)(3) - (-4)(-5)$

34. $(-7)(9) - (6)(5)$

35. $\dfrac{4(-7) - (3)(-2)}{-11}$

36. $\dfrac{(-4)(9) + (5)(-3)}{1 - 18}$

37. $3 - 2[4(-3 - 1)]$

38. $-6 - [3(-4 - 7)]$

39. A record high temperature of 125°F occurred in Laughlin, Nevada on June 29, 1994. A record low temperature of −50°F occurred in San Jacinto, Nevada on January 8, 1937. Find the difference between the record high and low temperatures.

40. In North America the highest elevation, which is on Mt. McKinley, Alaska, is 20,320 feet above sea level. The lowest elevation in North America, which is at Death Valley, California, is 282 feet below sea level. Find the absolute value of the difference in elevation between Mt. McKinley and Death Valley.

41. As a running back in a football game, Marquette carried the ball 7 times. On two plays he gained 6 yards each play; on another play he lost 4 yards; on the next three plays he gained 8 yards per play; and on the last play he lost 1 yard. Write a numerical expression that gives Marquette's overall yardage for the game, and simplify that expression.

42. Shelley started the month with \$3278 in her checking account. During the month she deposited \$175 each week for 4 weeks but had debit charges of \$50, \$189, \$160, \$20, and \$115. What is the balance in her checking account after these deposits and debits?

In Problems 43–54, simplify each algebraic expression by combining similar terms.

43. $12x + 3x - 7x$

44. $9y + 3 - 14y - 12$

45. $8x + 5y - 13x - y$

46. $9a + 11b + 4a - 17b$

47. $3ab - 4ab - 2a$

48. $5xy - 9xy + xy - y$

49. $3(x + 6) + 7(x + 8)$

50. $5(x - 4) - 3(x - 9)$

51. $-3(x - 2) - 4(x + 6)$

52. $-2x - 3(x - 4) + 2x$

53. $2(a - 1) - a - 3(a - 2)$

54. $-(a - 1) + 3(a - 2) - 4a + 1$

In Problems 55–68, evaluate each of the algebraic expressions for the given values of the variables.

55. $5x + 8y$ for $x = -7$ and $y = -3$

56. $7x - 9y$ for $x = -3$ and $y = 4$

57. $\dfrac{-5x - 2y}{-2x - 7}$ for $x = 6$ and $y = 4$

58. $\dfrac{-3x + 4y}{3x}$ for $x = -4$ and $y = -6$

59. $-2a + \frac{a - b}{a - 2}$ for $a = -5$ and $b = 9$

60. $\frac{2a + b}{b + 6} - 3b$ for $a = 3$ and $b = -4$

61. $5a + 6b - 7a - 2b$ for $a = -1$ and $b = 5$

62. $3x + 7y - 5x + y$ for $x = -4$ and $y = 3$

63. $2xy + 6 + 5xy - 8$ for $x = -1$ and $y = 1$

64. $7(x + 6) - 9(x + 1)$ for $x = -2$

65. $-3(x - 4) - 2(x + 8)$ for $x = 7$

66. $2(x - 1) - (x + 2) + 3(x - 4)$ for $x = -4$

67. $(a - b) - (a + b) - b$ for $a = -1$ and $b = -3$

68. $2ab - 3(a - b) + b + a$ for $a = 2$ and $b = -5$

Chapter 1 Test

For Problems 1–10, simplify each of the numerical expressions.

1. $6 + (-7) - 4 + 12$
2. $7 + 4(9) + 2$
3. $-4(2 - 8) + 14$
4. $5(-7) - (-3)(8)$
5. $8 \div (-4) + (-6)(9) - 2$
6. $(-8)(-7) + (-6) - (9)(12)$
7. $\dfrac{6(-4) - (-8)(-5)}{-16}$
8. $-14 + 23 - 17 - 19 + 26$
9. $(-14)(4) \div 4 + (-6)$
10. $6(-9) - (-8) - (-7)(4) + 11$
11. It was reported on the 5 o'clock weather show that the current temperature was 7°F. The forecast was for the temperature to drop 13 degrees by 6:00 A.M. If the forecast is correct, what will the temperature be at 6:00 A.M.?

For Problems 12–17, evaluate each of the algebraic expressions for the given values of the variables.

12. $7x - 9y$ for $x = -4$ and $y = -6$
13. $-4a - 6b$ for $a = -9$ and $b = 12$
14. $3xy - 8y + 5x$ for $x = 7$ and $y = -2$
15. $5(x - 4) - 6(x + 7)$ for $x = -5$
16. $3x - 2y - 4x - x + 7y$ for $x = 6$ and $y = -7$
17. $3(x - 2) - 5(x - 4) + 6(x - 1)$ for $x = -3$
18. Classify 79 as a prime or composite number.
19. Express 360 as a product of prime factors.
20. Find the greatest common factor of 36, 60, and 84.
21. Find the least common multiple of 9 and 24.
22. State the property of integers demonstrated by $[-3 + (-4)] + (-6) = -3 + [(-4) + (-6)]$.
23. State the property of integers demonstrated by $8(25 + 37) = 8(25) + 8(37)$.
24. Simplify $-7x + 9y - y + x - 2y - 7x$ by combining similar terms.
25. Simplify $-2(x - 4) - 5(x + 7) - 6(x - 1)$ by applying the distributive property and combining similar terms.

Real Numbers

AP/Wide World Photos

People that watch the stock market are familiar with rational numbers expressed in decimal form.

Caleb left an estate valued at \$750,000. His will states that three-fourths of the estate is to be divided equally among his three children. The numerical expression $\left(\frac{1}{3}\right)\left(\frac{3}{4}\right)(750{,}000)$ can be used to determine how much each of his three children should receive.

When the market opened on Monday morning, Garth bought some shares of a stock at \$13.25 per share. The rational numbers 0.75, -1.50, 2.25, -0.25, and -0.50 represent the daily changes in the market for that stock for the week. We use the numerical expression $13.25 + 0.75 + (-1.50) + 2.25 + (-0.25) + (-0.50)$ to determine the value of one share of Garth's stock when the market closed on Friday.

The width of a rectangle is w feet, and its length is four feet more than three times its width. The algebraic expression $2w + 2(3w + 4)$ represents the perimeter of the rectangle.

Again in this chapter we use the concepts of **numerical** and **algebraic expressions** to review some computational skills from arithmetic and to continue the transition from arithmetic to algebra. However, the set of **rational numbers** now becomes the primary focal point. We urge you to use this chapter to review and improve your arithmetic skills so that the algebraic concepts in subsequent chapters can build upon a solid foundation.

2.1 Rational Numbers: Multiplication and Division

Any number that can be written in the form $\frac{a}{b}$, where a and b are integers and b is not zero, we call a **rational number**. (We call the form $\frac{a}{b}$ a fraction or sometimes a common fraction.) The following are examples of rational numbers:

$$\frac{1}{2}, \quad \frac{7}{9}, \quad \frac{15}{7}, \quad \frac{-3}{4}, \quad \frac{5}{-7}, \quad \frac{-11}{-13}$$

All integers are rational numbers, because every integer can be expressed as the indicated quotient of two integers. Some examples follow.

$$6 = \frac{6}{1} = \frac{12}{2} = \frac{18}{3}, \quad \text{and so on}$$

$$27 = \frac{27}{1} = \frac{54}{2} = \frac{81}{3}, \quad \text{and so on}$$

$$0 = \frac{0}{1} = \frac{0}{2} = \frac{0}{3}, \quad \text{and so on}$$

Our work in Chapter 1 with division involving negative integers helps with the next three examples.

$$-4 = \frac{-4}{1} = \frac{-8}{2} = \frac{-12}{3}, \quad \text{and so on}$$

$$-6 = \frac{6}{-1} = \frac{12}{-2} = \frac{18}{-3}, \quad \text{and so on}$$

$$10 = \frac{10}{1} = \frac{-10}{-1} = \frac{-20}{-2}, \quad \text{and so on}$$

Observe the following general properties.

Property 2.1

$$\frac{-a}{b} = \frac{a}{-b} = -\frac{a}{b} \quad \text{and} \quad \frac{-a}{-b} = \frac{a}{b}$$

Therefore, a rational number such as $\frac{-2}{3}$ can also be written as $\frac{2}{-3}$ or $-\frac{2}{3}$. (However, we seldom express rational numbers with negative denominators.)

Multiplying Rational Numbers

We define multiplication of rational numbers in common fractional form as follows:

Definition 2.1

If a, b, c, and d are integers, and b and d are not equal to zero, then

$$\frac{a}{b} \cdot \frac{c}{d} = \frac{a \cdot c}{b \cdot d}$$

To multiply rational numbers in common fractional form we simply multiply numerators and multiply denominators. Furthermore, we see from the definition that the rational numbers are commutative and associative with respect to multiplication. We are free to rearrange and regroup factors as we do with integers. The following examples illustrate Definition 2.1:

$$\frac{1}{3} \cdot \frac{2}{5} = \frac{1 \cdot 2}{3 \cdot 5} = \frac{2}{15}$$

$$\frac{3}{4} \cdot \frac{5}{7} = \frac{3 \cdot 5}{4 \cdot 7} = \frac{15}{28}$$

$$\frac{-2}{3} \cdot \frac{7}{9} = \frac{-2 \cdot 7}{3 \cdot 9} = \frac{-14}{27} \quad \text{or} \quad -\frac{14}{27}$$

$$\frac{1}{5} \cdot \frac{9}{-11} = \frac{1 \cdot 9}{5(-11)} = \frac{9}{-55} \quad \text{or} \quad -\frac{9}{55}$$

$$-\frac{3}{4} \cdot \frac{7}{13} = \frac{-3}{4} \cdot \frac{7}{13} = \frac{-3 \cdot 7}{4 \cdot 13} = \frac{-21}{52} \quad \text{or} \quad -\frac{21}{52}$$

$$\frac{3}{5} \cdot \frac{5}{3} = \frac{3 \cdot 5}{5 \cdot 3} = \frac{15}{15} = 1$$

The last example is a very special case. **If the product of two numbers is 1, the numbers are said to be reciprocals of each other**.

Using Definition 2.1 and applying the multiplication property of one, the fraction $\frac{a \cdot k}{b \cdot k}$, where b and k are nonzero integers, simplifies as shown.

$$\frac{a \cdot k}{b \cdot k} = \frac{a}{b} \cdot \frac{k}{k} = \frac{a}{b} \cdot 1 = \frac{a}{b}$$

This result is stated as Property 2.2.

Property 2.2

If b and k are nonzero integers, and a is any integer, then

$$\frac{a \cdot k}{b \cdot k} = \frac{a}{b}$$

We often use Property 2.2 when we work with rational numbers. It is called the fundamental principle of fractions and provides the basis for equivalent fractions. In the following examples, the property will be used for, what is often called, reducing fractions to lowest terms or expressing fractions in simplest or reduced form.

EXAMPLE 1

Reduce $\frac{12}{18}$ to lowest terms.

Solution

$$\frac{12}{18} = \frac{2 \cdot 6}{3 \cdot 6} = \frac{2}{3}$$

■

EXAMPLE 2

Change $\frac{14}{35}$ to simplest form.

Solution

$$\frac{14}{35} = \frac{2 \cdot \not{7}}{5 \cdot \not{7}} = \frac{2}{5}$$

A common factor of 7 has been divided out of both numerator and denominator ■

EXAMPLE 3

Express $\frac{-24}{32}$ in reduced form.

Solution

$$\frac{-24}{32} = -\frac{3 \cdot 8}{4 \cdot 8} = -\frac{3}{4} \cdot \frac{8}{8} = -\frac{3}{4} \cdot 1 = -\frac{3}{4}$$

The multiplication property of 1 is being used ■

EXAMPLE 4

Reduce $-\frac{72}{90}$.

Solution

$$-\frac{72}{90} = -\frac{2 \cdot 2 \cdot 2 \cdot \not{3} \cdot \not{3}}{2 \cdot \not{3} \cdot \not{3} \cdot 5} = -\frac{4}{5}$$

The prime factored forms of the numerator and denominator may be used to help recognize common factors ■

The fractions may contain variables in the numerator or the denominator (or both), but this creates no great difficulty. Our thought processes remain the same, as these next examples illustrate. Variables appearing in the denominators represent **nonzero** integers.

EXAMPLE 5

Reduce $\frac{9x}{17x}$.

Solution

$$\frac{9x}{17x} = \frac{9 \cdot \cancel{x}}{17 \cdot \cancel{x}} = \frac{9}{17}$$ ■

EXAMPLE 6

Simplify $\frac{8x}{36y}$.

Solution

$$\frac{8x}{36y} = \frac{2 \cdot 2 \cdot 2 \cdot x}{2 \cdot 2 \cdot 3 \cdot 3 \cdot y} = \frac{2x}{9y}$$ ■

EXAMPLE 7

Express $\frac{-9xy}{30y}$ in reduced form.

Solution

$$\frac{-9xy}{30y} = -\frac{9xy}{30y} = -\frac{3 \cdot \cancel{3} \cdot x \cdot \cancel{y}}{2 \cdot \cancel{3} \cdot 5 \cdot \cancel{y}} = -\frac{3x}{10}$$ ■

EXAMPLE 8

Reduce $\frac{-7abc}{-9ac}$.

Solution

$$\frac{-7abc}{-9ac} = \frac{7abc}{9ac} = \frac{7\cancel{a}b\cancel{c}}{9\cancel{a}\cancel{c}} = \frac{7b}{9}$$ ■

We are now ready to consider multiplication problems with the understanding that the final answer should be expressed in reduced form. Study the following examples carefully; we use different methods to handle the problems.

EXAMPLE 9

Multiply $\frac{7}{9} \cdot \frac{5}{14}$.

Solution

$$\frac{7}{9} \cdot \frac{5}{14} = \frac{7 \cdot 5}{9 \cdot 14} = \frac{\cancel{7} \cdot 5}{3 \cdot 3 \cdot 2 \cdot \cancel{7}} = \frac{5}{18}$$ ■

EXAMPLE 10

Find the product of $\frac{8}{9}$ and $\frac{18}{24}$.

Solution

$$\frac{\overset{1}{\cancel{8}}}{\underset{1}{\cancel{9}}} \cdot \frac{\overset{2}{\cancel{18}}}{\underset{3}{\cancel{24}}} = \frac{2}{3}$$

A common factor of 8 has been divided out of 8 and 24, and a common factor of 9 has been divided out of 9 and 18 ■

EXAMPLE 11

Multiply $\left(-\frac{6}{8}\right)\left(\frac{14}{32}\right)$.

Solution

$$\left(-\frac{6}{8}\right)\left(\frac{14}{32}\right) = -\frac{\overset{3}{\cancel{6}} \cdot \overset{7}{\cancel{14}}}{\underset{4}{\cancel{8}} \cdot \underset{16}{\cancel{32}}} = -\frac{21}{64}$$

Divide a common factor of 2 out of 6 and 8, and a common factor of 2 out of 14 and 32 ■

EXAMPLE 12

Multiply $\left(-\frac{9}{4}\right)\left(-\frac{14}{15}\right)$.

Solution

$$\left(-\frac{9}{4}\right)\left(-\frac{14}{15}\right) = \frac{\cancel{3} \cdot 3 \cdot \cancel{2} \cdot 7}{\cancel{2} \cdot 2 \cdot \cancel{3} \cdot 5} = \frac{21}{10}$$

Immediately we recognize that *a negative times a negative is positive* ■

EXAMPLE 13

Multiply $\frac{9x}{7y} \cdot \frac{14y}{45}$.

Solution

$$\frac{9x}{7y} \cdot \frac{14y}{45} = \frac{\cancel{9} \cdot x \cdot \overset{2}{\cancel{14}} \cdot \cancel{y}}{\cancel{7} \cdot \cancel{y} \cdot \underset{5}{\cancel{45}}} = \frac{2x}{5}$$

■

EXAMPLE 14

Multiply $\frac{-6c}{7ab} \cdot \frac{14b}{5c}$.

Solution

$$\frac{-6c}{7ab} \cdot \frac{14b}{5c} = -\frac{2 \cdot 3 \cdot \cancel{c} \cdot 2 \cdot \cancel{7} \cdot \cancel{b}}{\cancel{7} \cdot a \cdot \cancel{b} \cdot 5 \cdot \cancel{c}} = -\frac{12}{5a}$$

■

Dividing Rational Numbers

The following example motivates a definition for division of rational numbers in fractional form.

$$\frac{\frac{3}{4}}{\frac{2}{3}} = \left(\frac{\frac{3}{4}}{\frac{2}{3}}\right)\left(\frac{\frac{3}{2}}{\frac{3}{2}}\right) = \frac{\left(\frac{3}{4}\right)\left(\frac{3}{2}\right)}{1} = \left(\frac{3}{4}\right)\left(\frac{3}{2}\right) = \frac{9}{8}$$

Notice that this is a form of 1, and $\frac{3}{2}$ is the reciprocal of $\frac{2}{3}$

In other words, $\frac{3}{4}$ divided by $\frac{2}{3}$ is equivalent to $\frac{3}{4}$ times $\frac{3}{2}$. The following definition for division should seem reasonable:

Definition 2.2

If b, c, and d are nonzero integers and a is any integer, then $\frac{a}{b} \div \frac{c}{d} = \frac{a}{b} \cdot \frac{d}{c}$

Notice that to divide $\frac{a}{b}$ by $\frac{c}{d}$, we multiply $\frac{a}{b}$ times the reciprocal of $\frac{c}{d}$, which is $\frac{d}{c}$. The following examples demonstrate the important steps of a division problem.

$$\frac{2}{3} \div \frac{1}{2} = \frac{2}{3} \cdot \frac{2}{1} = \frac{4}{3}$$

$$\frac{5}{6} \div \frac{3}{4} = \frac{5}{6} \cdot \frac{4}{3} = \frac{5 \cdot 4}{6 \cdot 3} = \frac{5 \cdot \cancel{2} \cdot 2}{\cancel{2} \cdot 3 \cdot 3} = \frac{10}{9}$$

$$-\frac{9}{12} \div \frac{3}{6} = -\frac{\overset{3}{\cancel{9}}}{\underset{2}{\cancel{12}}} \cdot \frac{\overset{1}{\cancel{6}}}{\underset{1}{\cancel{3}}} = -\frac{3}{2}$$

$$\left(-\frac{27}{56}\right) \div \left(-\frac{33}{72}\right) = \left(-\frac{27}{56}\right)\left(-\frac{72}{33}\right) = \frac{\overset{9}{\cancel{27}} \cdot \overset{9}{\cancel{72}}}{\underset{7}{\cancel{56}} \cdot \underset{11}{\cancel{33}}} = \frac{81}{77}$$

$$\frac{6}{7} \div 2 = \frac{6}{7} \cdot \frac{1}{2} = \frac{\overset{3}{\cancel{6}}}{7} \cdot \frac{1}{\underset{1}{\cancel{2}}} = \frac{3}{7}$$

$$\frac{5x}{7y} \div \frac{10}{28y} = \frac{5x}{7y} \cdot \frac{28y}{10} = \frac{\cancel{5} \cdot x \cdot \overset{2}{\cancel{\overset{4}{\cancel{28}}}} \cdot \cancel{y}}{\cancel{7} \cdot \cancel{y} \cdot \underset{2}{\cancel{10}}} = 2x$$

PROBLEM 1

Frank has purchased 50 candy bars to make s'mores for the Boy Scout troop. If he uses $\frac{2}{3}$ of a candy bar for each s'more, how many s'mores will he be able to make?

Solution

To find how many s'mores can be made, we need to divide 50 by $\frac{2}{3}$.

$$50 \div \frac{2}{3} = 50 \cdot \frac{3}{2} = \frac{50}{1} \cdot \frac{3}{2} = \frac{\overset{25}{\cancel{50}}}{1} \cdot \frac{3}{\underset{1}{\cancel{2}}} = \frac{75}{1} = 75$$

Frank can make 75 s'mores. ■

Problem Set 2.1

For Problems 1–24, reduce each fraction to lowest terms.

1. $\frac{8}{12}$
2. $\frac{12}{16}$
3. $\frac{16}{24}$
4. $\frac{18}{32}$
5. $\frac{15}{9}$
6. $\frac{48}{36}$
7. $\frac{-8}{48}$
8. $\frac{-3}{15}$
9. $\frac{27}{-36}$
10. $\frac{9}{-51}$
11. $\frac{-54}{-56}$
12. $\frac{-24}{-80}$
13. $\frac{24x}{44x}$
14. $\frac{15y}{25y}$
15. $\frac{9x}{21y}$
16. $\frac{4y}{30x}$
17. $\frac{14xy}{35y}$
18. $\frac{55xy}{77x}$
19. $\frac{-20ab}{52bc}$
20. $\frac{-23ac}{41c}$
21. $\frac{-56yz}{-49xy}$
22. $\frac{-21xy}{-14ab}$
23. $\frac{65abc}{91ac}$
24. $\frac{68xyz}{85yz}$

For Problems 25–58, multiply or divide as indicated, and express answers in reduced form.

25. $\frac{3}{4} \cdot \frac{5}{7}$
26. $\frac{4}{5} \cdot \frac{3}{11}$
27. $\frac{2}{7} \div \frac{3}{5}$
28. $\frac{5}{6} \div \frac{11}{13}$
29. $\frac{3}{8} \cdot \frac{12}{15}$
30. $\frac{4}{9} \cdot \frac{3}{2}$
31. $\frac{-6}{13} \cdot \frac{26}{9}$
32. $\frac{3}{4} \cdot \frac{-14}{12}$
33. $\frac{7}{9} \div \frac{5}{9}$
34. $\frac{3}{11} \div \frac{7}{11}$
35. $\frac{1}{4} \div \frac{-5}{6}$
36. $\frac{7}{8} \div \frac{14}{-16}$
37. $\left(-\frac{8}{10}\right)\left(-\frac{10}{32}\right)$
38. $\left(-\frac{6}{7}\right)\left(-\frac{21}{24}\right)$
39. $-9 \div \frac{1}{3}$
40. $-10 \div \frac{1}{4}$
41. $\frac{5x}{9y} \cdot \frac{7y}{3x}$
42. $\frac{4a}{11b} \cdot \frac{6b}{7a}$

43. $\frac{6a}{14b} \cdot \frac{16b}{18a}$

44. $\frac{5y}{8x} \cdot \frac{14z}{15y}$

45. $\frac{10x}{-9y} \cdot \frac{15}{20x}$

46. $\frac{3x}{4y} \cdot \frac{-8w}{9z}$

47. $ab \cdot \frac{2}{b}$

48. $3xy \cdot \frac{4}{x}$

49. $\left(-\frac{7x}{12y}\right)\left(-\frac{24y}{35x}\right)$

50. $\left(-\frac{10a}{15b}\right)\left(-\frac{45b}{65a}\right)$

51. $\frac{3}{x} \div \frac{6}{y}$

52. $\frac{6}{x} \div \frac{14}{y}$

53. $\frac{5x}{9y} \div \frac{13x}{36y}$

54. $\frac{3x}{5y} \div \frac{7x}{10y}$

55. $\frac{-7}{x} \div \frac{9}{x}$

56. $\frac{8}{y} \div \frac{28}{-y}$

57. $\frac{-4}{n} \div \frac{-18}{n}$

58. $\frac{-34}{n} \div \frac{-51}{n}$

For Problems 59–74, perform the operations as indicated, and express answers in lowest terms.

59. $\frac{3}{4} \cdot \frac{8}{9} \cdot \frac{12}{20}$

60. $\frac{5}{6} \cdot \frac{9}{10} \cdot \frac{8}{7}$

61. $\left(-\frac{3}{8}\right)\left(\frac{13}{14}\right)\left(-\frac{12}{9}\right)$

62. $\left(-\frac{7}{9}\right)\left(\frac{5}{11}\right)\left(-\frac{18}{14}\right)$

63. $\left(\frac{3x}{4y}\right)\left(\frac{8}{9x}\right)\left(\frac{12y}{5}\right)$

64. $\left(\frac{2x}{3y}\right)\left(\frac{5y}{x}\right)\left(\frac{9}{4x}\right)$

65. $\left(-\frac{2}{3}\right)\left(\frac{3}{4}\right) \div \frac{1}{8}$

66. $\frac{3}{4} \cdot \frac{4}{5} \div \frac{1}{6}$

67. $\frac{5}{7} \div \left(-\frac{5}{6}\right)\left(-\frac{6}{7}\right)$

68. $\left(-\frac{3}{8}\right) \div \left(-\frac{4}{5}\right)\left(\frac{1}{2}\right)$

69. $\left(-\frac{6}{7}\right) \div \left(\frac{5}{7}\right)\left(-\frac{5}{6}\right)$

70. $\left(-\frac{4}{3}\right) \div \left(\frac{4}{5}\right)\left(\frac{3}{5}\right)$

71. $\left(\frac{4}{9}\right)\left(-\frac{9}{8}\right) \div \left(-\frac{3}{4}\right)$

72. $\left(-\frac{7}{8}\right)\left(\frac{4}{7}\right) \div \left(-\frac{3}{2}\right)$

73. $\left(\frac{5}{2}\right)\left(\frac{2}{3}\right) \div \left(-\frac{1}{4}\right) \div (-3)$

74. $\frac{1}{3} \div \left(\frac{3}{4}\right)\left(\frac{1}{2}\right) \div 2$

75. Maria's department has $\frac{3}{4}$ of all of the accounts within the ABC Advertising Agency. Maria is personally responsible for $\frac{1}{3}$ of all accounts in her department. For what portion of all of the accounts at ABC is Maria personally responsible?

76. Pablo has a board that is $4\frac{1}{2}$ feet long, and he wants to cut it into three pieces of the same length (see Figure 2.1). Find the length of each of the three pieces.

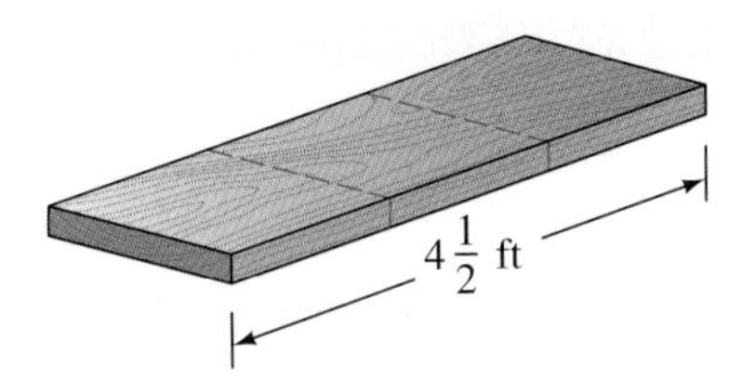

Figure 2.1

77. A recipe for a birthday cake calls for $\frac{3}{4}$ cup of sugar. How much sugar is needed to make 3 cakes?

78. Jonas left an estate valued at \$750,000. His will states that three-fourths of the estate is to be divided equally among his three children. How much should each receive?

79. One of Arlene's recipes calls for $3\frac{1}{2}$ cups of milk. If she wants to make one-half of the recipe, how much milk should she use?

80. The total length of the four sides of a square is $8\frac{2}{3}$ yards. How long is each side of the square?

81. If it takes $3\frac{1}{4}$ yards of material to make one dress, how much material is needed for 20 dresses?

82. If your calculator is equipped to handle rational numbers in $\frac{a}{b}$ form, check your answers for Problems 1–12 and 59–74.

THOUGHTS INTO WORDS

83. State in your own words the property

$$-\frac{a}{b} = \frac{-a}{b} = \frac{a}{-b}$$

84. Explain how you would reduce $\frac{72}{117}$ to lowest terms.

85. What mistake was made in the following simplification process?

$$\frac{1}{2} \div \left(\frac{2}{3}\right)\left(\frac{3}{4}\right) \div 3 = \frac{1}{2} \div \frac{1}{2} \div 3 = \frac{1}{2} \cdot 2 \cdot \frac{1}{3} = \frac{1}{3}$$

How would you correct the error?

FURTHER INVESTIGATIONS

86. The division problem $35 \div 7$ can be interpreted as "how many 7s are there in 35?" Likewise, a division problem such as $3 \div \frac{1}{2}$ can be interpreted as, "how many one-halves in 3?" Use this *how-many* interpretation to do the following division problems.

(a) $4 \div \frac{1}{2}$ **(b)** $3 \div \frac{1}{4}$

(c) $5 \div \frac{1}{8}$ **(d)** $6 \div \frac{1}{7}$

(e) $\frac{5}{6} \div \frac{1}{6}$ **(f)** $\frac{7}{8} \div \frac{1}{8}$

87. Estimation is important in mathematics. In each of the following, estimate whether the answer is larger than 1 or smaller than 1 by using the *how-many* idea from Problem 86.

(a) $\frac{3}{4} \div \frac{1}{2}$ **(b)** $1 \div \frac{7}{8}$

(c) $\frac{1}{2} \div \frac{3}{4}$ **(d)** $\frac{8}{7} \div \frac{7}{8}$

(e) $\frac{2}{3} \div \frac{1}{4}$ **(f)** $\frac{3}{5} \div \frac{3}{4}$

88. Reduce each of the following to lowest terms. Don't forget that we reviewed some divisibility rules in Problem Set 1.2.

(a) $\frac{99}{117}$ **(b)** $\frac{175}{225}$

(c) $\frac{-111}{123}$ **(d)** $\frac{-234}{270}$

(e) $\frac{270}{495}$ **(f)** $\frac{324}{459}$

(g) $\frac{91}{143}$ **(h)** $\frac{187}{221}$

2.2 Addition and Subtraction of Rational Numbers

Suppose that it is one-fifth of a mile between your dorm and the student center, and two-fifths of a mile between the student center and the library along a straight line as indicated in Figure 2.2. The total distance between your dorm and the library is three-fifths of a mile, and we write $\frac{1}{5} + \frac{2}{5} = \frac{3}{5}$.

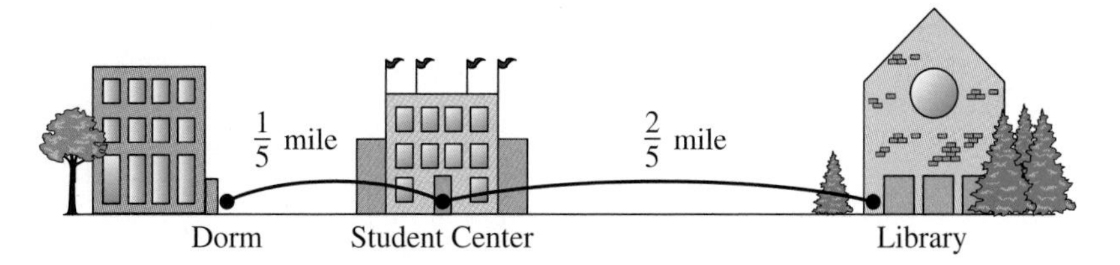

Figure 2.2

A pizza is cut into seven equal pieces and you eat two of the pieces. How much of the pizza (Figure 2.3) remains? We represent the whole pizza by $\frac{7}{7}$ and then conclude that $\frac{7}{7} - \frac{2}{7} = \frac{5}{7}$ of the pizza remains.

Figure 2.3

These examples motivate the following definition for addition and subtraction of rational numbers in $\frac{a}{b}$ form:

Definition 2.3

If a, b, and c are integers, and b is not zero, then

$$\frac{a}{b} + \frac{c}{b} = \frac{a+c}{b} \qquad \text{Addition}$$

$$\frac{a}{b} - \frac{c}{b} = \frac{a-c}{b} \qquad \text{Subtraction}$$

We say that rational numbers **with common denominators** can be added or subtracted by adding or subtracting the numerators and placing the results over the common denominator. Consider the following examples:

$$\frac{3}{7} + \frac{2}{7} = \frac{3+2}{7} = \frac{5}{7}$$

$$\frac{7}{8} - \frac{2}{8} = \frac{7-2}{8} = \frac{5}{8}$$

$$\frac{2}{6} + \frac{1}{6} = \frac{2+1}{6} = \frac{3}{6} = \frac{1}{2}$$ We agree to reduce the final answer

$$\frac{3}{11} - \frac{5}{11} = \frac{3-5}{11} = \frac{-2}{11} \quad \text{or} \quad -\frac{2}{11}$$

$$\frac{5}{x} + \frac{7}{x} = \frac{5+7}{x} = \frac{12}{x}$$

$$\frac{9}{y} - \frac{3}{y} = \frac{9-3}{y} = \frac{6}{y}$$

In the last two examples, the variables x and y cannot be equal to zero in order to exclude division by zero. It is always necessary to restrict denominators to nonzero values, although we will not take the time or space to list such restrictions for every problem.

How do we add or subtract if the fractions do not have a common denominator? We use the fundamental principle of fractions, $\frac{a}{b} = \frac{a \cdot k}{b \cdot k}$, and obtain equivalent fractions that have a common denominator. **Equivalent fractions** are fractions that name the same number. Consider the following example, which shows the details.

EXAMPLE 1

Add $\frac{1}{2} + \frac{1}{3}$.

Solution

$$\frac{1}{2} = \frac{1 \cdot 3}{2 \cdot 3} = \frac{3}{6}$$ $\frac{1}{2}$ and $\frac{3}{6}$ are equivalent fractions naming the same number

$$\frac{1}{3} = \frac{1 \cdot 2}{3 \cdot 2} = \frac{2}{6}$$ $\frac{1}{3}$ and $\frac{2}{6}$ are equivalent fractions naming the same number

$$\frac{1}{2} + \frac{1}{3} = \frac{3}{6} + \frac{2}{6} = \frac{3+2}{6} = \frac{5}{6}$$ ■

Notice that we chose 6 as the common denominator, and 6 is the least common multiple of the original denominators 2 and 3. (Recall that the least common multiple is the smallest nonzero whole number divisible by the given numbers.) In general, we use the least common multiple of the denominators of the fractions to be added or subtracted as a **least common denominator** (LCD).

Recall from Section 1.2 that the least common multiple may be found either by inspection or by using prime factorization forms of the numbers. Let's consider some examples involving these procedures.

EXAMPLE 2

Add $\frac{1}{4} + \frac{2}{5}$.

Solution

By inspection we see that the LCD is 20. Thus both fractions can be changed to equivalent fractions that have a denominator of 20.

$$\frac{1}{4} + \frac{2}{5} = \frac{1 \cdot 5}{4 \cdot 5} + \frac{2 \cdot 4}{5 \cdot 4} = \frac{5}{20} + \frac{8}{20} = \frac{13}{20}$$

Use of fundamental principle of fractions

■

EXAMPLE 3

Subtract $\frac{5}{8} - \frac{7}{12}$.

Solution

By inspection it is clear that the LCD is 24.

$$\frac{5}{8} - \frac{7}{12} = \frac{5 \cdot 3}{8 \cdot 3} - \frac{7 \cdot 2}{12 \cdot 2} = \frac{15}{24} - \frac{14}{24} = \frac{1}{24}$$

■

If the LCD is not obvious by inspection, then we can use the technique from Chapter 1 to find the least common multiple. We proceed as follows.

Step 1 Express each denominator as a product of prime factors.

Step 2 The LCD contains each different prime factor as many times as the *most* times it appears in any one of the factorizations from step 1.

EXAMPLE 4

Add $\frac{5}{18} + \frac{7}{24}$.

Solution

If we cannot find the LCD by inspection, then we can use the prime factorization forms.

$$\left.\begin{array}{l} 18 = 2 \cdot 3 \cdot 3 \\ 24 = 2 \cdot 2 \cdot 2 \cdot 3 \end{array}\right\} \longrightarrow \quad \text{LCD} = 2 \cdot 2 \cdot 2 \cdot 3 \cdot 3 = 72$$

$$\frac{5}{18} + \frac{7}{24} = \frac{5 \cdot 4}{18 \cdot 4} + \frac{7 \cdot 3}{24 \cdot 3} = \frac{20}{72} + \frac{21}{72} = \frac{41}{72}$$

■

EXAMPLE 5

Subtract $\frac{3}{14} - \frac{8}{35}$.

Solution

$$\left.\begin{array}{l} 14 = 2 \cdot 7 \\ 35 = 5 \cdot 7 \end{array}\right\} \longrightarrow \quad \text{LCD} = 2 \cdot 5 \cdot 7 = 70$$

$$\frac{3}{14} - \frac{8}{35} = \frac{3 \cdot 5}{14 \cdot 5} - \frac{8 \cdot 2}{35 \cdot 2} = \frac{15}{70} - \frac{16}{70} = \frac{-1}{70} \quad \text{or} \quad -\frac{1}{70}$$

■

EXAMPLE 6

Add $\frac{-5}{8} + \frac{3}{14}$.

Solution

$$\left.\begin{aligned} 8 &= 2 \cdot 2 \cdot 2 \\ 14 &= 2 \cdot 7 \end{aligned}\right\} \longrightarrow \text{LCD} = 2 \cdot 2 \cdot 2 \cdot 7 = 56$$

$$\frac{-5}{8} + \frac{3}{14} = \frac{-5 \cdot 7}{8 \cdot 7} + \frac{3 \cdot 4}{14 \cdot 4} = \frac{-35}{56} + \frac{12}{56} = \frac{-23}{56} \quad \text{or} \quad -\frac{23}{56}$$ ■

EXAMPLE 7

Add $-3 + \frac{2}{5}$.

Solution

$$-3 + \frac{2}{5} = \frac{-3 \cdot 5}{1 \cdot 5} + \frac{2}{5} = \frac{-15}{5} + \frac{2}{5} = \frac{-15 + 2}{5} = \frac{-13}{5} \quad \text{or} \quad -\frac{13}{5}$$ ■

Denominators that contain variables do not complicate the situation very much, as the next examples illustrate.

EXAMPLE 8

Add $\frac{2}{x} + \frac{3}{y}$.

Solution

By inspection, the LCD is xy.

$$\frac{2}{x} + \frac{3}{y} = \frac{2 \cdot y}{x \cdot y} + \frac{3 \cdot x}{y \cdot x} = \frac{2y}{xy} + \frac{3x}{xy} = \frac{2y + 3x}{xy}$$

Commutative property

■

EXAMPLE 9

Subtract $\frac{3}{8x} - \frac{5}{12y}$.

Solution

$$\left.\begin{aligned} 8x &= 2 \cdot 2 \cdot 2 \cdot x \\ 12y &= 2 \cdot 2 \cdot 3 \cdot y \end{aligned}\right\} \longrightarrow \text{LCD} = 2 \cdot 2 \cdot 2 \cdot 3 \cdot x \cdot y = 24xy$$

$$\frac{3}{8x} - \frac{5}{12y} = \frac{3 \cdot 3y}{8x \cdot 3y} - \frac{5 \cdot 2x}{12y \cdot 2x} = \frac{9y}{24xy} - \frac{10x}{24xy} = \frac{9y - 10x}{24xy}$$ ■

EXAMPLE 10 Add $\frac{7}{4a} + \frac{-5}{6bc}$.

Solution

$$\left.\begin{aligned} 4a &= 2 \cdot 2 \cdot a \\ 6bc &= 2 \cdot 3 \cdot b \cdot c \end{aligned}\right\} \longrightarrow \text{LCD} = 2 \cdot 2 \cdot 3 \cdot a \cdot b \cdot c = 12abc$$

$$\frac{7}{4a} + \frac{-5}{6bc} = \frac{7 \cdot 3bc}{4a \cdot 3bc} + \frac{-5 \cdot 2a}{6bc \cdot 2a} = \frac{21bc}{12abc} + \frac{-10a}{12abc} = \frac{21bc - 10a}{12abc}$$ ■

■ Simplifying Numerical Expressions

Let's now consider simplifying numerical expressions that contain rational numbers. As with integers, multiplications and divisions are done first, and then the additions and subtractions are performed. In these next examples only the major steps are shown, so be sure that you can fill in all of the other details.

EXAMPLE 11 Simplify $\frac{3}{4} + \frac{2}{3} \cdot \frac{3}{5} - \frac{1}{2} \cdot \frac{1}{5}$.

Solution

$$\frac{3}{4} + \frac{2}{3} \cdot \frac{3}{5} - \frac{1}{2} \cdot \frac{1}{5} = \frac{3}{4} + \frac{2}{5} - \frac{1}{10}$$ Perform the multiplications

$$= \frac{15}{20} + \frac{8}{20} - \frac{2}{20} = \frac{15 + 8 - 2}{20} = \frac{21}{20}$$ Change to equivalent fractions and combine numerators ■

EXAMPLE 12 Simplify $\frac{3}{5} \div \frac{8}{5} + \left(-\frac{1}{2}\right)\left(\frac{1}{3}\right) + \frac{5}{12}$.

Solution

$$\frac{3}{5} \div \frac{8}{5} + \left(-\frac{1}{2}\right)\left(\frac{1}{3}\right) + \frac{5}{12} = \frac{3}{5} \cdot \frac{5}{8} + \left(-\frac{1}{2}\right)\left(\frac{1}{3}\right) + \frac{5}{12}$$ Change division to multiply by the reciprocal

$$= \frac{3}{8} + \frac{-1}{6} + \frac{5}{12}$$

$$= \frac{9}{24} + \frac{-4}{24} + \frac{10}{24}$$

$$= \frac{9 + (-4) + 10}{24}$$

$$= \frac{15}{24} = \frac{5}{8}$$ Reduce! ■

The distributive property, $a(b + c) = ab + ac$, holds true for rational numbers and — as with integers — can be used to facilitate manipulation.

EXAMPLE 13

Simplify $12\left(\frac{1}{3} + \frac{1}{4}\right)$.

Solution

For help in this situation, let's change the form by applying the distributive property.

$$\begin{aligned} 12\left(\frac{1}{3} + \frac{1}{4}\right) &= 12\left(\frac{1}{3}\right) + 12\left(\frac{1}{4}\right) \\ &= 4 + 3 \\ &= 7 \end{aligned}$$

■

EXAMPLE 14

Simplify $\frac{5}{8}\left(\frac{1}{2} + \frac{1}{3}\right)$.

Solution

In this case it may be easier not to apply the distributive property but to work with the expression in its given form.

$$\begin{aligned} \frac{5}{8}\left(\frac{1}{2} + \frac{1}{3}\right) &= \frac{5}{8}\left(\frac{3}{6} + \frac{2}{6}\right) \\ &= \frac{5}{8}\left(\frac{5}{6}\right) \\ &= \frac{25}{48} \end{aligned}$$

■

Examples 13 and 14 emphasize a point we made in Chapter 1. Think first, and decide whether or not the properties can be used to make the manipulations easier. Example 15 illustrates how to combine similar terms that have fractional coefficients.

EXAMPLE 15

Simplify $\frac{1}{2}x + \frac{2}{3}x - \frac{3}{4}x$ by combining similar terms.

Solution

We can use the distributive property and our knowledge of adding and subtracting rational numbers to solve this type of problem.

$$\begin{aligned} \frac{1}{2}x + \frac{2}{3}x - \frac{3}{4}x &= \left(\frac{1}{2} + \frac{2}{3} - \frac{3}{4}\right)x \\ &= \left(\frac{6}{12} + \frac{8}{12} - \frac{9}{12}\right)x \\ &= \frac{5}{12}x \end{aligned}$$

■

PROBLEM 1

Brian brought 5 cups of flour along on a camping trip. He wants to make biscuits and cake for tonight's supper. It takes $\frac{3}{4}$ of a cup of flour for the biscuits and $2\frac{3}{4}$ cups of flour for the cake. How much flour will be left over for the rest of his camping trip?

Solution

Let's do this problem in two steps. First add the amounts of flour needed for the biscuits and cake.

$$\frac{3}{4} + 2\frac{3}{4} = \frac{3}{4} + \frac{11}{4} = \frac{14}{4} = \frac{7}{2}$$

Then to find the amount of flour left over, we will subtract $\frac{7}{2}$ from 5.

$$5 - \frac{7}{2} = \frac{10}{2} - \frac{7}{2} = \frac{3}{2} = 1\frac{1}{2}$$

So $1\frac{1}{2}$ cups of flour are left over. ■

Problem Set 2.2

For Problems 1–64, add or subtract as indicated, and express your answers in lowest terms.

1. $\frac{2}{7} + \frac{3}{7}$
2. $\frac{3}{11} + \frac{5}{11}$
3. $\frac{7}{9} - \frac{2}{9}$
4. $\frac{11}{13} - \frac{6}{13}$
5. $\frac{3}{4} + \frac{9}{4}$
6. $\frac{5}{6} + \frac{7}{6}$
7. $\frac{11}{12} - \frac{3}{12}$
8. $\frac{13}{16} - \frac{7}{16}$
9. $\frac{1}{8} - \frac{5}{8}$
10. $\frac{2}{9} - \frac{5}{9}$
11. $\frac{5}{24} + \frac{11}{24}$
12. $\frac{7}{36} + \frac{13}{36}$
13. $\frac{8}{x} + \frac{7}{x}$
14. $\frac{17}{y} + \frac{12}{y}$
15. $\frac{5}{3y} + \frac{1}{3y}$
16. $\frac{3}{8x} + \frac{1}{8x}$
17. $\frac{1}{3} + \frac{1}{5}$
18. $\frac{1}{6} + \frac{1}{8}$
19. $\frac{15}{16} - \frac{3}{8}$
20. $\frac{13}{12} - \frac{1}{6}$
21. $\frac{7}{10} + \frac{8}{15}$
22. $\frac{7}{12} + \frac{5}{8}$
23. $\frac{11}{24} + \frac{5}{32}$
24. $\frac{5}{18} + \frac{8}{27}$
25. $\frac{5}{18} - \frac{13}{24}$
26. $\frac{1}{24} - \frac{7}{36}$
27. $\frac{5}{8} - \frac{2}{3}$
28. $\frac{3}{4} - \frac{5}{6}$
29. $-\frac{2}{13} - \frac{7}{39}$
30. $-\frac{3}{11} - \frac{13}{33}$
31. $-\frac{3}{14} + \frac{1}{21}$
32. $-\frac{3}{20} + \frac{14}{25}$
33. $-4 - \frac{3}{7}$
34. $-2 - \frac{5}{6}$
35. $\frac{3}{4} - 6$
36. $\frac{5}{8} - 7$
37. $\frac{3}{x} + \frac{4}{y}$
38. $\frac{5}{x} + \frac{8}{y}$

39. $\frac{7}{a} - \frac{2}{b}$

40. $\frac{13}{a} - \frac{4}{b}$

41. $\frac{2}{x} + \frac{7}{2x}$

42. $\frac{5}{2x} + \frac{7}{x}$

43. $\frac{10}{3x} - \frac{2}{x}$

44. $\frac{13}{4x} - \frac{3}{x}$

45. $\frac{1}{x} - \frac{7}{5x}$

46. $\frac{2}{x} - \frac{17}{6x}$

47. $\frac{3}{2y} + \frac{5}{3y}$

48. $\frac{7}{3y} + \frac{9}{4y}$

49. $\frac{5}{12y} - \frac{3}{8y}$

50. $\frac{9}{4y} - \frac{5}{9y}$

51. $\frac{1}{6n} - \frac{7}{8n}$

52. $\frac{3}{10n} - \frac{11}{15n}$

53. $\frac{5}{3x} + \frac{7}{3y}$

54. $\frac{3}{2x} + \frac{7}{2y}$

55. $\frac{8}{5x} + \frac{3}{4y}$

56. $\frac{1}{5x} + \frac{5}{6y}$

57. $\frac{7}{4x} - \frac{5}{9y}$

58. $\frac{2}{7x} - \frac{11}{14y}$

59. $-\frac{3}{2x} - \frac{5}{4y}$

60. $-\frac{13}{8a} - \frac{11}{10b}$

61. $3 + \frac{2}{x}$

62. $\frac{5}{x} + 4$

63. $2 - \frac{3}{2x}$

64. $-1 - \frac{1}{3x}$

For Problems 65–80, simplify each numerical expression and express your answers in reduced form.

65. $\frac{1}{4} - \frac{3}{8} + \frac{5}{12} - \frac{1}{24}$

66. $\frac{3}{4} + \frac{2}{3} - \frac{1}{6} + \frac{5}{12}$

67. $\frac{5}{6} + \frac{2}{3} \cdot \frac{3}{4} - \frac{1}{4} \cdot \frac{2}{5}$

68. $\frac{2}{3} + \frac{1}{2} \cdot \frac{2}{5} - \frac{1}{3} \cdot \frac{1}{5}$

69. $\frac{3}{4} \cdot \frac{6}{9} - \frac{5}{6} \cdot \frac{8}{10} + \frac{2}{3} \cdot \frac{6}{8}$

70. $\frac{3}{5} \cdot \frac{5}{7} + \frac{2}{3} \cdot \frac{3}{5} - \frac{1}{7} \cdot \frac{2}{5}$

71. $4 - \frac{2}{3} \cdot \frac{3}{5} - 6$

72. $3 + \frac{1}{2} \cdot \frac{1}{3} - 2$

73. $\frac{4}{5} - \frac{10}{12} - \frac{5}{6} \div \frac{14}{8} + \frac{10}{21}$

74. $\frac{3}{4} \div \frac{6}{5} + \frac{8}{12} \cdot \frac{6}{9} - \frac{5}{12}$

75. $24\left(\frac{3}{4} - \frac{1}{6}\right)$ Don't forget the distributive property!

76. $18\left(\frac{2}{3} + \frac{1}{9}\right)$

77. $64\left(\frac{3}{16} + \frac{5}{8} - \frac{1}{4} + \frac{1}{2}\right)$

78. $48\left(\frac{5}{12} - \frac{1}{6} + \frac{3}{8}\right)$

79. $\frac{7}{13}\left(\frac{2}{3} - \frac{1}{6}\right)$

80. $\frac{5}{9}\left(\frac{1}{2} + \frac{1}{4}\right)$

For Problems 81–96, simplify each algebraic expression by combining similar terms.

81. $\frac{1}{3}x + \frac{2}{5}x$

82. $\frac{1}{4}x + \frac{2}{3}x$

83. $\frac{1}{3}a - \frac{1}{8}a$

84. $\frac{2}{5}a - \frac{2}{7}a$

85. $\frac{1}{2}x + \frac{2}{3}x + \frac{1}{6}x$

86. $\frac{1}{3}x + \frac{2}{5}x + \frac{5}{6}x$

87. $\frac{3}{5}n - \frac{1}{4}n + \frac{3}{10}n$

88. $\frac{2}{5}n - \frac{7}{10}n + \frac{8}{15}n$

89. $n + \frac{4}{3}n - \frac{1}{9}n$

90. $2n - \frac{6}{7}n + \frac{5}{14}n$

91. $-n - \frac{7}{9}n - \frac{5}{12}n$

92. $-\frac{3}{8}n - n - \frac{3}{14}n$

93. $\frac{3}{7}x + \frac{1}{4}y + \frac{1}{2}x + \frac{7}{8}y$

94. $\frac{5}{6}x + \frac{3}{4}y + \frac{4}{9}x + \frac{7}{10}y$

95. $\frac{2}{9}x + \frac{5}{12}y - \frac{7}{15}x - \frac{13}{15}y$

96. $-\frac{9}{10}x - \frac{3}{14}y + \frac{2}{25}x + \frac{5}{21}y$

97. Beth wants to make three sofa pillows for her new sofa. After consulting the chart provided by the fabric shop, she decides to make a 12″ round pillow, an 18″ square pillow, and a 12″ × 16″ rectangular pillow. According to the chart, how much fabric will Beth need to purchase?

Fabric Shop Chart

10″ round	$\frac{3}{8}$ yard
12″ round	$\frac{1}{2}$ yard
12″ square	$\frac{5}{8}$ yard
18″ square	$\frac{3}{4}$ yard
12″ × 16″ rectangular	$\frac{7}{8}$ yard

98. Marcus is decorating his room and plans on hanging three prints that are each $13\frac{3}{8}$ inches wide. He is going to hang the prints side by side with $2\frac{1}{4}$ inches between the prints. What width of wall space is needed to display the three prints?

99. From a board that is $12\frac{1}{2}$ feet long, a piece $1\frac{3}{4}$ feet long is cut off from one end. Find the length of the remaining piece of board.

100. Vinay has a board that is $6\frac{1}{2}$ feet long. If he cuts off a piece $2\frac{3}{4}$ feet long, how long is the remaining piece of board?

101. Mindy takes a daily walk of $2\frac{1}{2}$ miles. One day a thunderstorm forced her to stop her walk after $\frac{3}{4}$ of a mile. By how much was her walk shortened that day?

102. Blake Scott leaves $\frac{1}{4}$ of his estate to the Boy Scouts, $\frac{2}{5}$ to the local cancer fund, and the rest to his church. What fractional part of the estate does the church receive?

103. A triangular plot of ground measures $14\frac{1}{2}$ yards by $12\frac{1}{3}$ yards by $9\frac{5}{6}$ yards. How many yards of fencing are needed to enclose the plot?

104. For her exercise program, Lian jogs for $2\frac{1}{2}$ miles, then walks for $\frac{3}{4}$ of a mile, and finally jogs for another $1\frac{1}{4}$ miles. Find the total distance that Lian covers.

105. If your calculator handles rational numbers in $\frac{a}{b}$ form, check your answers for Problems 65–80.

THOUGHTS INTO WORDS

106. Give a step-by-step description of how to add the rational numbers $\frac{3}{8}$ and $\frac{5}{18}$.

107. Give a step-by-step description of how to add the fractions $\frac{5}{4x}$ and $\frac{7}{6x}$.

108. The will of a deceased collector of antique automobiles specified that his cars be left to his three children. Half were to go to his elder son, $\frac{1}{3}$ to his daughter, and $\frac{1}{9}$ to his younger son. At the time of his death, 17 cars were in the collection. The administrator of his estate borrowed a car to make 18. Then he distributed the cars as follows:

Elder son: $\frac{1}{2}(18) = 9$

Daughter: $\frac{1}{3}(18) = 6$

Younger son: $\frac{1}{9}(18) = 2$

This totaled 17 cars, so he then returned the borrowed car. Where is the error in this solution?

2.3 Real Numbers and Algebraic Expressions

We classify decimals — also called decimal fractions — as **terminating, repeating**, or **nonrepeating**. Here are examples of these classifications:

Terminating decimals	Repeating decimals	Nonrepeating decimals
0.3	0.333333 . . .	0.5918654279 . . .
0.26	0.5466666 . . .	0.26224222722229 . . .
0.347	0.14141414 . . .	0.1451172111931111148 . . .
0.9865	0.237237237 . . .	0.645751311 . . .

A repeating decimal has a block of digits that repeats indefinitely. This repeating block of digits may contain any number of digits and may or may not begin repeating immediately after the decimal point. Technically a terminating decimal can be thought of as repeating zeros after the last digit. For example, $0.3 = 0.30 = 0.300 = 0.3000$, etc.

In Section 2.1 we defined a rational number to be any number that can be written in the form $\frac{a}{b}$, where a and b are integers and b is not zero. **A rational number can also be defined as any number that has a terminating or repeating decimal representation**. Thus we can express rational numbers in either common-fraction form or decimal-fraction form, as the next examples illustrate. A repeating decimal can also be written by using a bar over the digits that repeat; for example, $0.\overline{14}$.

Terminating decimals	Repeating decimals
$\frac{3}{4} = 0.75$	$\frac{1}{3} = 0.3333\ldots$
$\frac{1}{8} = 0.125$	$\frac{2}{3} = 0.66666\ldots$

$$\frac{5}{16} = 0.3125 \qquad \frac{1}{6} = 0.166666\ldots$$

$$\frac{7}{25} = 0.28 \qquad \frac{1}{12} = 0.08333\ldots$$

$$\frac{2}{5} = 0.4 \qquad \frac{14}{99} = 0.14141414\ldots$$

The nonrepeating decimals are called **irrational numbers** and do appear in forms other than decimal form. For example, $\sqrt{2}$, $\sqrt{3}$, and π are irrational numbers; a partial representation for each of these follows.

$$\left.\begin{aligned} \sqrt{2} &= 1.414213562373\ldots \\ \sqrt{3} &= 1.73205080756887\ldots \\ \pi &= 3.14159265358979\ldots \end{aligned}\right\} \text{ Nonrepeating decimals}$$

(We will do more work with the irrationals in Chapter 9.)

The rational numbers together with the irrationals form the set of **real numbers**. The following tree diagram of the real number system is helpful for summarizing some basic ideas.

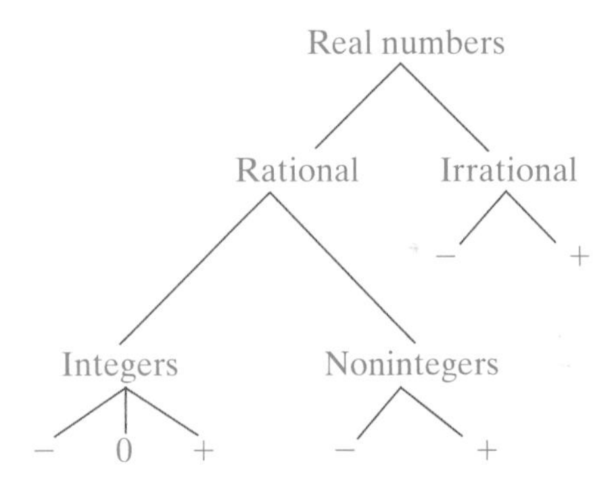

Any real number can be traced down through the diagram as follows.

5 is real, rational, an integer, and positive,

-4 is real, rational, an integer, and negative,

$\frac{3}{4}$ is real, rational, a noninteger, and positive,

0.23 is real, rational, a noninteger, and positive,

$-0.161616\ldots$ is real, rational, a noninteger, and negative,

$\sqrt{7}$ is real, irrational, and positive,

$-\sqrt{2}$ is real, irrational, and negative

In Section 1.3, we associated the set of integers with evenly spaced points on a line as indicated in Figure 2.4. This idea of associating numbers with points on a

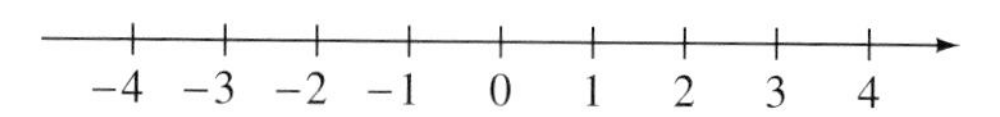

Figure 2.4

line can be extended so that there is a one-to-one correspondence between points on a line and the entire set of real numbers (as shown in Figure 2.5). That is to say, to each real number there corresponds one and only one point on the line, and to each point on the line there corresponds one and only one real number. The line is often referred to as the **real number line**, and the number associated with each point on the line is called the **coordinate** of the point.

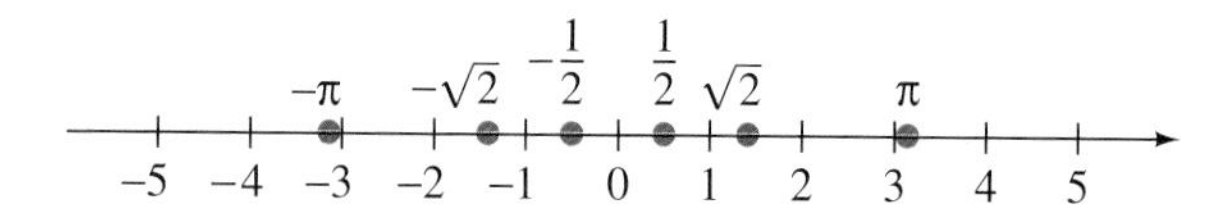

Figure 2.5

The properties we discussed in Section 1.5 pertaining to integers are true for all real numbers; we restate them here for your convenience. The multiplicative inverse property was added to the list; a discussion of that property follows.

Commutative Property of Addition

If a and b are real numbers, then

$$a + b = b + a$$

Commutative Property of Multiplication

If a and b are real numbers, then

$$ab = ba$$

Associative Property of Addition

If a, b, and c are real numbers, then

$$(a + b) + c = a + (b + c)$$

Associative Property of Multiplication

If a, b, and c are real numbers, then

$$(ab)c = a(bc)$$

Identity Property of Addition

If a is any real number, then

$$a + 0 = 0 + a = a$$

Identity Property of Multiplication

If a is any real number, then

$$a(1) = 1(a) = a$$

Additive Inverse Property

For every real number a, there exists a real number $-a$, such that

$$a + (-a) = (-a) + a = 0$$

Multiplication Property of Zero

If a is any real number, then

$$a(0) = 0(a) = 0$$

Multiplicative Property of Negative One

If a is any real number, then

$$a(-1) = -1(a) = -a$$

Multiplicative Inverse Property

For every nonzero real number a, there exists a real number $\frac{1}{a}$, such that

$$a\left(\frac{1}{a}\right) = \frac{1}{a}(a) = 1$$

Distributive Property

If a, b, and c are real numbers, then

$$a(b + c) = ab + ac$$

The number $\frac{1}{a}$ is called the **multiplicative inverse** or the **reciprocal of *a***. For example, the reciprocal of 2 is $\frac{1}{2}$ and $2\left(\frac{1}{2}\right) = \frac{1}{2}(2) = 1$. Likewise, the reciprocal of $\frac{1}{2}$ is $\frac{1}{\frac{1}{2}} = 2$. Therefore, 2 and $\frac{1}{2}$ are said to be reciprocals (or multiplicative inverses) of each other. Also, $\frac{2}{5}$ and $\frac{5}{2}$ are multiplicative inverses, and $\left(\frac{2}{5}\right)\left(\frac{5}{2}\right) = 1$. Since division by zero is undefined, zero does not have a reciprocal.

Basic Operations with Decimals

The basic operations with decimals may be related to the corresponding operation with common fractions. For example, $0.3 + 0.4 = 0.7$ because $\frac{3}{10} + \frac{4}{10} = \frac{7}{10}$, and $0.37 - 0.24 = 0.13$ because $\frac{37}{100} - \frac{24}{100} = \frac{13}{100}$. In general, to add or subtract decimals, we add or subtract the hundredths, the tenths, the ones, the tens, and so on. To keep place values aligned, we line up the decimal points.

Addition		Subtraction	
1	1 11	6 16	8 11 13
2.14	5.214	7.6	9.235
3.12	3.162	4.9	6.781
5.16	7.218	2.7	2.454
10.42	8.914		
	24.508		

The following examples can be used to formulate a general rule for multiplying decimals.

$$\text{Because } \frac{7}{10} \cdot \frac{3}{10} = \frac{21}{100}, \text{ then } (0.7)(0.3) = 0.21.$$

$$\text{Because } \frac{9}{10} \cdot \frac{23}{100} = \frac{207}{1000}, \text{ then } (0.9)(0.23) = 0.207.$$

$$\text{Because } \frac{11}{100} \cdot \frac{13}{100} = \frac{143}{10{,}000}, \text{ then } (0.11)(0.13) = 0.0143.$$

In general, to multiply decimals we (1) multiply the numbers and ignore the decimal points, and then (2) insert the decimal point in the product so that the number of digits to the right of the decimal point in the product is equal to the sum of the number of digits to the right of the decimal point in each factor.

(0.7)	$\times$	(0.3)	$=$	0.21
↑		↑		↑
One digit to right	+	One digit to right	=	Two digits to right
(0.9)		(0.23)	$=$	0.207
↑		↑		↑
One digit to right	+	Two digits to right	=	Three digits to right
(0.11)		(0.13)	$=$	0.0143
↑		↑		↑
Two digits to right	+	Two digits to right	=	Four digits to right

We frequently use the vertical format when multiplying decimals.

$$\begin{array}{rl} 41.2 & \text{One digit to right} \\ \underline{0.13} & \underline{\text{Two digits to right}} \\ 1236 & \\ \underline{412} & \\ 5.356 & \text{Three digits to right} \end{array} \qquad \begin{array}{rl} 0.021 & \text{Three digits to right} \\ \underline{0.03} & \underline{\text{Two digits to right}} \\ 0.00063 & \text{Five digits to right} \end{array}$$

Notice that in the last example we actually multiplied $3 \cdot 21$ and then inserted three 0s to the left so that there would be five digits to the right of the decimal point.

Once again let's look at some links between common fractions and decimals.

$$\text{Because } \frac{6}{10} \div 2 = \frac{\overset{3}{\cancel{6}}}{10} \cdot \frac{1}{\cancel{2}} = \frac{3}{10}, \quad \text{then } 2\overset{0.3}{\overline{)0.6}}$$

$$\text{Because } \frac{39}{100} \div 13 = \frac{\overset{3}{\cancel{39}}}{100} \cdot \frac{1}{\cancel{13}} = \frac{3}{100}, \quad \text{then } 13\overset{0.03}{\overline{)0.39}}$$

$$\text{Because } \frac{85}{100} \div 5 = \frac{\overset{17}{\cancel{85}}}{100} \cdot \frac{1}{\cancel{5}} = \frac{17}{100}, \quad \text{then } 5\overset{0.17}{\overline{)0.85}}$$

In general, to divide a decimal by a nonzero whole number we (1) place the decimal point in the quotient directly above the decimal point in the dividend

$$\left(\text{Divisor}\overset{\text{Quotient}}{\overline{)\text{Dividend}}}\right),$$

and then (2) divide as with whole numbers, except that in the division process, zeros are placed in the quotient immediately to the right of the decimal point in order to show the correct place value.

$$\begin{array}{r} 0.121 \\ 4\overline{)0.484} \end{array} \qquad \begin{array}{r} 0.24 \\ 32\overline{)7.68} \\ \underline{6\,4} \\ 1\,28 \\ \underline{1\,28} \end{array} \qquad \begin{array}{r} 0.019 \\ 12\overline{)0.228} \\ \underline{12} \\ 108 \\ \underline{108} \end{array}$$

Zero needed to show the correct place value

Don't forget that division can be checked by multiplication. For example, since $(12)(0.019) = 0.228$ we know that our last division example is correct.

Problems involving division by a decimal are easier to handle if we change the problem to an equivalent problem that has a whole number divisor. Consider the following examples in which the original division problem was changed to fractional form to show the reasoning involved in the procedure.

$$0.6\overline{)0.24} \rightarrow \frac{0.24}{0.6} = \left(\frac{0.24}{0.6}\right)\left(\frac{10}{10}\right) = \frac{2.4}{6} \rightarrow \begin{array}{r} 0.4 \\ 6\overline{)2.4} \end{array},$$

$$0.12\overline{)0.156} \rightarrow \frac{0.156}{0.12} = \left(\frac{0.156}{0.12}\right)\left(\frac{100}{100}\right) = \frac{15.6}{12} \rightarrow \begin{array}{r} 1.3 \\ 12\overline{)15.6} \\ \underline{12} \\ 36 \\ \underline{36} \end{array},$$

$$1.3\overline{)0.026} \rightarrow \frac{0.026}{1.3} = \left(\frac{0.026}{1.3}\right)\left(\frac{10}{10}\right) = \frac{0.26}{13} \rightarrow \begin{array}{r} 0.02 \\ 13\overline{)0.26} \\ \underline{26} \end{array}$$

The format commonly used with such problems is as follows.

$$\begin{array}{r} 5.6 \\ {}_{x}21.\overline{)1_{x}17.6} \\ \underline{1\,05} \\ 12\,6 \\ \underline{12\,6} \end{array}$$

The arrows indicate that the divisor and dividend were multiplied by 100, which changes the divisor to a whole number

$$\begin{array}{r} 0.04 \\ 3_{x}7.\overline{)_{x}1.48} \\ \underline{1\,48} \end{array}$$

The divisor and dividend were multiplied by 10

Our agreements for operating with positive and negative integers extend to all real numbers. For example, the product of two negative real numbers is a positive real number. Make sure that you agree with the following results. (You may need to do some work on scratch paper since the steps are not shown.)

$0.24 + (-0.18) = 0.06,$ $\qquad (-0.4)(0.8) = -0.32,$

$-7.2 + 5.1 = -2.1,$ $\qquad (-0.5)(-0.13) = 0.065,$

$-0.6 + (-0.8) = -1.4,$ $\qquad (1.4) \div (-0.2) = -7,$

$2.4 - 6.1 = -3.7,$ $\qquad (-0.18) \div (0.3) = -0.6,$

$0.31 - (-0.52) = 0.83,$ $\qquad (-0.24) \div (-4) = 0.06$

$(0.2)(-0.3) = -0.06,$

Numerical and algebraic expressions may contain the decimal form as well as the fractional form of rational numbers. We continue to follow the agreement that multiplications and divisions are done *first* and then the additions and subtractions unless parentheses indicate otherwise. The following examples illustrate a variety of situations that involve both the decimal form and fractional form of rational numbers.

EXAMPLE 1

Simplify $6.3 \div 7 + (4)(2.1) - (0.24) \div (-0.4)$.

Solution

$$\begin{aligned} 6.3 \div 7 + (4)(2.1) - (0.24) \div (-0.4) &= 0.9 + 8.4 - (-0.6) \\ &= 0.9 + 8.4 + 0.6 \\ &= 9.9 \end{aligned}$$

■

EXAMPLE 2

Evaluate $\frac{3}{5}a - \frac{1}{7}b$ for $a = \frac{5}{2}$ and $b = -1$.

Solution

$$\begin{aligned} \frac{3}{5}a - \frac{1}{7}b &= \frac{3}{5}\left(\frac{5}{2}\right) - \frac{1}{7}(-1) \quad \text{for } a = \frac{5}{2} \text{ and } b = -1 \\ &= \frac{3}{2} + \frac{1}{7} \\ &= \frac{21}{14} + \frac{2}{14} \\ &= \frac{23}{14} \end{aligned}$$

■

EXAMPLE 3

Evaluate $\frac{1}{2}x + \frac{2}{3}x - \frac{1}{5}x$ for $x = -\frac{3}{4}$.

Solution

First, let's combine similar terms by using the distributive property.

$$\begin{aligned} \frac{1}{2}x + \frac{2}{3}x - \frac{1}{5}x &= \left(\frac{1}{2} + \frac{2}{3} - \frac{1}{5}\right)x \\ &= \left(\frac{15}{30} + \frac{20}{30} - \frac{6}{30}\right)x \\ &= \frac{29}{30}x \end{aligned}$$

Now we can evaluate.

$$\frac{29}{30}x = \frac{29}{30}\left(-\frac{3}{4}\right) \quad \text{when } x = -\frac{3}{4}$$

$$= \frac{29}{\underset{10}{\cancel{30}}}\left(-\frac{\overset{1}{\cancel{3}}}{4}\right) = -\frac{29}{40}$$

■

EXAMPLE 4

Evaluate $2x + 3y$ for $x = 1.6$ and $y = 2.7$.

Solution

$$2x + 3y = 2(1.6) + 3(2.7) \quad \text{when } x = 1.6 \text{ and } y = 2.7$$
$$= 3.2 + 8.1 = 11.3$$

■

EXAMPLE 5

Evaluate $0.9x + 0.7x - 0.4x + 1.3x$ for $x = 0.2$.

Solution

First, let's combine similar terms by using the distributive property.

$$0.9x + 0.7x - 0.4x + 1.3x = (0.9 + 0.7 - 0.4 + 1.3)x = 2.5x$$

Now we can evaluate.

$$2.5x = (2.5)(0.2) \quad \text{for } x = 0.2$$
$$= 0.5$$

■

PROBLEM 1

A layout artist is putting together a group of images. She has four images whose widths are 1.35 centimeters, 2.6 centimeters, 5.45 centimeters, and 3.2 centimeters. If the images are set side by side, what will be their combined width?

Solution

To find the combined width, we need to add the widths.

$$\begin{array}{r} 1.35 \\ 2.6 \\ 5.45 \\ +3.2 \\ \hline 12.60 \end{array}$$

So the combined width would be 12.6 centimeters. ■

Problem Set 2.3

For Problems 1–8, classify the real numbers by tracing down the diagram on p. 63.

1. -2

2. $1/3$

3. $\sqrt{5}$

4. $-0.09090909\ldots$

5. 0.16

6. $-\sqrt{3}$

7. $-8/7$

8. 0.125

For Problems 9–40, perform the indicated operations.

9. $0.37 + 0.25$

10. $7.2 + 4.9$

11. $2.93 - 1.48$

12. $14.36 - 5.89$

13. $(7.6) + (-3.8)$

14. $(6.2) + (-2.4)$

15. $(-4.7) + 1.4$

16. $(-14.1) + 9.5$

17. $-3.8 + 11.3$

18. $-2.5 + 14.8$

19. $6.6 - (-1.2)$

20. $18.3 - (-7.4)$

21. $-11.5 - (-10.6)$

22. $-14.6 - (-8.3)$

23. $-17.2 - (-9.4)$

24. $-21.4 - (-14.2)$

25. $(0.4)(2.9)$

26. $(0.3)(3.6)$

27. $(-0.8)(0.34)$

28. $(-0.7)(0.67)$

29. $(9)(-2.7)$

30. $(8)(-7.6)$

31. $(-0.7)(-64)$

32. $(-0.9)(-56)$

33. $(-0.12)(-0.13)$

34. $(-0.11)(-0.15)$

35. $1.56 \div 1.3$

36. $7.14 \div 2.1$

37. $5.92 \div (-0.8)$

38. $-2.94 \div 0.6$

39. $-0.266 \div (-0.7)$

40. $-0.126 \div (-0.9)$

For Problems 41–54, simplify each of the numerical expressions.

41. $16.5 - 18.7 + 9.4$

42. $17.7 + 21.2 - 14.6$

43. $0.34 - 0.21 - 0.74 + 0.19$

44. $-5.2 + 6.8 - 4.7 - 3.9 + 1.3$

45. $0.76(0.2 + 0.8)$

46. $9.8(1.8 - 0.8)$

47. $0.6(4.1) + 0.7(3.2)$

48. $0.5(74) - 0.9(87)$

49. $7(0.6) + 0.9 - 3(0.4) + 0.4$

50. $-5(0.9) - 0.6 + 4.1(6) - 0.9$

51. $(0.96) \div (-0.8) + 6(-1.4) - 5.2$

52. $(-2.98) \div 0.4 - 5(-2.3) + 1.6$

53. $5(2.3) - 1.2 - 7.36 \div 0.8 + 0.2$

54. $0.9(12) \div 0.4 - 1.36 \div 17 + 9.2$

For Problems 55–68, simplify each algebraic expression by combining similar terms.

55. $x - 0.4x - 1.8x$

56. $-2x + 1.7x - 4.6x$

57. $5.4n - 0.8n - 1.6n$

58. $6.2n - 7.8n - 1.3n$

59. $-3t + 4.2t - 0.9t + 0.2t$

60. $7.4t - 3.9t - 0.6t + 4.7t$

61. $3.6x - 7.4y - 9.4x + 10.2y$

62. $5.7x + 9.4y - 6.2x - 4.4y$

63. $0.3(x - 4) + 0.4(x + 6) - 0.6x$

64. $0.7(x + 7) - 0.9(x - 2) + 0.5x$

65. $6(x - 1.1) - 5(x - 2.3) - 4(x + 1.8)$

66. $4(x + 0.7) - 9(x + 0.2) - 3(x - 0.6)$

67. $5(x - 0.5) + 0.3(x - 2) - 0.7(x + 7)$

68. $-8(x - 1.2) + 6(x - 4.6) + 4(x + 1.7)$

For Problems 69–82, evaluate each algebraic expression for the given values of the variables. Don't forget that for some problems it might be helpful to combine similar terms first and then to evaluate.

69. $x + 2y + 3z$ for $x = \frac{3}{4}$, $y = \frac{1}{3}$, and $z = -\frac{1}{6}$

70. $2x - y - 3z$ for $x = -\frac{2}{5}$, $y = -\frac{3}{4}$, and $z = \frac{1}{2}$

71. $\frac{3}{5}y - \frac{2}{3}y - \frac{7}{15}y$ for $y = -\frac{5}{2}$

72. $\frac{1}{2}x + \frac{2}{3}x - \frac{3}{4}x$ for $x = \frac{7}{8}$

73. $-x - 2y + 4z$ for $x = 1.7$, $y = -2.3$, and $z = 3.6$

74. $-2x + y - 5z$ for $x = -2.9$, $y = 7.4$, and $z = -6.7$

75. $5x - 7y$ for $x = -7.8$ and $y = 8.4$

76. $8x - 9y$ for $x = -4.3$ and $y = 5.2$

77. $0.7x + 0.6y$ for $x = -2$ and $y = 6$

78. $0.8x + 2.1y$ for $x = 5$ and $y = -9$

79. $1.2x + 2.3x - 1.4x - 7.6x$ for $x = -2.5$

80. $3.4x - 1.9x + 5.2x$ for $x = 0.3$

81. $-3a - 1 + 7a - 2$ for $a = 0.9$

82. $5x - 2 + 6x + 4$ for $x = -1.1$

83. Tanya bought 400 shares of one stock at $14.78 per share, and 250 shares of another stock at $16.36 per share. How much did she pay for the 650 shares?

84. On a trip Brent bought the following amounts of gasoline: 9.7 gallons, 12.3 gallons, 14.6 gallons, 12.2 gallons, 13.8 gallons, and 15.5 gallons. How many gallons of gasoline did he purchase on the trip?

85. Kathrin has a piece of copper tubing that is 76.4 centimeters long. She needs to cut it into four pieces of equal length. Find the length of each piece.

86. On a trip Biance filled the gasoline tank and noted that the odometer read 24,876.2 miles. After the next filling the odometer read 25,170.5 miles. It took 13.5 gallons of gasoline to fill the tank. How many miles per gallon did she get on that tank of gasoline?

87. The total length of the four sides of a square is 18.8 centimeters. How long is each side of the square?

88. When the market opened on Monday morning, Garth bought some shares of a stock at $13.25 per share. The daily changes in the market for that stock for the week were 0.75, −1.50, 2.25, −0.25, and −0.50. What was the value of one share of that stock when the market closed on Friday afternoon?

89. Victoria bought two pounds of Gala apples at $1.79 per pound and three pounds of Fuji apples at $.99 per pound. How much did she spend for the apples?

90. In 2005 the average speed of the winner of the Daytona 500 was 135.173 miles per hour. In 1978 the average speed of the winner was 159.73 miles per hour. How much faster was the average speed of the winner in 1978 compared to the winner in 2005?

91. Andrea's automobile averages 25.4 miles per gallon. With this average rate of fuel consumption, what distance should she be able to travel on a 12.7-gallon tank of gasoline?

92. Use a calculator to check your answers for Problems 41–54.

THOUGHTS INTO WORDS

93. At this time how would you describe the difference between arithmetic and algebra?

94. How have the properties of the real numbers been used thus far in your study of arithmetic and algebra?

95. Do you think that $2\sqrt{2}$ is a rational or an irrational number? Defend your answer.

FURTHER INVESTIGATIONS

96. Without doing the actual dividing, defend the statement, "$\frac{1}{7}$ produces a repeating decimal." [*Hint*: Think about the possible remainders when dividing by 7.]

97. Express each of the following in repeating decimal form.

(a) $\frac{1}{7}$ **(b)** $\frac{2}{7}$

(c) $\frac{4}{9}$ **(d)** $\frac{5}{6}$

(e) $\frac{3}{11}$ **(f)** $\frac{1}{12}$

98. (a) How can we tell that $\frac{5}{16}$ will produce a terminating decimal?

(b) How can we tell that $\frac{7}{15}$ will not produce a terminating decimal?

(c) Determine which of the following will produce a terminating decimal: $\frac{7}{8}, \frac{11}{16}, \frac{5}{12}, \frac{7}{24}, \frac{11}{75}, \frac{13}{32}, \frac{17}{40}, \frac{11}{30}, \frac{9}{20}, \frac{3}{64}$.

2.4 Exponents

We use exponents to indicate repeated multiplication. For example, we can write $5 \cdot 5 \cdot 5$ as 5^3, where the 3 indicates that 5 is to be used as a factor 3 times. The following general definition is helpful:

Definition 2.4

If n is a positive integer, and b is any real number, then

$$b^n = \underbrace{bbb \cdots b}_{n \text{ factors of } b}$$

We refer to the b as the **base** and n as the **exponent**. The expression b^n can be read as "b to the nth **power**." We frequently associate the terms **squared** and **cubed** with exponents of 2 and 3, respectively. For example, b^2 is read as "b squared" and b^3 as "b cubed." An exponent of 1 is usually not written, so b^1 is written as b. The following examples further clarify the concept of an exponent.

$$2^3 = 2 \cdot 2 \cdot 2 = 8, \qquad (0.6)^2 = (0.6)(0.6) = 0.36,$$

$$3^5 = 3 \cdot 3 \cdot 3 \cdot 3 \cdot 3 = 243, \qquad \left(\frac{1}{2}\right)^4 = \frac{1}{2} \cdot \frac{1}{2} \cdot \frac{1}{2} \cdot \frac{1}{2} = \frac{1}{16},$$

$$(-5)^2 = (-5)(-5) = 25, \qquad -5^2 = -(5 \cdot 5) = -25$$

We especially want to call your attention to the last two examples. Notice that $(-5)^2$ means that -5 is the base, which is to be used as a factor twice. However, -5^2 means that 5 is the base, and after 5 is squared, we take the opposite of that result.

Exponents provide a way of writing algebraic expressions in compact form. Sometimes we need to change from the compact form to an expanded form as these next examples demonstrate.

$x^4 = x \cdot x \cdot x \cdot x,$ $\quad (2x)^3 = (2x)(2x)(2x),$

$2y^3 = 2 \cdot y \cdot y \cdot y,$ $\quad (-2x)^3 = (-2x)(-2x)(-2x),$

$-3x^5 = -3 \cdot x \cdot x \cdot x \cdot x \cdot x,$ $\quad -x^2 = -(x \cdot x),$

$a^2 + b^2 = a \cdot a + b \cdot b$

At other times we need to change from an expanded form to a more compact form using the exponent notation.

$$3 \cdot x \cdot x = 3x^2$$

$$2 \cdot 5 \cdot x \cdot x \cdot x = 10x^3$$

$$3 \cdot 4 \cdot x \cdot x \cdot y = 12x^2y$$

$$7 \cdot a \cdot a \cdot a \cdot b \cdot b = 7a^3b^2$$

$$(2x)(3y) = 2 \cdot x \cdot 3 \cdot y = 2 \cdot 3 \cdot x \cdot y = 6xy$$

$$(3a^2)(4a) = 3 \cdot a \cdot a \cdot 4 \cdot a = 3 \cdot 4 \cdot a \cdot a \cdot a = 12a^3$$

$$(-2x)(3x) = -2 \cdot x \cdot 3 \cdot x = -2 \cdot 3 \cdot x \cdot x = -6x^2$$

The commutative and associative properties for multiplication allowed us to rearrange and regroup factors in the last three examples.

The concept of *exponent* can be used to extend our work with combining similar terms, operating with fractions, and evaluating algebraic expressions. Study the following examples very carefully; they will help you pull together many ideas.

EXAMPLE 1

Simplify $4x^2 + 7x^2 - 2x^2$ by combining similar terms.

Solution

By applying the distributive property, we obtain

$$\begin{aligned} 4x^2 + 7x^2 - 2x^2 &= (4 + 7 - 2)x^2 \\ &= 9x^2 \end{aligned}$$

■

EXAMPLE 2

Simplify $-8x^3 + 9y^2 + 4x^3 - 11y^2$ by combining similar terms.

Solution

By rearranging terms and then applying the distributive property we obtain

$$\begin{aligned} -8x^3 + 9y^2 + 4x^3 - 11y^2 &= -8x^3 + 4x^3 + 9y^2 - 11y^2 \\ &= (-8 + 4)x^3 + (9 - 11)y^2 \\ &= -4x^3 - 2y^2 \end{aligned}$$

■

EXAMPLE 3 Simplify $-7x^2 + 4x + 3x^2 - 9x$.

Solution

$$\begin{aligned} -7x^2 + 4x + 3x^2 - 9x &= -7x^2 + 3x^2 + 4x - 9x \\ &= (-7 + 3)x^2 + (4 - 9)x \\ &= -4x^2 - 5x \end{aligned}$$

■

As soon as you feel comfortable with this process of combining similar terms, you may want to do some of the steps mentally—then your work may appear as follows.

$$9a^2 + 6a^2 - 12a^2 = 3a^2,$$

$$6x^2 + 7y^2 - 3x^2 - 11y^2 = 3x^2 - 4y^2,$$

$$7x^2y + 5xy^2 - 9x^2y + 10xy^2 = -2x^2y + 15xy^2,$$

$$2x^3 - 5x^2 - 10x - 7x^3 + 9x^2 - 4x = -5x^3 + 4x^2 - 14x$$

The next two examples illustrate the use of exponents when reducing fractions.

EXAMPLE 4 Reduce $\frac{8x^2y}{12xy}$.

Solution

$$\frac{8x^2y}{12xy} = \frac{2 \cdot 2 \cdot 2 \cdot \cancel{x} \cdot x \cdot \cancel{y}}{2 \cdot 2 \cdot 3 \cdot \cancel{x} \cdot \cancel{y}} = \frac{2x}{3}$$

■

EXAMPLE 5 Reduce $\frac{15a^2b^3}{25a^3b}$.

Solution

$$\frac{15a^2b^3}{25a^3b} = \frac{3 \cdot \cancel{5} \cdot \cancel{a} \cdot \cancel{a} \cdot \cancel{b} \cdot b \cdot b}{\cancel{5} \cdot 5 \cdot \cancel{a} \cdot \cancel{a} \cdot a \cdot \cancel{b}} = \frac{3b^2}{5a}$$

■

The next three examples show how exponents can be used when multiplying and dividing fractions.

EXAMPLE 6 Multiply $\left(\frac{4x}{6y}\right)\left(\frac{12y^2}{7x^2}\right)$ and express the answer in reduced form.

Solution

$$\left(\frac{4x}{6y}\right)\left(\frac{12y^2}{7x^2}\right) = \frac{4 \cdot \overset{2}{\cancel{12}} \cdot \cancel{x} \cdot \cancel{y} \cdot y}{\cancel{6} \cdot 7 \cdot \cancel{y} \cdot \cancel{x} \cdot x} = \frac{8y}{7x}$$

■

EXAMPLE 7 Multiply and simplify

$$\left(\frac{8a^3}{9b}\right)\left(\frac{12b^2}{16a}\right)$$

Solution

$$\left(\frac{8a^3}{9b}\right)\left(\frac{12b^2}{16a}\right) = \frac{8 \cdot \overset{2}{\cancel{\overset{4}{\cancel{12}}}} \cdot \cancel{a} \cdot a \cdot a \cdot \cancel{b} \cdot b}{\underset{3}{\cancel{9}} \cdot \underset{2}{\cancel{16}} \cdot \cancel{b} \cdot \cancel{a}} = \frac{2a^2b}{3}$$ ■

EXAMPLE 8 Divide and express in reduced form,

$$\frac{-2x^3}{3y^2} \div \frac{4}{9xy}$$

Solution

$$\frac{-2x^3}{3y^2} \div \frac{4}{9xy} = -\frac{2x^3}{3y^2} \cdot \frac{9xy}{4} = -\frac{\overset{1}{\cancel{2}} \cdot \overset{3}{\cancel{9}} \cdot x \cdot x \cdot x \cdot x \cdot \cancel{y}}{\cancel{3} \cdot \underset{2}{\cancel{4}} \cdot \cancel{y} \cdot y} = -\frac{3x^4}{2y}$$ ■

The next two examples demonstrate the use of exponents when adding and subtracting fractions.

EXAMPLE 9 Add $\frac{4}{x^2} + \frac{7}{x}$.

Solution

The LCD is x^2. Thus,

$$\frac{4}{x^2} + \frac{7}{x} = \frac{4}{x^2} + \frac{7 \cdot x}{x \cdot x} = \frac{4}{x^2} + \frac{7x}{x^2} = \frac{4 + 7x}{x^2}$$ ■

EXAMPLE 10 Subtract $\frac{3}{xy} - \frac{4}{y^2}$.

Solution

$$\left.\begin{aligned} xy &= x \cdot y \\ y^2 &= y \cdot y \end{aligned}\right\} \longrightarrow \text{The LCD is } xy^2.$$

$$\frac{3}{xy} - \frac{4}{y^2} = \frac{3 \cdot y}{xy \cdot y} - \frac{4 \cdot x}{y^2 \cdot x} = \frac{3y}{xy^2} - \frac{4x}{xy^2}$$

$$= \frac{3y - 4x}{xy^2}$$ ■

56. $-10x^2 + 4x + 4x^2 - 8x$

57. $x^2 - 2x - 4 + 6x^2 - x + 12$

58. $-3x^3 - x^2 + 7x - 2x^3 + 7x^2 - 4x$

For Problems 59–68, reduce each fraction to simplest form.

59. $\frac{9xy}{15x}$

60. $\frac{8x^2y}{14x}$

61. $\frac{22xy^2}{6xy^3}$

62. $\frac{18x^3y}{12xy^4}$

63. $\frac{7a^2b^3}{17a^3b}$

64. $\frac{9a^3b^3}{22a^4b^2}$

65. $\frac{-24abc^2}{32bc}$

66. $\frac{4a^2c^3}{-22b^2c^4}$

67. $\frac{-5x^4y^3}{-20x^2y}$

68. $\frac{-32xy^2z^4}{-48x^3y^3z}$

For Problems 69–86, perform the indicated operations and express your answers in reduced form.

69. $\left(\frac{7x^2}{9y}\right)\left(\frac{12y}{21x}\right)$

70. $\left(\frac{3x}{8y^2}\right)\left(\frac{14xy}{9y}\right)$

71. $\left(\frac{5c}{a^2b^2}\right) \div \left(\frac{12c}{ab}\right)$

72. $\left(\frac{13ab^2}{12c}\right) \div \left(\frac{26b}{14c}\right)$

73. $\frac{6}{x} + \frac{5}{y^2}$

74. $\frac{8}{y} - \frac{6}{x^2}$

75. $\frac{5}{x^4} - \frac{7}{x^2}$

76. $\frac{9}{x} - \frac{11}{x^3}$

77. $\frac{3}{2x^3} + \frac{6}{x}$

78. $\frac{5}{3x^2} + \frac{6}{x}$

79. $\frac{-5}{4x^2} + \frac{7}{3x^2}$

80. $\frac{-8}{5x^3} + \frac{10}{3x^3}$

81. $\frac{11}{a^2} - \frac{14}{b^2}$

82. $\frac{9}{x^2} + \frac{8}{y^2}$

83. $\frac{1}{2x^3} - \frac{4}{3x^2}$

84. $\frac{2}{3x^3} - \frac{5}{4x}$

85. $\frac{3}{x} - \frac{4}{y} - \frac{5}{xy}$

86. $\frac{5}{x} + \frac{7}{y} - \frac{1}{xy}$

For Problems 87–100, evaluate each algebraic expression for the given values of the variables.

87. $4x^2 + 7y^2$ for $x = -2$ and $y = -3$

88. $5x^2 + 2y^3$ for $x = -4$ and $y = -1$

89. $3x^2 - y^2$ for $x = \frac{1}{2}$ and $y = -\frac{1}{3}$

90. $x^2 - 2y^2$ for $x = -\frac{2}{3}$ and $y = \frac{3}{2}$

91. $x^2 - 2xy + y^2$ for $x = -\frac{1}{2}$ and $y = 2$

92. $x^2 + 2xy + y^2$ for $x = -\frac{3}{2}$ and $y = -2$

93. $-x^2$ for $x = -8$

94. $-x^3$ for $x = 5$

95. $-x^2 - y^2$ for $x = -3$ and $y = -4$

96. $-x^2 + y^2$ for $x = -2$ and $y = 6$

97. $-a^2 - 3b^3$ for $a = -6$ and $b = -1$

98. $-a^3 + 3b^2$ for $a = -3$ and $b = -5$

99. $y^2 - 3xy$ for $x = 0.4$ and $y = -0.3$

100. $x^2 + 5xy$ for $x = -0.2$ and $y = -0.6$

101. Use a calculator to check your answers for Problems 1–34.

THOUGHTS INTO WORDS

102. Your friend keeps getting an answer of 16 when simplifying -2^4. What mistake is he making and how would you help him?

103. Explain how you would simplify $\frac{12x^2y}{18xy}$.

2.5 Translating from English to Algebra

In order to use the tools of algebra for solving problems, we must be able to translate back and forth between the English language and the language of algebra. In this section we want to translate algebraic expressions to English phrases (word phrases) and English phrases to algebraic expressions. Let's begin by translating some algebraic expressions to word phrases.

Algebraic expression	Word phrase
$x + y$	The sum of x and y
$x - y$	The difference of x and y
$y - x$	The difference of y and x
xy	The product of x and y
$\frac{x}{y}$	The quotient of x and y
$3x$	The product of 3 and x
$x^2 + y^2$	The sum of x squared and y squared
$2xy$	The product of 2, x, and y
$2(x + y)$	Two times the quantity x plus y
$x - 3$	Three less than x

Now let's consider the reverse process, translating some word phrases to algebraic expressions. Part of the difficulty in translating from English to algebra is that different word phrases translate into the same algebraic expression. So we need to become familiar with different ways of saying the same thing, especially when referring to the four fundamental operations. The following examples should help to acquaint you with some of the phrases used in the basic operations.

The sum of x and 4
x plus 4
x increased by 4
4 added to x
4 more than x → $x + 4$

The difference of n and 5
n minus 5
n less 5
n decreased by 5
5 less than n
Subtract 5 from n → $n - 5$

$\left[\begin{array}{l}\text{The product of 4 and } y \\ \text{4 times } y \\ y \text{ multiplied by 4}\end{array}\right] \rightarrow 4y$

$\left[\begin{array}{l}\text{The quotient of } n \text{ and 6} \\ n \text{ divided by 6} \\ \text{6 divided into } n\end{array}\right] \rightarrow \dfrac{n}{6}$

Often a word phrase indicates more than one operation. Furthermore, the standard vocabulary of sum, difference, product, and quotient may be replaced by other terminology. Study the following translations very carefully. Also remember that the commutative property holds for addition and multiplication but not for subtraction and division. Therefore the phrase "x plus y" can be written as $x + y$ or $y + x$. However the phrase "x minus y" means that y must be subtracted from x, and the phrase is written as $x - y$. So be very careful of phrases that involve subtraction or division.

Word phrase	Algebraic expression
The sum of two times x and three times y	$2x + 3y$
The sum of the squares of a and b	$a^2 + b^2$
Five times x divided by y	$\dfrac{5x}{y}$
Two more than the square of x	$x^2 + 2$
Three less than the cube of b	$b^3 - 3$
Five less than the product of x and y	$xy - 5$
Nine minus the product of x and y	$9 - xy$
Four times the sum of x and 2	$4(x + 2)$
Six times the quantity w minus 4	$6(w - 4)$

Suppose you are told that the sum of two numbers is 12, and one of the numbers is 8. What is the other number? The other number is $12 - 8$, which equals 4. Now suppose you are told that the product of two numbers is 56, and one of the numbers is 7. What is the other number? The other number is $56 \div 7$, which equals 8. The following examples illustrate the use of these addition-subtraction and multiplication-division relationships in a more general setting.

EXAMPLE 1

The sum of two numbers is 83, and one of the numbers is x. What is the other number?

Solution

Using the addition and subtraction relationship, we can represent the other number by $83 - x$. ■

EXAMPLE 2

The difference of two numbers is 14. The smaller number is n. What is the larger number?

Solution

Since the smaller number plus the difference must equal the larger number, we can represent the larger number by $n + 14$. ■

EXAMPLE 3

The product of two numbers is 39, and one of the numbers is y. Represent the other number.

Solution

Using the multiplication and division relationship, we can represent the other number by $\frac{39}{y}$. ■

The English statement may not contain key words such as sum, difference, product, or quotient; instead, the statement may describe a physical situation—from this description you need to deduce the operations involved. We make some suggestions for handling such situations in the following examples.

EXAMPLE 4

Arlene can type 70 words per minute. How many words can she type in m minutes?

Solution

In 10 minutes she would type $70(10) = 700$ words. In 50 minutes she would type $70(50) = 3500$ words. Thus, in m minutes she would type $70m$ words. ■

Notice the use of some specific examples $70(10) = 700$ and $70(50) = 3500$, to help formulate the general expression. This technique of first formulating some specific examples and then generalizing can be very effective.

EXAMPLE 5

Lynn has n nickels and d dimes. Express, in cents, this amount of money.

Solution

Three nickels and 8 dimes are $5(3) + 10(8) = 95$ cents. Thus n nickels and d dimes are $5n + 10d$ cents. ■

EXAMPLE 6

A train travels at the rate of r miles per hour. How far will it travel in 8 hours?

Solution

Suppose that a train travels at 50 miles per hour. Using the formula *distance equals rate times time*, it would travel $50 \cdot 8 = 400$ miles. Therefore, at r miles per hour, it would travel $r \cdot 8$ miles. We usually write the expression $r \cdot 8$ as $8r$. ■

EXAMPLE 7

The cost of a 5-pound box of candy is d dollars. How much is the cost per pound for the candy?

Solution

The price per pound is figured by dividing the total cost by the number of pounds. Therefore, the price per pound is represented by $\frac{d}{5}$. ■

The English statement to be translated to algebra may contain some geometric ideas. For example, suppose that we want to express in inches the length of a line segment that is f feet long. Since 1 foot = 12 inches, we can represent f feet by 12 times f, written as $12f$ inches.

Tables 2.1 and 2.2 list some of the basic relationships pertaining to linear measurements in the English and metric systems, respectively. (Additional listings of both systems are located on the inside back cover.)

Table 2.1

English system
12 inches = 1 foot
3 feet = 36 inches = 1 yard
5280 feet = 1760 yards = 1 mile

Table 2.2

Metric system
1 kilometer = 1000 meters
1 hectometer = 100 meters
1 dekameter = 10 meters
1 decimeter = 0.1 meter
1 centimeter = 0.01 meter
1 millimeter = 0.001 meter

EXAMPLE 8

The distance between two cities is k kilometers. Express this distance in meters.

Solution

Since 1 kilometer equals 1000 meters, we need to multiply k by 1000. Therefore, the distance in meters is represented by $1000k$. ■

EXAMPLE 9

The length of a line segment is i inches. Express that length in yards.

Solution

To change from inches to yards, we must divide by 36. Therefore $\frac{i}{36}$ represents, in yards, the length of the line segment. ■

EXAMPLE 10

The width of a rectangle is w centimeters, and the length is 5 centimeters less than twice the width. What is the length of the rectangle? What is the perimeter of the rectangle? What is the area of the rectangle?

Solution

We can represent the length of the rectangle by $2w - 5$. Now we can sketch a rectangle as in Figure 2.6 and record the given information. The perimeter of a rectangle is the sum of the lengths of the four sides. Therefore, the perimeter is given by $2w + 2(2w - 5)$, which can be written as $2w + 4w - 10$ and then simplified to $6w - 10$. The area of a rectangle is the product of the length and width. Therefore, the area in square centimeters is given by $w(2w - 5) = w \cdot 2w + w(-5) = 2w^2 - 5w$.

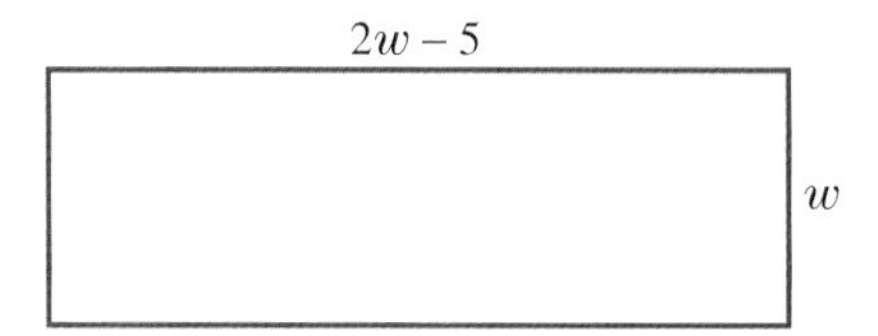

Figure 2.6 ■

EXAMPLE 11

The length of a side of a square is x feet. Express the length of a side in inches. What is the area of the square in square inches?

Solution

Because 1 foot equals 12 inches, we need to multiply x by 12. Therefore, $12x$ represents the length of a side in inches. The area of a square is the length of a side squared. So the area in square inches is given by $(12x)^2 = (12x)(12x) = 12 \cdot 12 \cdot x \cdot x = 144x^2$. ■

Problem Set 2.5

For Problems 1–12, write a word phrase for each of the algebraic expressions. For example, lw can be expressed as "the product of l and w."

1. $a - b$

2. $x + y$

3. $\frac{1}{3}Bh$

4. $\frac{1}{2}bh$

5. $2(l + w)$

6. πr^2

7. $\frac{A}{w}$

8. $\frac{C}{\pi}$

9. $\frac{a + b}{2}$

10. $\frac{a - b}{4}$

11. $3y + 2$

12. $3(x - y)$

For Problems 13–36, translate each word phrase into an algebraic expression. For example, "the sum of x and 14" translates into $x + 14$.

13. The sum of l and w

14. The difference of x and y

15. The product of a and b

16. The product of $\frac{1}{3}$, B, and h

17. The quotient of d and t

18. r divided into d

19. The product of l, w, and h

20. The product of π and the square of r

21. x subtracted from y

22. The difference "x subtract y"

23. Two larger than the product of x and y

24. Six plus the cube of x

25. Seven minus the square of y

26. The quantity, x minus 2, cubed

27. The quantity, x minus y, divided by four

28. Eight less than x

29. Ten less x

30. Nine times the quantity, n minus 4

31. Ten times the quantity, n plus 2

32. The sum of four times x and five times y

33. Seven subtracted from the product of x and y

34. Three times the sum of n and 2

35. Twelve less than the product of x and y

36. Twelve less the product of x and y

For Problems 37–68, answer the question with an algebraic expression.

37. The sum of two numbers is 35, and one of the numbers is n. What is the other number?

38. The sum of two numbers is 100, and one of the numbers is x. What is the other number?

39. The difference of two numbers is 45, and the smaller number is n. What is the other number?

40. The product of two numbers is 25, and one of the numbers is x. What is the other number?

41. Janet is y years old. How old will she be in 10 years?

42. Hector is y years old. How old was he 5 years ago?

43. Debra is x years old, and her mother is 3 years less than twice as old as Debra. How old is Debra's mother?

44. Jack is x years old, and Dudley is 1 year more than three times as old as Jack. How old is Dudley?

45. Donna has d dimes and q quarters in her bank. How much money in cents does she have?

46. Andy has c cents, which is all in dimes. How many dimes does he have?

47. A car travels d miles in t hours. How fast is the car traveling per hour (i.e., what is the rate)?

48. If g gallons of gas cost d dollars, what is the price per gallon?

49. If p pounds of candy cost d dollars, what is the price per pound?

50. Sue can type x words per minute. How many words can she type in 1 hour?

51. Larry's annual salary is d dollars. What is his monthly salary?

52. Nancy's monthly salary is d dollars. What is her annual salary?

53. If n represents a whole number, what is the next larger whole number?

54. If n represents an even number, what is the next larger even number?

55. If n represents an odd number, what is the next larger odd number?

56. Maria is y years old, and her sister is twice as old. What is the sum of their ages?

57. Willie is y years old, and his father is 2 years less than twice Willie's age. What is the sum of their ages?

58. Harriet has p pennies, n nickels, and d dimes. How much money in cents does she have?

59. The perimeter of a rectangle is y yards and f feet. What is the perimeter in inches?

60. The perimeter of a triangle is m meters and c centimeters. What is the perimeter in centimeters?

61. A rectangular plot of ground is f feet long. What is its length in yards?

62. The height of a telephone pole is f feet. What is the height in yards?

63. The width of a rectangle is w feet, and its length is three times the width. What is the perimeter of the rectangle in feet?

64. The width of a rectangle is w feet, and its length is 1 foot more than twice its width. What is the perimeter of the rectangle in feet?

65. The length of a rectangle is l inches, and its width is 2 inches less than one-half of its length. What is the perimeter of the rectangle in inches?

66. The length of a rectangle is l inches, and its width is 3 inches more than one-third of its length. What is the perimeter of the rectangle in inches?

67. The first side of a triangle is f feet long. The second side is 2 feet longer than the first side. The third side is twice as long as the second side. What is the perimeter of the triangle in inches?

68. The first side of a triangle is y yards long. The second side is 3 yards shorter than the first side. The third side is 3 times as long as the second side. What is the perimeter of the triangle in feet?

69. The width of a rectangle is w yards, and the length is twice the width. What is the area of the rectangle in square yards?

70. The width of a rectangle is w yards and the length is 4 yards more than the width. What is the area of the rectangle in square yards?

71. The length of a side of a square is s yards. What is the area of the square in square feet?

72. The length of a side of a square is y centimeters. What is the area of the square in square millimeters?

THOUGHTS INTO WORDS

73. What does the phrase "translating from English to algebra" mean to you?

74. Your friend is having trouble with Problems 61 and 62. For example, for Problem 61 she doesn't know if the answer should be $3f$ or $\frac{f}{3}$. What can you do to help her?

Chapter 2 Summary

(2.1) The property $\dfrac{a \cdot k}{b \cdot k} = \dfrac{a}{b}$ is used to express fractions in reduced form.

To **multiply** rational numbers in common fractional form we multiply numerators, multiply denominators, and express the result in reduced form.

To **divide** rational numbers in common fractional form, we multiply by the reciprocal of the divisor.

(2.2) Addition and **subtraction** of rational numbers in common fractional form are based on the following.

$$\frac{a}{b} + \frac{c}{b} = \frac{a + c}{b} \qquad \text{Addition}$$

$$\frac{a}{b} - \frac{c}{b} = \frac{a - c}{b} \qquad \text{Subtraction}$$

To add or subtract fractions that do not have a common denominator, we use the fundamental principle of fractions, $\dfrac{a}{b} = \dfrac{a \cdot k}{b \cdot k}$, and obtain equivalent fractions that have a common denominator.

(2.3) To **add** or **subtract decimals**, we write the numbers in a column so that the decimal points are lined up, and then we add or subtract as we do with integers.

To **multiply decimals** we (1) multiply the numbers and ignore the decimal points, and then (2) insert the decimal point in the product so that the number of digits to the right of the decimal point in the product is equal to the sum of the numbers of digits to the right of the decimal point in each factor.

To **divide a decimal by a nonzero whole number** we (1) place the decimal point in the quotient directly above the decimal point in the dividend, and then (2) divide as with whole numbers, except that in the division process, we place zeros in the quotient immediately to the right of the decimal point (if necessary) to show the correct place value.

To **divide by a decimal** we change to an equivalent problem that has a whole number divisor.

(2.4) Expressions of the form b^n, where

$$b^n = b \cdot b \cdot b \cdots b \qquad n \text{ factors of } b$$

are read as "b to the nth power"; b is the **base** and n is the **exponent**.

(2.5) To translate English phrases into algebraic expressions, we must be familiar with the standard vocabulary of **sum, difference, product**, and **quotient** as well as other terms used to express the same ideas.

Chapter 2 Review Problem Set

For Problems 1–14, find the value of each of the following.

1. 2^6

2. $(-3)^3$

3. -4^2

4. 5^3

5. $-\left(\dfrac{1}{2}\right)^2$

6. $\left(\dfrac{3}{4}\right)^2$

7. $\left(\dfrac{1}{2} + \dfrac{2}{3}\right)^2$

8. $(0.6)^3$

9. $(0.12)^2$

10. $(0.06)^2$

11. $\left(-\dfrac{2}{3}\right)^3$

12. $\left(-\dfrac{1}{2}\right)^4$

13. $\left(\dfrac{1}{4} - \dfrac{1}{2}\right)^3$

14. $\left(\dfrac{1}{2} + \dfrac{1}{3} - \dfrac{1}{6}\right)^2$

For Problems 15–24, perform the indicated operations and express your answers in reduced form.

15. $\dfrac{3}{8} + \dfrac{5}{12}$

16. $\dfrac{9}{14} - \dfrac{3}{35}$

17. $\dfrac{2}{3} + \dfrac{-3}{5}$

18. $\dfrac{7}{x} + \dfrac{9}{2y}$

19. $\frac{5}{xy} - \frac{8}{x^2}$

20. $\left(\frac{7y}{8x}\right)\left(\frac{14x}{35}\right)$

21. $\left(\frac{6xy}{9y^2}\right) \div \left(\frac{15y}{18x^2}\right)$

22. $\left(\frac{-3x}{12y}\right)\left(\frac{8y}{-7x}\right)$

23. $\left(\frac{-4y}{3x}\right)\left(-\frac{3x}{4y}\right)$

24. $\left(\frac{6n}{7}\right)\left(\frac{9n}{8}\right)$

For Problems 25–36, simplify each of the following numerical expressions.

25. $\frac{1}{6} + \frac{2}{3} \cdot \frac{3}{4} - \frac{5}{6} \div \frac{8}{6}$

26. $\frac{3}{4} \cdot \frac{1}{2} - \frac{4}{3} \cdot \frac{3}{2}$

27. $\frac{7}{9} \cdot \frac{3}{5} + \frac{7}{9} \cdot \frac{2}{5}$

28. $\frac{4}{5} \div \frac{1}{5} \cdot \frac{2}{3} - \frac{1}{4}$

29. $\frac{2}{3} \cdot \frac{1}{4} \div \frac{1}{2} + \frac{2}{3} \cdot \frac{1}{4}$

30. $0.48 + 0.72 - 0.35 - 0.18$

31. $0.81 + (0.6)(0.4) - (0.7)(0.8)$

32. $1.28 \div 0.8 - 0.81 \div 0.9 + 1.7$

33. $(0.3)^2 + (0.4)^2 - (0.6)^2$

34. $(1.76)(0.8) + (1.76)(0.2)$

35. $(2^2 - 2 - 2^3)^2$

36. $1.92(0.9 + 0.1)$

For Problems 37–42, simplify each of the following algebraic expressions by combining similar terms. Express your answers in reduced form when working with common fractions.

37. $\frac{3}{8}x^2 - \frac{2}{5}y^2 - \frac{2}{7}x^2 + \frac{3}{4}y^2$

38. $0.24ab + 0.73bc - 0.82ab - 0.37bc$

39. $\frac{1}{2}x + \frac{3}{4}x - \frac{5}{6}x + \frac{1}{24}x$

40. $1.4a - 1.9b + 0.8a + 3.6b$

41. $\frac{2}{5}n + \frac{1}{3}n - \frac{5}{6}n$

42. $n - \frac{3}{4}n + 2n - \frac{1}{5}n$

For Problems 43–48, evaluate the following algebraic expressions for the given values of the variables.

43. $\frac{1}{4}x - \frac{2}{5}y$ for $x = \frac{2}{3}$ and $y = -\frac{5}{7}$

44. $a^3 + b^2$ for $a = -\frac{1}{2}$ and $b = \frac{1}{3}$

45. $2x^2 - 3y^2$ for $x = 0.6$ and $y = 0.7$

46. $0.7w + 0.9z$ for $w = 0.4$ and $z = -0.7$

47. $\frac{3}{5}x - \frac{1}{3}x + \frac{7}{15}x - \frac{2}{3}x$ for $x = \frac{15}{17}$

48. $\frac{1}{3}n + \frac{2}{7}n - n$ for $n = 21$

For Problems 49–56, answer each of the following questions with an algebraic expression.

49. The sum of two numbers is 72, and one of the numbers is n. What is the other number?

50. Joan has p pennies and d dimes. How much money in cents does she have?

51. Ellen types x words in an hour. What is her typing rate per minute?

52. Harry is y years old. His brother is 3 years less than twice as old as Harry. How old is Harry's brother?

53. Larry chose a number n. Cindy chose a number 3 more than 5 times the number chosen by Larry. What number did Cindy choose?

54. The height of a file cabinet is y yards and f feet. How tall is the file cabinet in inches?

55. The length of a rectangular room is m meters. How long in centimeters is the room?

56. Corinne has n nickels, d dimes, and q quarters. How much money in cents does she have?

For Problems 57–66, translate each word phrase into an algebraic expression.

57. Five less than n

58. Five less n

59. Ten times the quantity, x minus 2

60. Ten times x minus 2

61. x minus three

62. d divided by r

63. x squared plus nine

64. x plus nine, the quantity squared

65. The sum of the cubes of x and y

66. Four less than the product of x and y

Chapter 2 Test

1. Find the value of each expression.
   (a) $(-3)^4$ **(b)** -2^6 **(c)** $(0.2)^3$

2. Express $\frac{42}{54}$ in reduced form.

3. Simplify $\frac{18xy^2}{32y}$.

For Problems 4–7, simplify each numerical expression.

4. $5.7 - 3.8 + 4.6 - 9.1$

5. $0.2(0.4) - 0.6(0.9) + 0.5(7)$

6. $-0.4^2 + 0.3^2 - 0.7^2$

7. $\left(\frac{1}{3} - \frac{1}{4} + \frac{1}{6}\right)^4$

For Problems 8–11, perform the indicated operations, and express your answers in reduced form.

8. $\frac{5}{12} \div \frac{15}{8}$

9. $-\frac{2}{3} - \frac{1}{2}\left(\frac{3}{4}\right) + \frac{5}{6}$

10. $3\left(\frac{2}{5}\right) - 4\left(\frac{5}{6}\right) + 6\left(\frac{7}{8}\right)$

11. $4\left(\frac{1}{2}\right)^3 - 3\left(\frac{2}{3}\right)^2 + 9\left(\frac{1}{4}\right)^2$

For Problems 12–17, perform the indicated operations, and express your answers in reduced form.

12. $\frac{8x}{15y} \cdot \frac{9y^2}{6x}$

13. $\frac{6xy}{9} \div \frac{y}{3x}$

14. $\frac{4}{x} - \frac{5}{y^2}$

15. $\frac{3}{2x} + \frac{7}{6x}$

16. $\frac{5}{3y} + \frac{9}{7y^2}$

17. $\left(\frac{15a^2b}{12a}\right)\left(\frac{8ab}{9b}\right)$

For Problems 18 and 19, simplify each algebraic expression by combining similar terms.

18. $3x - 2xy - 4x + 7xy$

19. $-2a^2 + 3b^2 - 5b^2 - a^2$

For Problems 20–23, evaluate each of the algebraic expressions for the given values of the variables.

20. $x^2 - xy + y^2$ for $x = \frac{1}{2}$ and $y = -\frac{2}{3}$

21. $0.2x - 0.3y - xy$ for $x = 0.4$ and $y = 0.8$

22. $\frac{3}{4}x - \frac{2}{3}y$ for $x = -\frac{1}{2}$ and $y = \frac{3}{5}$

23. $3x - 2y + xy$ for $x = 0.5$ and $y = -0.9$

24. David has n nickels, d dimes, and q quarters. How much money, in cents, does he have?

25. Hal chose a number n. Sheila chose a number 3 less than 4 times the number that Hal chose. Express the number that Sheila chose in terms of n.

For Problems 1–12, simply each of the numerical expressions.

1. $16 - 18 - 14 + 21 - 14 + 19$
2. $7(-6) - 8(-6) + 4(-9)$
3. $6 - [3 - (10 - 12)]$
4. $-9 - 2[4 - (-10 + 6)] - 1$
5. $\dfrac{-7(-4) - 5(-6)}{-2}$
6. $\dfrac{5(-3) + (-4)(6) - 3(4)}{-3}$
7. $\dfrac{3}{4} + \dfrac{1}{3} \div \dfrac{4}{3} - \dfrac{1}{2}$
8. $\left(\dfrac{2}{3}\right)\left(-\dfrac{3}{4}\right) - \left(\dfrac{5}{6}\right)\left(\dfrac{4}{5}\right)$
9. $\left(\dfrac{1}{2} - \dfrac{2}{3}\right)^2$
10. -4^3
11. $\dfrac{0.0046}{0.000023}$
12. $(0.2)^2 - (0.3)^3 + (0.4)^2$

For Problems 13–20, evaluate each algebraic expression for the given values of the variables.

13. $3xy - 2x - 4y$ for $x = -6$ and $y = 7$
14. $-4x^2y - 2xy^2 + xy$ for $x = -2$ and $y = -4$
15. $\dfrac{5x - 2y}{3x}$ for $x = \dfrac{1}{2}$ and $y = -\dfrac{1}{3}$
16. $0.2x - 0.3y + 2xy$ for $x = 0.1$ and $y = 0.3$
17. $-7x + 4y + 6x - 9y + x - y$ for $x = -0.2$ and $y = 0.4$
18. $\dfrac{2}{3}x - \dfrac{3}{5}y + \dfrac{3}{4}x - \dfrac{1}{2}y$ for $x = \dfrac{6}{5}$ and $y = -\dfrac{1}{4}$
19. $\dfrac{1}{5}n - \dfrac{1}{3}n + n - \dfrac{1}{6}n$ for $n = \dfrac{1}{5}$
20. $-ab + \dfrac{1}{5}a - \dfrac{2}{3}b$ for $a = -2$ and $b = \dfrac{3}{4}$

For Problems 21–24, express each of the numbers as a product of prime factors.

21. 54
22. 78
23. 91
24. 153

For Problems 25–28, find the greatest common factor of the given numbers.

25. 42 and 70
26. 63 and 81
27. 28, 36, and 52
28. 48, 66, and 78

For Problems 29–32, find the least common multiple of the given numbers.

29. 20 and 28
30. 40 and 100
31. 12, 18, and 27
32. 16, 20, and 80

For Problems 33–38, simplify each algebraic expression by combining similar terms.

33. $\dfrac{2}{3}x - \dfrac{1}{4}y - \dfrac{3}{4}x - \dfrac{2}{3}y$
34. $-n - \dfrac{1}{2}n + \dfrac{3}{5}n + \dfrac{5}{6}n$
35. $3.2a - 1.4b - 6.2a + 3.3b$
36. $-(n - 1) + 2(n - 2) - 3(n - 3)$
37. $-x + 4(x - 1) - 3(x + 2) - (x + 5)$
38. $2a - 5(a + 3) - 2(a - 1) - 4a$

For Problems 39–46, perform the indicated operations and express your answers in reduced form.

39. $\dfrac{5}{12} - \dfrac{3}{16}$
40. $\dfrac{3}{4} - \dfrac{5}{6} - \dfrac{7}{9}$
41. $\dfrac{5}{xy} - \dfrac{2}{x} + \dfrac{3}{y}$
42. $-\dfrac{7}{x^2} + \dfrac{9}{xy}$

43. $\left(\frac{7x}{9y}\right)\left(\frac{12y}{14}\right)$

44. $\left(-\frac{5a}{7b^2}\right)\left(-\frac{8ab}{15}\right)$

45. $\left(\frac{6x^2y}{11}\right) \div \left(\frac{9y^2}{22}\right)$

46. $\left(-\frac{9a}{8b}\right) \div \left(\frac{12a}{18b}\right)$

For Problems 47–50, answer the question with an algebraic expression.

47. Hector has p pennies, n nickels, and d dimes. How much money in cents does he have?

48. Ginny chose a number n. Penny chose a number 5 less than 4 times the number chosen by Ginny. What number did Penny choose?

49. The height of a flagpole is y yards, f feet, and i inches. How tall is the flagpole in inches?

50. A rectangular room is x meters by y meters. What is its perimeter in centimeters?

3

Equations, Inequalities, and Problem Solving

The inequality $\frac{95 + 82 + 93 + 84 + 5}{5} \geq 90$ *can be used to determine that Ashley needs a 96 or higher on her fifth exam to have an average of 90 or higher for the five exams if she got 95, 82, 93, and 84 on her first four exams.*

Tracy received a cell phone bill for \$136.74. Included in the \$136.74 were a monthly-plan charge of \$39.99 and a charge for 215 extra minutes. How much is Tracy being charged for each extra minute? If we let c represent the charge per minute, then the equation $39.99 + 215c = 136.74$ can be used to determine that the charge for each extra minute is \$0.45.

John was given a graduation present of \$25,000 to purchase a car. What price range should John be looking at if the \$25,000 has to include the sales tax of 6% and registration fees of \$850? If we let p represent the price of the car, then the inequality $p + 0.06p + 850 \leq 25{,}000$ can be used to determine that the car can cost \$22,783 or less.

Throughout this book, you will develop new skills to help solve equations and inequalities, and you will use equations and inequalities to solve applied problems. In this chapter you will use the skills you developed in the first chapter to solve equations and inequalities and then move on to work with applied problems.

3.1 Solving First-Degree Equations

These are examples of **numerical statements**:

$$3 + 4 = 7 \qquad 5 - 2 = 3 \qquad 7 + 1 = 12$$

The first two are true statements, and the third one is a false statement.

When we use x as a variable, the statements

$$x + 3 = 4, \qquad 2x - 1 = 7, \qquad \text{and} \qquad x^2 = 4$$

are called **algebraic equations** in x. We call the number a a **solution** or **root** of an equation if a true numerical statement is formed when a is substituted for x. (We also say that a satisfies the equation.) For example, 1 is a solution of $x + 3 = 4$ because substituting 1 for x produces the true numerical statement $1 + 3 = 4$. We call the set of all solutions of an equation its **solution set**. Thus the solution set of $x + 3 = 4$ is $\{1\}$. Likewise the solution set of $2x - 1 = 7$ is $\{4\}$ and the solution set of $x^2 = 4$ is $\{-2, 2\}$. **Solving an equation** refers to the process of determining the solution set. Remember that a set that consists of no elements is called the **empty** or **null set** and is denoted by $\varnothing$. Thus we say that the solution set of $x = x + 1$ is $\varnothing$; that is, there are no real numbers that satisfy $x = x + 1$.

In this chapter, we will consider techniques for solving **first-degree equations of one variable**. This means that the equations contain only one variable, and this variable has an exponent of one. Here are some examples of first-degree equations of one variable:

$$3x + 4 = 7 \qquad 8w + 7 = 5w - 4$$

$$\frac{1}{2}y + 2 = 9 \qquad 7x + 2x - 1 = 4x - 1$$

Equivalent equations are equations that have the same solution set. The following equations are all equivalent:

$$5x - 4 = 3x + 8$$

$$2x = 12$$

$$x = 6$$

You can verify the equivalence by showing that 6 is the solution for all three equations.

As we work with equations, we can use the properties of equality.

Property 3.1 Properties of Equality

For all real numbers, a, b, and c,

1. $a = a$ Reflexive property
2. If $a = b$, then $b = a$. Symmetric property
3. If $a = b$ and $b = c$, then $a = c$. Transitive property
4. If $a = b$, then a may be replaced by b, or b may be replaced by a, in any statement without changing the meaning of the statement.
   Substitution property

The general procedure for solving an equation is to continue replacing the given equation with equivalent but simpler equations until we obtain an equation of the form **variable = constant** or **constant = variable**. Thus, in the earlier example, $5x - 4 = 3x + 8$ was simplified to $2x = 12$, which was further simplified to $x = 6$, from which the solution of 6 is obvious. The exact procedure for simplifying equations becomes our next concern.

Two properties of equality play an important role in the process of solving equations. The first of these is the **addition-subtraction property of equality**.

Property 3.2 Addition-Subtraction Property of Equality

For all real numbers, a, b, and c,

1. $a = b$ if and only if $a + c = b + c$.
2. $a = b$ if and only if $a - c = b - c$.

Property 3.2 states that any number can be added to or subtracted from both sides of an equation and an equivalent equation is produced. Consider the use of this property in the next four examples.

EXAMPLE 1

Solve $x - 8 = 3$.

Solution

$$x - 8 = 3$$

$$x - 8 + 8 = 3 + 8 \quad \text{Add 8 to both sides}$$

$$x = 11$$

The solution set is $\{11\}$. ■

Remark: It is true that a simple equation such as Example 1 can be solved *by inspection*; for instance, "Some number minus 8 produces 3" yields an obvious answer of 11. However, as the equations become more complex, the technique of solving by inspection becomes ineffective. So it is necessary to develop more formal techniques for solving equations. Therefore, we will begin developing such techniques with very simple types of equations.

EXAMPLE 2

Solve $x + 14 = -8$.

Solution

$$x + 14 = -8$$

$$x + 14 - 14 = -8 - 14 \qquad \text{Subtract 14 from both sides}$$

$$x = -22$$

The solution set is $\{-22\}$. ■

EXAMPLE 3

Solve $n - \frac{1}{3} = \frac{1}{4}$.

Solution

$$n - \frac{1}{3} = \frac{1}{4}$$

$$n - \frac{1}{3} + \frac{1}{3} = \frac{1}{4} + \frac{1}{3} \qquad \text{Add } \frac{1}{3} \text{ to both sides}$$

$$n = \frac{3}{12} + \frac{4}{12} \qquad \text{Change to equivalent fractions with a denominator of 12}$$

$$n = \frac{7}{12}$$

The solution set is $\left\{\frac{7}{12}\right\}$. ■

EXAMPLE 4

Solve $0.72 = y + 0.35$.

Solution

$$0.72 = y + 0.35$$

$$0.72 - 0.35 = y + 0.35 - 0.35 \qquad \text{Subtract 0.35 from both sides}$$

$$0.37 = y$$

The solution set is $\{0.37\}$. ■

Note in Example 4 that the final equation is $0.37 = y$ instead of $y = 0.37$. Technically, the **symmetric property of equality** (if $a = b$, then $b = a$) permits us to change from $0.37 = y$ to $y = 0.37$, but such a change is not necessary to determine that the solution is 0.37. You should also realize that the symmetric property could be applied to the original equation. Thus, $0.72 = y + 0.35$ becomes $y + 0.35 = 0.72$, and subtracting 0.35 from both sides produces $y = 0.37$.

One other comment that pertains to Property 3.2 should be made at this time. Because subtracting a number is equivalent to adding its opposite, we can state Property 3.2 only in terms of addition. Thus to solve an equation such as Example 4, we add -0.35 to both sides rather than subtract 0.35 from both sides.

The other important property for solving equations is the **multiplication-division property of equality**.

Property 3.3 Multiplication-Division Property of Equality

For all real numbers, a, b, and c, where $c \neq 0$,

1. $a = b$ if and only if $ac = bc$.

2. $a = b$ if and only if $\frac{a}{c} = \frac{b}{c}$.

Property 3.3 states that an equivalent equation is obtained whenever both sides of a given equation are multiplied or divided by the same nonzero real number. The next examples illustrate how we use this property.

EXAMPLE 5

Solve $\frac{3}{4}x = 6$.

Solution

$$\frac{3}{4}x = 6$$

$$\frac{4}{3}\left(\frac{3}{4}x\right) = \frac{4}{3}(6) \qquad \text{Multiply both sides by } \frac{4}{3} \text{ because } \left(\frac{4}{3}\right)\left(\frac{3}{4}\right) = 1$$

$$x = 8$$

The solution set is $\{8\}$. ■

EXAMPLE 6

Solve $5x = 27$.

Solution

$$5x = 27$$

$$\frac{5x}{5} = \frac{27}{5} \qquad \text{Divide both sides by 5}$$

$$x = \frac{27}{5} \qquad \frac{27}{5} \text{ can be expressed as } 5\frac{2}{5} \text{ or } 5.4$$

The solution set is $\left\{\frac{27}{5}\right\}$. ■

EXAMPLE 7

Solve $-\frac{2}{3}p = \frac{1}{2}$.

Solution

$$-\frac{2}{3}p = \frac{1}{2}$$

$$\left(-\frac{3}{2}\right)\left(-\frac{2}{3}p\right) = \left(-\frac{3}{2}\right)\left(\frac{1}{2}\right) \qquad \text{Multiply both sides by } -\frac{3}{2} \text{ because } \left(-\frac{3}{2}\right)\left(-\frac{2}{3}\right) = 1$$

$$p = -\frac{3}{4}$$

The solution set is $\left\{-\frac{3}{4}\right\}$. ■

EXAMPLE 8

Solve $26 = -6x$.

Solution

$$26 = -6x$$

$$\frac{26}{-6} = \frac{-6x}{-6} \qquad \text{Divide both sides by } -6$$

$$-\frac{26}{6} = x \qquad \frac{26}{-6} = -\frac{26}{6}$$

$$-\frac{13}{3} = x \qquad \text{Don't forget to reduce!}$$

The solution set is $\left\{-\frac{13}{3}\right\}$. ■

Look back at Examples 5–8, and you will notice that we divided both sides of the equation by the coefficient of the variable whenever the coefficient was an integer; otherwise, we used the multiplication part of Property 3.3. Technically, because dividing by a number is equivalent to multiplying by its reciprocal, Property 3.3 could be stated only in terms of multiplication. Thus to solve an equation such as $5x = 27$, we could multiply both sides by $\frac{1}{5}$ instead of dividing both sides by 5.

EXAMPLE 9 Solve $0.2n = 15$.

Solution

$$0.2n = 15$$

$$\frac{0.2n}{0.2} = \frac{15}{0.2} \quad \text{Divide both sides by 0.2}$$

$$n = 75$$

The solution set is $\{75\}$.

Problem Set 3.1

For Problems 1–72, use the properties of equality to help solve each equation.

1. $x + 9 = 17$

2. $x + 7 = 21$

3. $x + 11 = 5$

4. $x + 13 = 2$

5. $-7 = x + 2$

6. $-12 = x + 4$

7. $8 = n + 14$

8. $6 = n + 19$

9. $21 + y = 34$

10. $17 + y = 26$

11. $x - 17 = 31$

12. $x - 22 = 14$

13. $14 = x - 9$

14. $17 = x - 28$

15. $-26 = n - 19$

16. $-34 = n - 15$

17. $y - \frac{2}{3} = \frac{3}{4}$

18. $y - \frac{2}{5} = \frac{1}{6}$

19. $x + \frac{3}{5} = \frac{1}{3}$

20. $x + \frac{5}{8} = \frac{2}{5}$

21. $b + 0.19 = 0.46$

22. $b + 0.27 = 0.74$

23. $n - 1.7 = -5.2$

24. $n - 3.6 = -7.3$

25. $15 - x = 32$

26. $13 - x = 47$

27. $-14 - n = 21$

28. $-9 - n = 61$

29. $7x = -56$

30. $9x = -108$

31. $-6x = 102$

32. $-5x = 90$

33. $5x = 37$

34. $7x = 62$

35. $-18 = 6n$

36. $-52 = 13n$

37. $-26 = -4n$

38. $-56 = -6n$

39. $\frac{t}{9} = 16$

40. $\frac{t}{12} = 8$

41. $\frac{n}{-8} = -3$

42. $\frac{n}{-9} = -5$

43. $-x = 15$

44. $-x = -17$

45. $\frac{3}{4}x = 18$

46. $\frac{2}{3}x = 32$

47. $-\frac{2}{5}n = 14$

48. $-\frac{3}{8}n = 33$

49. $\frac{2}{3}n = \frac{1}{5}$

50. $\frac{3}{4}n = \frac{1}{8}$

51. $\frac{5}{6}n = -\frac{3}{4}$

52. $\frac{6}{7}n = -\frac{3}{8}$

53. $\frac{3x}{10} = \frac{3}{20}$

54. $\frac{5x}{12} = \frac{5}{36}$

55. $\frac{-y}{2} = \frac{1}{6}$

56. $\frac{-y}{4} = \frac{1}{9}$

57. $-\frac{4}{3}x = -\frac{9}{8}$

58. $-\frac{6}{5}x = -\frac{10}{14}$

59. $-\frac{5}{12} = \frac{7}{6}x$

60. $-\frac{7}{24} = \frac{3}{8}x$

61. $-\frac{5}{7}x = 1$

62. $-\frac{11}{12}x = -1$

63. $-4n = \frac{1}{3}$

64. $-6n = \frac{3}{4}$

65. $-8n = \frac{6}{5}$

66. $-12n = \frac{8}{3}$

67. $1.2x = 0.36$

68. $2.5x = 17.5$

69. $30.6 = 3.4n$

70. $2.1 = 4.2n$

71. $-3.4x = 17$

72. $-4.2x = 50.4$

THOUGHTS INTO WORDS

73. Describe the difference between a numerical statement and an algebraic equation.

74. Are the equations $6 = 3x + 1$ and $1 + 3x = 6$ equivalent equations? Defend your answer.

3.2 Equations and Problem Solving

We often need more than one property of equality to help find the solution of an equation. Consider the following examples.

EXAMPLE 1

Solve $3x + 1 = 7$.

Solution

$$3x + 1 = 7$$

$$3x + 1 - 1 = 7 - 1 \quad \text{Subtract 1 from both sides}$$

$$3x = 6$$

$$\frac{3x}{3} = \frac{6}{3} \quad \text{Divide both sides by 3}$$

$$x = 2$$

The potential solution can be *checked* by substituting it into the original equation to see whether a true numerical statement results.

✔ **Check**

$$3x + 1 = 7$$

$$3(2) + 1 \stackrel{?}{=} 7$$

$$6 + 1 \stackrel{?}{=} 7$$

$$7 = 7$$

Now we know that the solution set is $\{2\}$. ■

EXAMPLE 2 Solve $5x - 6 = 14$.

Solution

$$5x - 6 = 14$$

$$5x - 6 + 6 = 14 + 6 \quad \text{Add 6 to both sides}$$

$$5x = 20$$

$$\frac{5x}{5} = \frac{20}{5} \quad \text{Divide both sides by 5}$$

$$x = 4$$

✔ **Check**

$$5x - 6 = 14$$

$$5(4) - 6 \stackrel{?}{=} 14$$

$$20 - 6 \stackrel{?}{=} 14$$

$$14 = 14$$

The solution set is $\{4\}$. ■

EXAMPLE 3 Solve $4 - 3a = 22$.

Solution

$$4 - 3a = 22$$

$$4 - 3a - 4 = 22 - 4 \quad \text{Subtract 4 from both sides}$$

$$-3a = 18$$

$$\frac{-3a}{-3} = \frac{18}{-3} \quad \text{Divide both sides by } -3$$

$$a = -6$$

✔ **Check**

$$4 - 3a = 22$$

$$4 - 3(-6) \stackrel{?}{=} 22$$

$$4 + 18 \stackrel{?}{=} 22$$

$$22 = 22$$

The solution set is $\{-6\}$. ■

Notice that in Examples 1, 2, and 3, we used the addition-subtraction property first, and then we used the multiplication-division property. In general, this sequence of steps provides the easiest method for solving such equations. Perhaps you should convince yourself of that fact by doing Example 1 again, but this time use the multiplication-division property first and then the addition-subtraction property.

EXAMPLE 4

Solve $19 = 2n + 4$.

Solution

$$19 = 2n + 4$$

$$19 - 4 = 2n + 4 - 4 \quad \text{Subtract 4 from both sides}$$

$$15 = 2n$$

$$\frac{15}{2} = \frac{2n}{2} \quad \text{Divide both sides by 2}$$

$$\frac{15}{2} = n$$

✔ **Check**

$$19 = 2n + 4$$

$$19 \stackrel{?}{=} 2\left(\frac{15}{2}\right) + 4$$

$$19 \stackrel{?}{=} 15 + 4$$

$$19 = 19$$

The solution set is $\left\{\frac{15}{2}\right\}$. ■

■ Word Problems

In the last section of Chapter 2, we translated English phrases into algebraic expressions. We are now ready to expand that concept and translate English sentences into algebraic equations. Such translations allow us to use the concepts of algebra to solve word problems. Let's consider some examples.

PROBLEM 1

A certain number added to 17 yields a sum of 29. What is the number?

Solution

Let n represent the number to be found. The sentence "A certain number added to 17 yields a sum of 29" translates into the algebraic equation $17 + n = 29$. We can solve this equation.

$$17 + n = 29$$

$$17 + n - 17 = 29 - 17$$

$$n = 12$$

The solution is 12, which is the number asked for in the problem. ■

We often refer to the statement "let n represent the number to be found" as **declaring the variable**. We need to choose a letter to use as a variable and indicate what it represents for a specific problem. This may seem like an unnecessary exercise, but as the problems become more complex, the process of declaring the variable becomes even more important. We could solve a problem like Problem 1 without setting up an algebraic equation; however, as problems increase in difficulty, the translation from English into an algebraic equation becomes a key issue. Therefore, even with these relatively simple problems we need to concentrate on the translation process.

PROBLEM 2

Six years ago Bill was 13 years old. How old is he now?

Solution

Let y represent Bill's age now; therefore, $y - 6$ represents his age 6 years ago. Thus,

$$y - 6 = 13$$

$$y - 6 + 6 = 13 + 6$$

$$y = 19$$

Bill is presently 19 years old. ■

PROBLEM 3

Betty worked 8 hours Saturday and earned \$60. How much did she earn per hour?

Solution A

Let x represent the amount Betty earned per hour. The number of hours worked times the wage per hour yields the total earnings. Thus,

$$8x = 60$$

$$\frac{8x}{8} = \frac{60}{8}$$

$$x = 7.50$$

Betty earned \$7.50 per hour.

Solution B

Let y represent the amount Betty earned per hour. The wage per hour equals the total wage divided by the number of hours. Thus,

$$y = \frac{60}{8}$$

$$y = 7.50$$

Betty earned \$7.50 per hour. ■

Sometimes we can use more than one equation to solve a problem. In Solution A, we set up the equation in terms of multiplication; whereas in Solution B, we were thinking in terms of division.

PROBLEM 4

Kendall paid \$244 for a CD player and six compact discs. The CD player cost ten times as much as one compact disc. Find the cost of the CD player and the cost of one compact disc.

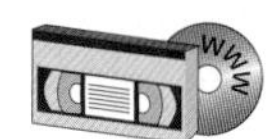

Solution

Let d represent the cost of one compact disc. Then the cost of the CD player is represented by $10d$, and the cost of six compact discs is represented by $6d$. The total cost is \$244, so we can proceed as follows:

Cost of CD player + Cost of six compact discs = \$244

$$10d + 6d = 244$$

Solving this equation, we have

$$16d = 244$$

$$d = 15.25$$

The cost of one compact disc is \$15.25, and the cost of the CD player is 10(15.25) or \$152.50. ■

PROBLEM 5

The cost of a five-day vacation cruise package was \$534. This cost included \$339 for the cruise and an amount for 2 nights of lodging on shore. Find the cost per night of the lodging.

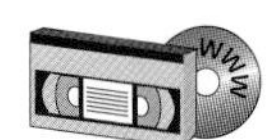

Solution

Let n represent the cost for one night of lodging; then $2n$ represents the total cost of lodging. Thus the cost for the cruise and lodging is the total cost of \$534. We can proceed as follows

Cost of cruise + Cost of lodging = \$534

$$339 + 2n = 534$$

We can solve this equation:

$$339 + 2n = 534$$

$$2n = 195$$

$$\frac{2n}{2} = \frac{195}{2}$$

$$n = 97.50$$

The cost of lodging per night is $97.50. ■

Problem Set 3.2

For Problems 1–40, solve each equation.

1. $2x + 5 = 13$
2. $3x + 4 = 19$
3. $5x + 2 = 32$
4. $7x + 3 = 24$
5. $3x - 1 = 23$
6. $2x - 5 = 21$
7. $4n - 3 = 41$
8. $5n - 6 = 19$
9. $6y - 1 = 16$
10. $4y - 3 = 14$
11. $2x + 3 = 22$
12. $3x + 1 = 21$
13. $10 = 3t - 8$
14. $17 = 2t + 5$
15. $5x + 14 = 9$
16. $4x + 17 = 9$
17. $18 - n = 23$
18. $17 - n = 29$
19. $-3x + 2 = 20$
20. $-6x + 1 = 43$
21. $7 + 4x = 29$
22. $9 + 6x = 23$
23. $16 = -2 - 9a$
24. $18 = -10 - 7a$
25. $-7x + 3 = -7$
26. $-9x + 5 = -18$
27. $17 - 2x = -19$
28. $18 - 3x = -24$
29. $-16 - 4x = 9$
30. $-14 - 6x = 7$
31. $-12t + 4 = 88$
32. $-16t + 3 = 67$
33. $14y + 15 = -33$
34. $12y + 13 = -15$
35. $32 - 16n = -8$
36. $-41 = 12n - 19$
37. $17x - 41 = -37$
38. $19y - 53 = -47$
39. $29 = -7 - 15x$
40. $49 = -5 - 14x$

For each of the following problems (a) choose a variable and indicate what it represents in the problem, (b) set up an equation that represents the situation described, and (c) solve the equation.

41. Twelve added to a certain number is 21. What is the number?
42. A certain number added to 14 is 25. Find the number.
43. Nine subtracted from a certain number is 13. Find the number.
44. A certain number subtracted from 32 is 15. What is the number?
45. Suppose that two items cost $43. If one of the items costs $25, what is the cost of the other item?
46. Eight years ago Rosa was 22 years old. Find Rosa's present age.
47. Six years from now, Nora will be 41 years old. What is her present age?
48. Chris bought eight pizzas for a total of $50. What was the price per pizza?
49. Chad worked 6 hours Saturday for a total of $39. How much per hour did he earn?
50. Jill worked 8 hours Saturday at $5.50 per hour. How much did she earn?
51. If 6 is added to three times a certain number, the result is 24. Find the number.
52. If 2 is subtracted from five times a certain number, the result is 38. Find the number.

53. Nineteen is 4 larger than three times a certain number. Find the number.

54. If nine times a certain number is subtracted from 7, the result is 52. Find the number.

55. Dress socks cost $2.50 a pair more than athletic socks. Randall purchased one pair of dress socks and six pairs of athletic socks for $21.75. Find the price of a pair of dress socks.

56. Together, a calculator and a mathematics textbook cost $85 in the college bookstore. The textbook price is $45 more than the price of the calculator. Find the price of the textbook.

57. The rainfall in June was 11.2 inches. This was 1 inch less than twice the rainfall in July. Find the amount of rainfall in inches for July.

58. Lunch at Joe's Hamburger Stand costs $1.75 less than lunch at Jodi's Taco Palace. A student spent his weekly lunch money, $24.50, eating four times at Jodi's and one time at Joe's. Find the cost of lunch at Jodi's Taco Palace.

59. If eight times a certain number is subtracted from 27, the result is 3. Find the number.

60. Twenty is 22 less than six times a certain number. Find the number.

61. A jeweler has priced a diamond ring at $550. This price represents $50 less than twice the cost of the ring to the jeweler. Find the cost of the ring to the jeweler.

62. Todd is following a 1750-calorie-per-day diet plan. This plan permits 650 calories less than twice the number of calories permitted by Lerae's diet plan. How many calories are permitted by Lerae's plan?

63. The length of a rectangular floor is 18 meters. This length is 2 meters less than five times the width of the floor. Find the width of the floor.

64. An executive is earning $45,000 per year. This is $15,000 less than twice her salary 4 years ago. Find her salary 4 years ago.

65. In the year 2000 it was estimated that there were 874 million speakers of the Chinese Mandarin language. This was 149 million less than three times the speakers of the English language. By this estimate how many million speakers of the English language were there in the year 2000?

66. A bill from the limousine company was $510. This included $150 for the service and $80 for each hour of use. Find the number of hours that the limousine was used.

67. Robin paid a $454 bill for a car DVD system. This included $379 for the DVD player and $60 an hour for installation. Find the number of hours it took to install the DVD system.

68. Tracy received a bill with cell phone use charges of $136.74. Included in the $136.74 were a charge of $39.99 for the monthly plan and a charge for 215 extra minutes. How much is Tracy being charged for each extra minute?

A selection of additional word problems is in Appendix B. All Appendix problems referenced as (3.2) are appropriate for this section.

THOUGHTS INTO WORDS

69. Give a step-by-step description of how you would solve the equation $17 = -3x + 2$.

70. What does the phrase "declare a variable" mean when solving a word problem?

71. Suppose that you are helping a friend with his homework, and he solves the equation $19 = 14 - x$ like this:

$$19 = 14 - x$$

$$19 + x = 14 - x + x$$

$$19 + x = 14$$

$$19 + x - 19 = 14 - 19$$

$$x = -5$$

The solution set is $\{-5\}$.

Does he have the correct solution set? What would you say to him about his method of solving the equation?

3.3 More on Solving Equations and Problem Solving

As equations become more complex, we need additional tools to solve them. So we need to organize our work carefully to minimize the chances for error. We will begin this section with some suggestions for solving equations, and then we will illustrate a *solution format* that is effective.

We can summarize the process of solving first-degree equations of one variable with the following three steps.

Step 1 Simplify both sides of the equation as much as possible.

Step 2 Use the addition or subtraction property of equality to isolate a term that contains the variable on one side and a constant on the other side of the equation.

Step 3 Use the multiplication or division property of equality to make the coefficient of the variable one.

The next examples illustrate this step-by-step process for solving equations. Study these examples carefully and be sure that you understand each step taken in the solving process.

EXAMPLE 1

Solve $5y - 4 + 3y = 12$.

Solution

$$5y - 4 + 3y = 12$$

$$8y - 4 = 12 \quad \text{Combine similar terms on the left side}$$

$$8y - 4 + 4 = 12 + 4 \quad \text{Add 4 to both sides}$$

$$8y = 16$$

$$\frac{8y}{8} = \frac{16}{8} \quad \text{Divide both sides by 8}$$

$$y = 2$$

The solution set is $\{2\}$. You can do the check alone now! ■

EXAMPLE 2

Solve $7x - 2 = 3x + 9$.

Solution

Notice that both sides of the equation are in simplified form; thus we can begin by applying the subtraction property of equality.

$$7x - 2 = 3x + 9$$

$$7x - 2 - 3x = 3x + 9 - 3x \quad \text{Subtract } 3x \text{ from both sides}$$

$$4x - 2 = 9$$

$$4x - 2 + 2 = 9 + 2 \quad \text{Add 2 to both sides}$$

$$4x = 11$$

$$\frac{4x}{4} = \frac{11}{4} \quad \text{Divide both sides by 4}$$

$$x = \frac{11}{4}$$

The solution set is $\left\{\frac{11}{4}\right\}$. ■

EXAMPLE 3 Solve $5n + 12 = 9n - 16$.

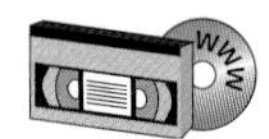

Solution

$$5n + 12 = 9n - 16$$

$$5n + 12 - 9n = 9n - 16 - 9n \quad \text{Subtract } 9n \text{ from both sides}$$

$$-4n + 12 = -16$$

$$-4n + 12 - 12 = -16 - 12 \quad \text{Subtract 12 from both sides}$$

$$-4n = -28$$

$$\frac{-4n}{-4} = \frac{-28}{-4} \quad \text{Divide both sides by } -4$$

$$n = 7$$

The solution set is $\{7\}$. ■

EXAMPLE 4 Solve $3x + 8 = 3x - 2$.

Solution

$$3x + 8 = 3x - 2$$

$$3x + 8 - 3x = 3x - 2 - 3x \quad \text{Subtract } 3x \text{ from both sides}$$

$$8 = -2 \quad \text{False statement}$$

Since we obtained an equivalent equation that is a false statement, there is no value of x that will make the equation a true statement. When the equation is not true under any condition, then the equation is called a **contradiction**. The solution set for an equation that is a contradiction is the empty or null set and is symbolized by $\varnothing$. ■

EXAMPLE 5 Solve $4x + 6 - x = 3x + 6$.

Solution

$$4x + 6 - x = 3x + 6$$

$$3x + 6 = 3x + 6 \quad \text{Combine similar terms on the left side}$$

$$3x + 6 - 3x = 3x + 6 - 3x \quad \text{Subtract } 3x \text{ from both sides}$$

$$6 = 6 \quad \text{True statement}$$

Since we obtained an equivalent equation that is a true statement, any value of x will make the equation a true statement. When an equation is true for any value of the variable, the equation is called an **identity**. The solution set for an equation that is an identity is the set of all real numbers. We will denote the set of all real numbers as {All reals}.

Word Problems

As we expand our skills for solving equations, we also expand our capabilities for solving word problems. No one definite procedure will ensure success at solving word problems, but the following suggestions can be helpful.

Suggestions for Solving Word Problems

1. Read the problem carefully, and make sure that you understand the meanings of all the words. Be especially alert for any technical terms in the statement of the problem.
2. Read the problem a second time (perhaps even a third time) to get an overview of the situation being described and to determine the known facts as well as what is to be found.
3. Sketch any figure, diagram, or chart that might be helpful in analyzing the problem.
4. Choose a meaningful variable to represent an unknown quantity in the problem (perhaps t if time is an unknown quantity); represent any other unknowns in terms of that variable.
5. Look for a **guideline** that you can use to set up an equation. A guideline might be a formula such as *distance equals rate times time* or a statement of a relationship such as *the sum of the two numbers* is 28. A guideline may also be indicated by a figure or diagram that you sketch for a particular problem.
6. Form an equation that contains the variable and that translates the conditions of the guideline from English into algebra.

(continued)

7. Solve the equation and use the solution to determine all facts requested in the problem.

8. **Check all answers by going back to the original statement of the problem and verifying that the answers make sense.**

If you decide not to check an answer, at least use the reasonableness-of-answer idea as a partial check. That is to say, ask yourself the question: Is this answer reasonable? For example, if the problem involves two investments that total \$10,000, then an answer of \$12,000 for one investment is certainly not reasonable.

Now let's consider some problems and use these suggestions.

Consecutive Number Problems

Some problems involve consecutive numbers or consecutive even or odd numbers. For example, 7, 8, 9, and 10 are consecutive numbers. To solve these applications, you must know how to represent consecutive numbers with variables. Let n represent the first number. For consecutive numbers, the next number is 1 more and is represented by $n + 1$. To continue, we add 1 to each preceding expression, obtaining the representations shown here:

$$\begin{array}{cccc} 7 & 8 & 9 & 10 \\ \downarrow & \downarrow & \downarrow & \downarrow \\ n & n+1 & n+2 & n+3 \end{array}$$

The pattern is somewhat different for consecutive even or odd numbers. For example, 2, 4, 6, and 8 are consecutive even numbers. Let n represent the first even number; then $n + 2$ represents the next even number. To continue, we add 2 to each preceding expression, obtaining these representations:

$$\begin{array}{cccc} 2 & 4 & 6 & 8 \\ \downarrow & \downarrow & \downarrow & \downarrow \\ n & n+2 & n+4 & n+6 \end{array}$$

Consecutive odd numbers have the same pattern of adding 2 to each preceding expression because consecutive odd numbers are two odd numbers with one and only one whole number between them.

PROBLEM 1

Find two consecutive even numbers whose sum is 74.

Solution

Let n represent the first number; then $n + 2$ represents the next even number. Since their sum is 74, we can set up and solve the following equation:

$$n + (n + 2) = 74$$

$$2n + 2 = 74$$

$$2n + 2 - 2 = 74 - 2$$

$$2n = 72$$

$$\frac{2n}{2} = \frac{72}{2}$$

$$n = 36$$

If $n = 36$, then $n + 2 = 38$; thus, the numbers are 36 and 38.

Check

To check your answers for Problem 1, determine whether the numbers satisfy the conditions stated in the original problem. Because 36 and 38 are two consecutive even numbers, and $36 + 38 = 74$ (their sum is 74), we know that the answers are correct. ■

The fifth entry in our list of problem-solving suggestions is to look for a *guideline* that can be used to set up an equation. The guideline may not be stated explicitly in the problem but may be implied by the nature of the problem. Consider the following example.

PROBLEM 2

Barry sells bicycles on a salary-plus-commission basis. He receives a monthly salary of \$300 and a commission of \$15 for each bicycle that he sells. How many bicycles must he sell in a month to earn a total monthly salary of \$750?

Solution

Let b represent the number of bicycles to be sold in a month. Then $15b$ represents his commission for those bicycles. The guideline "fixed salary plus commission equals total monthly salary" generates the following equation:

Fixed salary + Commission = Total monthly salary

$$\$300 + 15b = \$750$$

Solving this equation yields

$$300 + 15b - 300 = 750 - 300$$

$$15b = 450$$

$$\frac{15b}{15} = \frac{450}{15}$$

$$b = 30$$

Barry must sell 30 bicycles per month. (Does this number check?) ■

Geometric Problems

Sometimes the guideline for setting up an equation to solve a problem is based on a geometric relationship. Several basic geometric relationships pertain to angle measure. Let's state some of these relationships and then consider some problems.

1. Two angles for which the sum of their measure is 90° (the symbol ° indicates degrees) are called **complementary angles**.
2. Two angles for which the sum of their measure is 180° are called **supplementary angles**.
3. The sum of the measures of the three angles of a triangle is 180°.

PROBLEM 3

One of two complementary angles is 14° larger than the other. Find the measure of each of the angles.

Solution

If we let a represent the measure of the smaller angle, then $a + 14$ represents the measure of the larger angle. Since they are complementary angles, their sum is 90°, and we can proceed as follows:

$$a + (a + 14) = 90$$
$$2a + 14 = 90$$
$$2a + 14 - 14 = 90 - 14$$
$$2a = 76$$
$$\frac{2a}{2} = \frac{76}{2}$$
$$a = 38$$

If $a = 38$, then $a + 14 = 52$, and the angles measure 38° and 52°. ■

PROBLEM 4

Find the measures of the three angles of a triangle if the second is three times the first and the third is twice the second.

Solution

If we let a represent the measure of the smallest angle, then $3a$ and $2(3a)$ represent the measures of the other two angles. Therefore, we can set up and solve the following equation:

$$a + 3a + 2(3a) = 180$$
$$a + 3a + 6a = 180$$
$$10a = 180$$

$$\frac{10a}{10} = \frac{180}{10}$$

$$a = 18$$

If $a = 18$, then $3a = 54$ and $2(3a) = 108$. So the angles have measures of 18°, 54°, and 108°.

Problem Set 3.3

For Problems 1–32, solve each equation.

1. $2x + 7 + 3x = 32$
2. $3x + 9 + 4x = 30$
3. $7x - 4 - 3x = -36$
4. $8x - 3 - 2x = -45$
5. $3y - 1 + 2y - 3 = 4$
6. $y + 3 + 2y - 4 = 6$
7. $5n - 2 - 8n = 31$
8. $6n - 1 - 10n = 51$
9. $-2n + 1 - 3n + n - 4 = 7$
10. $-n + 7 - 2n + 5n - 3 = -6$
11. $3x + 4 = 2x - 5$
12. $5x - 2 = 4x + 6$
13. $5x - 7 = 6x - 9$
14. $7x - 3 = 8x - 13$
15. $6x + 1 = 3x - 8$
16. $4x - 10 = x + 17$
17. $7y - 3 = 5y + 10$
18. $8y + 4 = 5y - 4$
19. $8n - 2 = 8n - 7$
20. $7n - 10 = 9n - 13$
21. $-2x - 7 = -3x + 10$
22. $-4x + 6 = -5x - 9$
23. $-3x + 5 = -5x - 8$
24. $-4x + 7 = -4x + 4$
25. $-7 - 6x = 9 - 9x$
26. $-10 - 7x = 14 - 12x$
27. $2x - 1 - x = x - 1$
28. $3x - 4 - 4x = -5x + 4x - 4$
29. $5n - 4 - n = -3n - 6 + n$
30. $4x - 3 + 2x = 8x - 3 - x$
31. $-7 - 2n - 6n = 7n - 5n + 12$
32. $-3n + 6 + 5n = 7n - 8n - 9$

Solve each word problem by setting up and solving an algebraic equation.

33. The sum of a number plus four times the number is 85. What is the number?
34. A number subtracted from three times the number yields 68. Find the number.
35. Find two consecutive odd numbers whose sum is 72.
36. Find two consecutive even numbers whose sum is 94.
37. Find three consecutive even numbers whose sum is 114.
38. Find three consecutive odd numbers whose sum is 159.
39. Two more than three times a certain number is the same as 4 less than seven times the number. Find the number.
40. One more than five times a certain number is equal to eleven less than nine times the number. What is the number?
41. The sum of a number and five times the number equals eighteen less than three times the number. Find the number.
42. One of two supplementary angles is five times as large as the other. Find the measure of each angle.

43. One of two complementary angles is 6° smaller than twice the other angle. Find the measure of each angle.

44. If two angles are complementary and the difference between their measures is 62°, find the measure of each angle.

45. If two angles are supplementary and the larger angle is 20° less than three times the smaller angle, find the measure of each angle.

46. Find the measures of the three angles of a triangle if the largest is 14° less than three times the smallest, and the other angle is 4° larger than the smallest

47. One of the angles of a triangle has a measure of 40°. Find the measures of the other two angles if the difference between their measures is 10°.

48. Jesstan worked as a telemarketer on a salary-plus-commission basis. He was paid a salary of $300 a week and a $12 commission for each sale. If his earnings for the week were $960, how many sales did he make?

49. Marci sold an antique vase in an online auction for $69.00. This was $15 less than twice the amount she paid for it. What price did she pay for the vase?

50. A set of wheels sold in an online auction for $560. This was $35 more than three times the opening bid. How much was the opening bid?

51. Suppose that Bob is paid two times his normal hourly rate for each hour he works over 40 hours in a week. Last week he earned $504 for 48 hours of work. What is his hourly wage?

52. Last week on an algebra test, the highest grade was 9 points less than three times the lowest grade. The sum of the two grades was 135. Find the lowest and highest grades on the test.

53. At a university-sponsored concert, there were three times as many women as men. A total of 600 people attended the concert. How many men and how many women attended?

54. Suppose that a triangular lot is enclosed with 135 yards of fencing (see Figure 3.1). The longest side of the lot is 5 yards longer than twice the length of the shortest side. The other side is 10 yards longer than the shortest side. Find the lengths of the three sides of the lot.

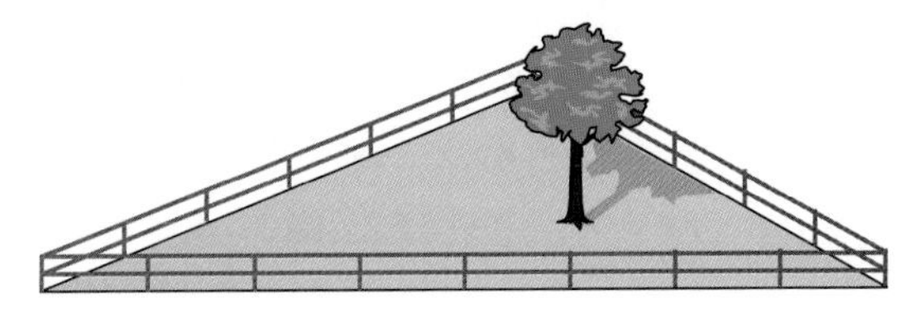

Figure 3.1

55. The textbook for a biology class cost $15 more than twice the cost of a used textbook for college algebra. If the cost of the two books together is $129, find the cost of the biology book.

56. A nutrition plan counts grams of fat, carbohydrates, and fiber. The grams of carbohydrates are to be 15 more than twice the grams of fat. The grams of fiber are to be three less than the grams of fat. If the grams of carbohydrate, fat, and fiber must total 48 grams for a dinner meal, how many grams of each would be in the meal?

57. At a local restaurant, $275 in tips is to be shared between the server, bartender, and busboy. The server gets $25 more than three times the amount the busboy receives. The bartender gets $50 more than the amount the busboy receives. How much will the server receive?

A selection of additional word problems is in Appendix B. All Appendix problems referenced as (3.3) are appropriate for this section.

THOUGHTS INTO WORDS

58. Give a step-by-step description of how you would solve the equation $3x + 4 = 5x - 2$.

59. Suppose your friend solved the problem *find two consecutive odd integers whose sum is 28* as follows:

$$x + (x + 1) = 28$$

$$2x = 27$$

$$x = \frac{27}{2} = 13\frac{1}{2}$$

She claims that $13\frac{1}{2}$ will check in the equation. Where has she gone wrong, and how would you help her?

FURTHER INVESTIGATIONS

60. Solve each of these equations.

(a) $7x - 3 = 4x - 3$

(b) $-x - 4 + 3x = 2x - 7$

(c) $-3x + 9 - 2x = -5x + 9$

(d) $5x - 3 = 6x - 7 - x$

(e) $7x + 4 = -x + 4 + 8x$

(f) $3x - 2 - 5x = 7x - 2 - 5x$

(g) $-6x - 8 = 6x + 4$

(h) $-8x + 9 = -8x + 5$

61. Make up an equation whose solution set is the null set and explain why.

62. Make up an equation whose solution set is the set of all real numbers and explain why.

3.4 Equations Involving Parentheses and Fractional Forms

We will use the distributive property frequently in this section as we add to our techniques for solving equations. Recall that in symbolic form the distributive property states that $a(b + c) = ab + ac$. Consider the following examples, which illustrate the use of this property to remove parentheses. Pay special attention to the last two examples, which involve a negative number in front of the parentheses.

$$3(x + 2) = 3 \cdot x + 3 \cdot 2 = 3x + 6$$

$$5(y - 3) = 5 \cdot y - 5 \cdot 3 = 5y - 15 \qquad a(b - c) = ab - ac$$

$$2(4x + 7) = 2(4x) + 2(7) = 8x + 14$$

$$-1(n + 4) = (-1)(n) + (-1)(4) = -n - 4$$

$$-6(x - 2) = (-6)(x) - (-6)(2) = -6x + 12$$

Do this step mentally!

It is often necessary to solve equations in which the variable is part of an expression enclosed in parentheses. We use the distributive property to remove the parentheses, and then we proceed in the usual way. Consider the next examples. (Notice that we are beginning to show only the major steps when solving an equation.)

EXAMPLE 1

Solve $3(x + 2) = 23$.

Solution

$$3(x + 2) = 23$$

$$3x + 6 = 23 \qquad \text{Applied distributive property to left side}$$

$$3x = 17 \qquad \text{Subtracted 6 from both sides}$$

$$x = \frac{17}{3} \qquad \text{Divided both sides by 3}$$

The solution set is $\left\{\frac{17}{3}\right\}$. ■

EXAMPLE 2 Solve $4(x + 3) = 2(x - 6)$.

Solution

$$4(x + 3) = 2(x - 6)$$

$$4x + 12 = 2x - 12 \qquad \text{Applied distributive property on each side}$$

$$2x + 12 = -12 \qquad \text{Subtracted } 2x \text{ from both sides}$$

$$2x = -24 \qquad \text{Subtracted 12 from both sides}$$

$$x = -12 \qquad \text{Divided both sides by 2}$$

The solution set is $\{-12\}$. ■

It may be necessary to remove more than one set of parentheses and then to use the distributive property again to combine similar terms. Consider the following two examples.

EXAMPLE 3 Solve $5(w + 3) + 3(w + 1) = 14$.

Solution

$$5(w + 3) + 3(w + 1) = 14$$

$$5w + 15 + 3w + 3 = 14 \qquad \text{Applied distributive property}$$

$$8w + 18 = 14 \qquad \text{Combined similar terms}$$

$$8w = -4 \qquad \text{Subtracted 18 from both sides}$$

$$w = -\frac{4}{8} \qquad \text{Divided both sides by 8}$$

$$w = -\frac{1}{2} \qquad \text{Reduced}$$

The solution set is $\left\{-\frac{1}{2}\right\}$. ■

EXAMPLE 4 Solve $6(x - 7) - 2(x - 4) = 13$.

Solution

$$6(x - 7) - 2(x - 4) = 13$$ Be careful with this sign!

$$6x - 42 - 2x + 8 = 13$$ Distributive property

$$4x - 34 = 13$$ Combined similar terms

$$4x = 47$$ Added 34 to both sides

$$x = \frac{47}{4}$$ Divided both sides by 4

The solution set is $\left\{\frac{47}{4}\right\}$. ■

In a previous section, we solved equations like $x - \frac{2}{3} = \frac{3}{4}$ by adding $\frac{2}{3}$ to both sides. If an equation contains several fractions, then it is usually easier to clear the equation of all fractions by multiplying both sides by the least common denominator of all the denominators. Perhaps several examples will clarify this idea.

EXAMPLE 5 Solve $\frac{1}{2}x + \frac{2}{3} = \frac{5}{6}$.

Solution

$$\frac{1}{2}x + \frac{2}{3} = \frac{5}{6}$$

$$6\left(\frac{1}{2}x + \frac{2}{3}\right) = 6\left(\frac{5}{6}\right)$$ 6 is the LCD of 2, 3, and 6

$$6\left(\frac{1}{2}x\right) + 6\left(\frac{2}{3}\right) = 6\left(\frac{5}{6}\right)$$ Distributive property

$$3x + 4 = 5$$ Note how the equation has been *cleared of all fractions*

$$3x = 1$$

$$x = \frac{1}{3}$$

The solution set is $\left\{\frac{1}{3}\right\}$. ■

EXAMPLE 6

Solve $\frac{5n}{6} - \frac{1}{4} = \frac{3}{8}$.

Solution

$$\frac{5n}{6} - \frac{1}{4} = \frac{3}{8}$$

$$24\left(\frac{5n}{6} - \frac{1}{4}\right) = 24\left(\frac{3}{8}\right) \qquad \text{24 is the LCD of 6, 4, and 8}$$

$$24\left(\frac{5n}{6}\right) - 24\left(\frac{1}{4}\right) = 24\left(\frac{3}{8}\right) \qquad \text{Distributive property}$$

$$20n - 6 = 9$$

$$20n = 15$$

$$n = \frac{15}{20} = \frac{3}{4}$$

The solution set is $\left\{\frac{3}{4}\right\}$. ■

We use many of the ideas presented in this section to help solve the equations in the next examples. Study the solutions carefully and be sure that you can supply reasons for each step. It might be helpful to cover up the solutions and try to solve the equations on your own.

EXAMPLE 7

Solve $\frac{x + 3}{2} + \frac{x + 4}{5} = \frac{3}{10}$.

Solution

$$\frac{x + 3}{2} + \frac{x + 4}{5} = \frac{3}{10}$$

$$10\left(\frac{x + 3}{2} + \frac{x + 4}{5}\right) = 10\left(\frac{3}{10}\right) \qquad \text{10 is the LCD of 2, 5, and 10}$$

$$10\left(\frac{x + 3}{2}\right) + 10\left(\frac{x + 4}{5}\right) = 10\left(\frac{3}{10}\right) \qquad \text{Distributive property}$$

$$5(x + 3) + 2(x + 4) = 3$$

$$5x + 15 + 2x + 8 = 3$$

$$7x + 23 = 3$$

$$7x = -20$$

$$x = -\frac{20}{7}$$

The solution set is $\left\{-\frac{20}{7}\right\}$. ■

EXAMPLE 8

Solve $\frac{x-1}{4} - \frac{x-2}{6} = \frac{2}{3}$.

Solution

$$\frac{x-1}{4} - \frac{x-2}{6} = \frac{2}{3}$$

$$12\left(\frac{x-1}{4} - \frac{x-2}{6}\right) = 12\left(\frac{2}{3}\right) \quad \text{12 is the LCD of 4, 6, and 3}$$

$$12\left(\frac{x-1}{4}\right) - 12\left(\frac{x-2}{6}\right) = 12\left(\frac{2}{3}\right) \quad \text{Distributive property}$$

$$3(x-1) - 2(x-2) = 8$$

$$3x - 3 - 2x + 4 = 8 \quad \text{Be careful with this sign!}$$

$$x + 1 = 8$$

$$x = 7$$

The solution set is $\{7\}$. ■

■ Word Problems

We are now ready to solve some word problems using equations of the different types presented in this section. Again, it might be helpful for you to attempt to solve the problems on your own before looking at the book's approach.

PROBLEM 1

Loretta has 19 coins (quarters and nickels) that amount to \$2.35. How many coins of each kind does she have?

Solution

Let q represent the number of quarters. Then $19 - q$ represents the number of nickels. We can use the following guideline to help set up an equation:

Value of quarters in cents + Value of nickels in cents = Total value in cents

$$25q \quad + \quad 5(19 - q) \quad = \quad 235$$

Solving the equation, we obtain

$$25q + 95 - 5q = 235$$
$$20q + 95 = 235$$
$$20q = 140$$
$$q = 7$$

If $q = 7$, then $19 - q = 12$, so she has 7 quarters and 12 nickels. ■

PROBLEM 2

Find a number such that 4 less than two-thirds the number is equal to one-sixth the number.

Solution

Let n represent the number. Then $\frac{2}{3}n - 4$ represents 4 less than two-thirds the number, and $\frac{1}{6}n$ represents one-sixth the number.

$$\frac{2}{3}n - 4 = \frac{1}{6}n$$
$$6\left(\frac{2}{3}n - 4\right) = 6\left(\frac{1}{6}n\right)$$
$$4n - 24 = n$$
$$3n - 24 = 0$$
$$3n = 24$$
$$n = 8$$

The number is 8. ■

PROBLEM 3

Lance is paid $1\frac{1}{2}$ times his normal hourly rate for each hour he works over 40 hours in a week. Last week he worked 50 hours and earned \$462. What is his normal hourly rate?

Solution

Let x represent Lance's normal hourly rate. Then $\frac{3}{2}x$ represents $1\frac{1}{2}$ times his normal hourly rate. We can use the following guideline to set up the equation:

Regular wages for first 40 hours + Wages for 10 hours of overtime = Total wages

$$40x + 10\left(\frac{3}{2}x\right) = 462$$

Solving this equation, we obtain

$$40x + 15x = 462$$
$$55x = 462$$
$$x = 8.40$$

Lance's normal hourly rate is \$8.40. ■

PROBLEM 4

Find three consecutive whole numbers such that the sum of the first plus twice the second plus three times the third is 134.

Solution

Let n represent the first whole number. Then $n + 1$ represents the second whole number and $n + 2$ represents the third whole number.

$$n + 2(n + 1) + 3(n + 2) = 134$$
$$n + 2n + 2 + 3n + 6 = 134$$
$$6n + 8 = 134$$
$$6n = 126$$
$$n = 21$$

The numbers are 21, 22, and 23. ■

Keep in mind that the problem-solving suggestions we offered in Section 3.3 simply outline a general algebraic approach to solving problems. You will add to this list throughout this course and in any subsequent mathematics courses that you take. Furthermore, you will be able to pick up additional problem-solving ideas from your instructor and from fellow classmates as problems are discussed in class. Always be on the alert for any ideas that might help you become a better problem solver.

Problem Set 3.4

Solve each equation.

1. $7(x + 2) = 21$

2. $4(x + 4) = 24$

3. $5(x - 3) = 35$

4. $6(x - 2) = 18$

5. $-3(x + 5) = 12$

6. $-5(x - 6) = -15$

7. $4(n - 6) = 5$

8. $3(n + 4) = 7$

9. $6(n + 7) = 8$

10. $8(n - 3) = 12$

11. $-10 = -5(t - 8)$

12. $-16 = -4(t + 7)$

13. $5(x - 4) = 4(x + 6)$

14. $6(x - 4) = 3(2x + 5)$

15. $8(x + 1) = 9(x - 2)$

16. $4(x - 7) = 5(x + 2)$

17. $8(t + 5) = 4(2t + 10)$

18. $7(t - 5) = 5(t + 3)$

19. $2(6t + 1) = 4(3t - 1)$

20. $6(t + 5) = 2(3t + 15)$

21. $-2(x - 6) = -(x - 9)$

22. $-(x + 7) = -2(x + 10)$

23. $-3(t - 4) - 2(t + 4) = 9$

24. $5(t - 4) - 3(t - 2) = 12$

25. $3(n - 10) - 5(n + 12) = -86$

26. $4(n + 9) - 7(n - 8) = 83$

27. $3(x + 1) + 4(2x - 1) = 5(2x + 3)$

28. $4(x - 1) + 5(x + 2) = 3(x - 8)$

29. $-(x + 2) + 2(x - 3) = -2(x - 7)$

30. $-2(x + 6) + 3(3x - 2) = -3(x - 4)$

31. $5(2x - 1) - (3x + 4) = 4(x + 3) - 27$

32. $3(4x + 1) - 2(2x + 1) = -2(x - 1) - 1$

33. $-(a - 1) - (3a - 2) = 6 + 2(a - 1)$

34. $3(2a - 1) - 2(5a + 1) = 4(3a + 4)$

35. $3(x - 1) + 2(x - 3) = -4(x - 2) + 10(x + 4)$

36. $-2(x - 4) - (3x - 2) = -2 + (-6x + 2)$

37. $3 - 7(x - 1) = 9 - 6(2x + 1)$

38. $8 - 5(2x + 1) = 2 - 6(x - 3)$

39. $\frac{3}{4}x - \frac{2}{3} = \frac{5}{6}$

40. $\frac{1}{2}x - \frac{4}{3} = -\frac{5}{6}$

41. $\frac{5}{6}x + \frac{1}{4} = -\frac{9}{4}$

42. $\frac{3}{8}x + \frac{1}{6} = -\frac{7}{12}$

43. $\frac{1}{2}x - \frac{3}{5} = \frac{3}{4}$

44. $\frac{1}{4}x - \frac{2}{5} = \frac{5}{6}$

45. $\frac{n}{3} + \frac{5n}{6} = \frac{1}{8}$

46. $\frac{n}{6} + \frac{3n}{8} = \frac{5}{12}$

47. $\frac{5y}{6} - \frac{3}{5} = \frac{2y}{3}$

48. $\frac{3y}{7} + \frac{1}{2} = \frac{y}{4}$

49. $\frac{h}{6} + \frac{h}{8} = 1$

50. $\frac{h}{4} + \frac{h}{3} = 1$

51. $\frac{x + 2}{3} + \frac{x + 3}{4} = \frac{13}{3}$

52. $\frac{x - 1}{4} + \frac{x + 2}{5} = \frac{39}{20}$

53. $\frac{x - 1}{5} - \frac{x + 4}{6} = -\frac{13}{15}$

54. $\frac{x + 1}{7} - \frac{x - 3}{5} = \frac{4}{5}$

55. $\frac{x + 8}{2} - \frac{x + 10}{7} = \frac{3}{4}$

56. $\frac{x + 7}{3} - \frac{x + 9}{6} = \frac{5}{9}$

57. $\frac{x - 2}{8} - 1 = \frac{x + 1}{4}$

58. $\frac{x - 4}{2} + 3 = \frac{x - 2}{4}$

59. $\frac{x + 1}{4} = \frac{x - 3}{6} + 2$

60. $\frac{x + 3}{5} = \frac{x - 6}{2} + 1$

Solve each word problem by setting up and solving an appropriate algebraic equation.

61. Find two consecutive whole numbers such that the smaller number plus four times the larger number equals 39.

62. Find two consecutive whole numbers such that the smaller number subtracted from five times the larger number equals 57.

63. Find three consecutive whole numbers such that twice the sum of the two smallest numbers is 10 more than three times the largest number.

64. Find four consecutive whole numbers such that the sum of the first three numbers equals the fourth number.

65. The sum of two numbers is 17. If twice the smaller number is 1 more than the larger number, find the numbers.

66. The sum of two numbers is 53. If three times the smaller number is 1 less than the larger number, find the numbers.

67. Find a number such that 20 more than one-third of the number equals three-fourths of the number.

68. The sum of three-eighths of a number and five-sixths of the same number is 29. Find the number.

69. Mrs. Nelson had to wait 4 minutes in line at her bank's automated teller machine. This was 3 minutes less than one-half of the time she waited in line at the grocery store. How long in minutes did she wait in line at the grocery store?

70. Raoul received a $30 tip for waiting on a large party. This was $5 more than one-fourth of the tip the headwaiter received. How much did the headwaiter get for a tip?

71. Suppose that a board 20 feet long is cut into two pieces. Four times the length of the shorter piece is 4 feet less than three times the length of the longer piece. Find the length of each piece.

72. Ellen is paid time and a half for each hour over 40 hours she works in a week. Last week she worked 44 hours and earned $391. What is her normal hourly rate?

73. Lucy has 35 coins consisting of nickels and quarters amounting to $5.75. How many coins of each kind does she have?

74. Suppose that Julian has 44 coins consisting of pennies and nickels. If the number of nickels is two more than twice the number of pennies, find the number of coins of each kind.

75. Max has a collection of 210 coins consisting of nickels, dimes, and quarters. He has twice as many dimes as nickels, and 10 more quarters than dimes. How many coins of each kind does he have?

76. Ginny has a collection of 425 coins consisting of pennies, nickels, and dimes. She has 50 more nickels than pennies and 25 more dimes than nickels. How many coins of each kind does she have?

77. Maida has 18 coins consisting of dimes and quarters amounting to $3.30. How many coins of each kind does she have?

78. Ike has some nickels and dimes amounting to $2.90. The number of dimes is one less than twice the number of nickels. How many coins of each kind does he have?

79. Mario has a collection of 22 specimens in his aquarium consisting of crabs, fish, and plants. There are three times as many fish as crabs. There are two more plants than crabs. How many specimens of each kind are in the collection?

80. Tickets for a concert were priced at $8 for students and $10 for nonstudents. There were 1500 tickets sold for a total of $12,500. How many student tickets were sold?

Figure 3.2

81. The supplement of an angle is 30° larger than twice its complement. Find the measure of the angle.

82. The sum of the measure of an angle and three times its complement is 202°. Find the measure of the angle.

83. In triangle ABC, the measure of angle A is 2° less than one-fifth of the measure of angle C. The measure of angle B is 5° less than one-half of the measure of angle C. Find the measures of the three angles of the triangle.

84. If one-fourth of the complement of an angle plus one-fifth of the supplement of the angle equals 36°, find the measure of the angle.

85. The supplement of an angle is 10° smaller than three times its complement. Find the size of the angle.

86. In triangle ABC, the measure of angle C is eight times the measure of angle A, and the measure of angle B is 10° more than the measure of angle C. Find the measure of each angle of the triangle.

A selection of additional word problems is in Appendix B. All Appendix problems referenced as (3.4) are appropriate for this section.

THOUGHTS INTO WORDS

87. Discuss how you would solve the equation

$3(x - 2) - 5(x + 3) = -4(x + 9)$.

88. Why must potential answers to word problems be checked back into the original statement of the problem?

89. Consider these two solutions:

$$3(x + 2) = 9 \qquad 3(x - 4) = 7$$

$$\frac{3(x + 2)}{3} = \frac{9}{3} \qquad \frac{3(x - 4)}{3} = \frac{7}{3}$$

$$x + 2 = 3 \qquad x - 4 = \frac{7}{3}$$

$$x = 1 \qquad x = \frac{19}{3}$$

Are both of these solutions correct? Comment on the effectiveness of the two different approaches.

FURTHER INVESTIGATIONS

90. Solve each equation.

(a) $-2(x - 1) = -2x + 2$

(b) $3(x + 4) = 3x - 4$

(c) $5(x - 1) = -5x - 5$

(d) $\frac{x - 3}{3} + 4 = 3$

(e) $\frac{x + 2}{3} + 1 = \frac{x - 2}{3}$

(f) $\frac{x - 1}{5} - 2 = \frac{x - 11}{5}$

(g) $4(x - 2) - 2(x + 3) = 2(x + 6)$

(h) $5(x + 3) - 3(x - 5) = 2(x + 15)$

(i) $7(x - 1) + 4(x - 2) = 15(x - 1)$

91. Find three consecutive integers such that the sum of the smallest integer and the largest integer is equal to twice the middle integer.

3.5 Inequalities

Just as we use the symbol = to represent *is equal to*, we also use the symbols < and > to represent *is less than* and *is greater than*, respectively. Here are some examples of **statements of inequality**. Notice that the first four are true statements and the last two are false.

$6 + 4 > 7$	True
$8 - 2 < 14$	True
$4 \cdot 8 > 4 \cdot 6$	True
$5 \cdot 2 < 5 \cdot 7$	True

$5 + 8 > 19$ False

$9 - 2 < 3$ False

Algebraic inequalities contain one or more variables. These are examples of algebraic inequalities:

$x + 3 > 4$

$2x - 1 < 6$

$x^2 + 2x - 1 > 0$

$2x + 3y < 7$

$7ab < 9$

An algebraic inequality such as $x + 1 > 2$ is neither true nor false as it stands; it is called an **open sentence**. Each time a number is substituted for x, the algebraic inequality $x + 1 > 2$ becomes a numerical statement that is either true or false. For example, if $x = 0$, then $x + 1 > 2$ becomes $0 + 1 > 2$, which is false. If $x = 2$, then $x + 1 > 2$ becomes $2 + 1 > 2$, which is true. **Solving an inequality** refers to the process of finding the numbers that make an algebraic inequality a true numerical statement. We say that such numbers, called the **solutions of the inequality**, satisfy the inequality. The set of all solutions of an inequality is called its **solution set**. We often state solution sets for inequalities with **set builder notation**. For example, the solution set for $x + 1 > 2$ is the set of real numbers greater than 1, expressed as $\{x|x > 1\}$. The set builder notation $\{x|x > 1\}$ is read as "the set of all x such that x is greater than 1." We sometimes graph solution sets for inequalities on a number line; the solution set for $\{x|x > 1\}$ is pictured in Figure 3.3.

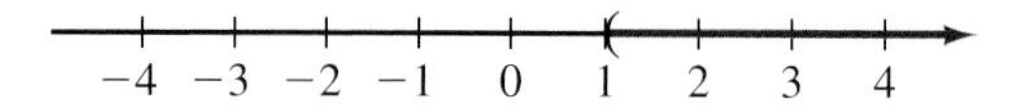

Figure 3.3

The left-hand parenthesis at 1 indicates that 1 is *not* a solution, and the red part of the line to the right of 1 indicates that all real numbers greater than 1 are solutions. We refer to the red portion of the number line as the *graph* of the solution set $\{x|x > 1\}$.

The solution set for $x + 1 \leq 3$ ($\leq$ is read "less than or equal to") is the set of real numbers less than or equal to 2, expressed as $\{x|x \leq 2\}$. The graph of the solution set for $\{x|x \leq 2\}$ is pictured in Figure 3.4. The right-hand bracket at 2 indicates that 2 *is included* in the solution set.

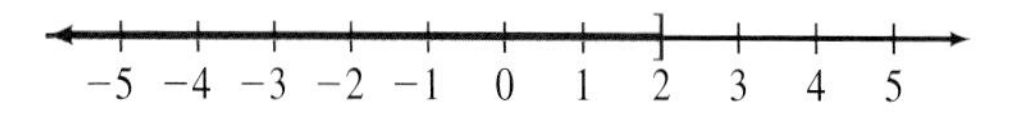

Figure 3.4

It is also convenient to express solution sets of inequalities using **interval notation**. The solution set $\{x|x > 6\}$ is written as $(6, \infty)$ using interval notation. In interval notation parentheses are used to indicate exclusion of the endpoint. The $>$ and $<$ symbols in inequalities also indicate the exclusion of the endpoint. So when the inequality has a $>$ or $<$ symbol, the interval notation uses a parenthesis. This is consistent with the use of parentheses on the number line.

In this same example, $\{x|x > 6\}$, the solution set has no upper endpoint, so the infinity symbol, ∞, is used to indicate that the interval continues indefinitely. The solution set for $\{x|x < 3\}$ is written as $(-\infty, 3)$ in interval notation. Here the solution set has no lower endpoint, so a negative sign precedes the infinity symbol because the interval is extending indefinitely in the opposite direction. The infinity symbol always has a parenthesis in interval notation because there is no actual endpoint to include.

The solution set $\{x|x \geq 5\}$ is written as $[5, \infty)$ using interval notation. In interval notation square brackets are used to indicate inclusion of the endpoint. The $\geq$ and $\leq$ symbols in inequalities also indicate the inclusion of the endpoint. So when the inequality has a $\geq$ or $\leq$ symbol, the interval notation uses a square bracket. Again the use of a bracket in interval notation is consistent with the use of a bracket on the number line.

The examples in the table below contain some simple algebraic inequalities, their solution sets, graphs of the solution sets, and the solution sets written in interval notation. Look them over very carefully to be sure you understand the symbols.

Algebraic inequality	Solution set	Graph of solution set	Interval notation
$x < 2$	$\{x\|x < 2\}$	−5 −4 −3 −2 −1 0 1 2 3 4 5	$(-\infty, 2)$
$x > -1$	$\{x\|x > -1\}$	−5 −4 −3 −2 −1 0 1 2 3 4 5	$(-1, \infty)$
$3 < x$	$\{x\|x > 3\}$	−5 −4 −3 −2 −1 0 1 2 3 4 5	$(3, \infty)$
$x \geq 1$ ($\geq$ is read "greater than or equal to")	$\{x\|x \geq 1\}$	−5 −4 −3 −2 −1 0 1 2 3 4 5	$[1, \infty)$
$x \leq 2$ ($\leq$ is read "less than or equal to")	$\{x\|x \leq 2\}$	−5 −4 −3 −2 −1 0 1 2 3 4 5	$(-\infty, 2]$
$1 \geq x$	$\{x\|x \leq 1\}$	−5 −4 −3 −2 −1 0 1 2 3 4 5	$(-\infty, 1]$

Figure 3.5

The general process for solving inequalities closely parallels that for solving equations. We continue to replace the given inequality with equivalent, but simpler inequalities. For example,

$$2x + 1 > 9 \quad (1)$$

$$2x > 8 \quad (2)$$

$$x > 4 \quad (3)$$

are all equivalent inequalities; that is, they have the same solutions. Thus, to solve (1), we can solve (3), which is obviously all numbers greater than 4. The exact procedure for simplifying inequalities is based primarily on two properties, and they become our topics of discussion at this time. The first of these is the **addition-subtraction property of inequality**.

Property 3.4 Addition-Subtraction Property of Inequality

For all real numbers a, b, and c,

1. $a > b$ if and only if $a + c > b + c$.

2. $a > b$ if and only if $a - c > b - c$.

Property 3.4 states that any number can be added to or subtracted from both sides of an inequality, and an equivalent inequality is produced. The property is stated in terms of $>$, but analogous properties exist for $<$, $\geq$, and $\leq$. Consider the use of this property in the next three examples.

EXAMPLE 1

Solve $x - 3 > -1$ and graph the solutions.

Solution

$$x - 3 > -1$$

$$x - 3 + 3 > -1 + 3 \quad \text{Add 3 to both sides}$$

$$x > 2$$

The solution set is $\{x|x > 2\}$, and it can be graphed as shown in Figure 3.6. The solution written in interval notation is $(2, \infty)$.

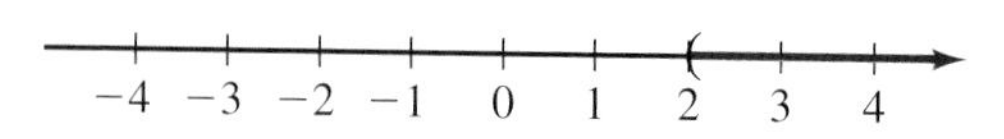

Figure 3.6

EXAMPLE 2

Solve $x + 4 \leq 5$ and graph the solutions.

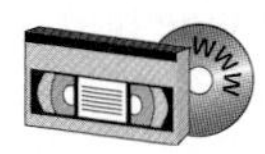

Solution

$$x + 4 \leq 5$$

$$x + 4 - 4 \leq 5 - 4 \quad \text{Subtract 4 from both sides}$$

$$x \leq 1$$

The solution set is $\{x|x \leq 1\}$, and it can be graphed as shown in Figure 3.7. The solution written in interval notation is $(-\infty, 1]$.

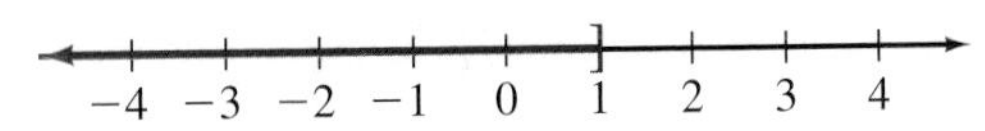

■

Figure 3.7

EXAMPLE 3

Solve $5 > 6 + x$ and graph the solutions.

Solution

$$5 > 6 + x$$

$$5 - 6 > 6 + x - 6 \quad \text{Subtract 6 from both sides}$$

$$-1 > x$$

Because $-1 > x$ is equivalent to $x < -1$, the solution set is $\{x|x < -1\}$, and it can be graphed as shown in Figure 3.8. The solution written in interval notation is $(-\infty, -1)$.

■

Figure 3.8

Now let's look at some numerical examples to see what happens when both sides of an inequality are multiplied or divided by some number.

$$4 > 3 \rightarrow 5(4) > 5(3) \rightarrow 20 > 15$$

$$-2 > -3 \rightarrow 4(-2) > 4(-3) \rightarrow -8 > -12$$

$$6 > 4 \rightarrow \frac{6}{2} > \frac{4}{2} \rightarrow 3 > 2$$

$$8 > -2 \rightarrow \frac{8}{4} > \frac{-2}{4} \rightarrow 2 > -\frac{1}{2}$$

Notice that multiplying or dividing both sides of an inequality by a positive number produces an inequality of the same sense. This means that if the original inequality is *greater than*, then the new inequality is *greater than*, and if the original is *less than*, then the resulting inequality is *less than*.

Now note what happens when we multiply or divide both sides by a negative number:

$$3 < 5 \rightarrow -2(3) > -2(5) \rightarrow -6 > -10$$

$$-4 < 1 \rightarrow -5(-4) > -5(1) \rightarrow 20 > -5$$

$$14 > 2 \rightarrow \frac{14}{-2} < \frac{2}{-2} \rightarrow -7 < -1$$

$$-3 > -6 \rightarrow \frac{-3}{-3} < \frac{-6}{-3} \rightarrow 1 < 2$$

Multiplying or dividing both sides of an inequality by a *negative number reverses the sense of the inequality*. Property 3.5 summarizes these ideas.

Property 3.5 Multiplication-Division Property of Inequality

(a) For all real numbers, a, b, and c, with $c > 0$,

1. $a > b$ if and only if $ac > bc$.

2. $a > b$ if and only if $\frac{a}{c} > \frac{b}{c}$.

(b) For all real numbers, a, b, and c, with $c < 0$,

1. $a > b$ if and only if $ac < bc$.

2. $a > b$ if and only if $\frac{a}{c} < \frac{b}{c}$.

Similar properties hold if each inequality is reversed or if $>$ is replaced with $\geq$, and $<$ is replaced with $\leq$. For example, if $a \leq b$ and $c < 0$, then $ac \geq bc$ and $\frac{a}{c} \geq \frac{b}{c}$.

Observe the use of Property 3.5 in the next three examples.

EXAMPLE 4

Solve $2x > 4$.

Solution

$$2x > 4$$

$$\frac{2x}{2} > \frac{4}{2} \quad \text{Divide both sides by 2}$$

$$x > 2$$

The solution set is $\{x|x > 2\}$ or $(2, \infty)$ in interval notation. ■

EXAMPLE 5

Solve $\frac{3}{4}x \leq \frac{1}{5}$.

Solution

$$\frac{3}{4}x \leq \frac{1}{5}$$

$$\frac{4}{3}\left(\frac{3}{4}x\right) \leq \frac{4}{3}\left(\frac{1}{5}\right) \quad \text{Multiply both sides by } \frac{4}{3}$$

$$x \leq \frac{4}{15}$$

The solution set is $\left\{x \mid x \leq \frac{4}{15}\right\}$ or $\left(-\infty, \frac{4}{15}\right]$ in interval notation. ■

EXAMPLE 6

Solve $-3x > 9$.

Solution

$$-3x > 9$$

$$\frac{-3x}{-3} < \frac{9}{-3} \quad \text{Divide both sides by } -3\text{, which reverses the inequality}$$

$$x < -3$$

The solution set is $\{x \mid x < -3\}$ or $(-\infty, -3)$ in interval notation. ■

As we mentioned earlier, many of the same techniques used to solve equations may be used to solve inequalities. However, you must be extremely careful when you apply Property 3.5. Study the next examples and notice the similarities between solving equations and solving inequalities.

EXAMPLE 7

Solve $4x - 3 > 9$.

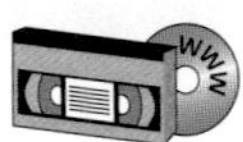

Solution

$$4x - 3 > 9$$

$$4x - 3 + 3 > 9 + 3 \quad \text{Add 3 to both sides}$$

$$4x > 12$$

$$\frac{4x}{4} > \frac{12}{4} \quad \text{Divide both sides by 4}$$

$$x > 3$$

The solution set is $\{x \mid x > 3\}$ or $(3, \infty)$ in interval notation. ■

EXAMPLE 8 Solve $-3n + 5 < 11$.

Solution

$$-3n + 5 < 11$$

$$-3n + 5 - 5 < 11 - 5 \quad \text{Subtract 5 from both sides}$$

$$-3n < 6$$

$$\frac{-3n}{-3} > \frac{6}{-3} \quad \text{Divide both sides by } -3 \text{, which reverses the inequality}$$

$$n > -2$$

The solution set is $\{n|n > -2\}$ or $(-2, \infty)$ in interval notation. ■

Checking the solutions for an inequality presents a problem. Obviously we cannot check all of the infinitely many solutions for a particular inequality. However, by checking at least one solution, especially when the multiplication-division property was used, we might catch the common mistake of forgetting to reverse the sense of the inequality. In Example 8 we are claiming that all numbers greater than -2 will satisfy the original inequality. Let's check one such number in the original inequality—say, -1.

$$-3n + 5 < 11$$

$$-3(-1) + 5 \stackrel{?}{<} 11$$

$$3 + 5 \stackrel{?}{<} 11$$

$$8 < 11$$

Thus -1 satisfies the original inequality. If we had forgotten to reverse the sense of the inequality when we divided both sides by -3, our answer would have been $n < -2$, and the check would have detected the error.

Problem Set 3.5

For Problems 1–10, determine whether each numerical inequality is true or false.

1. $2(3) - 4(5) < 5(3) - 2(-1) + 4$
2. $5 + 6(-3) - 8(-4) > 17$
3. $\frac{2}{3} - \frac{3}{4} + \frac{1}{6} > \frac{1}{5} + \frac{3}{4} - \frac{7}{10}$
4. $\frac{1}{2} + \frac{1}{3} < \frac{1}{3} + \frac{1}{4}$
5. $\left(-\frac{1}{2}\right)\left(\frac{4}{9}\right) > \left(\frac{3}{5}\right)\left(-\frac{1}{3}\right)$
6. $\left(\frac{5}{6}\right)\left(\frac{8}{12}\right) < \left(\frac{3}{7}\right)\left(\frac{14}{15}\right)$
7. $\frac{3}{4} + \frac{2}{3} \div \frac{1}{5} > \frac{2}{3} + \frac{1}{2} \div \frac{3}{4}$
8. $1.9 - 2.6 - 3.4 < 2.5 - 1.6 - 4.2$

9. $0.16 + 0.34 > 0.23 + 0.17$

10. $(0.6)(1.4) > (0.9)(1.2)$

For Problems 11–22, state the solution set and graph it on a number line.

11. $x > -2$
12. $x > -4$
13. $x \leq 3$
14. $x \leq 0$
15. $2 < x$
16. $-3 \leq x$
17. $-2 \geq x$
18. $1 > x$
19. $-x > 1$
20. $-x < 2$
21. $-2 < -x$
22. $-1 > -x$

For Problems 23–60, solve each inequality.

23. $x + 6 < -14$
24. $x + 7 > -15$
25. $x - 4 \geq -13$
26. $x - 3 \leq -12$
27. $4x > 36$
28. $3x < 51$
29. $6x < 20$
30. $8x > 28$
31. $-5x > 40$
32. $-4x < 24$
33. $-7n \leq -56$
34. $-9n \geq -63$
35. $48 > -14n$
36. $36 < -8n$
37. $16 < 9 + n$
38. $19 > 27 + n$
39. $3x + 2 > 17$
40. $2x + 5 < 19$
41. $4x - 3 \leq 21$
42. $5x - 2 \geq 28$
43. $-2x - 1 \geq 41$
44. $-3x - 1 \leq 35$
45. $6x + 2 < 18$
46. $8x + 3 > 25$
47. $3 > 4x - 2$
48. $7 < 6x - 3$
49. $-2 < -3x + 1$
50. $-6 > -2x + 4$
51. $-38 \geq -9t - 2$
52. $36 \geq -7t + 1$
53. $5x - 4 - 3x > 24$
54. $7x - 8 - 5x < 38$
55. $4x + 2 - 6x < -1$
56. $6x + 3 - 8x > -3$
57. $-5 \geq 3t - 4 - 7t$
58. $6 \leq 4t - 7t - 10$
59. $-x - 4 - 3x > 5$
60. $-3 - x - 3x < 10$

THOUGHTS INTO WORDS

61. Do the greater-than and less-than relationships possess the symmetric property? Explain your answer.

62. Is the solution set for $x < 3$ the same as for $3 > x$? Explain your answer.

63. How would you convince someone that it is necessary to reverse the inequality symbol when multiplying both sides of an inequality by a negative number?

FURTHER INVESTIGATIONS

Solve each inequality.

64. $x + 3 < x - 4$
65. $x - 4 < x + 6$
66. $2x + 4 > 2x - 7$
67. $5x + 2 > 5x + 7$
68. $3x - 4 - 3x > 6$
69. $-2x + 7 + 2x > 1$
70. $-5 \leq -4x - 1 + 4x$
71. $-7 \geq 5x - 2 - 5x$

3.6 Inequalities, Compound Inequalities, and Problem Solving

Inequalities

Let's begin this section by solving three inequalities with the same basic steps we used with equations. Again, be careful when applying the multiplication and division properties of inequality.

EXAMPLE 1

Solve $5x + 8 \leq 3x - 10$.

Solution

$$5x + 8 \leq 3x - 10$$

$$5x + 8 - 3x \leq 3x - 10 - 3x \qquad \text{Subtract } 3x \text{ from both sides}$$

$$2x + 8 \leq -10$$

$$2x + 8 - 8 \leq -10 - 8 \qquad \text{Subtract 8 from both sides}$$

$$2x \leq -18$$

$$\frac{2x}{2} \leq \frac{-18}{2} \qquad \text{Divide both sides by 2}$$

$$x \leq -9$$

The solution set is $\{x|x \leq -9\}$ or $(-\infty, -9]$ in interval notation. ■

EXAMPLE 2

Solve $4(x + 3) + 3(x - 4) \geq 2(x - 1)$.

Solution

$$4(x + 3) + 3(x - 4) \geq 2(x - 1)$$

$$4x + 12 + 3x - 12 \geq 2x - 2 \qquad \text{Distributive property}$$

$$7x \geq 2x - 2 \qquad \text{Combine similar terms}$$

$$7x - 2x \geq 2x - 2 - 2x \qquad \text{Subtract } 2x \text{ from both sides}$$

$$5x \geq -2$$

$$\frac{5x}{5} \geq \frac{-2}{5} \qquad \text{Divide both sides by 5}$$

$$x \geq -\frac{2}{5}$$

The solution set is $\left\{x|x \geq -\frac{2}{5}\right\}$ or $\left[-\frac{2}{5}, \infty\right)$ in interval notation. ■

EXAMPLE 3 Solve $-\frac{3}{2}n + \frac{1}{6}n < \frac{3}{4}$.

Solution

$$-\frac{3}{2}n + \frac{1}{6}n < \frac{3}{4}$$

$$12\left(-\frac{3}{2}n + \frac{1}{6}n\right) < 12\left(\frac{3}{4}\right)$$ Multiply both sides by 12, the LCD of all denominators

$$12\left(-\frac{3}{2}n\right) + 12\left(\frac{1}{6}n\right) < 12\left(\frac{3}{4}\right)$$ Distributive property

$$-18n + 2n < 9$$

$$-16n < 9$$

$$\frac{-16n}{-16} > \frac{9}{-16}$$ Divide both sides by -16, which reverses the inequality

$$n > -\frac{9}{16}$$

The solution set is $\left\{n|n > -\frac{9}{16}\right\}$ or $\left(-\frac{9}{16}, \infty\right)$ in interval notation. ■

In Example 3 we are claiming that all numbers greater than $-\frac{9}{16}$ will satisfy the original inequality. Let's check one number—say, 0.

$$-\frac{3}{2}n + \frac{1}{6}n < \frac{3}{4}$$

$$-\frac{3}{2}(0) + \frac{1}{6}(0) \stackrel{?}{<} \frac{3}{4}$$

$$0 < \frac{3}{4}$$

The check resulted in a true statement, which means that 0 is in the solution set. Had we forgotten to reverse the inequality sign when we divided both sides by -16, then the solution set would have been $\left\{n|n < -\frac{9}{16}\right\}$. Zero would not have been a member of that solution set, and we would have detected the error by the check.

■ Compound Inequalities

The words "and" and "or" are used in mathematics to form compound statements. We use "and" and "or" to join two inequalities to form a compound inequality.

Consider the compound inequality

$$x > 2 \quad \text{and} \quad x < 5$$

For the solution set, we must find values of x that make both inequalities true statements. The solution set of a compound inequality formed by the word "and" is the **intersection** of the solution sets of the two inequalities. The intersection of two sets, denoted by $\cap$, contains the elements that are common to both sets. For example, if $A = \{1, 2, 3, 4, 5, 6\}$ and $B = \{0, 2, 4, 6, 8, 10\}$, then $A \cap B = \{2, 4, 6\}$. So to find the solution set of the compound inequality $x > 2$ and $x < 5$, we find the solution set for each inequality and then determine the solutions that are common to both solution sets.

EXAMPLE 4

Graph the solution set for the compound inequality $x > 2$ and $x < 5$, and write the solution set in interval notation.

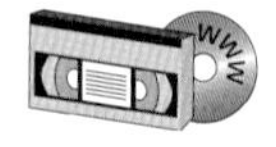

Solution

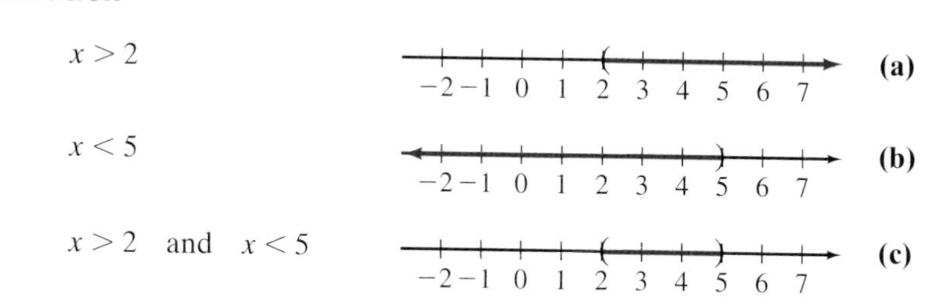

Figure 3.9

Thus all numbers greater than 2 and less than 5 are included in the solution set, and the graph is shown in Figure 3.9(c). In interval notation the solution set is $(2, 5)$. ■

EXAMPLE 5

Graph the solution set for the compound inequality $x \leq 1$ and $x \leq 4$, and write the solution set in interval notation.

Solution

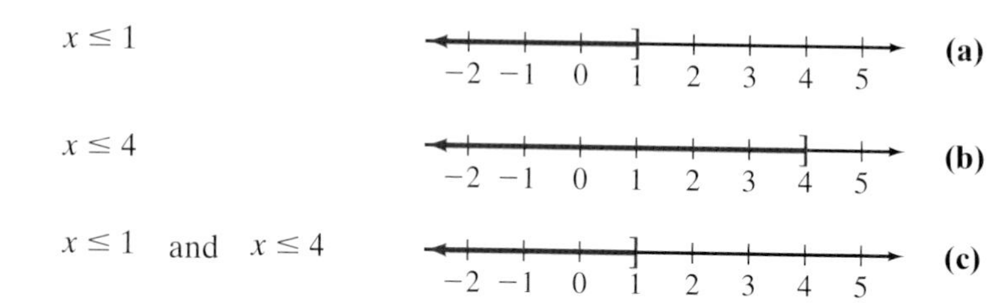

Figure 3.10

The intersection of the two solution sets is $x \leq 1$. The solution set $x \leq 1$ contains all the numbers that are less than or equal to 1, and the graph is shown in Figure 3.10(c). In interval notation the solution set is $(-\infty, 1]$. ■

The solution set of a compound inequality formed by the word "or" is the **union** of the solution sets of the two inequalities. The union of two sets, denoted by $\cup$, contains all the elements in both sets. For example, if $A = \{0, 1, 2\}$ and $B = \{1, 2, 3, 4\}$, then $A \cup B = \{0, 1, 2, 3, 4\}$. Note that even though 1 and 2 are in both set A and set B, there is no need to write them twice in $A \cup B$.

To find the solution set of the compound inequality

$$x > 1 \qquad \text{or} \qquad x > 3$$

we find the solution set for each inequality and then take all the values that satisfy either inequality or both.

EXAMPLE 6

Graph the solution set for $x > 1$ or $x > 3$ and write the solution in interval notation.

Solution

$x > 1$ (a)

$x > 3$ (b)

$x > 1$ or $x > 3$ (c)

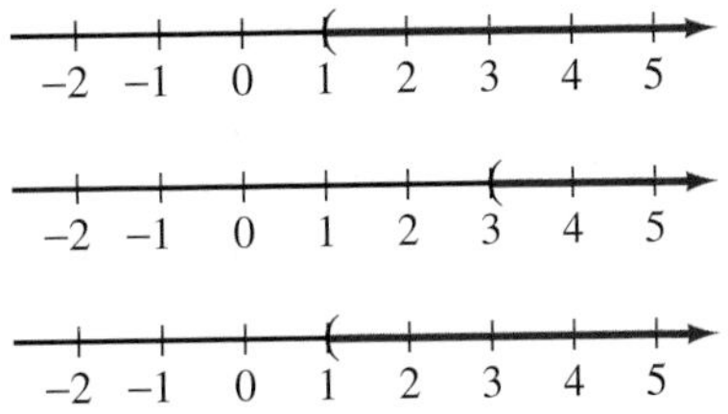

Figure 3.11

Thus all numbers greater than 1 are included in the solution set, and the graph is shown in Figure 3.11(c). The solution set is written as $(1, \infty)$ in interval notation. ■

EXAMPLE 7

Graph the solution set for $x \le 0$ or $x \ge 2$ and write the solution in interval notation.

Solution

$x \le 0$ (a)

$x \ge 2$ (b)

$x \le 0$ or $x \ge 2$ (c)

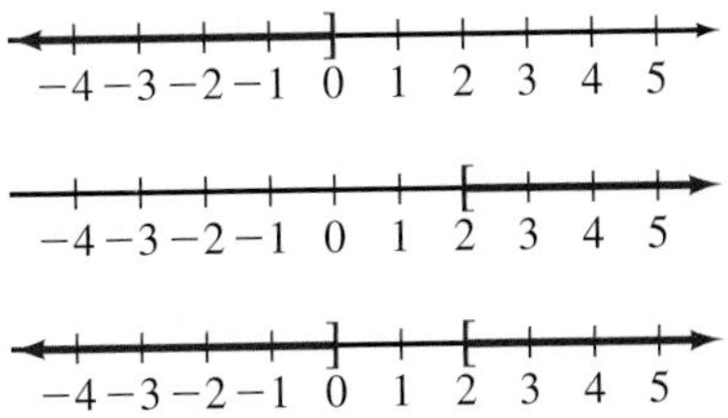

Figure 3.12

Thus all numbers less than or equal to 0 and all numbers greater than or equal to 2 are included in the solution set, and the graph is shown in Figure 3.12(c). Since the solution set contains two intervals that are not continuous, a $\cup$ symbol is used in the interval notation. The solution set is written as $(-\infty, 0] \cup [2, \infty)$ in interval notation. ■

■ Back to Problem Solving

Let's consider some word problems that translate into inequality statements. We gave suggestions for solving word problems in Section 3.3, which apply here except that the situation described in the problem will translate into an inequality instead of an equation.

PROBLEM 1

Ashley had scores of 95, 82, 93, and 84 on her first four exams of the semester. What score must she get on the fifth exam to have an average of 90 or higher for the five exams?

Solution

Let s represent the score needed on the fifth exam. Since the average is computed by adding all five scores and dividing by 5 (the number of scores), we have the following inequality to solve:

$$\frac{95 + 82 + 93 + 84 + s}{5} \geq 90$$

Solving this inequality, we obtain

$$\frac{354 + s}{5} \geq 90 \qquad \text{Simplify numerator of the left side}$$

$$5\left(\frac{354 + s}{5}\right) \geq 5(90) \qquad \text{Multiply both sides by 5}$$

$$354 + s \geq 450$$

$$354 + s - 354 \geq 450 - 354 \qquad \text{Subtract 354 from both sides}$$

$$s \geq 96$$

She must receive a score of 96 or higher on the fifth exam. ■

PROBLEM 2

The Cubs have won 40 baseball games and have lost 62 games. They will play 60 more games. To win more than 50% of all their games, how many of the 60 games remaining must they win?

Solution

Let w represent the number of games the Cubs must win out of the 60 games remaining. Since they are playing a total of $40 + 62 + 60 = 162$ games, to win more than 50% of their games, they need to win more than 81 games. Thus, we have the inequality

$$w + 40 > 81$$

Solving this yields

$$w > 41$$

The Cubs need to win at least 42 of the 60 games remaining. ■

Problem Set 3.6

For Problems 1–50, solve each inequality.

1. $3x + 4 > x + 8$
2. $5x + 3 < 3x + 11$
3. $7x - 2 < 3x - 6$
4. $8x - 1 > 4x - 21$
5. $6x + 7 > 3x - 3$
6. $7x + 5 < 4x - 12$
7. $5n - 2 \leq 6n + 9$
8. $4n - 3 \geq 5n + 6$
9. $2t + 9 \geq 4t - 13$
10. $6t + 14 \leq 8t - 16$
11. $-3x - 4 < 2x + 7$
12. $-x - 2 > 3x - 7$
13. $-4x + 6 > -2x + 1$
14. $-6x + 8 < -4x + 5$
15. $5(x - 2) \leq 30$
16. $4(x + 1) \geq 16$
17. $2(n + 3) > 9$
18. $3(n - 2) < 7$
19. $-3(y - 1) < 12$
20. $-2(y + 4) > 18$
21. $-2(x + 6) > -17$
22. $-3(x - 5) < -14$
23. $3(x - 2) < 2(x + 1)$
24. $5(x + 3) > 4(x - 2)$
25. $4(x + 3) > 6(x - 5)$
26. $6(x - 1) < 8(x + 5)$
27. $3(x - 4) + 2(x + 3) < 24$
28. $2(x + 1) + 3(x + 2) > -12$
29. $5(n + 1) - 3(n - 1) > -9$
30. $4(n - 5) - 2(n - 1) < 13$
31. $\frac{1}{2}n - \frac{2}{3}n \geq -7$
32. $\frac{3}{4}n + \frac{1}{6}n \leq 1$
33. $\frac{3}{4}n - \frac{5}{6}n < \frac{3}{8}$
34. $\frac{2}{3}n - \frac{1}{2}n > \frac{1}{4}$
35. $\frac{3x}{5} - \frac{2}{3} > \frac{x}{10}$
36. $\frac{5x}{4} + \frac{3}{8} < \frac{7x}{12}$
37. $n \geq 3.4 + 0.15n$
38. $x \geq 2.1 + 0.3x$
39. $0.09t + 0.1(t + 200) > 77$
40. $0.07t + 0.08(t + 100) > 38$
41. $0.06x + 0.08(250 - x) \geq 19$
42. $0.08x + 0.09(2x) \leq 130$
43. $\frac{x - 1}{2} + \frac{x + 3}{5} > \frac{1}{10}$

44. $\frac{x+3}{4} + \frac{x-5}{7} < \frac{1}{28}$

45. $\frac{x+2}{6} - \frac{x+1}{5} < -2$

46. $\frac{x-6}{8} - \frac{x+2}{7} > -1$

47. $\frac{n+3}{3} + \frac{n-7}{2} > 3$

48. $\frac{n-4}{4} + \frac{n-2}{3} < 4$

49. $\frac{x-3}{7} - \frac{x-2}{4} \leq \frac{9}{14}$

50. $\frac{x-1}{5} - \frac{x+2}{6} \geq \frac{7}{15}$

For Problems 51–66, graph the solution set for each compound inequality.

51. $x > -1$ and $x < 2$

52. $x > 1$ and $x < 4$

53. $x < -2$ or $x > 1$

54. $x < 0$ or $x > 3$

55. $x > -2$ and $x \leq 2$

56. $x \geq -1$ and $x < 3$

57. $x > -1$ and $x > 2$

58. $x < -2$ and $x < 3$

59. $x > -4$ or $x > 0$

60. $x < 2$ or $x < 4$

61. $x > 3$ and $x < -1$

62. $x < -3$ and $x > 6$

63. $x \leq 0$ or $x \geq 2$

64. $x \leq -2$ or $x \geq 1$

65. $x > -4$ or $x < 3$

66. $x > -1$ or $x < 2$

For Problems 67–78, solve each problem by setting up and solving an appropriate inequality.

67. Five more than three times a number is greater than 26. Find all of the numbers that satisfy this relationship.

68. Fourteen increased by twice a number is less than or equal to three times the number. Find the numbers that satisfy this relationship.

69. Suppose that the perimeter of a rectangle is to be no greater than 70 inches, and the length of the rectangle must be 20 inches. Find the largest possible value for the width of the rectangle.

70. One side of a triangle is three times as long as another side. The third side is 15 centimeters long. If the perimeter of the triangle is to be no greater than 75 centimeters, find the greatest lengths that the other two sides can be.

71. Sue bowled 132 and 160 in her first two games. What must she bowl in the third game to have an average of at least 150 for the three games?

72. Mike has scores of 87, 81, and 74 on his first three algebra tests. What score must he get on the fourth test to have an average of 85 or higher for the four tests?

73. This semester Lance has scores of 96, 90, and 94 on his first three algebra exams. What must he average on the last two exams to have an average higher than 92 for all five exams?

74. The Mets have won 45 baseball games and lost 55 games. They have 62 more games to play. To win more than 50% of all their games, how many of the 62 games remaining must they win?

75. An Internet business has costs of \$4000 plus \$32 per sale. The business receives revenue of \$48 per sale. What possible values for sales would ensure that the revenues exceed the costs?

76. The average height of the two forwards and the center of a basketball team is 6 feet, 8 inches. What must the average height of the two guards be so that the team's average height is at least 6 feet, 4 inches?

77. Scott shot rounds of 82, 84, 78, and 79 on the first four days of the golf tournament. What must he shoot on the fifth day of the tournament to average 80 or less for the 5 days?

78. Sydney earns $2300 a month. To qualify for a mortgage, her monthly payments must be less than 35% of her monthly income. Her monthly mortgage payments must be less than what amount in order to qualify for the mortgage?

A selection of additional word problems is in Appendix B. All Appendix problems referenced as (3.6) are appropriate for this section.

THOUGHTS INTO WORDS

79. Give an example of a compound statement using the word "and" outside the field of mathematics.

80. Give an example of a compound statement using the word "or" outside the field of mathematics.

81. Give a step-by-step description of how you would solve the inequality $3x - 2 > 4(x + 6)$.

Chapter 3 Summary

(3.1) Numerical equations may be true or false. **Algebraic equations** (open sentences) contain one or more variables. **Solving an equation** refers to the process of finding the number (or numbers) that makes an algebraic equation a true statement. A **first-degree equation of one variable** is an equation that contains only one variable, and this variable has an exponent of one.

Properties 3.1, 3.2, and 3.3 provide the basis for solving equations. Be sure that you can use these properties to solve the variety of equations presented in this chapter.

(3.2) It is often necessary to use both the addition and multiplication properties of equality to solve an equation.

Be sure to **declare the variable** as you translate English sentences into algebraic equations.

(3.3) Keep the following suggestions in mind as you solve word problems.

1. Read the problem carefully.
2. Sketch any figure, diagram, or chart that might be helpful.
3. Choose a meaningful variable.
4. Look for a *guideline*.
5. Form an equation or inequality.
6. Solve the equation or inequality.
7. Check your answers.

(3.4) The **distributive property** is used to *remove parentheses*.

If an equation contains several fractions, then it is usually advisable to *clear the equation of all fractions* by multiplying both sides by the least common denominator of all the denominators in the equation.

(3.5) Properties 3.4 and 3.5 provide the basis for solving inequalities. Be sure that you can use these properties to solve the variety of inequalities presented in this chapter.

We can use many of the same techniques used to solve equations to solve inequalities, *but* we must be very careful when multiplying or dividing both sides of an inequality by the same number. Don't forget that multiplying or dividing both sides of an inequality by a negative number reverses the sense of the inequality.

(3.6) The words "and" and "or" are used to form compound inequalities.

The solution set of a compound inequality formed by the word "and" is the **intersection** of the solution sets of the two inequalities. The solution set of a compound inequality formed by the word "or" is the **union** of the solution sets of the two inequalities.

To solve inequalities involving "or" we must satisfy one or more of the conditions. Thus the compound inequality $x < -1$ **or** $x > 2$ is satisfied by (a) all numbers less than -1, or (b) all numbers greater than 2, or (c) both (a) and (b).

Chapter 3 Review Problem Set

In Problems 1–20, solve each of the equations.

1. $9x - 2 = -29$
2. $-3 = -4y + 1$
3. $7 - 4x = 10$
4. $6y - 5 = 4y + 13$
5. $4n - 3 = 7n + 9$
6. $7(y - 4) = 4(y + 3)$
7. $2(x + 1) + 5(x - 3) = 11(x - 2)$

8. $-3(x + 6) = 5x - 3$

9. $\frac{2}{5}n - \frac{1}{2}n = \frac{7}{10}$

10. $\frac{3n}{4} + \frac{5n}{7} = \frac{1}{14}$

11. $\frac{x - 3}{6} + \frac{x + 5}{8} = \frac{11}{12}$

12. $\frac{n}{2} - \frac{n - 1}{4} = \frac{3}{8}$

13. $-2(x - 4) = -3(x + 8)$

14. $3x - 4x - 2 = 7x - 14 - 9x$

15. $5(n - 1) - 4(n + 2) = -3(n - 1) + 3n + 5$

16. $\frac{x - 3}{9} = \frac{x + 4}{8}$

17. $\frac{x - 1}{-3} = \frac{x + 2}{-4}$

18. $-(t - 3) - (2t + 1) = 3(t + 5) - 2(t + 1)$

19. $\frac{2x - 1}{3} = \frac{3x + 2}{2}$

20. $3(2t - 4) + 2(3t + 1) = -2(4t + 3) - (t - 1)$

For Problems 21–36, solve each inequality.

21. $3x - 2 > 10$

22. $-2x - 5 < 3$

23. $2x - 9 \geq x + 4$

24. $3x + 1 \leq 5x - 10$

25. $6(x - 3) > 4(x + 13)$

26. $2(x + 3) + 3(x - 6) < 14$

27. $\frac{2n}{5} - \frac{n}{4} < \frac{3}{10}$

28. $\frac{n + 4}{5} + \frac{n - 3}{6} > \frac{7}{15}$

29. $-16 < 8 + 2y - 3y$

30. $-24 > 5x - 4 - 7x$

31. $-3(n - 4) > 5(n + 2) + 3n$

32. $-4(n - 2) - (n - 1) < -4(n + 6)$

33. $\frac{3}{4}n - 6 \leq \frac{2}{3}n + 4$

34. $\frac{1}{2}n - \frac{1}{3}n - 4 \geq \frac{3}{5}n + 2$

35. $-12 > -4(x - 1) + 2$

36. $36 < -3(x + 2) - 1$

For Problems 37–40, graph the solution set for each of the compound inequalities.

37. $x > -3$ and $x < 2$

38. $x < -1$ or $x > 4$

39. $x < 2$ or $x > 0$

40. $x > 1$ and $x > 0$

Set up an equation or an inequality to solve Problems 41–52.

41. Three-fourths of a number equals 18. Find the number.

42. Nineteen is 2 less than three times a certain number. Find the number.

43. The difference of two numbers is 21. If 12 is the smaller number, find the other number.

44. One subtracted from nine times a certain number is the same as 15 added to seven times the number. Find the number.

45. Monica has scores of 83, 89, 78, and 86 on her first four exams. What score must she receive on the fifth exam so that her average for all five exams is 85 or higher?

46. The sum of two numbers is 40. Six times the smaller number equals four times the larger. Find the numbers.

47. Find a number such that 2 less than two-thirds of the number is 1 more than one-half of the number.

48. Ameya's average score for her first three psychology exams is 84. What must she get on the fourth exam so that her average for the four exams is 85 or higher?

49. Miriam has 30 coins (nickels and dimes) that amount to \$2.60. How many coins of each kind does she have?

50. Suppose that Khoa has a bunch of nickels, dimes, and quarters amounting to \$15.40. The number of dimes is 1 more than three times the number of nickels, and the number of quarters is twice the number of dimes. How many coins of each kind does he have?

51. The supplement of an angle is 14° more than three times the complement of the angle. Find the measure of the angle.

52. Pam rented a car from a rental agency that charges \$25 a day and \$0.20 per mile. She kept the car for 3 days and her bill was \$215. How many miles did she drive during that 3-day period?

If you have not already done so, you may want to use the word problems in Appendix B for more practice with word problems. All Appendix problems that have a chapter 3 reference would be appropriate.

Chapter 3 Test

For Problems 1–12, solve each of the equations.

1. $7x - 3 = 11$
2. $-7 = -3x + 2$
3. $4n + 3 = 2n - 15$
4. $3n - 5 = 8n + 20$
5. $4(x - 2) = 5(x + 9)$
6. $9(x + 4) = 6(x - 3)$
7. $5(y - 2) + 2(y + 1) = 3(y - 6)$
8. $\frac{3}{5}x - \frac{2}{3} = \frac{1}{2}$
9. $\frac{x - 2}{4} = \frac{x + 3}{6}$
10. $\frac{x + 2}{3} + \frac{x - 1}{2} = 2$
11. $\frac{x - 3}{6} - \frac{x - 1}{8} = \frac{13}{24}$
12. $-5(n - 2) = -3(n + 7)$

For Problems 13–18, solve each of the inequalities.

13. $3x - 2 < 13$
14. $-2x + 5 \geq 3$
15. $3(x - 1) \leq 5(x + 3)$
16. $-4 > 7(x - 1) + 3$
17. $-2(x - 1) + 5(x - 2) < 5(x + 3)$
18. $\frac{1}{2}n + 2 \leq \frac{3}{4}n - 1$

For Problems 19 and 20, graph the solution set for each compound inequality.

19. $x \geq -2$ and $x \leq 4$
20. $x < 1$ or $x > 3$

For Problems 21–25, set up an equation or an inequality and solve each problem.

21. Jean-Paul received a cell phone bill for \$98.24. Included in the \$98.24 were a monthly-plan charge of \$29.99 and a charge for 195 extra minutes. How much is Jean-Paul being charged for each extra minute?
22. Suppose that a triangular plot of ground is enclosed with 70 meters of fencing. The longest side of the lot is two times the length of the shortest side, and the third side is 10 meters longer than the shortest side. Find the length of each side of the plot.
23. Tina had scores of 86, 88, 89, and 91 on her first four history exams. What score must she get on the fifth exam to have an average of 90 or higher for the five exams?
24. Sean has 103 coins consisting of nickels, dimes, and quarters. The number of dimes is 1 less than twice the number of nickels, and the number of quarters is 2 more than three times the number of nickels. How many coins of each kind does he have?
25. In triangle *ABC*, the measure of angle *C* is one-half the measure of angle *A*, and the measure of angle *B* is 30° more than the measure of angle *A*. Find the measure of each angle of the triangle.

4

Formulas and Problem Solving

The equation $s = 3 + 0.6s$ can be used to determine how much the owner of a pizza parlor must charge for a pizza if it costs \$3 to make the pizza, and he wants to make a profit of 60% based on the selling price.

© Michael Newman/PhotoEdit

Kirk starts jogging at 5 miles per hour. One-half hour later, Consuela starts jogging on the same route at 7 miles per hour. How long will it take Consuela to catch Kirk? If we let t represent the time that Consuela jogs, then $t + \frac{1}{2}$ represents Kirk's time. We can use the equation $7t = 5\left(t + \frac{1}{2}\right)$ to determine that Consuela should catch Kirk in $1\frac{1}{4}$ hours.

We used the formula *distance equals rate times time*, which is usually expressed as $d = rt$, to set up the equation $7t = 5\left(t + \frac{1}{2}\right)$. Throughout this chapter, we will use a variety of formulas to solve problems that connect algebraic and geometric concepts.

4.1 Ratio, Proportion, and Percent

Ratio and Proportion

In Figure 4.1, as gear A revolves 4 times, gear B will revolve 3 times. We say that the gear ratio of A to B is 4 to 3, or the gear ratio of B to A is 3 to 4. Mathematically, a **ratio** is the comparison of two numbers by division. We can write the gear ratio of A to B as

$$4 \text{ to } 3 \qquad \text{or} \qquad 4:3 \qquad \text{or} \qquad \frac{4}{3}$$

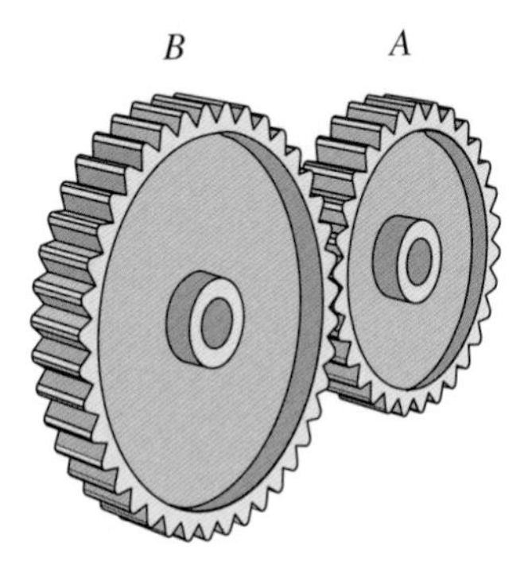

Figure 4.1

We express ratios as fractions in reduced form. For example, if there are 7500 women and 5000 men at a certain university, then the ratio of women to men is $\frac{7500}{5000} = \frac{3}{2}$.

A statement of equality between two ratios is called a **proportion**. For example,

$$\frac{2}{3} = \frac{8}{12}$$

is a proportion that states that the ratios $\frac{2}{3}$ and $\frac{8}{12}$ are equal. In the general proportion

$$\frac{a}{b} = \frac{c}{d}, \qquad b \neq 0 \text{ and } d \neq 0$$

if we multiply both sides of the equation by the common denominator, bd, we obtain

$$(bd)\left(\frac{a}{b}\right) = (bd)\left(\frac{c}{d}\right)$$

$$ad = bc$$

These products ad and bc are called *cross products* and are equal to each other. Let's state this as a property of proportions.

$$\frac{a}{b} = \frac{c}{d} \text{ if and only if } ad = bc, \quad \text{where } b \neq 0 \text{ and } d \neq 0$$

EXAMPLE 1

Solve $\frac{x}{20} = \frac{3}{4}$.

Solution

$$\frac{x}{20} = \frac{3}{4}$$

$$4x = 60 \qquad \text{Cross products are equal}$$

$$x = 15$$

The solution set is $\{15\}$. ■

EXAMPLE 2

Solve $\frac{x-3}{5} = \frac{x+2}{4}$.

Solution

$$\frac{x-3}{5} = \frac{x+2}{4}$$

$$4(x-3) = 5(x+2) \qquad \text{Cross products are equal}$$

$$4x - 12 = 5x + 10 \qquad \text{Distributive property}$$

$$-12 = x + 10 \qquad \text{Subtracted } 4x \text{ from both sides}$$

$$-22 = x \qquad \text{Subtracted 10 from both sides}$$

The solution set is $\{-22\}$. ■

If a variable appears in one or both of the denominators, then a proper restriction should be made to avoid division by zero, as the next example illustrates.

EXAMPLE 3

Solve $\frac{7}{a-2} = \frac{4}{a+3}$.

Solution

$$\frac{7}{a-2} = \frac{4}{a+3}, \qquad a \neq 2 \text{ and } a \neq -3$$

$$7(a+3) = 4(a-2) \qquad \text{Cross products are equal}$$

$$7a + 21 = 4a - 8 \qquad \text{Distributive property}$$

$$3a + 21 = -8 \qquad \text{Subtracted } 4a \text{ from both sides}$$

$$3a = -29 \quad \text{Subtracted 21 from both sides}$$

$$a = -\frac{29}{3} \quad \text{Divided both sides by 3}$$

The solution set is $\left\{-\frac{29}{3}\right\}$. ■

EXAMPLE 4

Solve $\frac{x}{4} + 3 = \frac{x}{5}$.

Solution

This is *not* a proportion, so we can multiply both sides by 20 to clear the equation of all fractions.

$$\frac{x}{4} + 3 = \frac{x}{5}$$

$$20\left(\frac{x}{4} + 3\right) = 20\left(\frac{x}{5}\right) \quad \text{Multiply both sides by 20}$$

$$20\left(\frac{x}{4}\right) + 20(3) = 20\left(\frac{x}{5}\right) \quad \text{Distributive property}$$

$$5x + 60 = 4x$$

$$x + 60 = 0 \quad \text{Subtracted } 4x \text{ from both sides}$$

$$x = -60 \quad \text{Subtracted 60 from both sides}$$

The solution set is $\{-60\}$. ■

Remark: Example 4 demonstrates the importance of thinking first before pushing the pencil. Since the equation was not in the form of a proportion, we needed to revert to a previous technique for solving such equations.

■ Problem Solving Using Proportions

Some word problems can be conveniently set up and solved using the concepts of ratio and proportion. Consider the following examples.

PROBLEM 1

On the map in Figure 4.2, 1 inch represents 20 miles. If two cities are $6\frac{1}{2}$ inches apart on the map, find the number of miles between the cities.

Solution

Let m represent the number of miles between the two cities. Now let's set up a proportion where one ratio compares distances in inches on the map, and the other ratio compares *corresponding* distances in miles on land:

$$\frac{1}{6\frac{1}{2}} = \frac{20}{m}$$

Figure 4.2

To solve this equation, we equate the cross products:

$$m(1) = \left(6\frac{1}{2}\right)(20)$$

$$m = \left(\frac{13}{2}\right)(20) = 130$$

The distance between the two cities is 130 miles. ■

PROBLEM 2

A sum of \$1750 is to be divided between two people in the ratio of 3 to 4. How much does each person receive?

Solution

Let d represent the amount of money to be received by one person. Then $1750 - d$ represents the amount for the other person. We set up this proportion:

$$\frac{d}{1750 - d} = \frac{3}{4}$$

$$4d = 3(1750 - d)$$

$$4d = 5250 - 3d$$

$$7d = 5250$$

$$d = 750$$

If $d = 750$, then $1750 - d = 1000$; therefore, one person receives \$750, and the other person receives \$1000. ■

Percent

The word **percent** means "per one hundred," and we use the symbol % to express it. For example, we write 7 percent as 7%, which means $\frac{7}{100}$ or 0.07. In other words, percent is a special kind of ratio — namely, one in which the denominator is always 100. Proportions provide a convenient basis for changing common fractions to percents. Consider the next examples.

EXAMPLE 5

Express $\frac{7}{20}$ as a percent.

Solution

We are asking "What number compares to 100 as 7 compares to 20?" Therefore, if we let n represent that number, we can set up the following proportion:

$$\frac{n}{100} = \frac{7}{20}$$

$$20n = 700$$

$$n = 35$$

Thus, $\frac{7}{20} = \frac{35}{100} = 35\%$. ■

EXAMPLE 6

Express $\frac{5}{6}$ as a percent.

Solution

$$\frac{n}{100} = \frac{5}{6}$$

$$6n = 500$$

$$n = \frac{500}{6} = \frac{250}{3} = 83\frac{1}{3}$$

Therefore, $\frac{5}{6} = 83\frac{1}{3}\%$. ■

Some Basic Percent Problems

What is 8% of 35? Fifteen percent of what number is 24? Twenty-one is what percent of 70? These are the three basic types of percent problems. We can solve each of these problems easily by translating into and solving a simple algebraic equation.

PROBLEM 3

What is 8% of 35?

Solution

Let n represent the number to be found. The word "is" refers to equality, and the word "of" means multiplication. Thus the question translates into

$$n = (8\%)(35)$$

which can be solved as follows:

$$\begin{aligned} n &= (0.08)(35) \\ &= 2.8 \end{aligned}$$

Therefore, 2.8 is 8% of 35. ■

PROBLEM 4

Fifteen percent of what number is 24?

Solution

Let n represent the number to be found.

$$\begin{aligned} (15\%)(n) &= (24) \\ 0.15n &= 24 \\ 15n &= 2400 && \text{Multiplied both sides by 100} \\ n &= 160 \end{aligned}$$

Therefore, 15% of 160 is 24. ■

PROBLEM 5

Twenty-one is what percent of 70?

Solution

Let r represent the percent to be found.

$$\begin{aligned} 21 &= r(70) \\ \frac{21}{70} &= r \\ \frac{3}{10} &= r && \text{Reduce!} \\ \frac{30}{100} &= r && \text{Changed } \frac{3}{10} \text{ to } \frac{30}{100} \\ 30\% &= r \end{aligned}$$

Therefore, 21 is 30% of 70. ■

PROBLEM 6

Seventy-two is what percent of 60?

Solution

Let r represent the percent to be found.

$$72 = r(60)$$

$$\frac{72}{60} = r$$

$$\frac{6}{5} = r$$

$$\frac{120}{100} = r \qquad \text{Changed } \frac{6}{5} \text{ to } \frac{120}{100}$$

$$120\% = r$$

Therefore, 72 is 120% of 60. ■

Again, it is helpful to get into the habit of checking answers for *reasonableness*. We also suggest that you alert yourself to a potential computational error by estimating the answer before you actually do the problem. For example, prior to doing Problem 6, you may have estimated as follows: Since 72 is greater than 60, you know that the answer has to be greater than 100%. Furthermore, 1.5 (or 150%) times 60 equals 90. Therefore, you can estimate the answer to be somewhere between 100% and 150%. That may seem like a rather rough estimate, but many times such an estimate will reveal a computational error.

Problem Set 4.1

For Problems 1–32, solve each of the equations.

1. $\frac{x}{6} = \frac{3}{2}$

2. $\frac{x}{9} = \frac{5}{3}$

3. $\frac{5}{12} = \frac{n}{24}$

4. $\frac{7}{8} = \frac{n}{16}$

5. $\frac{x}{3} = \frac{5}{2}$

6. $\frac{x}{7} = \frac{4}{3}$

7. $\frac{x-2}{4} = \frac{x+4}{3}$

8. $\frac{x-6}{7} = \frac{x+9}{8}$

9. $\frac{x+1}{6} = \frac{x+2}{4}$

10. $\frac{x-2}{6} = \frac{x-6}{8}$

11. $\frac{h}{2} - \frac{h}{3} = 1$

12. $\frac{h}{5} + \frac{h}{4} = 2$

13. $\frac{x+1}{3} - \frac{x+2}{2} = 4$

14. $\frac{x-2}{5} - \frac{x+3}{6} = -4$

15. $\frac{-4}{x+2} = \frac{-3}{x-7}$

16. $\frac{-9}{x+1} = \frac{-8}{x+5}$

17. $\frac{-1}{x-7} = \frac{5}{x-1}$

18. $\frac{3}{x-10} = \frac{-2}{x+6}$

19. $\frac{3}{2x-1} = \frac{2}{3x+2}$

20. $\frac{1}{4x+3} = \frac{2}{5x-3}$

21. $\frac{n+1}{n} = \frac{8}{7}$

22. $\frac{5}{6} = \frac{n}{n+1}$

23. $\frac{x-1}{2} - 1 = \frac{3}{4}$

24. $-2 + \frac{x+3}{4} = \frac{5}{6}$

25. $-3 - \frac{x+4}{5} = \frac{3}{2}$

26. $\frac{x-5}{3} + 2 = \frac{5}{9}$

27. $\frac{n}{150-n} = \frac{1}{2}$

28. $\frac{n}{200-n} = \frac{3}{5}$

29. $\frac{300-n}{n} = \frac{3}{2}$

30. $\frac{80-n}{n} = \frac{7}{9}$

31. $\frac{-1}{5x-1} = \frac{-2}{3x+7}$

32. $\frac{-3}{2x-5} = \frac{-4}{x-3}$

For Problems 33–44, use proportions to change each common fraction to a percent.

33. $\frac{11}{20}$

34. $\frac{17}{20}$

35. $\frac{3}{5}$

36. $\frac{7}{25}$

37. $\frac{1}{6}$

38. $\frac{5}{7}$

39. $\frac{3}{8}$

40. $\frac{1}{16}$

41. $\frac{3}{2}$

42. $\frac{5}{4}$

43. $\frac{12}{5}$

44. $\frac{13}{6}$

For Problems 45–56, answer the question by setting up and solving an appropriate equation.

45. What is 7% of 38?

46. What is 35% of 52?

47. 15% of what number is 6.3?

48. 55% of what number is 38.5?

49. 76 is what percent of 95?

50. 72 is what percent of 120?

51. What is 120% of 50?

52. What is 160% of 70?

53. 46 is what percent of 40?

54. 26 is what percent of 20?

55. 160% of what number is 144?

56. 220% of what number is 66?

For Problems 57–73, solve each problem using a proportion.

57. A blueprint has a scale where 1 inch represents 6 feet. Find the dimensions of a rectangular room that measures $2\frac{1}{2}$ inches by $3\frac{1}{4}$ inches on the blueprint.

58. On a certain map, 1 inch represents 15 miles. If two cities are 7 inches apart on the map, find the number of miles between the cities.

59. Suppose that a car can travel 264 miles using 12 gallons of gasoline. How far will it go on 15 gallons?

60. Jesse used 10 gallons of gasoline to drive 170 miles. How much gasoline will he need to travel 238 miles?

61. If the ratio of the length of a rectangle to its width is $\frac{5}{2}$, and the width is 24 centimeters, find its length.

62. If the ratio of the width of a rectangle to its length is $\frac{4}{5}$, and the length is 45 centimeters, find the width.

63. A saltwater solution is made by dissolving 3 pounds of salt in 10 gallons of water. At this rate, how many pounds of salt are needed for 25 gallons of water? (See Figure 4.3.)

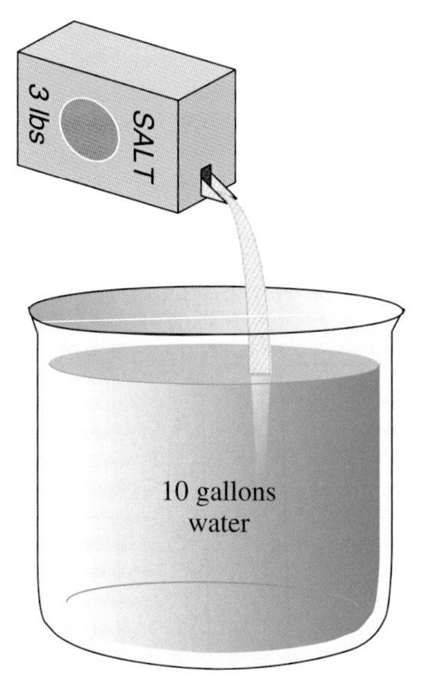

Figure 4.3

64. A home valued at $50,000 is assessed $900 in real estate taxes. At the same rate, how much are the taxes on a home valued at $60,000?

65. If 20 pounds of fertilizer will cover 1500 square feet of lawn, how many pounds are needed for 2500 square feet?

66. It was reported that a flu epidemic is affecting six out of every ten college students in a certain part of the country. At this rate, how many students will be affected at a university of 15,000 students?

67. A preelection poll indicated that three out of every seven eligible voters were going to vote in an upcoming election. At this rate, how many people are expected to vote in a city of 210,000?

68. A board 28 feet long is cut into two pieces whose lengths are in the ratio of 2 to 5. Find the lengths of the two pieces.

69. In a nutrition plan the ratio of calories to grams of carbohydrates is 16 to 1. According to this ratio, how many grams of carbohydrates would be in a plan that has 2200 calories?

70. The ratio of male students to female students at a certain university is 5 to 4. If there is a total of 6975 students, find the number of male students and the number of female students.

71. An investment of $500 earns $45 in a year. At the same rate, how much additional money must be invested to raise the earnings to $72 per year?

72. A sum of $1250 is to be divided between two people in the ratio of 2 to 3. How much does each person receive?

73. An inheritance of $180,000 is to be divided between a child and the local cancer fund in the ratio of 5 to 1. How much money will the child receive?

A selection of additional word problems is in Appendix B. All Appendix problems referenced as (4.1) are appropriate for this section.

THOUGHTS INTO WORDS

74. Explain the difference between a ratio and a proportion.

75. What is wrong with the following procedure? Explain how it should be done.

$$\frac{x}{2} + 4 = \frac{x}{6}$$

$$6\left(\frac{x}{2} + 4\right) = 2(x)$$

$$3x + 24 = 2x$$

$$x = -24$$

76. Estimate an answer for each of the following problems. Also explain how you arrived at your estimate. Then work out the problem to see how well you estimated.

(a) The ratio of female students to male students at a small private college is 5 to 3. If there is a total of 1096 students, find the number of male students.

(b) If 15 pounds of fertilizer will cover 1200 square feet of lawn, how many pounds are needed for 3000 square feet?

(c) An investment of $5000 earns $300 interest in a year. At the same rate, how much money must be invested to earn $450?

(d) If the ratio of the length of a rectangle to its width is 5 to 3 and the length is 70 centimeters, find its width.

FURTHER INVESTIGATIONS

Solve each of the following equations. Don't forget that division by zero is undefined.

77. $\frac{3}{x-2} = \frac{6}{2x-4}$

78. $\frac{8}{2x+1} = \frac{4}{x-3}$

79. $\frac{5}{x-3} = \frac{10}{x-6}$

80. $\frac{6}{x-1} = \frac{5}{x-1}$

81. $\frac{x-2}{2} = \frac{x}{2} - 1$

82. $\frac{x+3}{x} = 1 + \frac{3}{x}$

4.2 More on Percents and Problem Solving

We can solve the equation $x + 0.35 = 0.72$ by subtracting 0.35 from both sides of the equation. Another technique for solving equations that contain decimals is to clear the equation of all decimals by multiplying both sides by an appropriate power of 10. The following examples demonstrate both techniques in a variety of situations.

EXAMPLE 1 Solve $0.5x = 14$.

Solution

$$0.5x = 14$$

$$5x = 140 \quad \text{Multiplied both sides by 10}$$

$$x = 28 \quad \text{Divided both sides by 5}$$

The solution set is $\{28\}$. ■

EXAMPLE 2 Solve $x + 0.04x = 5.2$.

Solution

$$x + 0.04x = 5.2$$

$$1.04x = 5.2 \quad \text{Combined similar terms}$$

$$x = \frac{5.2}{1.04}$$

$$x = 5$$

The solution set is $\{5\}$. ■

EXAMPLE 3 Solve $0.08y + 0.09y = 3.4$.

Solution

$$0.08y + 0.09y = 3.4$$

$$0.17y = 3.4 \quad \text{Combined similar terms}$$

$$y = \frac{3.4}{0.17}$$

$$y = 20$$

The solution set is $\{20\}$. ■

EXAMPLE 4

Solve $0.10t = 560 - 0.12(t + 1000)$.

Solution

$$\begin{aligned} 0.10t &= 560 - 0.12(t + 1000) \\ 10t &= 56{,}000 - 12(t + 1000) && \text{Multiplied both sides by 100} \\ 10t &= 56{,}000 - 12t - 12{,}000 && \text{Distributive property} \\ 22t &= 44{,}000 \\ t &= 2000 \end{aligned}$$

The solution set is $\{2000\}$. ■

■ Problems Involving Percents

Many consumer problems can be solved with an equation approach. For example, we have this general guideline regarding discount sales:

Original selling price − Discount = Discount sale price

Next we consider some examples using algebraic techniques along with this basic guideline.

PROBLEM 1

Amy bought a dress at a 30% discount sale for \$35. What was the original price of the dress?

Solution

Let p represent the original price of the dress. We can use the basic discount guideline to set up an algebraic equation.

Original selling price − Discount = Discount sale price

$$(100\%)(p) \quad - \quad (30\%)(p) \quad = \quad \$35$$

Solving this equation, we obtain

$$\begin{aligned} (100\%)(p) - (30\%)(p) &= 35 \\ 1.00p - 0.30p &= 35 && \text{Changed percents to decimals} \end{aligned}$$

$$0.7p = 35$$
$$7p = 350$$
$$p = 50$$

The original price of the dress was \$50. ■

Don't forget that if an item is on sale for 30% off, then you are going to pay $100\% - 30\% = 70\%$ of the original price. So at a 30% discount sale, a \$50 dress can be purchased for $(70\%)(\$50) = (0.70)(\$50) = \$35$. (Note that we just checked our answer for Problem 1.)

PROBLEM 2

Find the cost of a \$60 pair of jogging shoes on sale for 20% off (see Figure 4.4).

Figure 4.4

Solution

Let x represent the discount sale price. Since the shoes are on sale for 20% off, we must pay 80% of the original price.

$$x = (80\%)(60)$$
$$= (0.8)(60) = 48$$

The sale price is \$48. ■

Here is another equation that we can use in consumer problems:

Selling price = Cost + Profit

Profit (also called "markup, markon, margin, and margin of profit") may be stated in different ways. It may be stated as a percent of the selling price, a percent of the cost, or simply in terms of dollars and cents. Let's consider some problems where the profit is either a percent of the selling price or a percent of the cost.

PROBLEM 3

A retailer has some shirts that cost him \$20 each. He wants to sell them at a profit of 60% of the cost. What selling price should be marked on the shirts?

Solution

Let s represent the selling price. The basic relationship *selling price equals cost plus profit* can be used as a guideline.

$$\begin{array}{ccccc} \text{Selling price} & = & \text{Cost} & + & \text{Profit (\% of cost)} \\ \downarrow & & \downarrow & & \downarrow \\ s & = & \$20 & + & (60\%)(20) \end{array}$$

Solving this equation, we obtain

$$s = 20 + (60\%)(20)$$

$$s = 20 + (0.6)(20) \qquad \text{Changed percent to decimal}$$

$$s = 20 + 12$$

$$s = 32$$

The selling price should be \$32. ■

PROBLEM 4

Kathrin bought a painting for \$120 and later decided to resell it. She made a profit of 40% of the selling price. How much did she receive for the painting?

Solution

We can use the same basic relationship as a guideline, except that this time the profit is a percent of the selling price. Let s represent the selling price.

$$\begin{array}{ccccc} \text{Selling price} & = & \text{Cost} & + & \text{Profit (\% of selling price)} \\ \downarrow & & \downarrow & & \downarrow \\ s & = & 120 & + & (40\%)(s) \end{array}$$

Solving this equation, we obtain

$$s = 120 + (40\%)(s)$$

$$s = 120 + 0.4s$$

$$0.6s = 120 \qquad \text{Subtracted } 0.4s \text{ from both sides}$$

$$s = \frac{120}{0.6} = 200$$

She received \$200 for the painting. ■

Certain types of investment problems can be translated into algebraic equations. In some of these problems, we use the simple interest formula $i = Prt$, where i represents the amount of interest earned by investing P dollars at a yearly rate of r percent for t years.

PROBLEM 5

John invested \$9300 for 2 years and received \$1395 in interest. Find the annual interest rate John received on his investment.

Solution

$$i = Prt$$
$$1395 = 9300r(2)$$
$$1395 = 18600r$$
$$\frac{1395}{18600} = r$$
$$0.075 = r$$

The annual interest rate is 7.5%. ■

PROBLEM 6

How much principal must be invested to receive \$1500 in interest when the investment is made for 3 years at an annual interest rate of 6.25%?

Solution

$$i = Prt$$
$$1500 = P(0.0625)(3)$$
$$1500 = P(0.1875)$$
$$\frac{1500}{0.1875} = P$$
$$8000 = P$$

The principal must be \$8000. ■

PROBLEM 7

How much monthly interest will be charged on a credit card bill with a balance of \$754 when the credit card company charges an 18% annual interest rate?

Solution

$$i = Prt$$
$$i = 754(0.18)\left(\frac{1}{12}\right)$$ Remember, 1 month is $\frac{1}{12}$ of a year
$$i = 11.31$$

The interest charge would be \$11.31. ■

Problem Set 4.2

For Problems 1–22, solve each of the equations.

1. $x - 0.36 = 0.75$
2. $x - 0.15 = 0.42$
3. $x + 7.6 = 14.2$
4. $x + 11.8 = 17.1$
5. $0.62 - y = 0.14$
6. $7.4 - y = 2.2$
7. $0.7t = 56$
8. $1.3t = 39$
9. $x = 3.36 - 0.12x$
10. $x = 5.3 - 0.06x$
11. $s = 35 + 0.3s$
12. $s = 40 + 0.5s$
13. $s = 42 + 0.4s$
14. $s = 24 + 0.6s$
15. $0.07x + 0.08(x + 600) = 78$
16. $0.06x + 0.09(x + 200) = 63$
17. $0.09x + 0.1(2x) = 130.5$
18. $0.11x + 0.12(3x) = 188$
19. $0.08x + 0.11(500 - x) = 50.5$
20. $0.07x + 0.09(2000 - x) = 164$
21. $0.09x = 550 - 0.11(5400 - x)$
22. $0.08x = 580 - 0.1(6000 - x)$

For Problems 23–38, set up an equation and solve each problem.

23. Tom bought an electric drill at a 30% discount sale for \$35. What was the original price of the drill?
24. Magda bought a dress for \$140, which represents a 20% discount of the original price. What was the original price of the dress?
25. Find the cost of a \$4800 wide-screen plasma television that is on sale for 25% off.
26. Byron purchased a computer monitor at a 10% discount sale for \$121.50. What was the original price of the monitor?
27. Suppose that Jack bought a \$32 putter on sale for 35% off. How much did he pay for the putter?
28. Swati bought a 13-inch portable color TV for 20% off of the list price. The list price was \$229.95. What did she pay for the TV?
29. Pierre bought a coat for \$126 that was listed for \$180. What rate of discount did he receive?
30. Phoebe paid \$32 for a pair of sandals that was listed for \$40. What rate of discount did she receive?
31. A retailer has some toe rings that cost him \$5 each. He wants to sell them at a profit of 70% of the cost. What should be the selling price of the toe rings?
32. A retailer has some video games that cost her \$25 each. She wants to sell them at a profit of 80% of the cost. What price should she charge for the video games?
33. The owner of a pizza parlor wants to make a profit of 55% of the cost for each pizza sold. If it costs \$8 to make a pizza, at what price should it be sold?
34. Produce in a food market usually has a high markup because of loss due to spoilage. If a head of lettuce costs a retailer \$0.50, at what price should it be sold to realize a profit of 130% of the cost?
35. Jewelry has a very high markup rate. If a ring costs a jeweler \$400, at what price should it be sold to gain a profit of 60% of the selling price?
36. If a box of candy costs a retailer \$2.50 and he wants to make a profit of 50% based on the selling price, what price should he charge for the candy?
37. If the cost of a pair of shoes for a retailer is \$32 and he sells them for \$44.80, what is his rate of profit based on the cost?
38. A retailer has some skirts that cost her \$24. If she sells them for \$31.20, find her rate of profit based on the cost.

For Problems 39–46, use the formula $i = Prt$ to reach a solution.

39. Find the annual interest rate if \$560 in interest is earned when \$3500 is invested for 2 years.
40. How much interest will be charged on a student loan if \$8000 is borrowed for 9 months at a 19.2% annual interest rate?
41. How much principal, invested at 8% annual interest for 3 years, is needed to earn \$1000?

42. How long will $2400 need to be invested at a 5.5% annual interest rate to earn $330?

43. What will be the interest earned on a $5000 certificate of deposit invested at 3.8% annual interest for 10 years?

44. One month a credit card company charged $38.15 in interest on a balance of $2725. What annual interest rate is the credit card company charging?

45. How much is a month's interest on a mortgage balance of $145,000 at a 6.5% annual interest rate?

46. For how many years must $2000 be invested at a 5.4% annual interest rate to earn $162?

A selection of additional word problems is in Appendix B. All Appendix problems referenced as (4.2) are appropriate for this section.

THOUGHTS INTO WORDS

47. What is wrong with the following procedure, and how should it be changed?

$$1.2x + 2 = 3.8$$
$$10(1.2x) + 2 = 10(3.8)$$
$$12x + 2 = 38$$
$$12x = 36$$
$$x = 3$$

48. From a consumer's viewpoint, would you prefer that a retailer figure profit based on the cost or on the selling price of an item? Explain your answer.

FURTHER INVESTIGATIONS

49. A retailer buys an item for $40, resells it for $50, and claims that she is making only a 20% profit. Is her claim correct?

50. A store has a special discount sale of 40% off on all items. It also advertises an additional 10% off on items bought in quantities of a dozen or more. How much will it cost to buy a dozen items of some particular kind that regularly sell for $5 per item? (Be careful, a 40% discount followed by a 10% discount is not equal to a 50% discount.)

51. Is a 10% discount followed by a 40% discount the same as a 40% discount followed by a 10% discount? Justify your answer.

52. Some people use the following formula for determining the selling price of an item when the profit is based on a percent of the selling price:

$$\text{Selling price} = \frac{\text{Cost}}{100\% - \text{Percent of profit}}$$

Show how to develop this formula.

Solve each of the following equations and express the solutions in decimal form. Your calculator might be of some help.

53. $2.4x + 5.7 = 9.6$

54. $-3.2x - 1.6 = 5.8$

55. $0.08x + 0.09(800 - x) = 68.5$

56. $0.10x + 0.12(720 - x) = 80$

57. $7x - 0.39 = 0.03$

58. $9x - 0.37 = 0.35$

59. $0.2(t + 1.6) = 3.4$

60. $0.4(t - 3.8) = 2.2$

4.3 Formulas

To find the distance traveled in 3 hours at a rate of 50 miles per hour, we multiply the rate by the time. Thus the distance is $50(3) = 150$ miles. We usually state the rule *distance equals rate times time* as a formula: $d = rt$. **Formulas** are simply rules we state in symbolic language and express as equations. Thus the formula $d = rt$ is an equation that involves three variables: d, r, and t.

As we work with formulas, it is often necessary to solve for a specific variable when we have numerical values for the remaining variables. Consider the following examples.

EXAMPLE 1

Solve $d = rt$ for r if $d = 330$ and $t = 6$.

Solution

Substitute 330 for d and 6 for t in the given formula to obtain

$$330 = r(6)$$

Solving this equation yields

$$330 = 6r$$

$$55 = r$$

■

EXAMPLE 2

Solve $C = \frac{5}{9}(F - 32)$ for F if $C = 10$. (This formula expresses the relationship between the Fahrenheit and Celsius temperature scales.)

Solution

Substitute 10 for C to obtain

$$10 = \frac{5}{9}(F - 32)$$

Solving this equation produces

$$\frac{9}{5}(10) = \frac{9}{5}\left(\frac{5}{9}\right)(F - 32) \qquad \text{Multiply both sides by } \frac{9}{5}$$

$$18 = F - 32$$

$$50 = F$$

■

Sometimes it may be convenient to change a formula's form by using the properties of equality. For example, the formula $d = rt$ can be changed as follows:

$$d = rt$$

$$\frac{d}{r} = \frac{rt}{r} \qquad \text{Divide both sides by } r$$

$$\frac{d}{r} = t$$

We say that the formula $d = rt$ has been **solved for the variable *t***. The formula can also be **solved for *r*** as follows:

$$d = rt$$

$$\frac{d}{t} = \frac{rt}{t} \qquad \text{Divide both sides by } t$$

$$\frac{d}{t} = r$$

Geometric Formulas

There are several formulas in geometry that we use quite often. Let's briefly review them at this time; they will be used periodically throughout the remainder of the text. These formulas (along with some others) and Figures 4.5–4.15 are also listed in the inside front cover of this text.

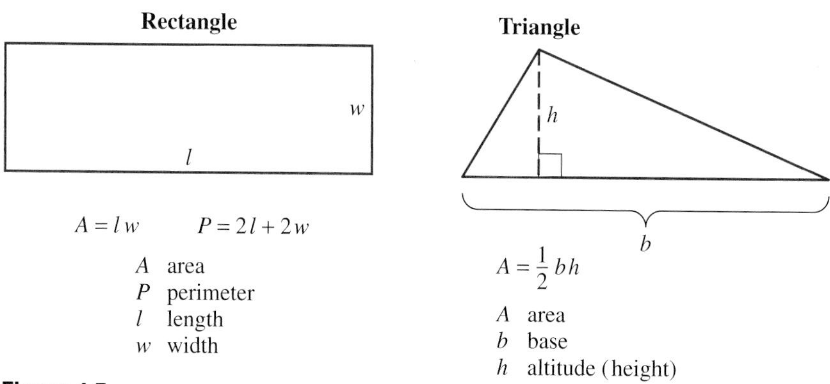

Figure 4.5

Figure 4.6

Trapezoid

b_1, h, b_2

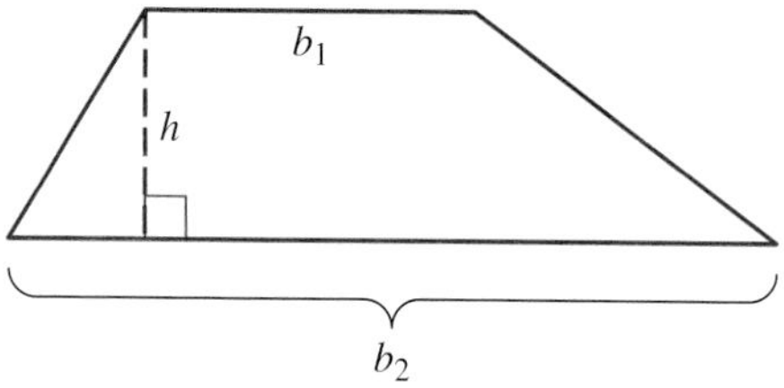

$$A = \frac{1}{2}h(b_1 + b_2)$$

A area
b_1, b_2 bases
h altitude

Figure 4.7

Parallelogram

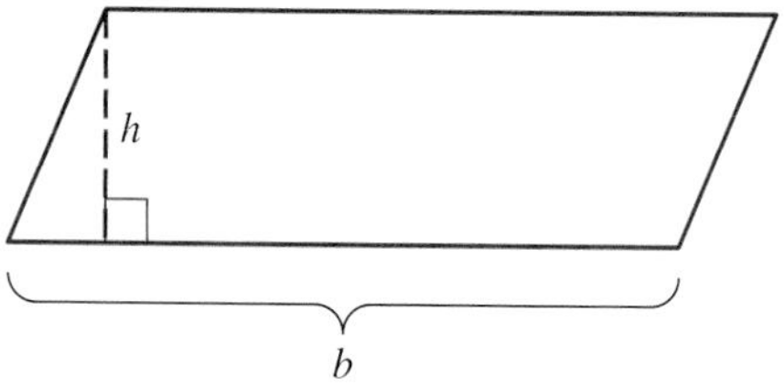

h, b

$$A = bh$$

A area
b base
h altitude (height)

Figure 4.8

Circle

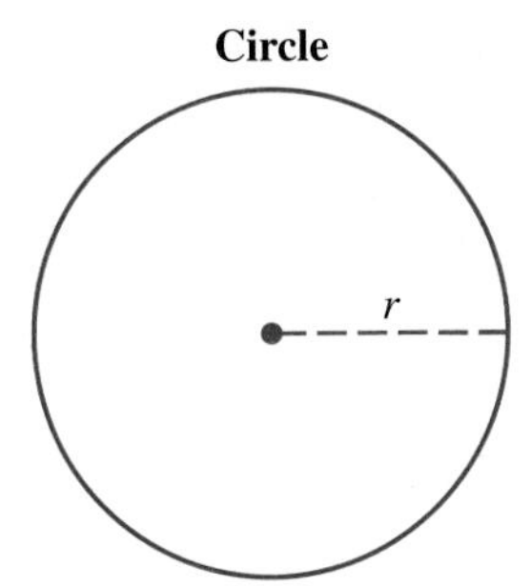

$A = \pi r^2 \qquad C = 2\pi r$

A area
C circumference
r radius

Figure 4.9

Sphere

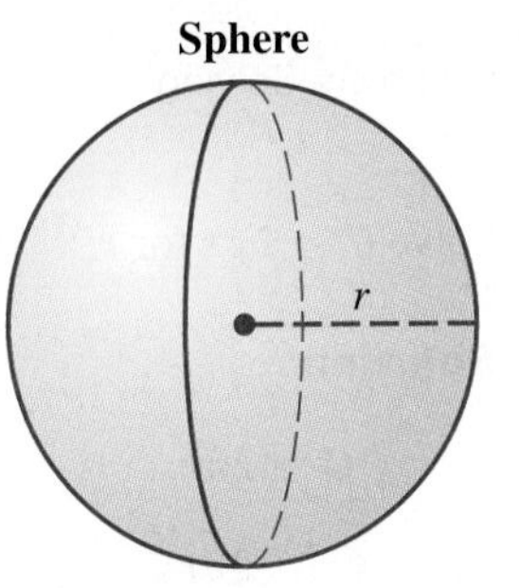

$V = \frac{4}{3}\pi r^3 \qquad S = 4\pi r^2$

S surface area
V volume
r radius

Figure 4.10

Prism

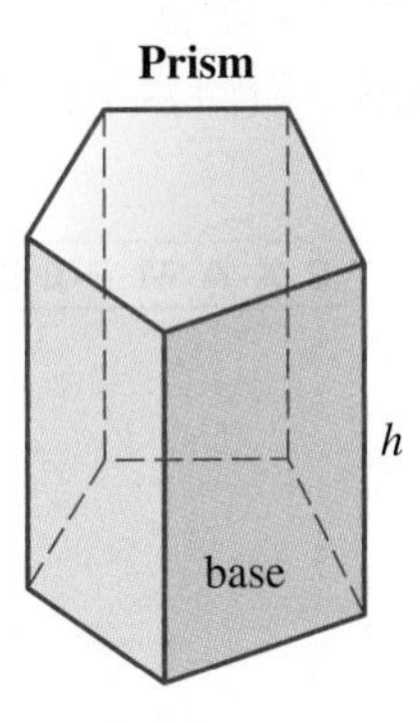

$V = Bh$

V volume
B area of base
h altitude (height)

Figure 4.11

Rectangular Prism

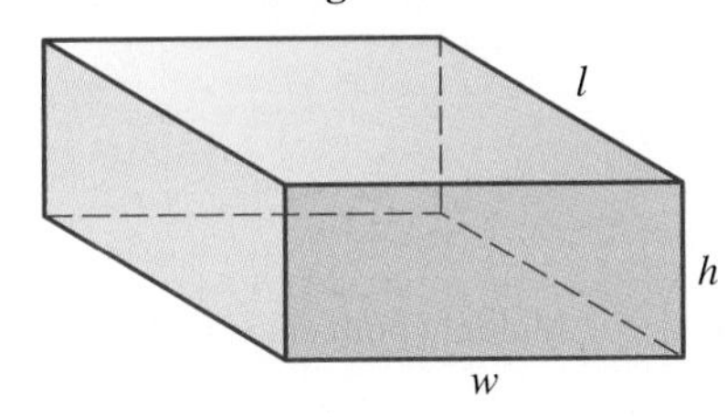

$V = lwh \qquad S = 2hw + 2hl + 2lw$

V volume
S total surface area
w width
l length
h altitude (height)

Figure 4.12

Pyramid

$V = \frac{1}{3}Bh$

V volume
B area of base
h altitude (height)

Figure 4.13

Right Circular Cylinder

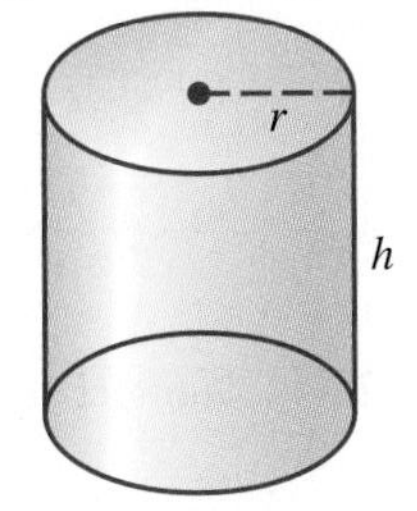

$V = \pi r^2 h \qquad S = 2\pi r^2 + 2\pi rh$

V volume
S total surface area
r radius
h altitude (height)

Figure 4.14

Right Circular Cone

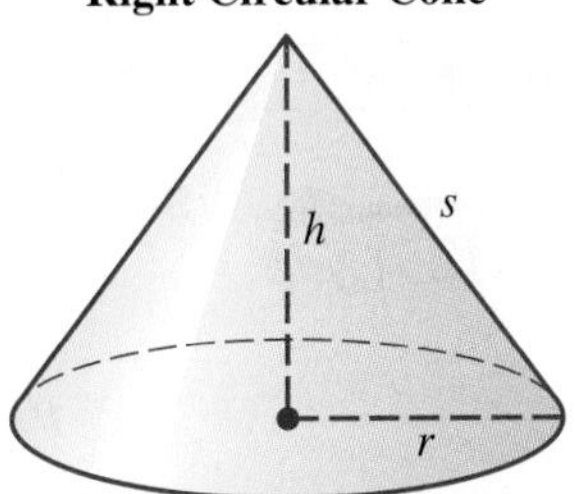

$V = \frac{1}{3}\pi r^2 h \qquad S = \pi r^2 + \pi rs$

V volume
S total surface area
r radius
h altitude (height)
s slant height

Figure 4.15

EXAMPLE 3 Solve $C = 2\pi r$ for r.

Solution

$$C = 2\pi r$$

$$\frac{C}{2\pi} = \frac{2\pi r}{2\pi} \quad \text{Divide both sides by}$$

$$\frac{C}{2\pi} = r$$

■

EXAMPLE 4 Solve $V = \frac{1}{3}Bh$ for h.

Solution

$$V = \frac{1}{3}Bh$$

$$3(V) = 3\left(\frac{1}{3}Bh\right) \quad \text{Multiply both sides by 3}$$

$$3V = Bh$$

$$\frac{3V}{B} = \frac{Bh}{B} \quad \text{Divide both sides by } B$$

$$\frac{3V}{B} = h$$

■

EXAMPLE 5 Solve $P = 2l + 2w$ for w.

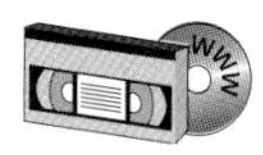

Solution

$$P = 2l + 2w$$

$$P - 2l = 2l + 2w - 2l \quad \text{Subtract } 2l \text{ from both sides}$$

$$P - 2l = 2w$$

$$\frac{P - 2l}{2} = \frac{2w}{2} \quad \text{Divide both sides by 2}$$

$$\frac{P - 2l}{2} = w$$

■

PROBLEM 1

Find the total surface area of a right circular cylinder that has a radius of 10 inches and a height of 14 inches.

Solution

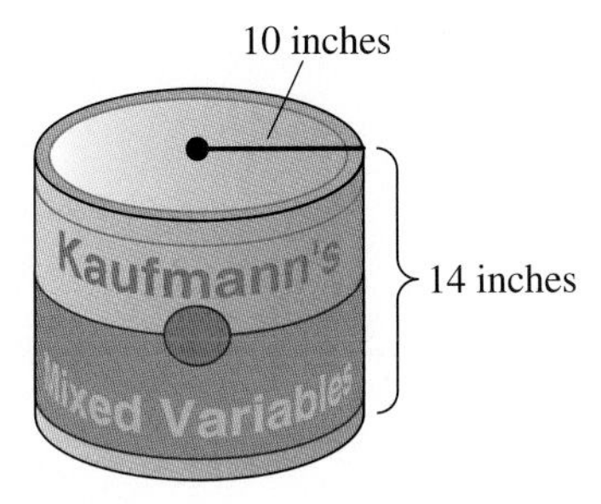

Figure 4.16

Let's sketch a right circular cylinder and record the given information as shown in Figure 4.16. Substitute 10 for r and 14 for h in the formula for the total surface area of a right circular cylinder to obtain

$$\begin{aligned} S &= 2\pi r^2 + 2\pi rh \\ &= 2\pi(10)^2 + 2\pi(10)(14) \\ &= 200\pi + 280\pi \\ &= 480\pi \end{aligned}$$

The total surface area is 480π square inches. ■

In Problem 1 we used the figure to record the given information, and it also served as a reminder of the geometric figure under consideration. Now let's consider a problem where the figure helps us to analyze the problem.

PROBLEM 2

A sidewalk 3 feet wide surrounds a rectangular plot of ground that measures 75 feet by 100 feet. Find the area of the sidewalk.

Solution

We can make a sketch and record the given information as in Figure 4.17. The area of the sidewalk can be found by subtracting the area of the rectangular plot from

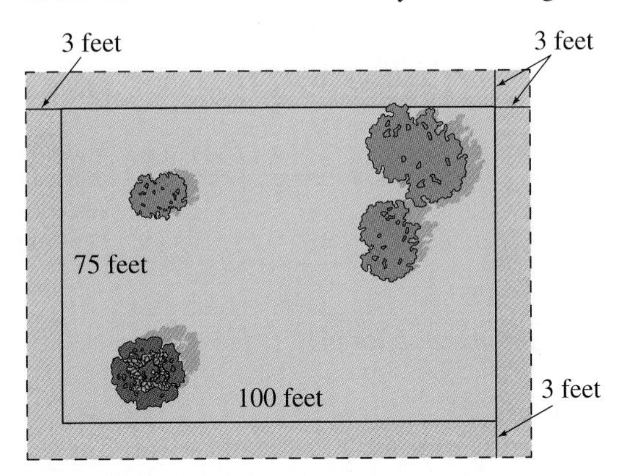

Figure 4.17

the area of the plot plus the sidewalk (the large dashed rectangle). The width of the large rectangle is $75 + 3 + 3 = 81$ feet, and its length is $100 + 3 + 3 = 106$ feet.

$$\begin{aligned} A &= (81)(106) - (75)(100) \\ &= 8586 - 7500 = 1086 \end{aligned}$$

The area of the sidewalk is 1086 square feet. ■

Changing Forms of Equations

In Chapter 8 you will be working with equations that contain two variables. At times you will need to solve for one variable in terms of the other variable—that is, change the form of the equation just like we have done with formulas. The next examples illustrate once again how we can use the properties of equality for such situations.

EXAMPLE 6 Solve $3x + y = 4$ for x.

Solution

$$3x + y = 4$$

$$3x + y - y = 4 - y \qquad \text{Subtract } y \text{ from both sides}$$

$$3x = 4 - y$$

$$\frac{3x}{3} = \frac{4 - y}{3} \qquad \text{Divide both sides by 3}$$

$$x = \frac{4 - y}{3}$$

EXAMPLE 7 Solve $4x - 5y = 7$ for y.

Solution

$$4x - 5y = 7$$

$$4x - 5y - 4x = 7 - 4x \qquad \text{Subtract } 4x \text{ from both sides}$$

$$-5y = 7 - 4x$$

$$\frac{-5y}{-5} = \frac{7 - 4x}{-5} \qquad \text{Divide both sides by } -5$$

$$y = \frac{7 - 4x}{-5}\left(\frac{-1}{-1}\right) \qquad \text{Multiply numerator and denominator of fraction on the right by } -1$$

We commonly do this so that the denominator is positive

$$y = \frac{4x - 7}{5}$$

EXAMPLE 8 Solve $y = mx + b$ for m.

Solution

$$y = mx + b$$

$$y - b = mx + b - b \qquad \text{Subtract } b \text{ from both sides}$$

$$y - b = mx$$

$$\frac{y - b}{x} = \frac{mx}{x} \qquad \text{Divide both sides by } x$$

$$\frac{y - b}{x} = m$$

Problem Set 4.3

For Problems 1–10, solve for the specified variable using the given facts.

1. Solve $d = rt$ for t if $d = 336$ and $r = 48$.
2. Solve $d = rt$ for r if $d = 486$ and $t = 9$.
3. Solve $i = Prt$ for P if $i = 200$, $r = 0.08$, and $t = 5$.
4. Solve $i = Prt$ for t if $i = 540$, $P = 750$, and $r = 0.09$.
5. Solve $\mathrm{F} = \frac{9}{5}\mathrm{C} + 32$ for C if F = 68.
6. Solve $\mathrm{C} = \frac{5}{9}(\mathrm{F} - 32)$ for F if C = 15.
7. Solve $V = \frac{1}{3}Bh$ for B if $V = 112$ and $h = 7$.
8. Solve $V = \frac{1}{3}Bh$ for h if $V = 216$ and $B = 54$.
9. Solve $A = P + Prt$ for t if $A = 652$, $P = 400$, and $r = 0.07$.
10. Solve $A = P + Prt$ for P if $A = 1032$, $r = 0.06$, and $t = 12$.

For Problems 11–32, use the geometric formulas given in this section to help solve the problems.

11. Find the perimeter of a rectangle that is 14 centimeters long and 9 centimeters wide.
12. If the perimeter of a rectangle is 80 centimeters and its length is 24 centimeters, find its width.
13. If the perimeter of a rectangle is 108 inches and its length is $3\frac{1}{4}$ feet, find its width in inches.
14. How many yards of fencing does it take to enclose a rectangular plot of ground that is 69 feet long and 42 feet wide?
15. A dirt path 4 feet wide surrounds a rectangular garden that is 38 feet long and 17 feet wide. Find the area of the dirt path.
16. Find the area of a cement walk 3 feet wide that surrounds a rectangular plot of ground 86 feet long and 42 feet wide.
17. Suppose that paint costs $8.00 per liter and that 1 liter will cover 9 square meters of surface. We are going to paint (on one side only) 50 rectangular pieces of wood of the same size that have a length of 60 centimeters and a width of 30 centimeters. What will the cost of the paint be?
18. A lawn is in the shape of a triangle with one side 130 feet long and the altitude to that side 60 feet long. Will one bag of fertilizer that covers 4000 square feet be enough to fertilize the lawn?
19. Find the length of an altitude of a trapezoid with bases of 8 inches and 20 inches and an area of 98 square inches.
20. A flower garden is in the shape of a trapezoid with bases of 6 yards and 10 yards. The distance between the bases is 4 yards. Find the area of the garden.
21. In Figure 4.18 you'll notice that the diameter of a metal washer is 4 centimeters. The diameter of the hole is 2 centimeters. How many square centimeters of metal are there in 50 washers? Express the answer in terms of π.

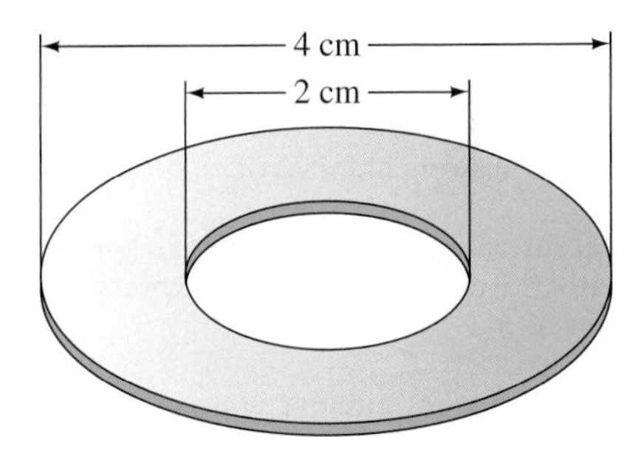

Figure 4.18

22. Find the area of a circular plot of ground that has a radius of length 14 meters. Use $3\frac{1}{7}$ as an approximation for π.
23. Find the area of a circular region that has a diameter of 1 yard. Express the answer in terms of π.
24. Find the area of a circular region if the circumference is 12π units. Express the answer in terms of π.

25. Find the total surface area and volume of a sphere that has a radius 9 inches long. Express the answers in terms of π.

26. A circular pool is 34 feet in diameter and has a flagstone walk around it that is 3 feet wide (see Figure 4.19). Find the area of the walk. Express the answer in terms of π.

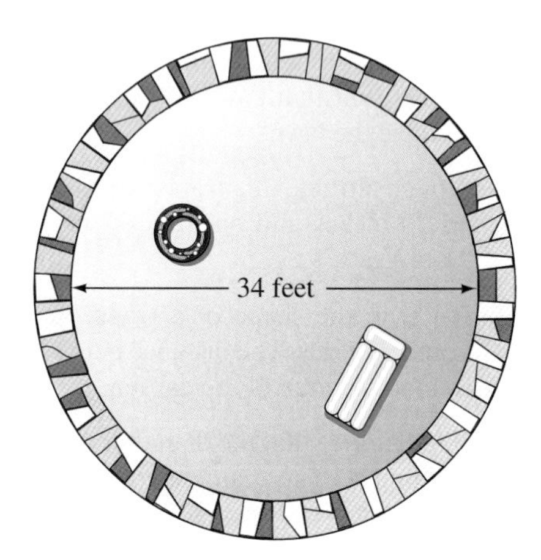

Figure 4.19

27. Find the volume and total surface area of a right circular cylinder that has a radius of 8 feet and a height of 18 feet. Express the answers in terms of π.

28. Find the total surface area and volume of a sphere that has a diameter 12 centimeters long. Express the answers in terms of π.

29. If the volume of a right circular cone is 324π cubic inches and a radius of the base is 9 inches long, find the height of the cone.

30. Find the volume and total surface area of a tin can if the radius of the base is 3 centimeters and the height of the can is 10 centimeters. Express the answers in terms of π.

31. If the total surface area of a right circular cone is 65π square feet, and a radius of the base is 5 feet long, find the slant height of the cone.

32. If the total surface area of a right circular cylinder is 104π square meters, and a radius of the base is 4 meters long, find the height of the cylinder.

For Problems 33–42, match the correct formula for each statement.

33. Area of a rectangle	A. $A = \pi r^2$
34. Circumference of a circle	B. $V = lwh$
35. Volume of a rectangular prism	C. $P = 2l + 2w$
36. Area of a triangle	D. $V = \frac{4}{3}\pi r^3$
37. Area of a circle	E. $A = lw$
38. Volume of a right circular cylinder	F. $A = bh$
39. Perimeter of a rectangle	G. $A = \frac{1}{2}h(b_1 + b_2)$
40. Volume of a sphere	H. $A = \frac{1}{2}bh$
41. Area of a parallelogram	I. $C = 2\pi r$
42. Area of a trapezoid	J. $V = \pi r^2 h$

For Problems 43–54, solve each formula for the indicated variable. (Before doing these problems, cover the right-hand column and see how many of these formulas you recognize!)

43. $V = Bh$ for h	Volume of a prism
44. $A = lw$ for l	Area of a rectangle
45. $V = \frac{1}{3}Bh$ for B	Volume of a pyramid
46. $A = \frac{1}{2}bh$ for h	Area of a triangle
47. $P = 2l + 2w$ for w	Perimeter of a rectangle
48. $V = \pi r^2 h$ for h	Volume of a cylinder
49. $V = \frac{1}{3}\pi r^2 h$ for h	Volume of a cone
50. $i = Prt$ for t	Simple interest formula
51. $\text{F} = \frac{9}{5}\text{C} + 32$ for C	Celsius to Fahrenheit
52. $A = P + Prt$ for t	Simple interest formula
53. $A = 2\pi r^2 + 2\pi rh$ for h	Surface area of a cylinder
54. $\text{C} = \frac{5}{9}(\text{F} - 32)$ for F	Fahrenheit to Celsius

For Problems 55–70, solve each equation for the indicated variable.

55. $3x + 7y = 9$ for x

56. $5x + 2y = 12$ for x

57. $9x - 6y = 13$ for y

58. $3x - 5y = 19$ for y

59. $-2x + 11y = 14$ for x

60. $-x + 14y = 17$ for x

61. $y = -3x - 4$ for x

62. $y = -7x + 10$ for x

63. $\frac{x - 2}{4} = \frac{y - 3}{6}$ for y

64. $\frac{x + 1}{3} = \frac{y - 5}{2}$ for y

65. $ax - by - c = 0$ for y

66. $ax + by = c$ for y

67. $\frac{x + 6}{2} = \frac{y + 4}{5}$ for x

68. $\frac{x - 3}{6} = \frac{y - 4}{8}$ for x

69. $m = \frac{y - b}{x}$ for y

70. $y = mx + b$ for x

THOUGHTS INTO WORDS

71. Suppose that both the length and width of a rectangle are doubled. How does this affect the perimeter of the rectangle? Defend your answer.

72. Suppose that the length of a radius of a circle is doubled. How does this affect the area of the circle? Defend your answer.

73. Some people subtract 32 and then divide by 2 to estimate the change from a Fahrenheit reading to a Celsius reading. Why does this give an estimate and how good is the estimate?

FURTHER INVESTIGATIONS

For each of the following problems, use 3.14 as an approximation for π. Your calculator should be of some help with these problems.

74. Find the area of a circular plot of ground that has a radius 16.3 meters long. Express your answer to the nearest tenth of a square meter.

75. Find the area, to the nearest tenth of a square centimeter, of the ring in Figure 4.20.

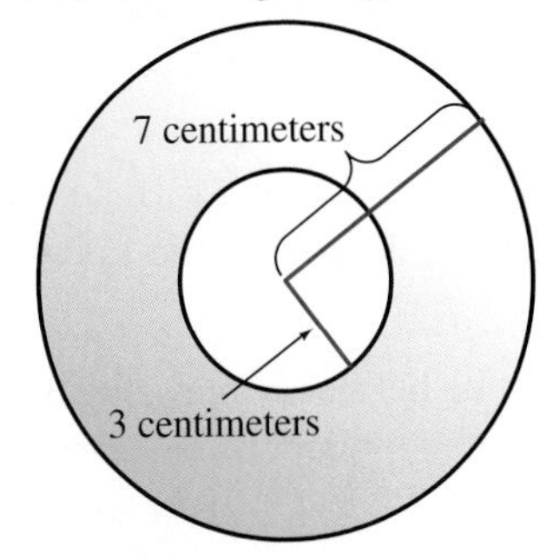

Figure 4.20

76. Find the area, to the nearest square inch, of each of these pizzas: 10-inch diameter, 12-inch diameter, 14-inch diameter.

77. Find the total surface area, to the nearest square centimeter, of the tin can shown in Figure 4.21.

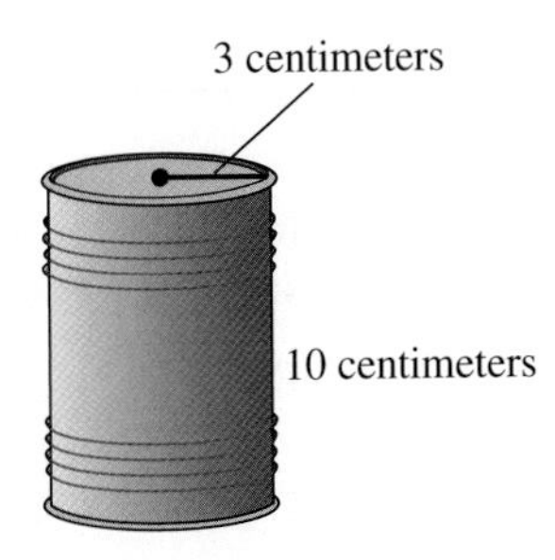

Figure 4.21

78. Find the total surface area, to the nearest square centimeter, of a baseball that has a radius of 4 centimeters (see Figure 4.22).

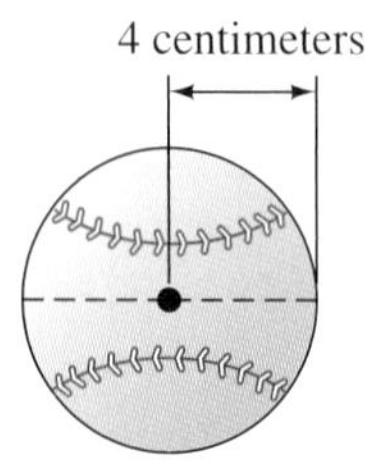

Figure 4.22

79. Find the volume, to the nearest cubic inch, of a softball that has a diameter of 5 inches (see Figure 4.23).

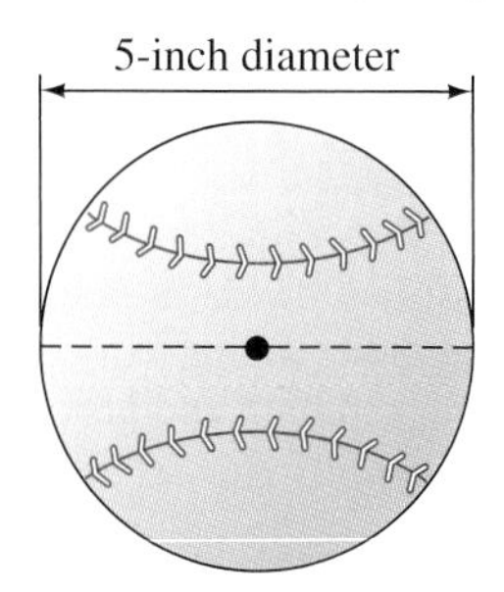

Figure 4.23

80. Find the volume, to the nearest cubic meter, of the rocket in Figure 4.24.

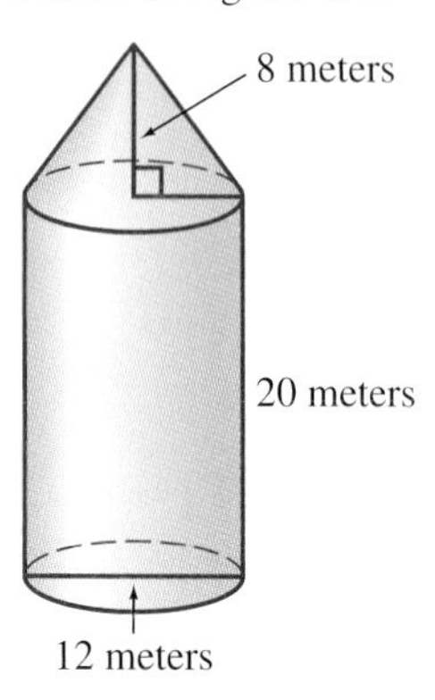

Figure 4.24

4.4 Problem Solving

We begin this section by restating the suggestions for solving word problems that we offered in Section 3.3.

Suggestions for Solving Word Problems

1. Read the problem carefully, and make sure that you understand the meanings of all the words. Be especially alert for any technical terms used in the statement of the problem.
2. Read the problem a second time (perhaps even a third time) to get an overview of the situation being described and to determine the known facts as well as what is to be found.
3. Sketch any figure, diagram, or chart that might be helpful in analyzing the problem.

(continued)

4. Choose a meaningful variable to represent an unknown quantity in the problem (perhaps t if time is the unknown quantity); represent any other unknowns in terms of that variable.
5. Look for a guideline that can be used to set up an equation. A guideline might be a formula such as *selling price equals cost plus profit*, or a relationship such as *interest earned from a 9% investment plus interest earned from a 10% investment equals total amount of interest earned*. A guideline may also be illustrated by a figure or diagram that you sketch for a particular problem.
6. Form an equation that contains the variable and that translates the conditions of the guideline from English into algebra.
7. Solve the equation, and use the solution to determine all the facts requested in the problem.
8. **Check all answers back into the original statement of the problem.**

Again we emphasize the importance of suggestion 5. Determining the guideline to follow when setting up the equation is a vital part of the analysis of a problem. Sometimes the guideline is a formula, such as one of the formulas we presented in the previous section and accompanying problem set. Let's consider a problem of that type.

PROBLEM 1

How long will it take \$500 to double itself if it is invested at 8% simple interest?

Solution

We can use the basic simple interest formula, $i = Prt$, where i represents interest, P is the principal (money invested), r is the rate (percent), and t is the time in years. For \$500 to double itself means that we want \$500 to earn another \$500 in interest. Thus, using $i = Prt$ as a guideline, we can proceed as follows:

$$i = Prt$$

$$500 = 500(8\%)(t)$$

Now let's solve this equation.

$$500 = 500(0.08)(t)$$

$$1 = 0.08t$$

$$100 = 8t$$

$$\frac{100}{8} = t$$

$$12\frac{1}{2} = t$$

It will take $12\frac{1}{2}$ years. ■

If the problem involves a geometric formula, then a sketch of the figure is helpful for recording the given information and analyzing the problem. The next example illustrates this idea.

PROBLEM 2

The length of a football field is 40 feet more than twice its width, and the perimeter of the field is 1040 feet. Find the length and width of the field.

Solution

Since the length is stated in terms of the width, we can let w represent the width, and then $2w + 40$ represents the length, as shown in Figure 4.25. A guideline for this problem is the perimeter formula $P = 2l + 2w$. Thus, the following equation can be set up and solved:

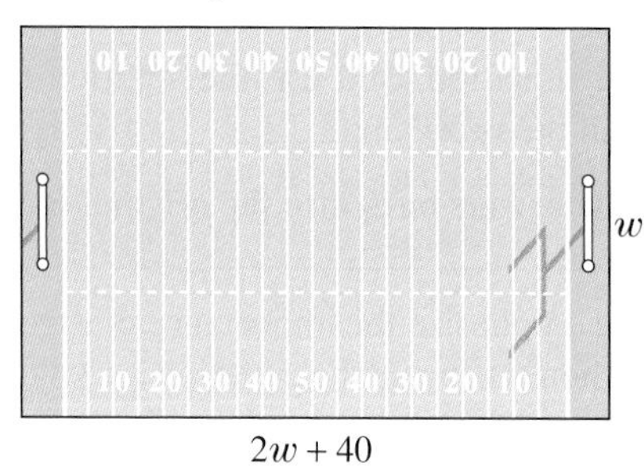

Figure 4.25

$$P = 2l + 2w$$

$$1040 = 2(2w + 40) + 2w$$

$$1040 = 4w + 80 + 2w$$

$$1040 = 6w + 80$$

$$960 = 6w$$

$$160 = w$$

If $w = 160$, then $2w + 40 = 2(160) + 40 = 360$. Thus, the football field is 360 feet long and 160 feet wide. ■

Sometimes the formulas we use when we are analyzing a problem are different than those we use as a guideline for setting up the equation. For example, uniform motion problems involve the formula $d = rt$, but the main guideline for setting up an equation for such problems is usually a statement about either *times*, *rates*, or *distances*. Let's consider an example.

PROBLEM 3

Pablo leaves city A on a moped traveling toward city B at 18 miles per hour. At the same time, Cindy leaves city B on a moped traveling toward city A at 23 miles per hour. The distance between the two cities is 123 miles. How long will it take before Pablo and Cindy meet on their mopeds?

Solution

First, let's sketch a diagram as in Figure 4.26. Let t represent the time that Pablo travels. Then t also represents the time that Cindy travels.

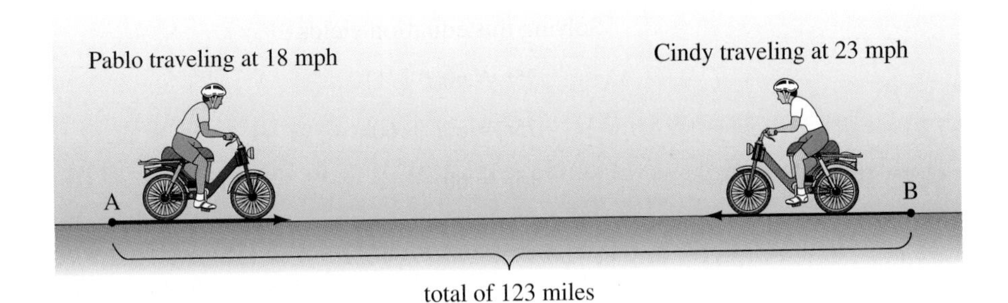

Figure 4.26

Distance Pablo travels + Distance Cindy travels = Total distance

$$18t + 23t = 123$$

Solving this equation yields

$$18t + 23t = 123$$
$$41t = 123$$
$$t = 3$$

They both travel for 3 hours. ■

Some people find it helpful to use a chart to organize the known and unknown facts in a uniform motion problem. We will illustrate with an example.

PROBLEM 4

A car leaves a town traveling at 60 kilometers per hour. How long will it take a second car traveling at 75 kilometers per hour to catch the first car if the second car leaves 1 hour later?

Solution

Let t represent the time of the second car. Then $t + 1$ represents the time of the first car because it travels 1 hour longer. We can now record the information of the problem in a chart.

	Rate	Time	Distance
First car	60	$t + 1$	$60(t + 1)$
Second car	75	t	$75t$

$d = rt$

Because the second car is to overtake the first car, the distances must be equal.

$$\underbrace{\text{Distance of second car}}_{75t} = \underbrace{\text{Distance of first car}}_{60(t+1)}$$

Solving this equation yields

$$75t = 60(t + 1)$$
$$75t = 60t + 60$$
$$15t = 60$$
$$t = 4$$

The second car should overtake the first car in 4 hours. (Check the answer!) ■

We would like to offer one bit of advice at this time. Don't become discouraged if solving word problems is giving you trouble. Problem solving is not a skill that can be developed overnight. It takes time, patience, hard work, and an open mind. Keep giving it your best shot, and gradually you should become more confident in your approach to such problems. Furthermore, we realize that some (perhaps many) of these problems may not seem "practical" to you. However, keep in mind that the real goal here is to develop problem-solving techniques. Finding and using a guideline, sketching a figure to record information and help in the analysis, estimating an answer before attempting to solve the problem, and using a chart to record information are some of the important tools we are trying to develop.

Problem Set 4.4

For Problems 1–12, solve each of the equations. These equations are the types you will be using in Problems 13–40.

1. $950(0.12)t = 950$

2. $1200(0.09)t = 1200$

3. $l + \frac{1}{4}l - 1 = 19$

4. $l + \frac{2}{3}l + 1 = 41$

5. $500(0.08)t = 1000$

6. $800(0.11)t = 1600$

7. $s + (2s - 1) + (3s - 4) = 37$

8. $s + (3s - 2) + (4s - 4) = 42$

9. $\frac{5}{2}r + \frac{5}{2}(r + 6) = 135$

10. $\frac{10}{3}r + \frac{10}{3}(r - 3) = 90$

11. $24\left(t - \frac{2}{3}\right) = 18t + 8$

12. $16t + 8\left(\frac{9}{2} - t\right) = 60$

Solve each of the following problems. Keep in mind the suggestions we offered in this section.

13. How long will it take \$4000 to double itself if it is invested at 8% simple interest?

14. How many years will it take \$1000 to double itself if it is invested at 5% simple interest?

15. How long will it take $8000 to triple itself if it is invested at 6% simple interest?

16. How many years will it take $500 to earn $750 in interest if it is invested at 6% simple interest?

17. The length of a rectangle is three times its width. If the perimeter of the rectangle is 112 inches, find its length and width.

18. The width of a rectangle is one-half of its length. If the perimeter of the rectangle is 54 feet, find its length and width.

19. Suppose that the length of a rectangle is 2 centimeters less than three times its width. The perimeter of the rectangle is 92 centimeters. Find the length and width of the rectangle.

20. Suppose that the length of a certain rectangle is 1 meter more than five times its width. The perimeter of the rectangle is 98 meters. Find the length and width of the rectangle.

21. The width of a rectangle is 3 inches less than one-half of its length. If the perimeter of the rectangle is 42 inches, find the area of the rectangle.

22. The width of a rectangle is 1 foot more than one-third of its length. If the perimeter of the rectangle is 74 feet, find the area of the rectangle.

23. The perimeter of a triangle is 100 feet. The longest side is 3 feet less than twice the shortest side, and the third side is 7 feet longer than the shortest side. Find the lengths of the sides of the triangle.

24. A triangular plot of ground has a perimeter of 54 yards. The longest side is twice the shortest side, and the third side is 2 yards longer than the shortest side. Find the lengths of the sides of the triangle.

25. The second side of a triangle is 1 centimeter longer than three times the first side. The third side is 2 centimeters longer than the second side. If the perimeter is 46 centimeters, find the length of each side of the triangle.

26. The second side of a triangle is 3 meters shorter than twice the first side. The third side is 4 meters longer than the second side. If the perimeter is 58 meters, find the length of each side of the triangle.

27. The perimeter of an equilateral triangle is 4 centimeters more than the perimeter of a square, and the length of a side of the triangle is 4 centimeters more than the length of a side of the square. Find the length of a side of the equilateral triangle. (An equilateral triangle has three sides of the same length.)

28. Suppose that a square and an equilateral triangle have the same perimeter. Each side of the equilateral triangle is 6 centimeters longer than each side of the square. Find the length of each side of the square. (An equilateral triangle has three sides of the same length.)

29. Suppose that the length of a radius of a circle is the same as the length of a side of a square. If the circumference of the circle is 15.96 centimeters longer than the perimeter of the square, find the length of a radius of the circle. (Use 3.14 as an approximation for π.)

30. The circumference of a circle is 2.24 centimeters more than six times the length of a radius. Find the radius of the circle. (Use 3.14 as an approximation for π.)

31. Sandy leaves a town traveling in her car at a rate of 45 miles per hour. One hour later, Monica leaves the same town traveling the same route at a rate of 50 miles per hour. How long will it take Monica to overtake Sandy?

32. Two cars start from the same place traveling in opposite directions. One car travels 4 miles per hour faster than the other car. Find their speeds if after 5 hours they are 520 miles apart.

33. The distance between Jacksonville and Miami is 325 miles. A freight train leaves Jacksonville and travels toward Miami at 40 miles per hour. At the same time, a passenger train leaves Miami and travels toward Jacksonville at 90 miles per hour. How long will it take the two trains to meet?

34. Kirk starts jogging at 5 miles per hour. One-half hour later, Nancy starts jogging on the same route at 7 miles per hour. How long will it take Nancy to catch Kirk?

35. A car leaves a town traveling at 40 miles per hour. Two hours later a second car leaves the town traveling the same route and overtakes the first car in 5 hours and 20 minutes. How fast was the second car traveling?

36. Two airplanes leave St. Louis at the same time and fly in opposite directions (see Figure 4.27). If one travels at 500 kilometers per hour, and the other at 600 kilometers per hour, how long will it take for them to be 1925 kilometers apart?

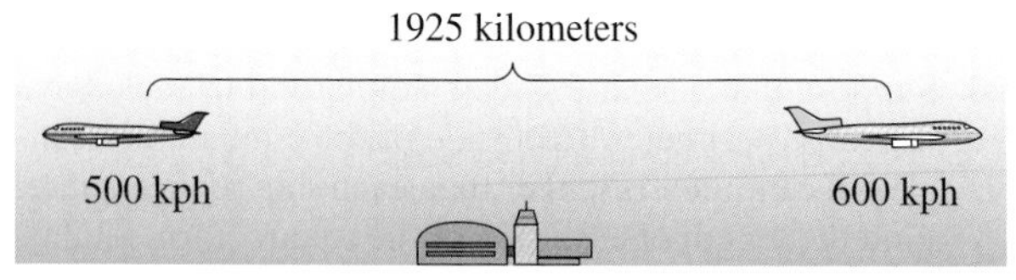

Figure 4.27

37. Two trains leave at the same time, one traveling east and the other traveling west. At the end of $9\frac{1}{2}$ hours they are 1292 miles apart. If the rate of the train traveling east is 8 miles per hour faster than the rate of the other train, find their rates.

38. Dawn starts on a 58-mile trip on her moped at 20 miles per hour. After a while the motor stops and she pedals the remainder of the trip at 12 miles per hour. The entire trip takes $3\frac{1}{2}$ hours. How far had Dawn traveled when the motor on the moped quit running?

39. Jeff leaves home and rides his bicycle out into the country for 3 hours. On his return trip along the same route, it takes him three-quarters of an hour longer. If his rate on the return trip was 2 miles per hour slower than on the trip out into the country, find the total roundtrip distance.

40. In $1\frac{1}{4}$ hours more time, Rita, riding her bicycle at 12 miles per hour, rode 2 miles farther than Sonya, who was riding her bicycle at 16 miles per hour. How long did each girl ride?

THOUGHTS INTO WORDS

41. Suppose that your friend analyzes Problem 31 as follows: Sandy has traveled 45 miles before Monica starts. Since Monica travels 5 miles per hour faster than Sandy, it will take her $\frac{45}{5} = 9$ hours to catch Sandy. How would you react to this analysis of the problem?

42. Summarize the ideas about problem solving that you have acquired thus far in this course.

4.5 More about Problem Solving

We begin this section with an important but often overlooked facet of problem solving: the importance of looking back over your solution and considering some of the following questions.

1. Is your answer to the problem a reasonable answer? Does it agree with the estimated answer you arrived at before doing the problem?
2. Have you checked your answer by substituting it back into the conditions stated in the problem?
3. Do you now see another plan that you can use to solve the problem? Perhaps there is even another guideline that you can use.
4. Do you now see that this problem is closely related to another problem that you have previously solved?

5. Have you "tucked away for future reference" the technique used to solve this problem?

Looking back over the solution of a newly solved problem can provide a foundation for solving problems in the future.

Now let's consider three problems that we often refer to as mixture problems. No basic formula applies for all of these problems, but the suggestion that you *think in terms of a pure substance* is often helpful in setting up a guideline. For example, a phrase such as "30% solution of acid" means that 30% of the amount of solution is acid and the remaining 70% is water.

PROBLEM 1

How many milliliters of pure acid must be added to 150 milliliters of a 30% solution of acid to obtain a 40% solution (see Figure 4.28)?

Remark: If a guideline is not apparent from reading the problem, it might help you to guess an answer and then check that guess. Suppose we guess that 30 milliliters of pure acid need to be added. To check, we must determine whether the final solution is 40% acid. Since we started with $0.30(150) = 45$ milliliters of pure acid and added our guess of 30 milliliters, the final solution will have $45 + 30 = 75$ milliliters of pure acid. The final amount of solution is $150 + 30 = 180$ milliliters. Thus, the final solution is $\frac{75}{180} = 41\frac{2}{3}\%$ pure acid.

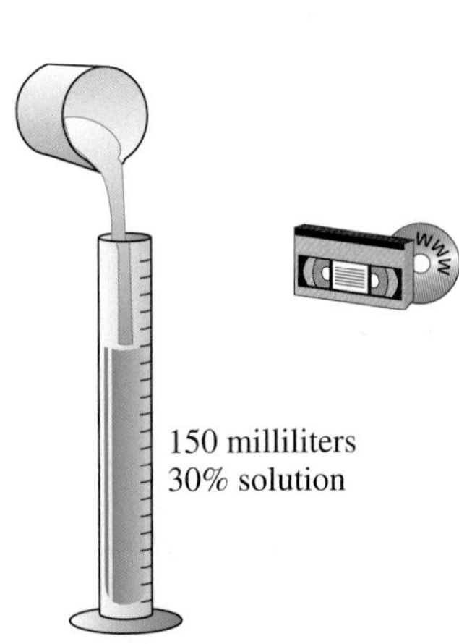

Figure 4.28

Solution

We hope that by guessing and checking the guess, you obtain the following guideline:

Amount of pure acid in original solution	+	Amount of pure acid to be added	=	Amount of pure acid in final solution

Let p represent the amount of pure acid to be added. Then using the guideline, we can form the following equation:

$$(30\%)(150) + p = 40\%(150 + p)$$

Now let's solve this equation to determine the amount of pure acid to be added.

$$(0.30)(150) + p = 0.40(150 + p)$$

$$45 + p = 60 + 0.4p$$

$$0.6p = 15$$

$$p = \frac{15}{0.6} = 25$$

We must add 25 milliliters of pure acid. (Perhaps you should check this answer.) ■

PROBLEM 2

Suppose we have a supply of a 30% solution of alcohol and a 70% solution. How many quarts of each should be mixed to produce a 20-quart solution that is 40% alcohol?

Solution

We can use a guideline similar to the one in Problem 1.

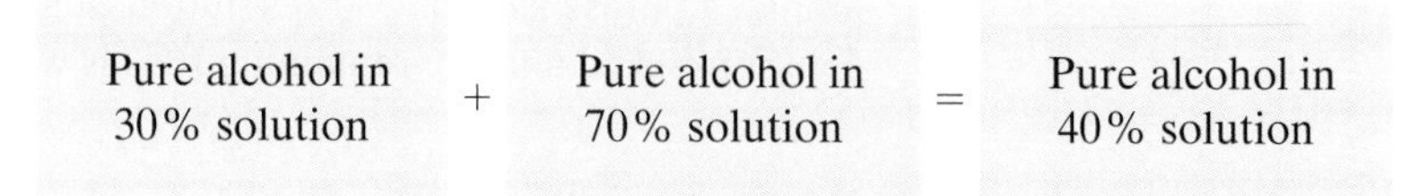

Let x represent the amount of 30% solution. Then $20 - x$ represents the amount of 70% solution. Now using the guideline, we translate into

$$(30\%)(x) + (70\%)(20 - x) = (40\%)(20)$$

Solving this equation, we obtain

$$\begin{aligned} 0.30x + 0.70(20 - x) &= 8 \\ 30x + 70(20 - x) &= 800 \qquad \text{Multiplied both sides by 100} \\ 30x + 1400 - 70x &= 800 \\ -40x &= -600 \\ x &= 15 \end{aligned}$$

Therefore, $20 - x = 5$. We should mix 15 quarts of the 30% solution with 5 quarts of the 70% solution. ■

PROBLEM 3

A 4-gallon radiator is full and contains a 40% solution of antifreeze. How much needs to be drained out and replaced with pure antifreeze to obtain a 70% solution?

Solution

This guideline can be used:

Pure antifreeze in the original solution − Pure antifreeze in the solution drained out + Pure antifreeze added = Pure antifreeze in the final solution

Let x represent the amount of pure antifreeze to be added. Then x also represents the amount of the 40% solution to be drained out. Thus, the guideline translates into the following equation:

$$(40\%)(4) - (40\%)(x) + x = (70\%)(4)$$

Solving this equation, we obtain

$$0.4(4) - 0.4x + x = 0.7(4)$$
$$1.6 + 0.6x = 2.8$$
$$0.6x = 1.2$$
$$x = 2$$

Therefore, we must drain out 2 gallons of the 40% solution and then add 2 gallons of pure antifreeze. (Checking this answer is a worthwhile exercise for you!) ■

PROBLEM 4

A woman invests a total of \$5000. Part of it is invested at 4% and the remainder at 6%. Her total yearly interest from the two investments is \$260. How much did she invest at each rate?

Solution

Let x represent the amount invested at 6%. Then $5000 - x$ represents the amount invested at 4%. Use the following guideline:

Interest earned from 6% investment	+	Interest earned from 4% investment	=	Total interest earned
↓		↓		↓
$(6\%)(x)$	+	$(4\%)(\$5000 - x)$	=	\$260

Solving this equation yields

$$(6\%)(x) + (4\%)(5000 - x) = 260$$
$$0.06x + 0.04(5000 - x) = 260$$
$$6x + 4(5000 - x) = 26{,}000 \quad \text{Multiplied both sides by 100}$$
$$6x + 20{,}000 - 4x = 26{,}000$$
$$2x + 20{,}000 = 26{,}000$$
$$2x = 6000$$
$$x = 3000$$

Therefore, $5000 - x = 2000$.

She invested \$3000 at 6% and \$2000 at 4%. ■

PROBLEM 5

An investor invests a certain amount of money at 3%. Then he finds a better deal and invests \$5000 more than that amount at 5%. His yearly income from the two investments is \$650. How much did he invest at each rate?

Solution

Let x represent the amount invested at 3%. Then $x + 5000$ represents the amount invested at 5%.

$$(3\%)(x) + (5\%)(x + 5000) = 650$$

$$0.03x + 0.05(x + 5000) = 650$$

$$3x + 5(x + 5000) = 65{,}000 \quad \text{Multiplied both sides by 100}$$

$$3x + 5x + 25{,}000 = 65{,}000$$

$$8x + 25{,}000 = 65{,}000$$

$$8x = 40{,}000$$

$$x = 5000$$

Therefore, $x + 5000 = 10{,}000$.

He invested \$5000 at 3% and \$10,000 at 5%. ■

Now let's consider a problem where the process of representing the various unknown quantities in terms of one variable is the key to solving the problem.

PROBLEM 6

Jody is 6 years younger than her sister Cathy, and in 7 years Jody will be three-fourths as old as Cathy. Find their present ages.

Solution

By letting c represent Cathy's present age we can represent all of the unknown quantities as follows:

c: Cathy's present age

$c - 6$: Jody's present age

$c + 7$: Cathy's age in 7 years

$c - 6 + 7$ or $c + 1$: Jody's age in 7 years

The statement that Jody's age in 7 years will be three-fourths of Cathy's age at that time serves as the guideline. So we can set up and solve the following equation.

$$c + 1 = \frac{3}{4}(c + 7)$$

$$4c + 4 = 3(c + 7) \quad \text{Multiplied both sides by 4}$$

$$4c + 4 = 3c + 21$$

$$c = 17$$

Therefore, Cathy's present age is 17, and Jody's present age is $17 - 6 = 11$. ■

43. $0.08t + 0.1(300 - t) > 28$

44. $-4 > 5x - 2 - 3x$

45. $\frac{2}{3}n - 2 \geq \frac{1}{2}n + 1$

46. $-3 < -2(x - 1) - x$

For Problems 47–54, set up an equation or an inequality and solve each problem.

47. Erin's salary this year is $32,000. This represents $2000 more than twice her salary 5 years ago. Find her salary 5 years ago.

48. One of two supplementary angles is 45° smaller than four times the other angle. Find the measure of each angle.

49. Jaamal has 25 coins (nickels and dimes) that amount to $2.10. How many coins of each kind does he have?

50. Hana bowled 144 and 176 in her first two games. What must she bowl in the third game to have an average of at least 150 for the three games?

51. A board 30 feet long is cut into two pieces whose lengths are in the ratio of 2 to 3. Find the lengths of the two pieces.

52. A retailer has some shoes that cost him $32 per pair. He wants to sell them at a profit of 20% of the selling price. What price should he charge for the shoes?

53. Two cars start from the same place traveling in opposite directions. One car travels 5 miles per hour faster than the other car. Find their speeds if after 6 hours they are 570 miles apart.

54. How many liters of pure alcohol must be added to 15 liters of a 20% solution to obtain a 40% solution?

5

Exponents and Polynomials

The average distance between the sun and the earth is approximately 93,000,000 miles. Using scientific notation, 93,000,000 can be written as $(9.3)(10^7)$.

A rectangular dock that measures 12 feet by 16 feet is treated with a uniform strip of nonslip coating along both sides and both ends. How wide is the strip if one-half of the dock is treated? If we let x represent the width of the strip, then we can use the equation $(16 - 2x)(12 - 2x) = \frac{1}{2}(12)(16)$ to determine that the width of the strip is 2 feet.

The equation we used to solve this problem is called a **quadratic equation.** Quadratic equations belong to a larger classification called **polynomial equations.** To solve problems involving polynomial equations, we need to develop some basic skills that pertain to polynomials. That is to say, we need to be able to add, subtract, multiply, divide, and factor polynomials. Chapters 5 and 6 will help you develop those skills as you work through problems that involve quadratic equations.

5.1 Addition and Subtraction of Polynomials

In earlier chapters, we called algebraic expressions such as $4x$, $5y$, $-6ab$, $7x^2$, and $-9xy^2z^3$ "terms." Recall that a term is an indicated product that may contain any number of factors. The variables in a term are called "literal factors," and the numerical factor is called the "numerical coefficient" of the term. Thus, in $-6ab$, a and b are literal factors and the numerical coefficient is -6. Terms that have the same literal factors are called "similar" or "like" terms.

Terms that contain variables with only whole numbers as exponents are called **monomials**. The previously listed terms, $4x$, $5y$, $-6ab$, $7x^2$, and $-9xy^2z^3$, are all monomials. (We will work with some algebraic expressions later, such as $7x^{-1}y^{-1}$ and $4a^{-2}b^{-3}$, which are not monomials.) The **degree of a monomial** is the sum of the exponents of the literal factors. Here are some examples:

$4xy$ is of degree 2.

$5x$ is of degree 1.

$14a^2b$ is of degree 3.

$-17xy^2z^3$ is of degree 6.

$-9y^4$ is of degree 4.

If the monomial contains only one variable, then the exponent of the variable is the degree of the monomial. Any nonzero constant term is said to be of degree zero.

A **polynomial** is a monomial or a finite sum (or difference) of monomials. The **degree of a polynomial** is the degree of the term with the highest degree in the polynomial. Some special classifications of polynomials are made according to the number of terms. We call a one-term polynomial a **monomial**, a two-term polynomial a **binomial**, and a three-term polynomial a **trinomial**. The following examples illustrate some of this terminology:

The polynomial $5x^3y^4$ is a monomial of degree 7.

The polynomial $4x^2y - 3xy$ is a binomial of degree 3.

The polynomial $5x^2 - 6x + 4$ is a trinomial of degree 2.

The polynomial $9x^4 - 7x^3 + 6x^2 + x - 2$ is given no special name but is of degree 4.

Adding Polynomials

In the preceding chapters, you have worked many problems involving the addition and subtraction of polynomials. For example, simplifying $4x^2 + 6x + 7x^2 - 2x$ to $11x^2 + 4x$ by combining similar terms can actually be considered the addition problem $(4x^2 + 6x) + (7x^2 - 2x)$. At this time we will simply review and extend some of those ideas.

EXAMPLE 1

Add $5x^2 + 7x - 2$ and $9x^2 - 12x + 13$.

Solution

We commonly use the horizontal format for such work. Thus,

$$\begin{aligned}(5x^2 + 7x - 2) + (9x^2 - 12x + 13) &= (5x^2 + 9x^2) + (7x - 12x) + (-2 + 13)\\ &= 14x^2 - 5x + 11\end{aligned}$$

■

The commutative, associative, and distributive properties provide the basis for rearranging, regrouping, and combining similar terms.

EXAMPLE 2

Add $5x - 1$, $3x + 4$, and $9x - 7$.

Solution

$$\begin{aligned}(5x - 1) + (3x + 4) + (9x - 7) &= (5x + 3x + 9x) + [-1 + 4 + (-7)]\\ &= 17x - 4\end{aligned}$$

■

EXAMPLE 3

Add $-x^2 + 2x - 1$, $2x^3 - x + 4$, and $-5x + 6$.

Solution

$$\begin{aligned}&(-x^2 + 2x - 1) + (2x^3 - x + 4) + (-5x + 6)\\ &\quad = (2x^3) + (-x^2) + (2x - x - 5x) + (-1 + 4 + 6)\\ &\quad = 2x^3 - x^2 - 4x + 9\end{aligned}$$

■

Subtracting Polynomials

Recall from Chapter 2 that $a - b = a + (-b)$. We define subtraction as *adding the opposite*. This same idea extends to polynomials in general. The opposite of a polynomial is formed by taking the opposite of each term. For example, the opposite of $(2x^2 - 7x + 3)$ is $-2x^2 + 7x - 3$. Symbolically, we express this as

$$-(2x^2 - 7x + 3) = -2x^2 + 7x - 3$$

Now consider some subtraction problems.

EXAMPLE 4

Subtract $2x^2 + 9x - 3$ from $5x^2 - 7x - 1$.

Solution

Use the horizontal format.

$$\begin{aligned}(5x^2 - 7x - 1) - (2x^2 + 9x - 3) &= (5x^2 - 7x - 1) + (-2x^2 - 9x + 3)\\ &= (5x^2 - 2x^2) + (-7x - 9x) + (-1 + 3)\\ &= 3x^2 - 16x + 2\end{aligned}$$

■

EXAMPLE 5

Subtract $-8y^2 - y + 5$ from $2y^2 + 9$.

Solution

$$\begin{aligned}(2y^2 + 9) - (-8y^2 - y + 5) &= (2y^2 + 9) + (8y^2 + y - 5)\\ &= (2y^2 + 8y^2) + (y) + (9 - 5)\\ &= 10y^2 + y + 4\end{aligned}$$

■

Later when dividing polynomials, you will need to use a vertical format to subtract polynomials. Let's consider two such examples.

EXAMPLE 6

Subtract $3x^2 + 5x - 2$ from $9x^2 - 7x - 1$.

Solution

$$\begin{array}{r} 9x^2 - 7x - 1 \\ \underline{3x^2 + 5x - 2} \end{array}$$

Notice which polynomial goes on the bottom and the alignment of similar terms in columns

Now we can mentally form the opposite of the bottom polynomial and add.

$$\begin{array}{r} 9x^2 - 7x - 1 \\ \underline{3x^2 + 5x - 2} \\ 6x^2 - 12x + 1 \end{array}$$

The opposite of $3x^2 + 5x - 2$ is $-3x^2 - 5x + 2$

■

EXAMPLE 7

Subtract $15y^3 + 5y^2 + 3$ from $13y^3 + 7y - 1$.

Solution

$$\begin{array}{rrrr} 13y^3 & & + 7y & - 1 \\ \underline{15y^3} & \underline{+ 5y^2} & \underline{} & \underline{+ 3} \\ -2y^3 & - 5y^2 & + 7y & - 4 \end{array}$$

Similar terms are arranged in columns

We mentally formed the opposite of the bottom polynomial and added

■

We can use the distributive property along with the properties $a = 1(a)$ and $-a = -1(a)$ when adding and subtracting polynomials. The next examples illustrate this approach.

EXAMPLE 8

Perform the indicated operations.

$$(3x - 4) + (2x - 5) - (7x - 1)$$

Solution

$$\begin{aligned}&(3x - 4) + (2x - 5) - (7x - 1)\\ &\quad = 1(3x - 4) + 1(2x - 5) - 1(7x - 1)\end{aligned}$$

$$= 1(3x) - 1(4) + 1(2x) - 1(5) - 1(7x) - 1(-1)$$
$$= 3x - 4 + 2x - 5 - 7x + 1$$
$$= 3x + 2x - 7x - 4 - 5 + 1$$
$$= -2x - 8$$

■

Certainly we can do some of the steps mentally; Example 9 gives a possible format.

EXAMPLE 9

Perform the indicated operations.

$$(-y^2 + 5y - 2) - (-2y^2 + 8y + 6) + (4y^2 - 2y - 5)$$

Solution

$$(-y^2 + 5y - 2) - (-2y^2 + 8y + 6) + (4y^2 - 2y - 5)$$
$$= -y^2 + 5y - 2 + 2y^2 - 8y - 6 + 4y^2 - 2y - 5$$
$$= -y^2 + 2y^2 + 4y^2 + 5y - 8y - 2y - 2 - 6 - 5$$
$$= 5y^2 - 5y - 13$$

■

When we use the horizontal format, as in Examples 8 and 9, we use parentheses to indicate a quantity. In Example 8 the quantities $(3x - 4)$ and $(2x - 5)$ are to be added; from this result we are to subtract the quantity $(7x - 1)$. Brackets, [ ], are also sometimes used as grouping symbols, especially if there is a need to indicate quantities within quantities. To remove the grouping symbols, perform the indicated operations, starting with the innermost set of symbols. Let's consider two examples of this type.

EXAMPLE 10

Perform the indicated operations.

$$3x - [2x + (3x - 1)]$$

Solution

First we need to add the quantities $2x$ and $(3x - 1)$.

$$3x - [2x + (3x - 1)] = 3x - [2x + 3x - 1]$$
$$= 3x - [5x - 1]$$

Now we need to subtract the quantity $[5x - 1]$ from $3x$.

$$3x - [5x - 1] = 3x - 5x + 1$$
$$= -2x + 1$$

■

EXAMPLE 11

Perform the indicated operations.

$$8 - \{7x - [2 + (x - 1)] + 4x\}$$

Solution

Start with the innermost set of grouping symbols (the parentheses) and proceed as follows:

$$\begin{aligned} 8 - \{7x - [2 + (x - 1)] + 4x\} &= 8 - \{7x - [x + 1] + 4x\} \\ &= 8 - (7x - x - 1 + 4x) \\ &= 8 - (10x - 1) \\ &= 8 - 10x + 1 \\ &= -10x + 9 \end{aligned}$$

■

For a final example in this section, we look at polynomials in a geometric setting.

EXAMPLE 12

Suppose that a parallelogram and a rectangle have dimensions as indicated in Figure 5.1. Find a polynomial that represents the sum of the areas of the two figures.

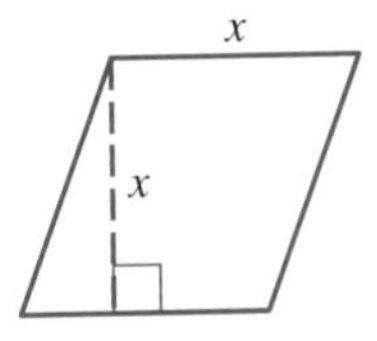

Figure 5.1

Solution

Using the area formulas $A = bh$ and $A = lw$ for parallelograms and rectangles, respectively, we can represent the sum of the areas of the two figures as follows:

Area of the parallelogram	$x(x) = x^2$
Area of the rectangle	$20(x) = 20x$

We can represent the total area by $x^2 + 20x$. ■

Problem Set 5.1

For Problems 1–8, determine the degree of each polynomial.

1. $7x^2y + 6xy$

2. $4xy - 7x$

3. $5x^2 - 9$

4. $8x^2y^2 - 2xy^2 - x$

5. $5x^3 - x^2 - x + 3$

6. $8x^4 - 2x^2 + 6$

7. $5xy$

8. $-7x + 4$

For Problems 9–22, add the polynomials.

9. $3x + 4$ and $5x + 7$

10. $3x - 5$ and $2x - 9$

11. $-5y - 3$ and $9y + 13$

12. $x^2 - 2x - 1$ and $-2x^2 + x + 4$

13. $-2x^2 + 7x - 9$ and $4x^2 - 9x - 14$

14. $3a^2 + 4a - 7$ and $-3a^2 - 7a + 10$

15. $5x - 2, 3x - 7$, and $9x - 10$

16. $-x - 4, 8x + 9$, and $-7x - 6$

17. $2x^2 - x + 4, -5x^2 - 7x - 2$, and $9x^2 + 3x - 6$

18. $-3x^2 + 2x - 6, 6x^2 + 7x + 3$, and $-4x^2 - 9$

19. $-4n^2 - n - 1$ and $4n^2 + 6n - 5$

20. $-5n^2 + 7n - 9$ and $-5n - 4$

21. $2x^2 - 7x - 10, -6x - 2$, and $-9x^2 + 5$

22. $7x - 11, -x^2 - 5x + 9$, and $-4x + 5$

For Problems 23–34, subtract the polynomials using a horizontal format.

23. $7x + 1$ from $12x + 6$

24. $10x + 3$ from $14x + 13$

25. $5x - 2$ from $3x - 7$

26. $7x - 2$ from $2x + 3$

27. $-x - 1$ from $-4x + 6$

28. $-3x + 2$ from $-x - 9$

29. $x^2 - 7x + 2$ from $3x^2 + 8x - 4$

30. $2x^2 + 6x - 1$ from $8x^2 - 2x + 6$

31. $-2n^2 - 3n + 4$ from $3n^2 - n + 7$

32. $3n^2 - 7n - 9$ from $-4n^2 + 6n + 10$

33. $-4x^3 - x^2 + 6x - 1$ from $-7x^3 + x^2 + 6x - 12$

34. $-4x^2 + 6x - 2$ from $-3x^3 + 2x^2 + 7x - 1$

For Problems 35–44, subtract the polynomials using a vertical format.

35. $3x - 2$ from $12x - 4$

36. $-4x + 6$ from $7x - 3$

37. $-5a - 6$ from $-3a + 9$

38. $7a - 11$ from $-2a - 1$

39. $8x^2 - x + 6$ from $6x^2 - x + 11$

40. $3x^2 - 2$ from $-2x^2 + 6x - 4$

41. $-2x^3 - 6x^2 + 7x - 9$ from $4x^3 + 6x^2 + 7x - 14$

42. $4x^3 + x - 10$ from $3x^2 - 6$

43. $2x^2 - 6x - 14$ from $4x^3 - 6x^2 + 7x - 2$

44. $3x - 7$ from $7x^3 + 6x^2 - 5x - 4$

For Problems 45–64, perform the indicated operations.

45. $(5x + 3) - (7x - 2) + (3x + 6)$

46. $(3x - 4) + (9x - 1) - (14x - 7)$

47. $(-x - 1) - (-2x + 6) + (-4x - 7)$

48. $(-3x + 6) + (-x - 8) - (-7x + 10)$

49. $(x^2 - 7x - 4) + (2x^2 - 8x - 9) - (4x^2 - 2x - 1)$

50. $(3x^2 + x - 6) - (8x^2 - 9x + 1) - (7x^2 + 2x - 6)$

51. $(-x^2 - 3x + 4) + (-2x^2 - x - 2)$
$- (-4x^2 + 7x + 10)$

52. $(-3x^2 - 2) + (7x^2 - 8) - (9x^2 - 2x - 4)$

53. $(3a - 2b) - (7a + 4b) - (6a - 3b)$

54. $(5a + 7b) + (-8a - 2b) - (5a + 6b)$

55. $(n - 6) - (2n^2 - n + 4) + (n^2 - 7)$

56. $(3n + 4) - (n^2 - 9n + 10) - (-2n + 4)$

57. $7x + [3x - (2x - 1)]$

58. $-6x + [-2x - (5x + 2)]$

59. $-7n - [4n - (6n - 1)]$

60. $9n - [3n - (5n + 4)]$

61. $(5a - 1) - [3a + (4a - 7)]$

62. $(-3a + 4) - [-7a + (9a - 1)]$

63. $13x - \{5x - [4x - (x - 6)]\}$

64. $-10x - \{7x - [3x - (2x - 3)]\}$

65. Subtract $5x - 3$ from the sum of $4x - 2$ and $7x + 6$.

66. Subtract $7x + 5$ from the sum of $9x - 4$ and $-3x - 2$.

67. Subtract the sum of $-2n - 5$ and $-n + 7$ from $-8n + 9$.

68. Subtract the sum of $7n - 11$ and $-4n - 3$ from $13n - 4$.

69. Find a polynomial that represents the perimeter of the rectangle in Figure 5.2.

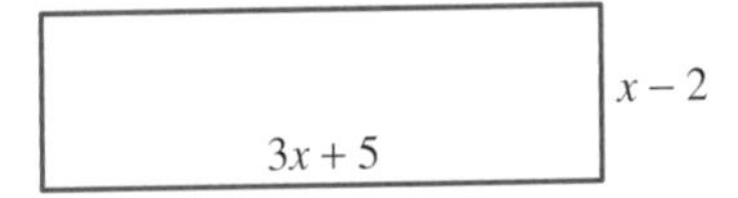

Figure 5.2

70. Find a polynomial that represents the area of the shaded region in Figure 5.3. The length of a radius of the larger circle is r units, and the length of a radius of the smaller circle is 4 units.

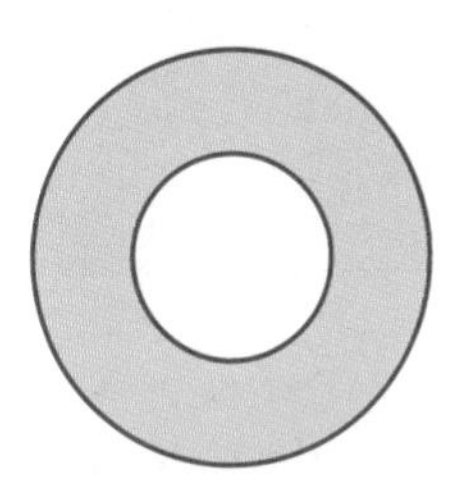

Figure 5.3

71. Find a polynomial that represents the sum of the areas of the rectangles and squares in Figure 5.4.

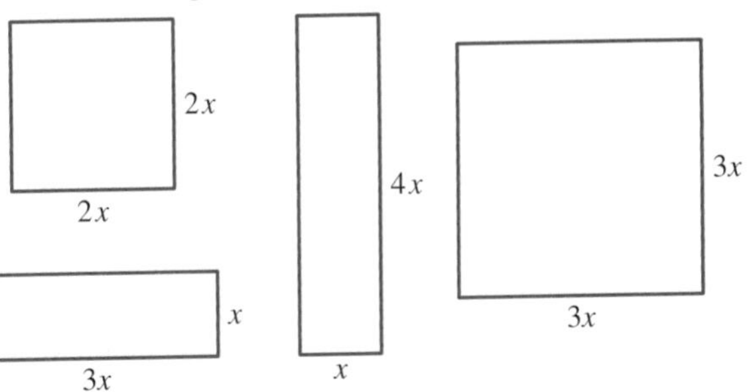

Figure 5.4

72. Find a polynomial that represents the total surface area of the rectangular solid in Figure 5.5.

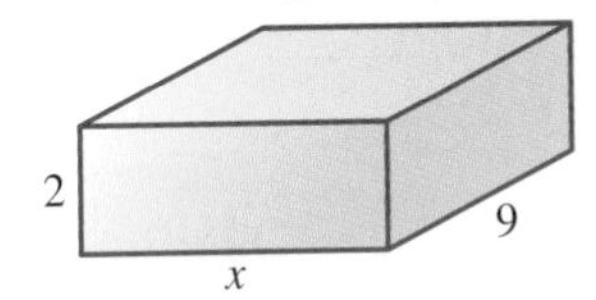

Figure 5.5

THOUGHTS INTO WORDS

73. Explain how to subtract the polynomial $3x^2 + 6x - 2$ from $4x^2 + 7$.

74. Is the sum of two binomials always another binomial? Defend your answer.

75. Is the sum of two binomials ever a trinomial? Defend your answer.

5.2 Multiplying Monomials

In Section 2.4, we used exponents and some of the basic properties of real numbers to simplify algebraic expressions into a more compact form; for example,

$$(3x)(4xy) = 3 \cdot 4 \cdot x \cdot x \cdot y = 12x^2y$$

Actually we were **multiplying monomials**, and it is this topic that we will pursue now. We can make multiplying monomials easier by using some basic properties of exponents. These properties are the direct result of the definition of an exponent. The following examples lead to the first property:

$$x^2 \cdot x^3 = (x \cdot x)(x \cdot x \cdot x) = x^5$$

$$a^3 \cdot a^4 = (a \cdot a \cdot a)(a \cdot a \cdot a \cdot a) = a^7$$

$$b \cdot b^2 = (b)(b \cdot b) = b^3$$

In general,

$$b^n \cdot b^m = \underbrace{(b \cdot b \cdot b \cdot \cdots \cdot b)}_{n \text{ factors of } b}\underbrace{(b \cdot b \cdot b \cdot \cdots \cdot b)}_{m \text{ factors of } b}$$

$$= \underbrace{b \cdot b \cdot b \cdot \cdots \cdot b}_{(n + m) \text{ factors of } b}$$

$$= b^{n+m}$$

Property 5.1

If b is any real number, and n and m are positive integers, then

$$b^n \cdot b^m = b^{n+m}$$

Property 5.1 states that when multiplying powers with the same base, add exponents.

EXAMPLE 1

Multiply.

(a) $x^4 \cdot x^3$ **(b)** $a^8 \cdot a^7$

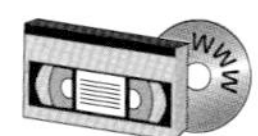

Solution

(a) $x^4 \cdot x^3 = x^{4+3} = x^7$ **(b)** $a^8 \cdot a^7 = a^{8+7} = a^{15}$ ■

Another property of exponents is demonstrated by these examples.

$$(x^2)^3 = x^2 \cdot x^2 \cdot x^2 = x^{2+2+2} = x^6$$

$$(a^3)^2 = a^3 \cdot a^3 = a^{3+3} = a^6$$

$$(b^3)^4 = b^3 \cdot b^3 \cdot b^3 \cdot b^3 = b^{3+3+3+3} = b^{12}$$

In general,

$$(b^n)^m = \underbrace{b^n \cdot b^n \cdot b^n \cdot \cdots \cdot b^n}_{m \text{ factors of } b^n}$$

$$= b^{\overbrace{n+n+n+\cdots+n}^{m \text{ of these } ns}}$$

$$= b^{mn}$$

Property 5.2

If b is any real number, and m and n are positive integers, then

$$(b^n)^m = b^{mn}$$

Property 5.2 states that when raising a power to a power, multiply exponents.

EXAMPLE 2

Raise each to the indicated power.

(a) $(x^4)^3$ **(b)** $(a^5)^6$

Solution

(a) $(x^4)^3 = x^{3 \cdot 4} = x^{12}$ **(b)** $(a^5)^6 = a^{6 \cdot 5} = a^{30}$ ■

The third property of exponents we will use in this section raises a monomial to a power.

$$(2x)^3 = (2x)(2x)(2x) = 2 \cdot 2 \cdot 2 \cdot x \cdot x \cdot x = 2^3 \cdot x^3$$

$$(3a^4)^2 = (3a^4)(3a^4) = 3 \cdot 3 \cdot a^4 \cdot a^4 = (3)^2(a^4)^2$$

$$(-2xy^5)^2 = (-2xy^5)(-2xy^5) = (-2)(-2)(x)(x)(y^5)(y^5) = (-2)^2(x)^2(y^5)^2$$

In general,

$$(ab)^n = \underbrace{ab \cdot ab \cdot ab \cdot \cdots \cdot ab}_{n \text{ factors of } ab}$$

$$= \underbrace{(a \cdot a \cdot a \cdot \cdots \cdot a)}_{n \text{ factors of } a}\underbrace{(b \cdot b \cdot b \cdot \cdots \cdot b)}_{n \text{ factors of } b}$$

$$= a^n b^n$$ ■

Property 5.3

If a and b are real numbers, and n is a positive integer, then

$$(ab)^n = a^n b^n$$

Property 5.3 states that when raising a monomial to a power, raise each factor to that power.

EXAMPLE 3

Raise each to the indicated power.

(a) $(2x^2y^3)^4$ **(b)** $(-3ab^5)^3$

Solution

(a) $(2x^2y^3)^4 = (2)^4(x^2)^4(y^3)^4 = 16x^8y^{12}$

(b) $(-3ab^5)^3 = (-3)^3(a^1)^3(b^5)^3 = -27a^3b^{15}$ ■

Consider the following examples in which we use the properties of exponents to help simplify the process of multiplying monomials.

1. $(3x^3)(5x^4) = 3 \cdot 5 \cdot x^3 \cdot x^4$

$= 15x^7$ $\quad x^3 \cdot x^4 = x^{3+4} = x^7$

2. $(-4a^2b^3)(6ab^2) = -4 \cdot 6 \cdot a^2 \cdot a \cdot b^3 \cdot b^2$

$= -24a^3b^5$

3. $(xy)(7xy^5) = 1 \cdot 7 \cdot x \cdot x \cdot y \cdot y^5$ $\quad$ The numerical coefficient of xy is 1

$= 7x^2y^6$

4. $\left(\frac{3}{4}x^2y^3\right)\left(\frac{1}{2}x^3y^5\right) = \frac{3}{4} \cdot \frac{1}{2} \cdot x^2 \cdot x^3 \cdot y^3 \cdot y^5$

$= \frac{3}{8}x^5y^8$

It is a simple process to raise a monomial to a power when using the properties of exponents. Study the next examples.

5. $(2x^3)^4 = (2)^4(x^3)^4$ by using $(ab)^n = a^nb^n$

$= (2)^4(x^{12})$ by using $(b^n)^m = b^{mn}$

$= 16x^{12}$

6. $(-2a^4)^5 = (-2)^5(a^4)^5$

$= -32a^{20}$

7. $\left(\frac{2}{5}x^2y^3\right)^3 = \left(\frac{2}{5}\right)^3(x^2)^3(y^3)^3$

$= \frac{8}{125}x^6y^9$

8. $(0.2a^6b^7)^2 = (0.2)^2(a^6)^2(b^7)^2$

$= 0.04a^{12}b^{14}$

Sometimes problems involve first raising monomials to a power and then multiplying the resulting monomials, as in the following examples.

9. $(3x^2)^3(2x^3)^2 = (3)^3(x^2)^3(2)^2(x^3)^2$
$= (27)(x^6)(4)(x^6)$
$= 108x^{12}$

10. $(-x^2y^3)^5(-2x^2y)^2 = (-1)^5(x^2)^5(y^3)^5(-2)^2(x^2)^2(y)^2$
$= (-1)(x^{10})(y^{15})(4)(x^4)(y^2)$
$= -4x^{14}y^{17}$

The distributive property along with the properties of exponents form a basis for finding the product of a monomial and a polynomial. The next examples illustrate these ideas.

11. $(3x)(2x^2 + 6x + 1) = (3x)(2x^2) + (3x)(6x) + (3x)(1)$
$= 6x^3 + 18x^2 + 3x$

12. $(5a^2)(a^3 - 2a^2 - 1) = (5a^2)(a^3) - (5a^2)(2a^2) - (5a^2)(1)$
$= 5a^5 - 10a^4 - 5a^2$

13. $(-2xy)(6x^2y - 3xy^2 - 4y^3)$
$= (-2xy)(6x^2y) - (-2xy)(3xy^2) - (-2xy)(4y^3)$
$= -12x^3y^2 + 6x^2y^3 + 8xy^4$

Once you feel comfortable with this process, you may want to perform most of the work mentally and then simply write down the final result. See whether you understand the following examples.

14. $3x(2x + 3) = 6x^2 + 9x$

15. $-4x(2x^2 - 3x - 1) = -8x^3 + 12x^2 + 4x$

16. $ab(3a^2b - 2ab^2 - b^3) = 3a^3b^2 - 2a^2b^3 - ab^4$

We conclude this section by making a connection between algebra and geometry.

EXAMPLE 4

Suppose that the dimensions of a rectangular solid are represented by x, $2x$, and $3x$ as shown in Figure 5.6. Express the volume and total surface area of the figure.

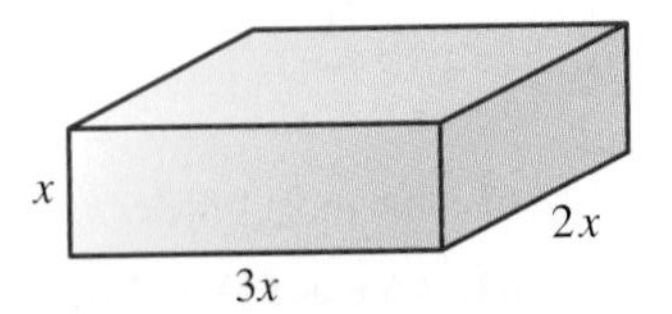

Figure 5.6

Solution

Using the formula $V = lwh$, we can express the volume of the rectangular solid as $(2x)(3x)(x)$, which equals $6x^3$. The total surface area can be described as follows:

Area of front and back rectangles: $2(x)(3x) = 6x^2$

Area of left side and right side: $2(2x)(x) = 4x^2$

Area of top and bottom: $2(2x)(3x) = 12x^2$

We can represent the total surface area by $6x^2 + 4x^2 + 12x^2$ or $22x^2$. ■

Problem Set 5.2

For Problems 1–30, multiply using the properties of exponents to help with the manipulation.

1. $(5x)(9x)$

2. $(7x)(8x)$

3. $(3x^2)(7x)$

4. $(9x)(4x^3)$

5. $(-3xy)(2xy)$

6. $(6xy)(-3xy)$

7. $(-2x^2y)(-7x)$

8. $(-5xy^2)(-4y)$

9. $(4a^2b^2)(-12ab)$

10. $(-3a^3b)(13ab^2)$

11. $(-xy)(-5x^3)$

12. $(-7y^2)(-x^2y)$

13. $(8ab^2c)(13a^2c)$

14. $(9abc^3)(14bc^2)$

15. $(5x^2)(2x)(3x^3)$

16. $(4x)(2x^2)(6x^4)$

17. $(4xy)(-2x)(7y^2)$

18. $(5y^2)(-3xy)(5x^2)$

19. $(-2ab)(-ab)(-3b)$

20. $(-7ab)(-4a)(-ab)$

21. $(6cd)(-3c^2d)(-4d)$

22. $(2c^3d)(-6d^3)(-5cd)$

23. $\left(\frac{2}{3}xy\right)\left(\frac{3}{5}x^2y^4\right)$

24. $\left(-\frac{5}{6}x\right)\left(\frac{8}{3}x^2y\right)$

25. $\left(-\frac{7}{12}a^2b\right)\left(\frac{8}{21}b^4\right)$

26. $\left(-\frac{9}{5}a^3b^4\right)\left(-\frac{15}{6}ab^2\right)$

27. $(0.4x^5)(0.7x^3)$

28. $(-1.2x^4)(0.3x^2)$

29. $(-4ab)(1.6a^3b)$

30. $(-6a^2b)(-1.4a^2b^4)$

For Problems 31–46, raise each monomial to the indicated power. Use the properties of exponents to help with the manipulation.

31. $(2x^4)^2$

32. $(3x^3)^2$

33. $(-3a^2b^3)^2$

34. $(-8a^4b^5)^2$

35. $(3x^2)^3$

36. $(2x^4)^3$

37. $(-4x^4)^3$

38. $(-3x^3)^3$

39. $(9x^4y^5)^2$

40. $(8x^6y^4)^2$

41. $(2x^2y)^4$

42. $(2x^2y^3)^5$

43. $(-3a^3b^2)^4$

44. $(-2a^4b^2)^4$

45. $(-x^2y)^6$

46. $(-x^2y^3)^7$

For Problems 47–60, multiply by using the distributive property.

47. $5x(3x + 2)$

48. $7x(2x + 5)$

49. $3x^2(6x - 2)$

50. $4x^2(7x - 2)$

51. $-4x(7x^2 - 4)$

52. $-6x(9x^2 - 5)$

53. $2x(x^2 - 4x + 6)$

54. $3x(2x^2 - x + 5)$

55. $-6a(3a^2 - 5a - 7)$

56. $-8a(4a^2 - 9a - 6)$

57. $7xy(4x^2 - x + 5)$

58. $5x^2y(3x^2 + 7x - 9)$

59. $-xy(9x^2 - 2x - 6)$

60. $xy^2(6x^2 - x - 1)$

For Problems 61–70, remove the parentheses by multiplying and then simplify by combining similar terms; for example,

$$\begin{aligned}3(x - y) + 2(x - 3y) &= 3x - 3y + 2x - 6y \\ &= 5x - 9y\end{aligned}$$

61. $5(x + 2y) + 4(2x + 3y)$

62. $3(2x + 5y) + 2(4x + y)$

63. $4(x - 3y) - 3(2x - y)$

64. $2(5x - 3y) - 5(x + 4y)$

65. $2x(x^2 - 3x - 4) + x(2x^2 + 3x - 6)$

66. $3x(2x^2 - x + 5) - 2x(x^2 + 4x + 7)$

67. $3[2x - (x - 2)] - 4(x - 2)$

68. $2[3x - (2x + 1)] - 2(3x - 4)$

69. $-4(3x + 2) - 5[2x - (3x + 4)]$

70. $-5(2x - 1) - 3[x - (4x - 3)]$

For Problems 71–80, perform the indicated operations and simplify.

71. $(3x)^2(2x^3)$

72. $(-2x)^3(4x^5)$

73. $(-3x)^3(-4x)^2$

74. $(3xy)^2(2x^2y)^4$

75. $(5x^2y)^2(xy^2)^3$

76. $(-x^2y)^3(6xy)^2$

77. $(-a^2bc^3)^3(a^3b)^2$

78. $(ab^2c^3)^4(-a^2b)^3$

79. $(-2x^2y^2)^4(-xy^3)^3$

80. $(-3xy)^3(-x^2y^3)^4$

81. Express in simplified form the sum of the areas of the two rectangles shown in Figure 5.7.

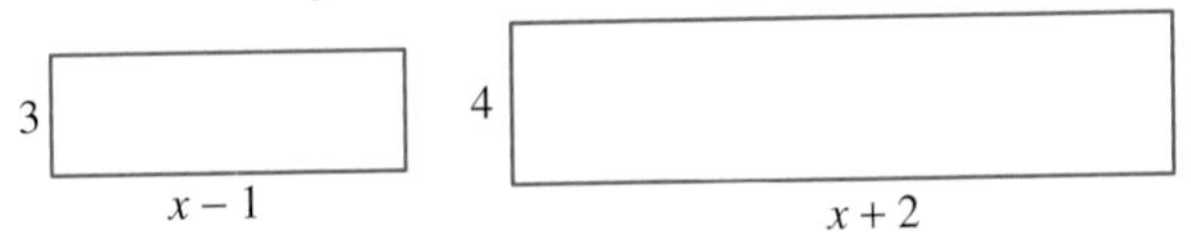

Figure 5.7

82. Express in simplified form the volume and the total surface area of the rectangular solid in Figure 5.8.

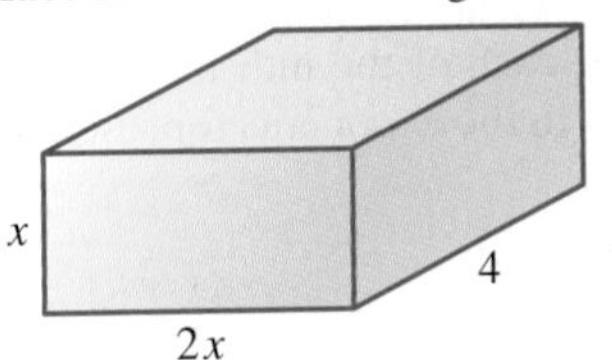

Figure 5.8

83. Represent the area of the shaded region in Figure 5.9. The length of a radius of the smaller circle is x, and the length of a radius of the larger circle is $2x$.

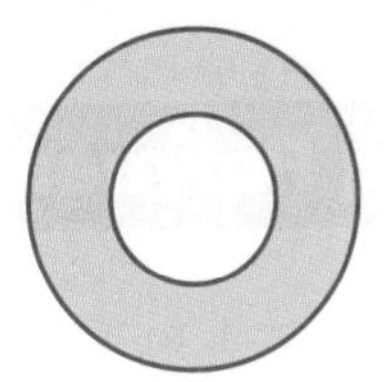

Figure 5.9

84. Represent the area of the shaded region in Figure 5.10.

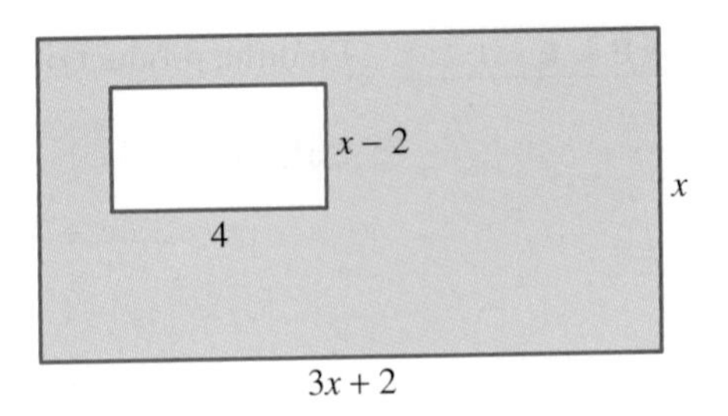

Figure 5.10

THOUGHTS INTO WORDS

85. How would you explain to someone why the product of x^3 and x^4 is x^7 and not x^{12}?

86. Suppose your friend was absent from class the day that this section was discussed. How would you help her understand why the property $(b^n)^m = b^{mn}$ is true?

87. How can Figure 5.11 be used to geometrically demonstrate that $x(x + 2) = x^2 + 2x$?

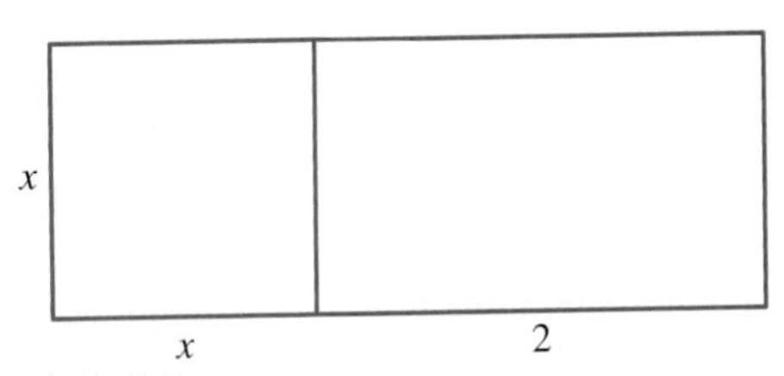

Figure 5.11

FURTHER INVESTIGATIONS

For Problems 88–97, find each of the indicated products. Assume that the variables in the exponents represent positive integers; for example,

$$(x^{2n})(x^{4n}) = x^{2n+4n} = x^{6n}$$

88. $(x^{n})(x^{3n})$

89. $(x^{2n})(x^{5n})$

90. $(x^{2n-1})(x^{3n+2})$

91. $(x^{5n+2})(x^{n-1})$

92. $(x^{3})(x^{4n-5})$

93. $(x^{6n-1})(x^{4})$

94. $(2x^{n})(3x^{2n})$

95. $(4x^{3n})(-5x^{7n})$

96. $(-6x^{2n+4})(5x^{3n-4})$

97. $(-3x^{5n-2})(-4x^{2n+2})$

5.3 Multiplying Polynomials

In general, to go from multiplying a monomial times a polynomial to multiplying two polynomials requires the use of the distributive property. Consider some examples.

EXAMPLE 1

Find the product of $(x + 3)$ and $(y + 4)$.

Solution

$$\begin{aligned}(x + 3)(y + 4) &= x(y + 4) + 3(y + 4)\\ &= x(y) + x(4) + 3(y) + 3(4)\\ &= xy + 4x + 3y + 12\end{aligned}$$

■

Notice that each term of the first polynomial is multiplied times each term of the second polynomial.

EXAMPLE 2

Find the product of $(x - 2)$ and $(y + z + 5)$.

Solution

$$\begin{aligned}(x - 2)(y + z + 5) &= x(y + z + 5) - 2(y + z + 5)\\ &= x(y) + x(z) + x(5) - 2(y) - 2(z) - 2(5)\\ &= xy + xz + 5x - 2y - 2z - 10\end{aligned}$$

■

Usually, multiplying polynomials will produce similar terms that can be combined to simplify the resulting polynomial.

EXAMPLE 3

Multiply $(x + 3)(x + 2)$.

Solution

$$\begin{aligned}(x + 3)(x + 2) &= x(x + 2) + 3(x + 2)\\ &= x^2 + 2x + 3x + 6\\ &= x^2 + 5x + 6 \qquad \text{Combine like terms}\end{aligned}$$

■

EXAMPLE 4

Multiply $(x - 4)(x + 9)$.

Solution

$$\begin{aligned}(x - 4)(x + 9) &= x(x + 9) - 4(x + 9)\\ &= x^2 + 9x - 4x - 36\\ &= x^2 + 5x - 36 \qquad \text{Combine like terms}\end{aligned}$$

■

EXAMPLE 5

Multiply $(x + 4)(x^2 + 3x + 2)$.

Solution

$$\begin{aligned}(x + 4)(x^2 + 3x + 2) &= x(x^2 + 3x + 2) + 4(x^2 + 3x + 2)\\ &= x^3 + 3x^2 + 2x + 4x^2 + 12x + 8\\ &= x^3 + 7x^2 + 14x + 8\end{aligned}$$

■

EXAMPLE 6

Multiply $(2x - y)(3x^2 - 2xy + 4y^2)$.

Solution

$$\begin{aligned}(2x - y)(3x^2 - 2xy + 4y^2) &= 2x(3x^2 - 2xy + 4y^2) - y(3x^2 - 2xy + 4y^2)\\ &= 6x^3 - 4x^2y + 8xy^2 - 3x^2y + 2xy^2 - 4y^3\\ &= 6x^3 - 7x^2y + 10xy^2 - 4y^3\end{aligned}$$

■

Perhaps the most frequently used type of multiplication problem is the product of two binomials. It will be a big help later if you can become proficient at multiplying binomials without showing all of the intermediate steps. This is quite easy to do if you use a three-step shortcut pattern demonstrated by the following examples.

EXAMPLE 7 Multiply $(x + 5)(x + 7)$.

Solution

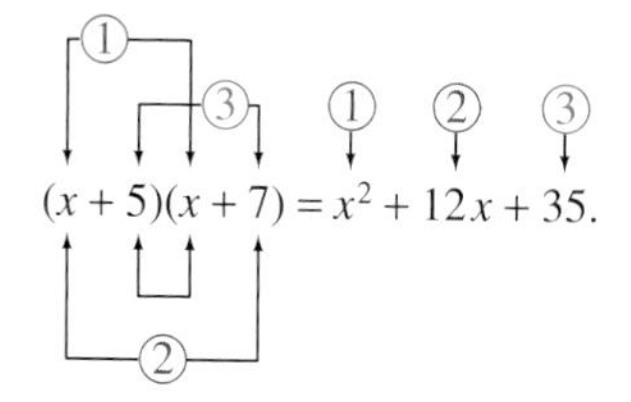

Figure 5.12

Step 1 Multiply $x \cdot x$.

Step 2 Multiply $5 \cdot x$ and $7 \cdot x$ and combine them.

Step 3 Multiply $5 \cdot 7$.

EXAMPLE 8 Multiply $(x - 8)(x + 3)$.

Solution

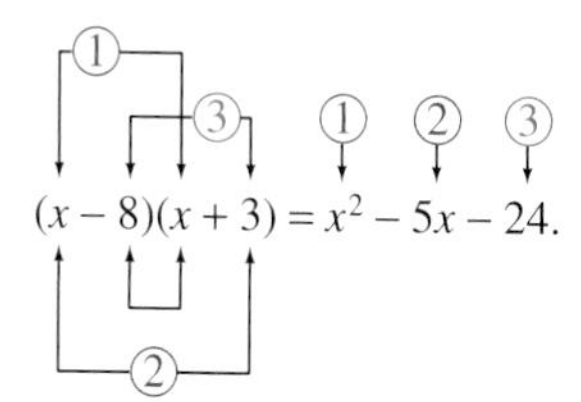

Figure 5.13

EXAMPLE 9 Multiply $(3x + 2)(2x - 5)$.

Solution

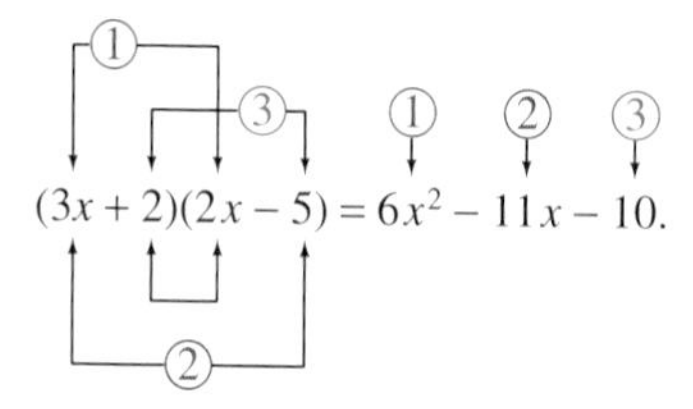

Figure 5.14

The mnemonic device FOIL is often used to remember the pattern for multiplying binomials. The letters in FOIL represent First, Outside, Inside, and Last. If you look back at the Examples 7 through 9, step 1 is to find the product of the first terms in each binomial; step 2 is to find the product of the outside terms and the inside terms; and step 3 is to find the product of the last terms in each binomial.

Now see whether you can use the pattern to find these products:

$$(x + 3)(x + 7)$$

$$(3x + 1)(2x + 5)$$

$$(x - 2)(x - 3)$$

$$(4x + 5)(x - 2)$$

Your answers should be $x^2 + 10x + 21$, $6x^2 + 17x + 5$, $x^2 - 5x + 6$, and $4x^2 - 3x - 10$.

Keep in mind that the shortcut pattern applies only to finding the product of two binomials. For other situations, such as finding the product of a binomial and a trinomial, we suggest showing the intermediate steps as follows:

$$\begin{aligned}(x + 3)(x^2 + 6x - 7) &= x(x^2) + x(6x) - x(7) + 3(x^2) + 3(6x) - 3(7)\\ &= x^3 + 6x^2 - 7x + 3x^2 + 18x - 21\\ &= x^3 + 9x^2 + 11x - 21\end{aligned}$$

Perhaps you could omit the first step, and shorten the form as follows:

$$\begin{aligned}(x - 4)(x^2 - 5x - 6) &= x^3 - 5x^2 - 6x - 4x^2 + 20x + 24\\ &= x^3 - 9x^2 + 14x + 24\end{aligned}$$

Remember that you are multiplying each term of the first polynomial times each term of the second polynomial and combining similar terms.

Exponents are also used to indicate repeated multiplication of polynomials. For example, we can write $(x + 4)(x + 4)$ as $(x + 4)^2$. Thus to square a binomial we simply write it as the product of two equal binomials and apply the shortcut pattern.

$$(x + 4)^2 = (x + 4)(x + 4) = x^2 + 8x + 16$$

$$(x - 5)^2 = (x - 5)(x - 5) = x^2 - 10x + 25$$

$$(2x + 3)^2 = (2x + 3)(2x + 3) = 4x^2 + 12x + 9$$

When you square binomials, be careful not to forget the middle term. That is to say, $(x + 3)^2 \neq x^2 + 3^2$; instead, $(x + 3)^2 = (x + 3)(x + 3) = x^2 + 6x + 9$.

The next example suggests a format to use when cubing a binomial.

$$\begin{aligned}(x + 4)^3 &= (x + 4)(x + 4)(x + 4)\\ &= (x + 4)(x^2 + 8x + 16)\\ &= x(x^2 + 8x + 16) + 4(x^2 + 8x + 16)\\ &= x^3 + 8x^2 + 16x + 4x^2 + 32x + 64\\ &= x^3 + 12x^2 + 48x + 64\end{aligned}$$

Special Product Patterns

When we multiply binomials, some special patterns occur that you should recognize. We can use these patterns to find products and later to factor polynomials. We will state each of the patterns in general terms followed by examples to illustrate the use of each pattern.

PATTERN

$$(a + b)^2 = (a + b)(a + b) = a^2 + 2ab + b^2$$

Square of the first term of the binomial + Twice the product of the two terms of the binomial + Square of the second term of the binomial

Examples

$$(x + 4)^2 = x^2 + 8x + 16$$
$$(2x + 3y)^2 = 4x^2 + 12xy + 9y^2$$
$$(5a + 7b)^2 = 25a^2 + 70ab + 49b^2$$

PATTERN

$$(a - b)^2 = (a - b)(a - b) = a^2 - 2ab + b^2$$

Square of the first term of the binomial − Twice the product of the two terms of the binomial + Square of the second term of the binomial

Examples

$$(x - 8)^2 = x^2 - 16x + 64$$
$$(3x - 4y)^2 = 9x^2 - 24xy + 16y^2$$
$$(4a - 9b)^2 = 16a^2 - 72ab + 81b^2$$

PATTERN

$$(a + b)(a - b) = a^2 - b^2$$

Square of the first term of the binomial − Square of the second term of the binomial

Examples

$$(x + 7)(x - 7) = x^2 - 49$$
$$(2x + y)(2x - y) = 4x^2 - y^2$$
$$(3a - 2b)(3a + 2b) = 9a^2 - 4b^2$$

As you might expect, there are geometric interpretations for many of the algebraic concepts presented in this section. We will give you the opportunity to make some of these connections between algebra and geometry in the next problem set. We conclude this section with a problem that allows us to use some algebra and geometry.

EXAMPLE 10

A rectangular piece of tin is 16 inches long and 12 inches wide as shown in Figure 5.15. From each corner a square piece x inches on a side is cut out. The flaps are then turned up to form an open box. Find polynomials that represent the volume and the exterior surface area of the box.

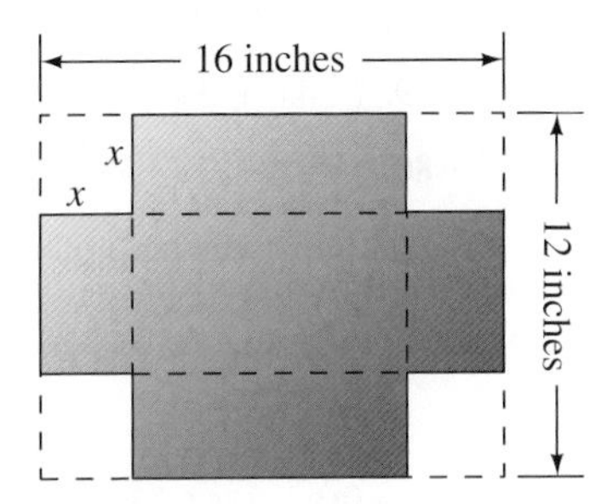

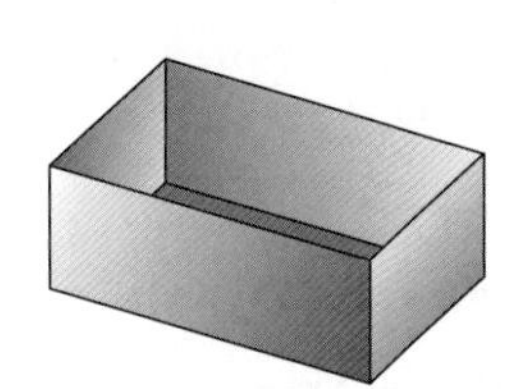

Figure 5.15

Solution

The length of the box is $16 - 2x$, the width is $12 - 2x$, and the height is x. From the volume formula $V = lwh$, the polynomial $(16 - 2x)(12 - 2x)(x)$, which simplifies to $4x^3 - 56x^2 + 192x$, represents the volume.

The outside surface area of the box is the area of the original piece of tin minus the four corners that were cut off. Therefore, the polynomial $16(12) - 4x^2$ or $192 - 4x^2$ represents the outside surface area of the box. ■

Problem Set 5.3

For Problems 1–10, find the indicated products by applying the distributive property; for example,

$$(x + 1)(y + 5) = x(y) + x(5) + 1(y) + 1(5)$$
$$= xy + 5x + y + 5$$

1. $(x + 2)(y + 3)$

2. $(x + 3)(y + 6)$

3. $(x - 4)(y + 1)$

4. $(x - 5)(y + 7)$

5. $(x - 5)(y - 6)$

6. $(x - 7)(y - 9)$

7. $(x + 2)(y + z + 1)$

8. $(x + 4)(y - z + 4)$

9. $(2x + 3)(3y + 1)$

10. $(3x - 2)(2y - 5)$

For Problems 11–36, find the indicated products by applying the distributive property and combining similar terms. Use the following format to show your work:

$$(x + 3)(x + 8) = x(x) + x(8) + 3(x) + 3(8)$$
$$= x^2 + 8x + 3x + 24$$
$$= x^2 + 11x + 24$$

11. $(x + 3)(x + 7)$

12. $(x + 4)(x + 2)$

13. $(x + 8)(x - 3)$

14. $(x + 9)(x - 6)$

15. $(x - 7)(x + 1)$

16. $(x - 10)(x + 8)$

17. $(n - 4)(n - 6)$ **18.** $(n - 3)(n - 7)$

19. $(3n + 1)(n + 6)$ **20.** $(4n + 3)(n + 6)$

21. $(5x - 2)(3x + 7)$ **22.** $(3x - 4)(7x + 1)$

23. $(x + 3)(x^2 + 4x + 9)$

24. $(x + 2)(x^2 + 6x + 2)$

25. $(x + 4)(x^2 - x - 6)$

26. $(x + 5)(x^2 - 2x - 7)$

27. $(x - 5)(2x^2 + 3x - 7)$

28. $(x - 4)(3x^2 + 4x - 6)$

29. $(2a - 1)(4a^2 - 5a + 9)$

30. $(3a - 2)(2a^2 - 3a - 5)$

31. $(3a + 5)(a^2 - a - 1)$

32. $(5a + 2)(a^2 + a - 3)$

33. $(x^2 + 2x + 3)(x^2 + 5x + 4)$

34. $(x^2 - 3x + 4)(x^2 + 5x - 2)$

35. $(x^2 - 6x - 7)(x^2 + 3x - 9)$

36. $(x^2 - 5x - 4)(x^2 + 7x - 8)$

For Problems 37–80, find the indicated products by using the shortcut pattern for multiplying binomials.

37. $(x + 2)(x + 9)$ **38.** $(x + 3)(x + 8)$

39. $(x + 6)(x - 2)$ **40.** $(x + 8)(x - 6)$

41. $(x + 3)(x - 11)$ **42.** $(x + 4)(x - 10)$

43. $(n - 4)(n - 3)$ **44.** $(n - 5)(n - 9)$

45. $(n + 6)(n + 12)$ **46.** $(n + 8)(n + 13)$

47. $(y + 3)(y - 7)$ **48.** $(y + 2)(y - 12)$

49. $(y - 7)(y - 12)$ **50.** $(y - 4)(y - 13)$

51. $(x - 5)(x + 7)$ **52.** $(x - 1)(x + 9)$

53. $(x - 14)(x + 8)$ **54.** $(x - 15)(x + 6)$

55. $(a + 10)(a - 9)$ **56.** $(a + 7)(a - 6)$

57. $(2a + 1)(a + 6)$ **58.** $(3a + 2)(a + 4)$

59. $(5x - 2)(x + 7)$ **60.** $(2x - 3)(x + 8)$

61. $(3x - 7)(2x + 1)$ **62.** $(5x - 6)(4x + 3)$

63. $(4a + 3)(3a - 4)$ **64.** $(5a + 4)(4a - 5)$

65. $(6n - 5)(2n - 3)$ **66.** $(4n - 3)(6n - 7)$

67. $(7x - 4)(2x + 3)$ **68.** $(8x - 5)(3x + 7)$

69. $(5 - x)(9 - 2x)$ **70.** $(4 - 3x)(2 + x)$

71. $(-2x + 3)(4x - 5)$ **72.** $(-3x + 1)(9x - 2)$

73. $(-3x - 1)(3x - 4)$ **74.** $(-2x - 5)(4x + 1)$

75. $(8n + 3)(9n - 4)$ **76.** $(6n + 5)(9n - 7)$

77. $(3 - 2x)(9 - x)$ **78.** $(5 - 4x)(4 - 5x)$

79. $(-4x + 3)(-5x - 2)$

80. $(-2x + 7)(-7x - 3)$

81. John, who has not mastered squaring a binomial, turns in the homework shown below. Identify any mistakes in John's work, and then write your answer for the problem.

(a) $(x + 4)^2 = x^2 + 16$

(b) $(x - 5)^2 = x^2 - 5x + 25$

(c) $(x - 6)^2 = x^2 - 12x - 36$

(d) $(3x + 2)^2 = 3x^2 + 12x + 4$

(e) $(x + 1)^2 = x^2 + 2x + 2$

(f) $(x - 3)^2 = x^2 + 6x + 9$

For Problems 82–112, use one of the appropriate patterns $(a + b)^2 = a^2 + 2ab + b^2$, $(a - b)^2 = a^2 - 2ab + b^2$, or $(a + b)(a - b) = a^2 - b^2$ to find the indicated products.

82. $(x + 3)^2$ **83.** $(x + 7)^2$

84. $(x + 9)^2$ **85.** $(5x - 2)(5x + 2)$

86. $(6x + 1)(6x - 1)$ **87.** $(x - 1)^2$

88. $(x - 4)^2$ **89.** $(3x + 7)^2$

90. $(2x + 9)^2$ **91.** $(2x - 3)^2$

92. $(4x - 5)^2$ **93.** $(2x + 3y)(2x - 3y)$

94. $(3a - b)(3a + b)$ **95.** $(1 - 5n)^2$

96. $(2 - 3n)^2$ **97.** $(3x + 4y)^2$

98. $(2x + 5y)^2$ **99.** $(3 + 4y)^2$

100. $(7 + 6y)^2$ **101.** $(1 + 7n)(1 - 7n)$

102. $(2 + 9n)(2 - 9n)$

103. $(4a - 7b)^2$

104. $(6a - b)^2$

105. $(x + 8y)^2$

106. $(x + 6y)^2$

107. $(5x - 11y)(5x + 11y)$

108. $(7x - 9y)(7x + 9y)$

109. $x(8x + 1)(8x - 1)$

110. $3x(5x + 7)(5x - 7)$

111. $-2x(4x + y)(4x - y)$

112. $-4x(2 - 3x)(2 + 3x)$

For Problems 113–120, find the indicated products. Don't forget that $(x + 2)^3$ means $(x + 2)(x + 2)(x + 2)$.

113. $(x + 2)^3$

114. $(x + 4)^3$

115. $(x - 3)^3$

116. $(x - 1)^3$

117. $(2n + 1)^3$

118. $(3n + 2)^3$

119. $(3n - 2)^3$

120. $(4n - 3)^3$

121. Explain how Figure 5.16 can be used to demonstrate geometrically that $(x + 3)(x + 5) = x^2 + 8x + 15$.

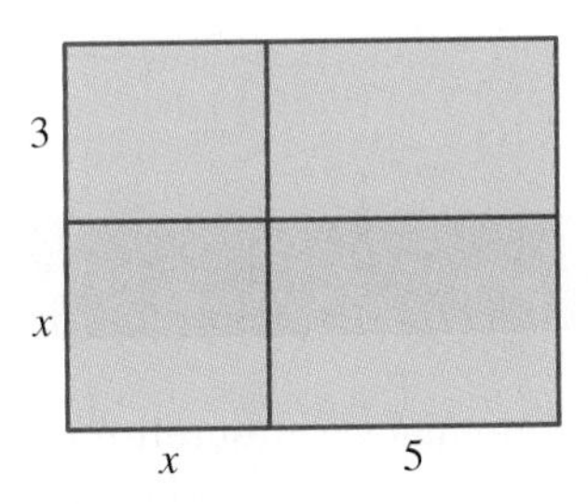

Figure 5.16

122. Explain how Figure 5.17 can be used to demonstrate geometrically that $(x + 5)(x - 3) = x^2 + 2x - 15$.

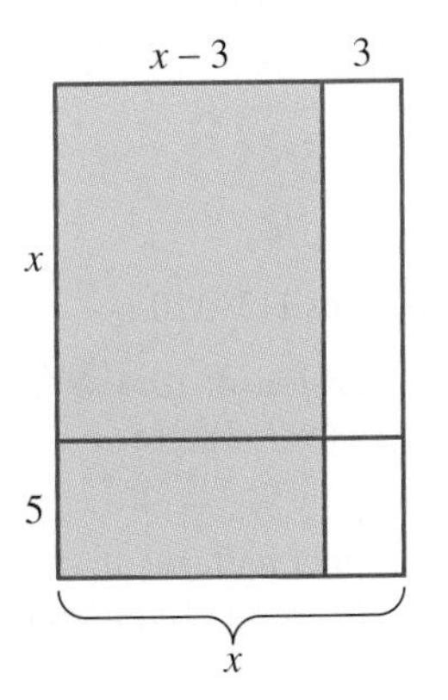

Figure 5.17

123. A square piece of cardboard is 14 inches long on each side. From each corner a square piece x inches on a side is cut out as shown in Figure 5.18. The flaps are then turned up to form an open box. Find polynomials that represent the volume and the exterior surface area of the box.

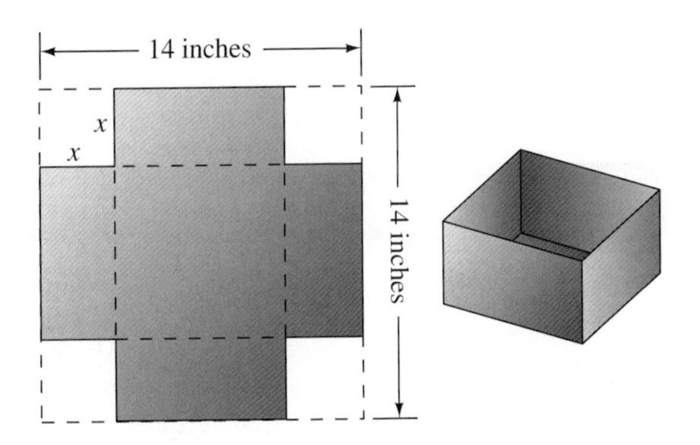

Figure 5.18

THOUGHTS INTO WORDS

124. Describe the process of multiplying two polynomials.

125. Illustrate as many uses of the distributive property as you can.

126. Determine the number of terms in the product of $(x + y + z)$ and $(a + b + c)$ without doing the multiplication. Explain how you arrived at your answer.

FURTHER INVESTIGATIONS

127. The following two patterns result from cubing binomials:

$$(a + b)^3 = a^3 + 3a^2b + 3ab^2 + b^3$$
$$(a - b)^3 = a^3 - 3a^2b + 3ab^2 - b^3$$

Use these patterns to redo Problems 113–120.

128. Find a pattern for the expansion of $(a + b)^4$. Then use the pattern to expand $(x + 2)^4$, $(x + 3)^4$, and $(2x + 1)^4$.

129. We can use some of the product patterns to do arithmetic computations mentally. For example, let's use the pattern $(a + b)^2 = a^2 + 2ab + b^2$ to mentally compute 31^2. Your thought process should be $31^2 = (30 + 1)^2 = 30^2 + 2(30)(1) + 1^2 = 961$. Compute each of the following numbers mentally and then check your answers.

(a) 21^2 **(b)** 41^2 **(c)** 71^2
(d) 32^2 **(e)** 52^2 **(f)** 82^2

130. Use the pattern $(a - b)^2 = a^2 - 2ab + b^2$ to compute each of the following numbers mentally and then check your answers.

(a) 19^2 **(b)** 29^2 **(c)** 49^2
(d) 79^2 **(e)** 38^2 **(f)** 58^2

131. Every whole number with a units digit of 5 can be represented by the expression $10x + 5$, where x is a whole number. For example, $35 = 10(3) + 5$ and $145 = 10(14) + 5$. Now observe the following pattern for squaring such a number:

$$(10x + 5)^2 = 100x^2 + 100x + 25$$
$$= \boxed{100x(x + 1) + 25}$$

The pattern inside the dashed box can be stated as "add 25 to the product of x, $x + 1$, and 100." Thus to mentally compute 35^2 we can think $35^2 = 3(4)(100) + 25 = 1225$. Compute each of the following numbers mentally and then check your answers.

(a) 15^2 **(b)** 25^2 **(c)** 45^2
(d) 55^2 **(e)** 65^2 **(f)** 75^2
(g) 85^2 **(h)** 95^2 **(i)** 105^2

5.4 Dividing by Monomials

To develop an effective process for dividing by a monomial we must rely on yet another property of exponents. This property is also a direct consequence of the definition of exponent and is illustrated by the following examples.

$$\frac{x^5}{x^2} = \frac{\cancel{x} \cdot \cancel{x} \cdot x \cdot x \cdot x}{\cancel{x} \cdot \cancel{x}} = x^3$$

$$\frac{a^4}{a^3} = \frac{\cancel{a} \cdot \cancel{a} \cdot \cancel{a} \cdot a}{\cancel{a} \cdot \cancel{a} \cdot \cancel{a}} = a$$

$$\frac{y^7}{y^3} = \frac{\cancel{y} \cdot \cancel{y} \cdot \cancel{y} \cdot y \cdot y \cdot y \cdot y}{\cancel{y} \cdot \cancel{y} \cdot \cancel{y}} = y^4$$

$$\frac{x^4}{x^4} = \frac{\cancel{x} \cdot \cancel{x} \cdot \cancel{x} \cdot \cancel{x}}{\cancel{x} \cdot \cancel{x} \cdot \cancel{x} \cdot \cancel{x}} = 1$$

$$\frac{y^3}{y^3} = \frac{\cancel{y} \cdot \cancel{y} \cdot \cancel{y}}{\cancel{y} \cdot \cancel{y} \cdot \cancel{y}} = 1$$

Property 5.4

If b is any nonzero real number, and n and m are positive integers, then

1. $\dfrac{b^n}{b^m} = b^{n-m}$ when $n > m$

2. $\dfrac{b^n}{b^m} = 1$ when $n = m$

(The situation $n < m$ will be discussed in a later section.)

Applying Property 5.4 to the previous examples yields these results:

$$\frac{x^5}{x^2} = x^{5-2} = x^3$$

$$\frac{a^4}{a^3} = a^{4-3} = a^1 \quad \text{Usually written as } a$$

$$\frac{y^7}{y^3} = y^{7-3} = y^4$$

$$\frac{x^4}{x^4} = 1$$

$$\frac{y^3}{y^3} = 1$$

Property 5.4 along with our knowledge of dividing integers provides the basis for dividing a monomial by another monomial. Consider the next examples.

$$\frac{16x^5}{2x^3} = 8x^{5-3} = 8x^2 \qquad \frac{-81a^{12}}{-9a^4} = 9a^{12-4} = 9a^8$$

$$\frac{-35x^9}{5x^4} = -7x^{9-4} = -7x^5 \qquad \frac{45x^4}{9x^4} = 5 \quad \frac{x^4}{x^4} = 1$$

$$\frac{56y^6}{-7y^2} = -8y^{6-2} = -8y^4 \qquad \frac{54x^3y^7}{-6xy^5} = -9x^{3-1}y^{7-5} = -9x^2y^2$$

Recall that $\dfrac{a+b}{c} = \dfrac{a}{c} + \dfrac{b}{c}$; this property serves as the basis for dividing a polynomial by a monomial. Consider these examples.

$$\frac{25x^3 + 10x^2}{5x} = \frac{25x^3}{5x} + \frac{10x^2}{5x} = 5x^2 + 2x$$

$$\frac{-35x^8 - 28x^6}{7x^3} = \frac{-35x^8}{7x^3} - \frac{28x^6}{7x^3} = -5x^5 - 4x^3 \qquad \frac{a-b}{c} = \frac{a}{c} - \frac{b}{c}$$

To divide a polynomial by a monomial, we simply divide each term of the polynomial by the monomial. Here are some additional examples.

$$\frac{12x^3y^2 - 14x^2y^5}{-2xy} = \frac{12x^3y^2}{-2xy} - \frac{14x^2y^5}{-2xy} = -6x^2y + 7xy^4$$

$$\frac{48ab^5 + 64a^2b}{-16ab} = \frac{48ab^5}{-16ab} + \frac{64a^2b}{-16ab} = -3b^4 - 4a$$

$$\frac{33x^6 - 24x^5 - 18x^4}{3x} = \frac{33x^6}{3x} - \frac{24x^5}{3x} - \frac{18x^4}{3x}$$

$$= 11x^5 - 8x^4 - 6x^3$$

As with many skills, once you feel comfortable with the process, you may want to perform some of the steps mentally. Your work could take on the following format.

$$\frac{24x^4y^5 - 56x^3y^9}{8x^2y^3} = 3x^2y^2 - 7xy^6$$

$$\frac{13a^2b - 12ab^2}{-ab} = -13a + 12b$$

Problem Set 5.4

For Problems 1–24, divide the monomials.

1. $\frac{x^{10}}{x^2}$
2. $\frac{x^{12}}{x^5}$
3. $\frac{4x^3}{2x}$
4. $\frac{8x^5}{4x^3}$
5. $\frac{-16n^6}{2n^2}$
6. $\frac{-54n^8}{6n^4}$
7. $\frac{72x^3}{-9x^3}$
8. $\frac{84x^5}{-7x^5}$
9. $\frac{65x^2y^3}{5xy}$
10. $\frac{70x^3y^4}{5x^2y}$
11. $\frac{-91a^4b^6}{-13a^3b^4}$
12. $\frac{-72a^5b^4}{-12ab^2}$
13. $\frac{18x^2y^6}{xy^2}$
14. $\frac{24x^3y^4}{x^2y^2}$
15. $\frac{32x^6y^2}{-x}$
16. $\frac{54x^5y^3}{-y^2}$
17. $\frac{-96x^5y^7}{12y^3}$
18. $\frac{-84x^4y^9}{14x^4}$
19. $\frac{-ab}{ab}$
20. $\frac{6ab}{-ab}$
21. $\frac{56a^2b^3c^5}{4abc}$
22. $\frac{60a^3b^2c}{15a^2c}$
23. $\frac{-80xy^2z^6}{-5xyz^2}$
24. $\frac{-90x^3y^2z^8}{-6xy^2z^4}$

For Problems 25–50, perform each division of polynomials by monomials.

25. $\frac{8x^4 + 12x^5}{2x^2}$
26. $\frac{12x^3 + 16x^6}{4x}$
27. $\frac{9x^6 - 24x^4}{3x^3}$
28. $\frac{35x^8 - 45x^6}{5x^4}$
29. $\frac{-28n^5 + 36n^2}{4n^2}$
30. $\frac{-42n^6 + 54n^4}{6n^4}$
31. $\frac{35x^6 - 56x^5 - 84x^3}{7x^2}$
32. $\frac{27x^7 - 36x^5 - 45x^3}{3x}$

33. $\dfrac{-24n^8 + 48n^5 - 78n^3}{-6n^3}$

34. $\dfrac{-56n^9 + 84n^6 - 91n^2}{-7n^2}$

35. $\dfrac{-60a^7 - 96a^3}{-12a}$

36. $\dfrac{-65a^8 - 78a^4}{-13a^2}$

37. $\dfrac{27x^2y^4 - 45xy^4}{-9xy^3}$

38. $\dfrac{-40x^4y^7 + 64x^5y^8}{-8x^3y^4}$

39. $\dfrac{48a^2b^2 + 60a^3b^4}{-6ab}$

40. $\dfrac{45a^3b^4 - 63a^2b^6}{-9ab^2}$

41. $\dfrac{12a^2b^2c^2 - 52a^2b^3c^5}{-4a^2bc}$

42. $\dfrac{48a^3b^2c + 72a^2b^4c^5}{-12ab^2c}$

43. $\dfrac{9x^2y^3 - 12x^3y^4}{-xy}$

44. $\dfrac{-15x^3y + 27x^2y^4}{xy}$

45. $\dfrac{-42x^6 - 70x^4 + 98x^2}{14x^2}$

46. $\dfrac{-48x^8 - 80x^6 + 96x^4}{16x^4}$

47. $\dfrac{15a^3b - 35a^2b - 65ab^2}{-5ab}$

48. $\dfrac{-24a^4b^2 + 36a^3b - 48a^2b}{-6ab}$

49. $\dfrac{-xy + 5x^2y^3 - 7x^2y^6}{xy}$

50. $\dfrac{-9x^2y^3 - xy + 14xy^4}{-xy}$

THOUGHTS INTO WORDS

51. How would you explain to someone why the quotient of x^8 and x^2 is x^6 and not x^4?

52. Your friend is having difficulty with problems such as $\dfrac{12x^2y}{xy}$ and $\dfrac{36x^3y^2}{-xy}$ where there appears to be no numerical coefficient in the denominator. What can you tell him that might help?

5.5 Dividing by Binomials

Perhaps the easiest way to explain the process of dividing a polynomial by a binomial is to work a few examples and describe the step-by-step procedure as we go along.

EXAMPLE 1

Divide $x^2 + 5x + 6$ by $x + 2$.

Solution

Step 1 Use the conventional long division format from arithmetic, and arrange both the dividend and the divisor in descending powers of the variable.

$$x + 2\overline{)x^2 + 5x + 6}$$

Step 2 Find the first term of the quotient by dividing the first term of the dividend by the first term of the divisor.

$$x + 2\overline{)x^2 + 5x + 6} \quad \text{quotient: } x \qquad \frac{x^2}{x} = x$$

Step 3 Multiply the entire divisor by the term of the quotient found in step 2, and position this product to be subtracted from the dividend.

$$\begin{array}{r} x \\ x + 2\overline{)x^2 + 5x + 6} \\ \underline{x^2 + 2x} \end{array} \qquad x(x + 2) = x^2 + 2x$$

Step 4 Subtract.

Remember to add the opposite!

$$\begin{array}{r} x \\ x + 2\overline{)x^2 + 5x + 6} \\ \underline{x^2 + 2x} \\ 3x + 6 \end{array}$$

Step 5 Repeat the process beginning with step 2; use the polynomial that resulted from the subtraction in step 4 as a new dividend.

$$\begin{array}{r} x + 3 \\ x + 2\overline{)x^2 + 5x + 6} \\ \underline{x^2 + 2x} \\ 3x + 6 \\ \underline{3x + 6} \end{array} \qquad \frac{3x}{x} = 3 \qquad 3x + 6$$

Thus $(x^2 + 5x + 6) \div (x + 2) = x + 3$, which can be checked by multiplying $(x + 2)$ and $(x + 3)$.

$$(x + 2)(x + 3) = x^2 + 5x + 6$$

■

A division problem such as $(x^2 + 5x + 6) \div (x + 2)$ can also be written as $\frac{x^2 + 5x + 6}{x + 2}$. Using this format, we can express the final result for Example 1 as $\frac{x^2 + 5x + 6}{x + 2} = x + 3$. (Technically, the restriction $x \neq -2$ should be made to avoid division by zero.)

In general, to check a division problem we can multiply the divisor times the quotient and add the remainder, which can be expressed as

Dividend = (Divisor)(Quotient) + Remainder

Sometimes the remainder is expressed as a fractional part of the divisor. The relationship then becomes

$$\frac{\text{Dividend}}{\text{Divisor}} = \text{Quotient} + \frac{\text{Remainder}}{\text{Divisor}}$$

EXAMPLE 2

Divide $2x^2 - 3x - 20$ by $x - 4$.

Solution

Step 1 $x - 4\overline{)2x^2 - 3x - 20}$

Step 2

$$\begin{array}{r} 2x \\ x - 4\overline{)2x^2 - 3x - 20} \end{array} \qquad \frac{2x^2}{x} = 2x$$

Step 3

$$\begin{array}{r} 2x \phantom{{}- 3x - 20} \\ x - 4\overline{)2x^2 - 3x - 20} \\ \underline{2x^2 - 8x} \phantom{{}- 20} \end{array} \qquad 2x(x - 4) = 2x^2 - 8x$$

Step 4

$$\begin{array}{r} 2x \phantom{{}- 3x - 20} \\ x - 4\overline{)2x^2 - 3x - 20} \\ \underline{2x^2 - 8x} \phantom{{}- 20} \\ 5x - 20 \end{array}$$

Step 5

$$\begin{array}{r} 2x + 5 \phantom{{}- 20} \\ x - 4\overline{)2x^2 - 3x - 20} \\ \underline{2x^2 - 8x} \phantom{{}- 20} \\ 5x - 20 \\ \underline{5x - 20} \end{array} \qquad \frac{5x}{x} = 5 \qquad 5(x - 4) = 5x - 20$$

Check

$$(x - 4)(2x + 5) = 2x^2 - 3x - 20$$

Therefore, $\dfrac{2x^2 - 3x - 20}{x - 4} = 2x + 5$. ■

Now let's continue to think in terms of the step-by-step division process but organize our work in the typical long division format.

EXAMPLE 3

Divide $12x^2 + x - 6$ by $3x - 2$.

Solution

$$\begin{array}{r} 4x + 3 \\ 3x - 2\overline{)12x^2 + x - 6} \\ \underline{12x^2 - 8x} \phantom{{}- 6} \\ 9x - 6 \\ \underline{9x - 6} \end{array}$$

Check

$$(3x - 2)(4x + 3) = 12x^2 + x - 6$$

Therefore, $\dfrac{12x^2 + x - 6}{3x - 2} = 4x + 3$. ■

Each of the next three examples illustrates another aspect of the division process. Study them carefully; then you should be ready to work the exercises in the next problem set.

EXAMPLE 4 Perform the division $(7x^2 - 3x - 4) \div (x - 2)$.

Solution

$$
\begin{array}{r}
7x + 11 \\
x - 2 \overline{) \, 7x^2 - 3x - 4} \\
\underline{7x^2 - 14x} \\
11x - 4 \\
\underline{11x - 22} \\
18
\end{array}
$$

← A remainder of 18

Check

Just as in arithmetic, we check by *adding* the remainder to the product of the divisor and quotient.

$$(x - 2)(7x + 11) + 18 \stackrel{?}{=} 7x^2 - 3x - 4$$
$$7x^2 - 3x - 22 + 18 \stackrel{?}{=} 7x^2 - 3x - 4$$
$$7x^2 - 3x - 4 = 7x^2 - 3x - 4$$

Therefore, $\dfrac{7x^2 - 3x - 4}{x - 2} = 7x + 11 + \dfrac{18}{x - 2}$. ■

EXAMPLE 5 Perform the division $\dfrac{x^3 - 8}{x - 2}$.

Solution

$$
\begin{array}{r}
x^2 + 2x + 4 \\
x - 2 \overline{) \, x^3 + 0x^2 + 0x - 8} \\
\underline{x^3 - 2x^2} \\
2x^2 + 0x - 8 \\
\underline{2x^2 - 4x} \\
4x - 8 \\
\underline{4x - 8}
\end{array}
$$

← Notice the insertion of x^2 and x terms with zero coefficients

Check

$$(x - 2)(x^2 + 2x + 4) \stackrel{?}{=} x^3 - 8$$
$$x^3 + 2x^2 + 4x - 2x^2 - 4x - 8 \stackrel{?}{=} x^3 - 8$$
$$x^3 - 8 = x^3 - 8$$

Therefore, $\dfrac{x^3 - 8}{x - 2} = x^2 + 2x + 4$. ■

EXAMPLE 6 Perform the division $\frac{x^3 + 5x^2 - 3x - 4}{x^2 + 2x}$.

Solution

$$\begin{array}{r l}
& x + 3 \\
x^2 + 2x \overline{)} & x^3 + 5x^2 - 3x - 4 \\
& \underline{x^3 + 2x^2} \\
& \quad 3x^2 - 3x - 4 \\
& \quad \underline{3x^2 + 6x} \\
& \qquad -9x - 4 \quad \leftarrow \text{A remainder of } -9x - 4
\end{array}$$

We stop the division process when the degree of the remainder is less than the degree of the divisor.

Check

$$(x^2 + 2x)(x + 3) + (-9x - 4) \stackrel{?}{=} x^3 + 5x^2 - 3x - 4$$

$$x^3 + 3x^2 + 2x^2 + 6x - 9x - 4 \stackrel{?}{=} x^3 + 5x^2 - 3x - 4$$

$$x^3 + 5x^2 - 3x - 4 = x^3 + 5x^2 - 3x - 4$$

Therefore, $\frac{x^3 + 5x^2 - 3x - 4}{x^2 + 2x} = x + 3 + \frac{-9x - 4}{x^2 + 2x}$. ■

Problem Set 5.5

For Problems 1–40, perform the divisions.

1. $(x^2 + 16x + 48) \div (x + 4)$
2. $(x^2 + 15x + 54) \div (x + 6)$
3. $(x^2 - 5x - 14) \div (x - 7)$
4. $(x^2 + 8x - 65) \div (x - 5)$
5. $(x^2 + 11x + 28) \div (x + 3)$
6. $(x^2 + 11x + 15) \div (x + 2)$
7. $(x^2 - 4x - 39) \div (x - 8)$
8. $(x^2 - 9x - 30) \div (x - 12)$
9. $(5n^2 - n - 4) \div (n - 1)$
10. $(7n^2 - 61n - 90) \div (n - 10)$
11. $(8y^2 + 53y - 19) \div (y + 7)$
12. $(6y^2 + 47y - 72) \div (y + 9)$
13. $(20x^2 - 31x - 7) \div (5x + 1)$
14. $(27x^2 + 21x - 20) \div (3x + 4)$
15. $(6x^2 + 25x + 8) \div (2x + 7)$
16. $(12x^2 + 28x + 27) \div (6x + 5)$
17. $(2x^3 - x^2 - 2x - 8) \div (x - 2)$
18. $(3x^3 - 7x^2 - 26x + 24) \div (x - 4)$
19. $(5n^3 + 11n^2 - 15n - 9) \div (n + 3)$
20. $(6n^3 + 29n^2 - 6n - 5) \div (n + 5)$
21. $(n^3 - 40n + 24) \div (n - 6)$
22. $(n^3 - 67n - 24) \div (n + 8)$
23. $(x^3 - 27) \div (x - 3)$
24. $(x^3 + 8) \div (x + 2)$

25. $\dfrac{27x^3 - 64}{3x - 4}$

26. $\dfrac{8x^3 + 27}{2x + 3}$

27. $\dfrac{1 + 3n^2 - 2n}{n + 2}$

28. $\dfrac{x + 5 + 12x^2}{3x - 2}$

29. $\dfrac{9t^2 + 3t + 4}{-1 + 3t}$

30. $\dfrac{4n^2 + 6n - 1}{4 + 2n}$

31. $\dfrac{6n^3 - 5n^2 - 7n + 4}{2n - 1}$

32. $\dfrac{21n^3 + 23n^2 - 9n - 10}{3n + 2}$

33. $\dfrac{4x^3 + 23x^2 - 30x + 32}{x + 7}$

34. $\dfrac{5x^3 - 12x^2 + 13x - 14}{x - 1}$

35. $(x^3 + 2x^2 - 3x - 1) \div (x^2 - 2x)$

36. $(x^3 - 6x^2 - 2x + 1) \div (x^2 + 3x)$

37. $(2x^3 - 4x^2 + x - 5) \div (x^2 + 4x)$

38. $(2x^3 - x^2 - 3x + 5) \div (x^2 + x)$

39. $(x^4 - 16) \div (x + 2)$

40. $(x^4 - 81) \div (x - 3)$

THOUGHTS INTO WORDS

41. Give a step-by-step description of how you would do the division problem $(2x^3 + 8x^2 - 29x - 30) \div (x + 6)$.

42. How do you know by inspection that the answer to the following division problem is incorrect?

$(3x^3 - 7x^2 - 22x + 8) \div (x - 4) = 3x^2 + 5x + 1$

5.6 Zero and Negative Integers as Exponents

Thus far in this text we have used only positive integers as exponents. The next definitions and properties serve as a basis for our work with exponents.

Definition 5.1

If n is a positive integer and b is any real number, then

$$b^n = \underbrace{bbb \cdots b}_{n \text{ factors of } b}$$

Property 5.5

If m and n are positive integers and a and b are real numbers, except $b \neq 0$ whenever it appears in a denominator, then

1. $b^n \cdot b^m = b^{n+m}$
2. $(b^n)^m = b^{mn}$

(continued)

3. $(ab)^n = a^n b^n$

4. $\left(\frac{a}{b}\right)^n = \frac{a^n}{b^n}$ Part 4 has not been stated previously

5. $\frac{b^n}{b^m} = b^{n-m}$ When $n > m$

$\frac{b^n}{b^m} = 1$ When $n = m$

Property 5.5 pertains to the use of positive integers as exponents. Zero and the negative integers can also be used as exponents. First let's consider the use of 0 as an exponent. We want to use 0 as an exponent in such a way that the basic properties of exponents will continue to hold. Consider the example $x^4 \cdot x^0$. If part 1 of Property 5.5 is to hold, then

$$x^4 \cdot x^0 = x^{4+0} = x^4$$

Note that x^0 acts like 1 because $x^4 \cdot x^0 = x^4$. This suggests the following definition.

Definition 5.2

If b is a nonzero real number, then

$$b^0 = 1$$

According to Definition 5.2 the following statements are all true.

$$4^0 = 1$$

$$(-628)^0 = 1$$

$$\left(\frac{4}{7}\right)^0 = 1$$

$$n^0 = 1, \qquad n \neq 0$$

$$(x^2y^5)^0 = 1, \qquad x \neq 0 \text{ and } y \neq 0$$

A similar line of reasoning indicates how negative integers should be used as exponents. Consider the example $x^3 \cdot x^{-3}$. If part 1 of Property 5.5 is to hold, then

$$x^3 \cdot x^{-3} = x^{3+(-3)} = x^0 = 1$$

Thus x^{-3} must be the reciprocal of x^3 because their product is 1; that is,

$$x^{-3} = \frac{1}{x^3}$$

This process suggests the following definition.

Definition 5.3

If n is a positive integer, and b is a nonzero real number, then

$$b^{-n} = \frac{1}{b^n}$$

According to Definition 5.3, the following statements are all true.

$$x^{-6} = \frac{1}{x^6}$$

$$2^{-3} = \frac{1}{2^3} = \frac{1}{8}$$

$$10^{-2} = \frac{1}{10^2} = \frac{1}{100} \quad \text{or} \quad 0.01$$

$$\frac{1}{x^{-4}} = \frac{1}{\frac{1}{x^4}} = x^4$$

$$\left(\frac{2}{3}\right)^{-2} = \frac{1}{\left(\frac{2}{3}\right)^2} = \frac{1}{\frac{4}{9}} = \frac{9}{4}$$

Remark: Note in the last example that $\left(\frac{2}{3}\right)^{-2} = \left(\frac{3}{2}\right)^2$. In other words, to raise a fraction to a negative power, we can invert the fraction and raise it to the corresponding positive power.

We can verify (we will not do so in this text) that all parts of Property 5.5 hold for *all integers*. In fact, we can replace part 5 with this statement.

Replacement for part 5 of Property 5.5

$$\frac{b^n}{b^m} = b^{n-m} \quad \text{for all integers } n \text{ and } m$$

The next examples illustrate the use of this new property. In each example, we simplify the original expression and use only positive exponents in the final result.

$$\frac{x^2}{x^5} = x^{2-5} = x^{-3} = \frac{1}{x^3}$$

$$\frac{a^{-3}}{a^{-7}} = a^{-3-(-7)} = a^{-3+7} = a^4$$

$$\frac{y^{-5}}{y^{-2}} = y^{-5-(-2)} = y^{-5+2} = y^{-3} = \frac{1}{y^3}$$

$$\frac{x^{-6}}{x^{-6}} = x^{-6-(-6)} = x^{-6+6} = x^0 = 1$$

The properties of exponents provide a basis for simplifying certain types of numerical expressions, as the following examples illustrate.

$$2^{-4} \cdot 2^6 = 2^{-4+6} = 2^2 = 4$$

$$10^5 \cdot 10^{-6} = 10^{5+(-6)} = 10^{-1} = \frac{1}{10} \quad \text{or} \quad 0.1$$

$$\frac{10^2}{10^{-2}} = 10^{2-(-2)} = 10^{2+2} = 10^4 = 10{,}000$$

$$(2^{-3})^{-2} = 2^{-3(-2)} = 2^6 = 64$$

Having the use of all integers as exponents also expands the type of work that we can do with algebraic expressions. In each of the following examples we simplify a given expression and use only positive exponents in the final result.

$$x^8x^{-2} = x^{8+(-2)} = x^6$$

$$a^{-4}a^{-3} = a^{-4+(-3)} = a^{-7} = \frac{1}{a^7}$$

$$(y^{-3})^4 = y^{-3(4)} = y^{-12} = \frac{1}{y^{12}}$$

$$(x^{-2}y^4)^{-3} = (x^{-2})^{-3}(y^4)^{-3} = x^6y^{-12} = \frac{x^6}{y^{12}}$$

$$\left(\frac{x^{-1}}{y^2}\right)^{-2} = \frac{(x^{-1})^{-2}}{(y^2)^{-2}} = \frac{x^2}{y^{-4}} = x^2y^4$$

$$(4x^{-2})(3x^{-1}) = 12x^{-2+(-1)} = 12x^{-3} = \frac{12}{x^3}$$

$$\left(\frac{12x^{-6}}{6x^{-2}}\right)^{-2} = (2x^{-6-(-2)})^{-2} = (2x^{-4})^{-2} \qquad \text{Divide the coefficients } \frac{12}{6} = 2$$

$$= (2)^{-2}(x^{-4})^{-2}$$

$$= \left(\frac{1}{2^2}\right)(x^8) = \frac{x^8}{4}$$

Scientific Notation

Many scientific applications of mathematics involve the use of very large and very small numbers. For example:

The speed of light is approximately 29,979,200,000 centimeters per second.

A light year—the distance light travels in 1 year—is approximately 5,865,696,000,000 miles.

A gigahertz equals 1,000,000,000 hertz.

The length of a typical virus cell equals 0.000000075 of a meter.

The length of a diameter of a water molecule is 0.0000000003 of a meter.

Working with numbers of this type in standard form is quite cumbersome. It is much more convenient to represent very small and very large numbers in **scientific notation**, sometimes called scientific form. A number is in scientific notation when it is written as the product of a number between 1 and 10 (including 1) and an integral power of 10. Symbolically, a number in scientific notation has the form $(N)(10^k)$, where $1 \leq N < 10$, and k is an integer. For example, 621 can be written as $(6.21)(10^2)$, and 0.0023 can be written as $(2.3)(10^{-3})$.

To switch from ordinary notation to scientific notation, you can use the following procedure.

> Write the given number as the product of a number greater than or equal to 1 and less than 10, and an integral power of 10. To determine the exponent of 10, count the number of places that the decimal point moved when going from the original number to the number greater than or equal to 1 and less than 10. This exponent is (a) negative if the original number is less than 1, (b) positive if the original number is greater than 10, and (c) zero if the original number itself is between 1 and 10.

Thus, we can write

$0.000179 = (1.79)(10^{-4})$ According to part (a) of the procedure

$8175 = (8.175)(10^3)$ According to part (b)

$3.14 = (3.14)(10^0)$ According to part (c)

We can express the applications given earlier in scientific notation as follows:

Speed of light: $29{,}979{,}200{,}000 = (2.99792)(10^{10})$ centimeters per second

Light year: $5{,}865{,}696{,}000{,}000 = (5.865696)(10^{12})$ miles

Gigahertz: $1{,}000{,}000{,}000 = (1)(10^9)$ hertz

Length of a virus cell: $0.000000075 = (7.5)(10^{-8})$ meter

Length of the diameter of a water molecule $= 0.0000000003$
$= (3)(10^{-10})$ meter

To switch from scientific notation to ordinary decimal notation you can use the following procedure.

> Move the decimal point the number of places indicated by the exponent of 10. The decimal point is moved to the right if the exponent is positive and to the left if it is negative.

Thus, we can write

$(4.71)(10^4) = 47{,}100$ Two zeros are needed for place value purposes

$(1.78)(10^{-2}) = 0.0178$ One zero is needed for place value purposes

The use of scientific notation along with the properties of exponents can make some arithmetic problems much easier to evaluate. The next examples illustrate this point.

EXAMPLE 1

Evaluate (4000)(0.000012).

Solution

$$\begin{aligned}(4000)(0.000012) &= (4)(10^3)(1.2)(10^{-5})\\ &= (4)(1.2)(10^3)(10^{-5})\\ &= (4.8)(10^{-2})\\ &= 0.048\end{aligned}$$

■

EXAMPLE 2

Evaluate $\dfrac{960{,}000}{0.032}$.

Solution

$$\begin{aligned}\frac{960{,}000}{0.032} &= \frac{(9.6)(10^5)}{(3.2)(10^{-2})}\\ &= (3)(10^7) \qquad \frac{10^5}{10^{-2}} = 10^{5-(-2)} = 10^7\\ &= 30{,}000{,}000\end{aligned}$$

■

EXAMPLE 3

Evaluate $\dfrac{(6000)(0.00008)}{(40{,}000)(0.006)}$.

Solution

$$\begin{aligned}\frac{(6000)(0.00008)}{(40{,}000)(0.006)} &= \frac{(6)(10^3)(8)(10^{-5})}{(4)(10^4)(6)(10^{-3})}\\ &= \frac{(48)(10^{-2})}{(24)(10^1)}\\ &= (2)(10^{-3}) \qquad \frac{10^{-2}}{10^1} = 10^{-2-1} = 10^{-3}\\ &= 0.002\end{aligned}$$

■

Problem Set 5.6

For Problems 1–30, evaluate each numerical expression.

1. 3^{-2}
2. 2^{-5}
3. 4^{-3}
4. 5^{-2}
5. $\left(\frac{3}{2}\right)^{-1}$
6. $\left(\frac{3}{4}\right)^{-2}$
7. $\frac{1}{2^{-4}}$
8. $\frac{1}{3^{-1}}$
9. $\left(-\frac{4}{3}\right)^{0}$
10. $\left(-\frac{1}{2}\right)^{-3}$
11. $\left(-\frac{2}{3}\right)^{-3}$
12. $(-16)^{0}$
13. $(-2)^{-2}$
14. $(-3)^{-2}$
15. $-(3^{-2})$
16. $-(2^{-2})$
17. $\frac{1}{\left(\frac{3}{4}\right)^{-3}}$
18. $\frac{1}{\left(\frac{3}{2}\right)^{-4}}$
19. $2^{6} \cdot 2^{-9}$
20. $3^{5} \cdot 3^{-2}$
21. $3^{6} \cdot 3^{-3}$
22. $2^{-7} \cdot 2^{2}$
23. $\frac{10^{2}}{10^{-1}}$
24. $\frac{10^{1}}{10^{-3}}$
25. $\frac{10^{-1}}{10^{2}}$
26. $\frac{10^{-2}}{10^{-2}}$
27. $(2^{-1} \cdot 3^{-2})^{-1}$
28. $(3^{-1} \cdot 4^{-2})^{-1}$
29. $\left(\frac{4^{-1}}{3}\right)^{-2}$
30. $\left(\frac{3}{2^{-1}}\right)^{-3}$

For Problems 31–84, simplify each algebraic expression and express your answers using positive exponents only.

31. $x^{6}x^{-1}$
32. $x^{-2}x^{7}$
33. $n^{-4}n^{2}$
34. $n^{-8}n^{3}$
35. $a^{-2}a^{-3}$
36. $a^{-4}a^{-6}$
37. $(2x^{3})(4x^{-2})$
38. $(5x^{-4})(6x^{7})$
39. $(3x^{-6})(9x^{2})$
40. $(8x^{-8})(4x^{2})$
41. $(5y^{-1})(-3y^{-2})$
42. $(-7y^{-3})(9y^{-4})$
43. $(8x^{-4})(12x^{4})$
44. $(-3x^{-2})(-6x^{2})$
45. $\frac{x^{7}}{x^{-3}}$
46. $\frac{x^{2}}{x^{-4}}$
47. $\frac{n^{-1}}{n^{3}}$
48. $\frac{n^{-2}}{n^{5}}$
49. $\frac{4n^{-1}}{2n^{-3}}$
50. $\frac{12n^{-2}}{3n^{-5}}$
51. $\frac{-24x^{-6}}{8x^{-2}}$
52. $\frac{56x^{-5}}{-7x^{-1}}$
53. $\frac{-52y^{-2}}{-13y^{-2}}$
54. $\frac{-91y^{-3}}{-7y^{-3}}$
55. $(x^{-3})^{-2}$
56. $(x^{-1})^{-5}$
57. $(x^{2})^{-2}$
58. $(x^{3})^{-1}$
59. $(x^{3}y^{4})^{-1}$
60. $(x^{4}y^{-2})^{-2}$
61. $(x^{-2}y^{-1})^{3}$
62. $(x^{-3}y^{-4})^{2}$
63. $(2n^{-2})^{3}$
64. $(3n^{-1})^{4}$
65. $(4n^{3})^{-2}$
66. $(2n^{2})^{-3}$
67. $(3a^{-2})^{4}$
68. $(5a^{-1})^{2}$
69. $(5x^{-1})^{-2}$
70. $(4x^{-2})^{-2}$
71. $(2x^{-2}y^{-1})^{-1}$
72. $(3x^{2}y^{-3})^{-2}$
73. $\left(\frac{x^{2}}{y}\right)^{-1}$
74. $\left(\frac{y^{2}}{x^{3}}\right)^{-2}$
75. $\left(\frac{a^{-1}}{b^{2}}\right)^{-4}$
76. $\left(\frac{a^{3}}{b^{-2}}\right)^{-3}$
77. $\left(\frac{x^{-1}}{y^{-3}}\right)^{-2}$
78. $\left(\frac{x^{-3}}{y^{-4}}\right)^{-1}$
79. $\left(\frac{x^{2}}{x^{3}}\right)^{-1}$
80. $\left(\frac{x^{4}}{x}\right)^{-2}$
81. $\left(\frac{2x^{-1}}{x^{-2}}\right)^{-3}$
82. $\left(\frac{3x^{-2}}{x^{-5}}\right)^{-1}$
83. $\left(\frac{18x^{-1}}{9x}\right)^{-2}$
84. $\left(\frac{35x^{2}}{7x^{-1}}\right)^{-1}$

85. The U.S. Social Security Administration pays approximately \$42,000,000,000 in monthly benefits to all beneficiaries. Write this number in scientific notation.

86. In 1998 approximately 10,200,000,000 pennies were made. Write, in scientific notation, the number of pennies made.

87. The thickness of a dollar bill is 0.0043 inches. Write this number in scientific notation.

88. The length of an Ebola virus cell is approximately 0.0000002 meters. Write this number in scientific notation.

89. The diameter of Jupiter is approximately 89,000 miles. Write this number in scientific notation.

90. Avogadro's Number, which has a value of approximately 602,200,000,000,000,000,000,000, refers to the calculated value of the number of molecules in a gram mole of any chemical substance. Write Avogadro's Number in scientific notation.

91. The cooling fan for a computer hard drive has a thickness of 0.025 meters. Write this number in scientific notation.

92. A sheet of 20-weight bond paper has a thickness of 0.097 millimeters. Write the thickness in scientific notation.

93. According to the 2004 annual report for Coca-Cola, the Web site CokePLAY.com was launched in Korea, and it was visited by 11,000,000 people in its first six months. Write the number of hits for the Web site in scientific notation.

94. In 2005, Wal-Mart reported that they served 138,000,000 customers worldwide each week. Write, in scientific notation, the number of customers for each week.

For Problems 95–106, write each number in standard decimal form; for example, $(1.4)(10^3) = 1400$.

95. $(8)(10^3)$

96. $(6)(10^2)$

97. $(5.21)(10^4)$

98. $(7.2)(10^3)$

99. $(1.14)(10^7)$

100. $(5.64)(10^8)$

101. $(7)(10^{-2})$

102. $(8.14)(10^{-1})$

103. $(9.87)(10^{-4})$

104. $(4.37)(10^{-5})$

105. $(8.64)(10^{-6})$

106. $(3.14)(10^{-7})$

For Problems 107–118, use scientific notation and the properties of exponents to evaluate each numerical expression.

107. $(0.007)(120)$

108. $(0.0004)(13)$

109. $(5{,}000{,}000)(0.00009)$

110. $(800{,}000)(0.0000006)$

111. $\dfrac{6000}{0.0015}$

112. $\dfrac{480}{0.012}$

113. $\dfrac{0.00086}{4300}$

114. $\dfrac{0.0057}{30{,}000}$

115. $\dfrac{0.00039}{0.0013}$

116. $\dfrac{0.0000082}{0.00041}$

117. $\dfrac{(0.0008)(0.07)}{(20{,}000)(0.0004)}$

118. $\dfrac{(0.006)(600)}{(0.00004)(30)}$

For Problems 119–122, convert the numbers to scientific notation and compute the answer.

119. The U.S. Social Security Administration pays approximately \$42,000,000,000 in monthly benefits to all beneficiaries. If there are 48,000,000 beneficiaries, find the average dollar amount each beneficiary receives.

120. In 1998 approximately 10,200,000,000 pennies were made. If the population was 270,000,000, find the average number of pennies produced per person. Round the answer to the nearest whole number.

121. The thickness of a dollar bill is 0.0043 inches. How tall will a stack of 1,000,000, dollars be? Express the answer to the nearest foot.

122. The diameter of Jupiter is approximately 11 times larger than the diameter of Earth. Find the diameter of Earth given that Jupiter has a diameter of approximately 89,000 miles. Express the answer to the nearest mile.

THOUGHTS INTO WORDS

123. Is the following simplification process correct?

$$(2^{-2})^{-1} = \left(\frac{1}{2^2}\right)^{-1} = \left(\frac{1}{4}\right)^{-1} = \frac{1}{\left(\frac{1}{4}\right)^1} = 4$$

Can you suggest a better way to do the problem?

124. Explain the importance of scientific notation.

FURTHER INVESTIGATIONS

125. Use your calculator to do Problems 1–16. Be sure that your answers are equivalent to the answers you obtained without the calculator.

126. Use your calculator to evaluate $(140{,}000)^2$. Your answer should be displayed in scientific notation; the format of the display depends on the particular calculator. For example, it may look like `1.96 10` or `1.96E + 10`. Thus in ordinary notation the answer is 19,600,000,000. Use your calculator to evaluate each expression. Express final answers in ordinary notation.

(a) $(9000)^3$ **(b)** $(4000)^3$

(c) $(150{,}000)^2$ **(d)** $(170{,}000)^2$

(e) $(0.012)^5$ **(f)** $(0.0015)^4$

(g) $(0.006)^3$ **(h)** $(0.02)^6$

127. Use your calculator to check your answers to Problems 107–118.

Chapter 5 Summary

(5.1) Terms that contain variables with only whole numbers as exponents are called **monomials**. A **polynomial** is a monomial or a finite sum (or difference) of monomials. Polynomials of one term, two terms, and three terms are called **monomials, binomials,** and **trinomials**, respectively.

Addition and subtraction of polynomials are based on using the distributive property and combining similar terms.

(5.2) and (5.3) The following properties of exponents serve as a basis for multiplying polynomials:

1. $b^n \cdot b^m = b^{n+m}$

2. $(b^n)^m = b^{mn}$

3. $(ab)^n = a^n b^n$

(5.4) The following properties of exponents serve as a basis for dividing monomials:

1. $\frac{b^n}{b^m} = b^{n-m}$ when $n > m$

2. $\frac{b^n}{b^m} = 1$ when $n = m$

Dividing a polynomial by a monomial is based on the property

$$\frac{a+b}{c} = \frac{a}{c} + \frac{b}{c}.$$

(5.5) To review the division of a polynomial by a binomial, turn to Section 5.5 and study the examples carefully.

(5.6) We use the following two definitions to expand our work with exponents to include zero and the negative integers.

Definition 5.2 If b is a nonzero real number, then $b^0 = 1$.

Definition 5.3 If n is a positive integer, and b is a non-zero real number, then $b^{-n} = \frac{1}{b^n}$.

The following properties of exponents are true for all integers:

1. $b^n \cdot b^m = b^{n+m}$

2. $(b^n)^m = b^{mn}$

3. $(ab)^n = a^n b^n$

4. $\left(\frac{a}{b}\right)^n = \frac{a^n}{b^n}$ $\quad$ $b \neq 0$ whenever it appears in a denominator

5. $\frac{b^n}{b^m} = b^{n-m}$

To represent a number in scientific notation, express it as the product of a number between 1 and 10 (including 1) and an integral power of 10.

Chapter 5 Review Problem Set

For Problems 1–4, perform the additions and subtractions.

1. $(5x^2 - 6x + 4) + (3x^2 - 7x - 2)$

2. $(7y^2 + 9y - 3) - (4y^2 - 2y + 6)$

3. $(2x^2 + 3x - 4) + (4x^2 - 3x - 6) - (3x^2 - 2x - 1)$

4. $(-3x^2 - 2x + 4) - (x^2 - 5x - 6) - (4x^2 + 3x - 8)$

For Problems 5–12, remove parentheses and combine similar terms.

5. $5(2x - 1) + 7(x + 3) - 2(3x + 4)$

6. $3(2x^2 - 4x - 5) - 5(3x^2 - 4x + 1)$

7. $6(y^2 - 7y - 3) - 4(y^2 + 3y - 9)$

8. $3(a - 1) - 2(3a - 4) - 5(2a + 7)$

9. $-(a + 4) + 5(-a - 2) - 7(3a - 1)$

10. $-2(3n - 1) - 4(2n + 6) + 5(3n + 4)$

11. $3(n^2 - 2n - 4) - 4(2n^2 - n - 3)$

12. $-5(-n^2 + n - 1) + 3(4n^2 - 3n - 7)$

For Problems 13–20, find the indicated products.

13. $(5x^2)(7x^4)$ **14.** $(-6x^3)(9x^5)$

15. $(-4xy^2)(-6x^2y^3)$ **16.** $(2a^3b^4)(-3ab^5)$

17. $(2a^2b^3)^3$ **18.** $(-3xy^2)^2$

19. $5x(7x + 3)$ **20.** $(-3x^2)(8x - 1)$

For Problems 21–40, find the indicated products. Be sure to simplify your answers.

21. $(x + 9)(x + 8)$ **22.** $(3x + 7)(x + 1)$

23. $(x - 5)(x + 2)$ **24.** $(y - 4)(y - 9)$

25. $(2x - 1)(7x + 3)$ **26.** $(4a - 7)(5a + 8)$

27. $(3a - 5)^2$ **28.** $(x + 6)(2x^2 + 5x - 4)$

29. $(5n - 1)(6n + 5)$ **30.** $(3n + 4)(4n - 1)$

31. $(2n + 1)(2n - 1)$ **32.** $(4n - 5)(4n + 5)$

33. $(2a + 7)^2$ **34.** $(3a + 5)^2$

35. $(x - 2)(x^2 - x + 6)$

36. $(2x - 1)(x^2 + 4x + 7)$

37. $(a + 5)^3$ **38.** $(a - 6)^3$

39. $(x^2 - x - 1)(x^2 + 2x + 5)$

40. $(n^2 + 2n + 4)(n^2 - 7n - 1)$

For Problems 41–48, perform the divisions.

41. $\dfrac{36x^4y^5}{-3xy^2}$ **42.** $\dfrac{-56a^5b^7}{-8a^2b^3}$

43. $\dfrac{-18x^4y^3 - 54x^6y^2}{6x^2y^2}$ **44.** $\dfrac{-30a^5b^{10} + 39a^4b^8}{-3ab}$

45. $\dfrac{56x^4 - 40x^3 - 32x^2}{4x^2}$

46. $(x^2 + 9x - 1) \div (x + 5)$

47. $(21x^2 - 4x - 12) \div (3x + 2)$

48. $(2x^3 - 3x^2 + 2x - 4) \div (x - 2)$

For Problems 49–60, evaluate each expression.

49. $3^2 + 2^2$ **50.** $(3 + 2)^2$

51. 2^{-4} **52.** $(-5)^0$

53. -5^0 **54.** $\dfrac{1}{3^{-2}}$

55. $\left(\dfrac{3}{4}\right)^{-2}$ **56.** $\dfrac{1}{\left(\dfrac{1}{4}\right)^{-1}}$

57. $\dfrac{1}{(-2)^{-3}}$ **58.** $2^{-1} + 3^{-2}$

59. $3^0 + 2^{-2}$ **60.** $(2 + 3)^{-2}$

For Problems 61–72, simplify each of the following, and express your answers using positive exponents only.

61. x^5x^{-8} **62.** $(3x^5)(4x^{-2})$

63. $\dfrac{x^{-4}}{x^{-6}}$ **64.** $\dfrac{x^{-6}}{x^{-4}}$

65. $\dfrac{24a^5}{3a^{-1}}$ **66.** $\dfrac{48n^{-2}}{12n^{-1}}$

67. $(x^{-2}y)^{-1}$ **68.** $(a^2b^{-3})^{-2}$

69. $(2x)^{-1}$ **70.** $(3n^2)^{-2}$

71. $(2n^{-1})^{-3}$ **72.** $(4ab^{-1})(-3a^{-1}b^2)$

For Problems 73–76, write each expression in standard decimal form.

73. $(6.1)(10^2)$ **74.** $(5.6)(10^4)$

75. $(8)(10^{-2})$ **76.** $(9.2)(10^{-4})$

For Problems 77–80, write each number in scientific notation.

77. 9000 **78.** 47

79. 0.047 **80.** 0.00021

For Problems 81–84, use scientific notation and the properties of exponents to evaluate each expression.

81. (0.00004)(12,000) **82.** (0.0021)(2000)

83. $\dfrac{0.0056}{0.0000028}$ **84.** $\dfrac{0.00078}{39{,}000}$

Chapter 5 Test

1. Find the sum of $-7x^2 + 6x - 2$ and $5x^2 - 8x + 7$.
2. Subtract $-x^2 + 9x - 14$ from $-4x^2 + 3x + 6$.
3. Remove parentheses and combine similar terms for the expression $3(2x - 1) - 6(3x - 2) - (x + 7)$.
4. Find the product $(-4xy^2)(7x^2y^3)$.
5. Find the product $(2x^2y)^2(3xy^3)$.

For Problems 6–12, find the indicated products and express answers in simplest form.

6. $(x - 9)(x + 2)$
7. $(n + 14)(n - 7)$
8. $(5a + 3)(8a + 7)$
9. $(3x - 7y)^2$
10. $(x + 3)(2x^2 - 4x - 7)$
11. $(9x - 5y)(9x + 5y)$
12. $(3x - 7)(5x - 11)$
13. Find the indicated quotient: $\dfrac{-96x^4y^5}{-12x^2y}$.
14. Find the indicated quotient: $\dfrac{56x^2y - 72xy^2}{-8xy}$.
15. Find the indicated quotient: $(2x^3 + 5x^2 - 22x + 15) \div (2x - 3)$.
16. Find the indicated quotient: $(4x^3 + 23x^2 + 36) \div (x + 6)$.
17. Evaluate $\left(\dfrac{2}{3}\right)^{-3}$.
18. Evaluate $4^{-2} + 4^{-1} + 4^0$.
19. Evaluate $\dfrac{1}{2^{-4}}$.
20. Find the product $(-6x^{-4})(4x^2)$ and express the answer using a positive exponent.
21. Simplify $\left(\dfrac{8x^{-1}}{2x^2}\right)^{-1}$ and express the answer using a positive exponent.
22. Simplify $(x^{-3}y^5)^{-2}$ and express the answer using positive exponents.
23. Write 0.00027 in scientific notation.
24. Express $(9.2)(10^6)$ in standard decimal form.
25. Evaluate $(0.000002)(3000)$.

Chapters

For Problems 1–10, evaluate each of the numerical expressions.

1. $5 + 3(2 - 7)^2 \div 3 \cdot 5$

2. $8 \div 2 \cdot (-1) + 3$

3. $7 - 2^2 \cdot 5 \div (-1)$

4. $4 + (-2) - 3(6)$

5. $(-3)^4$

6. -2^5

7. $\left(\frac{2}{3}\right)^{-1}$

8. $\frac{1}{4^{-2}}$

9. $\left(\frac{1}{2} - \frac{1}{3}\right)^{-2}$

10. $2^0 + 2^{-1} + 2^{-2}$

For Problems 11–16, evaluate each algebraic expression for the given values of the variables.

11. $\frac{2x + 3y}{x - y}$ for $x = \frac{1}{2}$ and $y = -\frac{1}{3}$

12. $\frac{2}{5}n - \frac{1}{3}n - n + \frac{1}{2}n$ for $n = -\frac{3}{4}$

13. $\frac{3a - 2b - 4a + 7b}{-a - 3a + b - 2b}$ for $a = -1$ and $b = -\frac{1}{3}$

14. $-2(x - 4) + 3(2x - 1) - (3x - 2)$ for $x = -2$

15. $(x^2 + 2x - 4) - (x^2 - x - 2) + (2x^2 - 3x - 1)$ for $x = -1$

16. $2(n^2 - 3n - 1) - (n^2 + n + 4) - 3(2n - 1)$ for $n = 3$

For Problems 17–29, find the indicated products.

17. $(3x^2y^3)(-5xy^4)$

18. $(-6ab^4)(-2b^3)$

19. $(-2x^2y^5)^3$

20. $-3xy(2x - 5y)$

21. $(5x - 2)(3x - 1)$

22. $(7x - 1)(3x + 4)$

23. $(-x - 2)(2x + 3)$

24. $(7 - 2y)(7 + 2y)$

25. $(x - 2)(3x^2 - x - 4)$

26. $(2x - 5)(x^2 + x - 4)$

27. $(2n + 3)^3$

28. $(1 - 2n)^3$

29. $(x^2 - 2x + 6)(2x^2 + 5x - 6)$

For Problems 30–34, perform the indicated divisions.

30. $\frac{-52x^3y^4}{13xy^2}$

31. $\frac{-126a^3b^5}{-9a^2b^3}$

32. $\frac{56xy^2 - 64x^3y - 72x^4y^4}{8xy}$

33. $(2x^3 + 2x^2 - 19x - 21) \div (x + 3)$

34. $(3x^3 + 17x^2 + 6x - 4) \div (3x - 1)$

For Problems 35–38, simplify each expression, and express your answers using positive exponents only.

35. $(-2x^3)(3x^{-4})$

36. $\frac{4x^{-2}}{2x^{-1}}$

37. $(3x^{-1}y^{-2})^{-1}$

38. $(xy^2z^{-1})^{-2}$

For Problems 39–41, use scientific notation and the properties of exponents to help evaluate each numerical expression.

39. $(0.00003)(4000)$

40. $(0.0002)(0.003)^2$

41. $\frac{0.00034}{0.0000017}$

For Problems 42–49, solve each of the equations.

42. $5x + 8 = 6x - 3$

43. $-2(4x - 1) = -5x + 3 - 2x$

44. $\frac{y}{2} - \frac{y}{3} = 8$

45. $6x + 8 - 4x = 10(3x + 2)$

46. $1.6 - 2.4x = 5x - 65$

47. $-3(x - 1) + 2(x + 3) = -4$

48. $\frac{3n + 1}{5} + \frac{n - 2}{3} = \frac{2}{15}$

49. $0.06x + 0.08(1500 - x) = 110$

For Problems 50–55, solve each of the inequalities.

50. $2x - 7 \leq -3(x + 4)$

51. $6x + 5 - 3x > 5$

52. $4(x - 5) + 2(3x + 6) < 0$

53. $-5x + 3 > -4x + 5$

54. $\dfrac{3x}{4} - \dfrac{x}{2} \leq \dfrac{5x}{6} - 1$

55. $0.08(700 - x) + 0.11x \geq 65$

For Problems 56–62, set up an equation and solve each problem.

56. The sum of 4 and three times a certain number is the same as the sum of the number and 10. Find the number.

57. Fifteen percent of some number is 6. Find the number.

58. Lou has 18 coins consisting of dimes and quarters. If the total value of the coins is \$3.30, how many coins of each denomination does he have?

59. A sum of \$1500 is invested, part of it at 8% interest and the remainder at 9%. If the total interest amounts to \$128, find the amount invested at each rate.

60. How many gallons of water must be added to 15 gallons of a 12% salt solution to change it to a 10% salt solution?

61. Two airplanes leave Atlanta at the same time and fly in opposite directions. If one travels at 400 miles per hour and the other at 450 miles per hour, how long will it take them to be 2975 miles apart?

62. The length of a rectangle is 1 meter more than twice its width. If the perimeter of the rectangle is 44 meters, find the length and width.

For additional word problems see Appendix B. All Appendix problems with references to chapters 3–5 would be appropriate.

6

Factoring, Solving Equations, and Problem Solving

Algebraic equations can be used to solve a large variety of problems dealing with geometric relationships.

A flower garden is in the shape of a right triangle with one leg 7 meters longer than the other leg and the hypotenuse 1 meter longer than the longer leg. Find the lengths of all three sides of the right triangle. A popular geometric formula, called the Pythagorean theorem, serves as a guideline for setting up an equation to solve this problem. We can use the equation $x^2 + (x + 7)^2 = (x + 8)^2$ to determine that the sides of the right triangle are 5 meters, 12 meters, and 13 meters long.

The distributive property has allowed us to combine similar terms and multiply polynomials. In this chapter, we will see yet another use of the distributive property as we learn how to **factor polynomials**. Factoring polynomials will allow us to solve other kinds of equations, which will in turn help us to solve a greater variety of word problems.

6.1 Factoring by Using the Distributive Property

In Chapter 1 we found the *greatest common factor* of two or more whole numbers by inspection or by using the prime factored form of the numbers. For example, by inspection we see that the greatest common factor of 8 and 12 is 4. This means that 4 is the largest whole number that is a factor of both 8 and 12. If it is difficult to determine the greatest common factor by inspection, then we can use the prime factorization technique as follows:

$$42 = 2 \cdot 3 \cdot 7$$

$$70 = 2 \cdot 5 \cdot 7$$

We see that $2 \cdot 7 = 14$ is the greatest common factor of 42 and 70.

It is meaningful to extend the concept of greatest common factor to monomials. Consider the next example.

EXAMPLE 1

Find the greatest common factor of $8x^2$ and $12x^3$.

Solution

$$8x^2 = 2 \cdot 2 \cdot 2 \cdot x \cdot x$$

$$12x^3 = 2 \cdot 2 \cdot 3 \cdot x \cdot x \cdot x$$

Therefore, the greatest common factor is $2 \cdot 2 \cdot x \cdot x = 4x^2$. ■

By "the greatest common factor of two or more monomials," we mean the monomial with the largest numerical coefficient and highest power of the variables that is a factor of the given monomials.

EXAMPLE 2

Find the greatest common factor of $16x^2y$, $24x^3y^2$, and $32xy$.

Solution

$$16x^2y = 2 \cdot 2 \cdot 2 \cdot 2 \cdot x \cdot x \cdot y$$

$$24x^3y^2 = 2 \cdot 2 \cdot 2 \cdot 3 \cdot x \cdot x \cdot x \cdot y \cdot y$$

$$32xy = 2 \cdot 2 \cdot 2 \cdot 2 \cdot 2 \cdot x \cdot y$$

Therefore, the greatest common factor is $2 \cdot 2 \cdot 2 \cdot x \cdot y = 8xy$. ■

We have used the distributive property to multiply a polynomial by a monomial; for example,

$$3x(x + 2) = 3x^2 + 6x$$

Suppose we start with $3x^2 + 6x$ and want to express it in factored form. We use the distributive property in the form $ab + ac = a(b + c)$.

$$3x^2 + 6x = 3x(x) + 3x(2) \quad \text{3x is the greatest common factor of } 3x^2 \text{ and } 6x$$
$$= 3x(x + 2) \quad \text{Use the distributive property}$$

The next four examples further illustrate this process of **factoring out the greatest common monomial factor**.

EXAMPLE 3

Factor $12x^3 - 8x^2$.

Solution

$$12x^3 - 8x^2 = 4x^2(3x) - 4x^2(2)$$
$$= 4x^2(3x - 2) \quad ab - ac = a(b - c)$$

■

EXAMPLE 4

Factor $12x^2y + 18xy^2$.

Solution

$$12x^2y + 18xy^2 = 6xy(2x) + 6xy(3y)$$
$$= 6xy(2x + 3y)$$

■

EXAMPLE 5

Factor $24x^3 + 30x^4 - 42x^5$.

Solution

$$24x^3 + 30x^4 - 42x^5 = 6x^3(4) + 6x^3(5x) - 6x^3(7x^2)$$
$$= 6x^3(4 + 5x - 7x^2)$$

■

EXAMPLE 6

Factor $9x^2 + 9x$.

Solution

$$9x^2 + 9x = 9x(x) + 9x(1)$$
$$= 9x(x + 1)$$

■

We want to emphasize the point made just before Example 3. It is important to realize that we are factoring out the *greatest* common monomial factor. We could factor an expression such as $9x^2 + 9x$ in Example 6 as $9(x^2 + x)$, $3(3x^2 + 3x)$, $3x(3x + 3)$, or even $\frac{1}{2}(18x^2 + 18x)$, but it is the form $9x(x + 1)$ that we want. We can accomplish this by factoring out the greatest common monomial factor; we sometimes refer to this process as **factoring completely**. A polynomial with integral coefficients is in completely factored form if these conditions are met:

1. It is expressed as a product of polynomials with integral coefficients.

2. No polynomial, other than a monomial, within the factored form can be further factored into polynomials with integral coefficients.

Thus $9(x^2 + x), 3(3x^2 + 3x)$, and $3x(3x + 2)$ are not completely factored because they violate condition 2. The form $\frac{1}{2}(18x^2 + 18x)$ violates both conditions 1 and 2.

Sometimes there may be a **common binomial factor** rather than a common monomial factor. For example, each of the two terms of $x(y + 2) + z(y + 2)$ has a binomial factor of $(y + 2)$. Thus we can factor $(y + 2)$ from each term and get

$$x(y + 2) + z(y + 2) = (y + 2)(x + z)$$

Consider a few more examples involving a common binomial factor.

$$a(b + c) - d(b + c) = (b + c)(a - d)$$

$$x(x + 2) + 3(x + 2) = (x + 2)(x + 3)$$

$$x(x + 5) - 4(x + 5) = (x + 5)(x - 4)$$

It may be that the original polynomial exhibits no apparent common monomial or binomial factor, which is the case with

$$ab + 3a + bc + 3c$$

However, by factoring a from the first two terms and c from the last two terms, we see that

$$ab + 3a + bc + 3c = a(b + 3) + c(b + 3)$$

Now a common binomial factor of $(b + 3)$ is obvious, and we can proceed as before.

$$a(b + 3) + c(b + 3) = (b + 3)(a + c)$$

This factoring process is called **factoring by grouping**. Let's consider two more examples of factoring by grouping.

$x^2 - x + 5x - 5 = x(x - 1) + 5(x - 1)$ Factor x from first two terms and 5 from last two terms

$= (x - 1)(x + 5)$ Factor common binomial factor of $(x - 1)$ from both terms

$6x^2 - 4x - 3x + 2 = 2x(3x - 2) - 1(3x - 2)$ Factor $2x$ from first two terms and -1 from last two terms

$= (3x - 2)(2x - 1)$ Factor common binomial factor of $(3x - 2)$ from both terms

Back to Solving Equations

Suppose we are told that the product of two numbers is 0. What do we know about the numbers? Do you agree with our conclusion that at least one of the numbers must be 0? The next property formalizes this idea.

Property 6.1

For all real numbers a and b,

$$ab = 0 \quad \text{if and only if } a = 0 \text{ or } b = 0$$

Property 6.1 provides us with another technique for solving equations.

EXAMPLE 7

Solve $x^2 + 6x = 0$.

Solution

To solve equations by applying Property 6.1, one side of the equation must be a product, and the other side of the equation must be zero. This equation already has zero on the right-hand side of the equation, but the left-hand side of this equation is a sum. We will factor the left-hand side, $x^2 + 6x$, to change the sum into a product.

$$x^2 + 6x = 0$$

$$x(x + 6) = 0 \qquad \text{Factor}$$

$$x = 0 \quad \text{or} \quad x + 6 = 0 \qquad ab = 0 \text{ if and only if } a = 0 \text{ or } b = 0$$

$$x = 0 \quad \text{or} \quad x = -6$$

The solution set is $\{-6, 0\}$. (Be sure to check both values in the original equation.) ■

EXAMPLE 8

Solve $x^2 = 12x$.

Solution

In order to solve this equation by Property 6.1, we will first get zero on the right-hand side of the equation by adding $-12x$ to each side. Then we factor the expression on the left-hand side of the equation.

$$x^2 = 12x$$

$$x^2 - 12x = 0 \qquad \text{Added } -12x \text{ to both sides}$$

$$x(x - 12) = 0$$

$$x = 0 \quad \text{or} \quad x - 12 = 0 \qquad ab = 0 \text{ if and only if } a = 0 \text{ or } b = 0$$

$$x = 0 \quad \text{or} \quad x = 12$$

The solution set is $\{0, 12\}$. ■

Remark: Notice in Example 8 that we did not divide both sides of the original equation by x. This would cause us to lose the solution of 0.

EXAMPLE 9

Solve $4x^2 - 3x = 0$.

Solution

$$4x^2 - 3x = 0$$

$$x(4x - 3) = 0$$

$$x = 0 \quad \text{or} \quad 4x - 3 = 0 \qquad ab = 0 \text{ if and only if } a = 0 \text{ or } b = 0$$

$$x = 0 \quad \text{or} \quad 4x = 3$$

$$x = 0 \quad \text{or} \quad x = \frac{3}{4}$$

The solution set is $\left\{0, \frac{3}{4}\right\}$. ■

EXAMPLE 10

Solve $x(x + 2) + 3(x + 2) = 0$.

Solution

In order to solve this equation by Property 6.1, we will factor the left-hand side of the equation. The greatest common factor of the terms is $(x + 2)$.

$$x(x + 2) + 3(x + 2) = 0$$

$$(x + 2)(x + 3) = 0$$

$$x + 2 = 0 \quad \text{or} \quad x + 3 = 0 \qquad ab = 0 \text{ if and only if } a = 0 \text{ or } b = 0$$

$$x = -2 \quad \text{or} \quad x = -3$$

The solution set is $\{-3, -2\}$. ■

Each time we expand our equation-solving capabilities, we also acquire more techniques for solving problems. Let's solve a geometric problem with the ideas we learned in this section.

PROBLEM 1

The area of a square is numerically equal to twice its perimeter. Find the length of a side of the square.

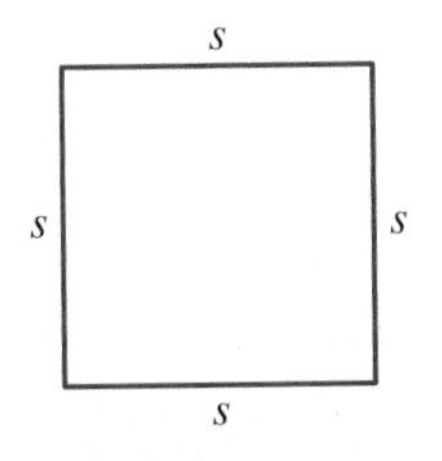

Figure 6.1

Solution

Sketch a square and let s represent the length of each side (see Figure 6.1). Then the area is represented by s^2 and the perimeter by $4s$. Thus,

$$s^2 = 2(4s)$$

$$s^2 = 8s$$

$$s^2 - 8s = 0$$
$$s(s - 8) = 0$$
$$s = 0 \quad \text{or} \quad s - 8 = 0$$
$$s = 0 \quad \text{or} \quad s = 8$$

Since 0 is not a reasonable answer to the problem, the solution is 8. (Be sure to check this solution in the original statement of the problem!) ■

Problem Set 6.1

For Problems 1–10, find the greatest common factor of the given expressions.

1. $24y$ and $30xy$

2. $32x$ and $40xy$

3. $60x^2y$ and $84xy^2$

4. $72x^3$ and $63x^2$

5. $42ab^3$ and $70a^2b^2$

6. $48a^2b^2$ and $96ab^4$

7. $6x^3$, $8x$, and $24x^2$

8. $72xy$, $36x^2y$, and $84xy^2$

9. $16a^2b^2$, $40a^2b^3$, and $56a^3b^4$

10. $70a^3b^3$, $42a^2b^4$, and $49ab^5$

For Problems 11–46, factor each polynomial completely.

11. $8x + 12y$

12. $18x + 24y$

13. $14xy - 21y$

14. $24x - 40xy$

15. $18x^2 + 45x$

16. $12x + 28x^3$

17. $12xy^2 - 30x^2y$

18. $28x^2y^2 - 49x^2y$

19. $36a^2b - 60a^3b^4$

20. $65ab^3 - 45a^2b^2$

21. $16xy^3 + 25x^2y^2$

22. $12x^2y^2 + 29x^2y$

23. $64ab - 72cd$

24. $45xy - 72zw$

25. $9a^2b^4 - 27a^2b$

26. $7a^3b^5 - 42a^2b^6$

27. $52x^4y^2 + 60x^6y$

28. $70x^5y^3 - 42x^8y^2$

29. $40x^2y^2 + 8x^2y$

30. $84x^2y^3 + 12xy^3$

31. $12x + 15xy + 21x^2$

32. $30x^2y + 40xy + 55y$

33. $2x^3 - 3x^2 + 4x$

34. $x^4 + x^3 + x^2$

35. $44y^5 - 24y^3 - 20y^2$

36. $14a - 18a^3 - 26a^5$

37. $14a^2b^3 + 35ab^2 - 49a^3b$

38. $24a^3b^2 + 36a^2b^4 - 60a^4b^3$

39. $x(y + 1) + z(y + 1)$

40. $a(c + d) + 2(c + d)$

41. $a(b - 4) - c(b - 4)$

42. $x(y - 6) - 3(y - 6)$

43. $x(x + 3) + 6(x + 3)$

44. $x(x - 7) + 9(x - 7)$

45. $2x(x + 1) - 3(x + 1)$

46. $4x(x + 8) - 5(x + 8)$

For Problems 47–60, use the process of *factoring by grouping* to factor each polynomial.

47. $5x + 5y + bx + by$

48. $7x + 7y + zx + zy$

49. $bx - by - cx + cy$

50. $2x - 2y - ax + ay$

51. $ac + bc + a + b$

52. $x + y + ax + ay$

53. $x^2 + 5x + 12x + 60$

54. $x^2 + 3x + 7x + 21$

55. $x^2 - 2x - 8x + 16$

56. $x^2 - 4x - 9x + 36$

57. $2x^2 + x - 10x - 5$

58. $3x^2 + 2x - 18x - 12$

59. $6n^2 - 3n - 8n + 4$

60. $20n^2 + 8n - 15n - 6$

For Problems 61–84, solve each equation.

61. $x^2 - 8x = 0$

62. $x^2 - 12x = 0$

63. $x^2 + x = 0$

64. $x^2 + 7x = 0$

65. $n^2 = 5n$

66. $n^2 = -2n$

67. $2y^2 - 3y = 0$

68. $4y^2 - 7y = 0$

69. $7x^2 = -3x$

70. $5x^2 = -2x$

71. $3n^2 + 15n = 0$

72. $6n^2 - 24n = 0$

73. $4x^2 = 6x$

74. $12x^2 = 8x$

75. $7x - x^2 = 0$

76. $9x - x^2 = 0$

77. $13x = x^2$

78. $15x = -x^2$

79. $5x = -2x^2$

80. $7x = -5x^2$

81. $x(x + 5) - 4(x + 5) = 0$

82. $x(3x - 2) - 7(3x - 2) = 0$

83. $4(x - 6) - x(x - 6) = 0$

84. $x(x + 9) = 2(x + 9)$

For Problems 85–91, set up an equation and solve each problem.

85. The square of a number equals nine times that number. Find the number.

86. Suppose that four times the square of a number equals 20 times that number. What is the number?

87. The area of a square is numerically equal to five times its perimeter. Find the length of a side of the square.

88. The area of a square is 14 times as large as the area of a triangle. One side of the triangle is 7 inches long, and the altitude to that side is the same length as a side of the square. Find the length of a side of the square. Also find the areas of both figures and be sure that your answer checks.

89. Suppose that the area of a circle is numerically equal to the perimeter of a square, and that the length of a radius of the circle is equal to the length of a side of the square. Find the length of a side of the square. Express your answer in terms of π.

90. One side of a parallelogram, an altitude to that side, and one side of a rectangle all have the same measure. If an adjacent side of the rectangle is 20 centimeters long, and the area of the rectangle is twice the area of the parallelogram, find the areas of both figures.

91. The area of a rectangle is twice the area of a square. If the rectangle is 6 inches long, and the width of the rectangle is the same as the length of a side of the square, find the dimensions of both the rectangle and the square.

THOUGHTS INTO WORDS

92. Suppose that your friend factors $24x^2y + 36xy$ like this:

$$\begin{aligned} 24x^2y + 36xy &= 4xy(6x + 9) \\ &= (4xy)(3)(2x + 3) \\ &= 12xy(2x + 3) \end{aligned}$$

Is this correct? Would you make any suggestions for changing her method?

93. The following solution is given for the equation $x(x - 10) = 0$:

$$x(x - 10) = 0$$
$$x^2 - 10x = 0$$
$$x(x - 10) = 0$$
$$x = 0 \quad \text{or} \quad x - 10 = 0$$
$$x = 0 \quad \text{or} \quad x = 10$$

The solution set is {0, 10}. Is this a correct solution? Would you suggest any changes to the method?

FURTHER INVESTIGATIONS

94. The total surface area of a right circular cylinder is given by the formula $A = 2\pi r^2 + 2\pi rh$, where r represents the radius of a base, and h represents the height of the cylinder. For computational purposes, it may be more convenient to change the form of the right side of the formula by factoring it.

$$A = 2\pi r^2 + 2\pi rh$$
$$= 2\pi r(r + h)$$

Use $A = 2\pi r(r + h)$ to find the total surface area of each of the following cylinders. Use $\frac{22}{7}$ as an approximation for π.

(a) $r = 7$ centimeters and $h = 12$ centimeters

(b) $r = 14$ meters and $h = 20$ meters

(c) $r = 3$ feet and $h = 4$ feet

(d) $r = 5$ yards and $h = 9$ yards

95. The formula $A = P + Prt$ yields the total amount of money accumulated (A) when P dollars is invested at r percent simple interest for t years. For computational purposes it may be convenient to change the right side of the formula by factoring.

$$A = P + Prt$$
$$= P(1 + rt)$$

Use $A = P(1 + rt)$ to find the total amount of money accumulated for each of the following investments.

(a) \$100 at 8% for 2 years

(b) \$200 at 9% for 3 years

(c) \$500 at 10% for 5 years

(d) \$1000 at 10% for 10 years

For Problems 96–99, solve each equation for the indicated variable.

96. $ax + bx = c$ for x

97. $b^2x^2 - cx = 0$ for x

98. $5ay^2 = by$ for y

99. $y + ay - by - c = 0$ for y

6.2 Factoring the Difference of Two Squares

In Section 5.3 we noted some special multiplication patterns. One of these patterns was

$$(a - b)(a + b) = a^2 - b^2$$

We can view this same pattern as follows:

Difference of Two Squares

$$a^2 - b^2 = (a - b)(a + b)$$

To apply the pattern is a fairly simple process, as these next examples illustrate. The steps inside the box are often performed mentally.

$x^2 - 36 =$	$(x)^2 - (6)^2$	$= (x - 6)(x + 6)$
$4x^2 - 25 =$	$(2x)^2 - (5)^2$	$= (2x - 5)(2x + 5)$
$9x^2 - 16y^2 =$	$(3x)^2 - (4y)^2$	$= (3x - 4y)(3x + 4y)$
$64 - y^2 =$	$(8)^2 - (y)^2$	$= (8 - y)(8 + y)$

Since multiplication is commutative, the order of writing the factors is not important. For example, $(x - 6)(x + 6)$ can also be written as $(x + 6)(x - 6)$.

You must be careful not to assume an analogous factoring pattern for the sum of two squares; it does not exist. For example, $x^2 + 4 \neq (x + 2)(x + 2)$ because $(x + 2)(x + 2) = x^2 + 4x + 4$. We say that the **sum of two squares is not factorable using integers**. The phrase "using integers" is necessary because $x^2 + 4$ could be written as $\frac{1}{2}(2x^2 + 8)$, but such factoring is of no help. Furthermore, we do not consider $(1)(x^2 + 4)$ as factoring $x^2 + 4$.

It is possible that both the technique of *factoring out a common monomial factor* and the pattern *difference of two squares* can be applied to the same polynomial. In general, it is best to look for a common monomial factor first.

EXAMPLE 1

Factor $2x^2 - 50$.

Solution

$$2x^2 - 50 = 2(x^2 - 25) \qquad \text{Common factor of 2}$$
$$= 2(x - 5)(x + 5) \qquad \text{Difference of squares}$$

■

In Example 1, by expressing $2x^2 - 50$ as $2(x - 5)(x + 5)$, we say that the algebraic expression has been **factored completely**. That means that the factors 2, $x - 5$, and $x + 5$ cannot be factored any further using integers.

EXAMPLE 2

Factor completely $18y^3 - 8y$.

Solution

$$18y^3 - 8y = 2y(9y^2 - 4) \qquad \text{Common factor of } 2y$$
$$= 2y(3y - 2)(3y + 2) \qquad \text{Difference of squares}$$

■

Sometimes it is possible to apply the difference-of-squares pattern more than once. Consider the next example.

EXAMPLE 3 Factor completely $x^4 - 16$.

Solution

$$x^4 - 16 = (x^2 + 4)(x^2 - 4)$$
$$= (x^2 + 4)(x + 2)(x - 2)$$

The following examples should help you to summarize the factoring ideas presented thus far.

$5x^2 + 20 = 5(x^2 + 4)$

$25 - y^2 = (5 - y)(5 + y)$

$3 - 3x^2 = 3(1 - x^2) = 3(1 + x)(1 - x)$

$36x^2 - 49y^2 = (6x - 7y)(6x + 7y)$

$a^2 + 9$ is not factorable using integers

$9x + 17y$ is not factorable using integers

Solving Equations

Each time we learn a new factoring technique, we also develop more power for solving equations. Let's consider how we can use the difference-of-squares factoring pattern to help solve certain kinds of equations.

EXAMPLE 4 Solve $x^2 = 25$.

Solution

$$x^2 = 25$$
$$x^2 - 25 = 0 \quad \text{Added } -25 \text{ to both sides}$$
$$(x + 5)(x - 5) = 0$$
$$x + 5 = 0 \quad \text{or} \quad x - 5 = 0 \quad \text{Remember: } ab = 0 \text{ if and only if } a = 0 \text{ or } b = 0$$
$$x = -5 \quad \text{or} \quad x = 5$$

The solution set is $\{-5, 5\}$. Check these answers!

EXAMPLE 5 Solve $9x^2 = 25$.

Solution

$$9x^2 = 25$$
$$9x^2 - 25 = 0$$
$$(3x + 5)(3x - 5) = 0$$

$$3x + 5 = 0 \quad \text{or} \quad 3x - 5 = 0$$

$$3x = -5 \quad \text{or} \quad 3x = 5$$

$$x = -\frac{5}{3} \quad \text{or} \quad x = \frac{5}{3}$$

The solution set is $\left\{-\frac{5}{3}, \frac{5}{3}\right\}$. ■

EXAMPLE 6

Solve $5y^2 = 20$.

Solution

$$5y^2 = 20$$

$$\frac{5y^2}{5} = \frac{20}{5} \qquad \text{Divide both sides by 5}$$

$$y^2 = 4$$

$$y^2 - 4 = 0$$

$$(y + 2)(y - 2) = 0$$

$$y + 2 = 0 \quad \text{or} \quad y - 2 = 0$$

$$y = -2 \quad \text{or} \quad y = 2$$

The solution set is $\{-2, 2\}$. Check it! ■

EXAMPLE 7

Solve $x^3 - 9x = 0$.

Solution

$$x^3 - 9x = 0$$

$$x(x^2 - 9) = 0$$

$$x(x - 3)(x + 3) = 0$$

$$x = 0 \quad \text{or} \quad x - 3 = 0 \quad \text{or} \quad x + 3 = 0$$

$$x = 0 \quad \text{or} \quad x = 3 \quad \text{or} \quad x = -3$$

The solution set is $\{-3, 0, 3\}$. ■

The more we know about solving equations, the more easily we can solve word problems.

PROBLEM 1

The combined area of two squares is 20 square centimeters. Each side of one square is twice as long as a side of the other square. Find the lengths of the sides of each square.

Solution

We can sketch two squares and label the sides of the smaller square s (see Figure 6.2). Then the sides of the larger square are $2s$. Since the sum of the areas of the two squares is 20 square centimeters, we can set up and solve the following equation:

$$s^2 + (2s)^2 = 20$$
$$s^2 + 4s^2 = 20$$
$$5s^2 = 20$$
$$s^2 = 4$$
$$s^2 - 4 = 0$$
$$(s + 2)(s - 2) = 0$$
$$s + 2 = 0 \quad \text{or} \quad s - 2 = 0$$
$$s = -2 \quad \text{or} \quad s = 2$$

s s $2s$ $2s$

Figure 6.2

Since s represents the length of a side of a square, we must disregard the solution -2. Thus one square has sides of length 2 centimeters and the other square has sides of length $2(2) = 4$ centimeters. ■

Problem Set 6.2

For Problems 1–12, use the difference-of-squares pattern to factor each polynomial.

1. $x^2 - 1$ **2.** $x^2 - 25$

3. $x^2 - 100$ **4.** $x^2 - 121$

5. $x^2 - 4y^2$ **6.** $x^2 - 36y^2$

7. $9x^2 - y^2$ **8.** $49y^2 - 64x^2$

9. $36a^2 - 25b^2$ **10.** $4a^2 - 81b^2$

11. $1 - 4n^2$ **12.** $4 - 9n^2$

For Problems 13–40, factor each polynomial completely. Indicate any that are not factorable using integers. Don't forget to look for a common monomial factor first.

13. $5x^2 - 20$ **14.** $7x^2 - 7$

15. $8x^2 + 32$ **16.** $12x^2 + 60$

17. $2x^2 - 18y^2$ **18.** $8x^2 - 32y^2$

19. $x^3 - 25x$ **20.** $2x^3 - 2x$

21. $x^2 + 9y^2$ **22.** $18x - 42y$

23. $45x^2 - 36xy$ **24.** $16x^2 + 25y^2$

25. $36 - 4x^2$ **26.** $75 - 3x^2$

27. $4a^4 + 16a^2$ **28.** $9a^4 + 81a^2$

29. $x^4 - 81$ **30.** $16 - x^4$

31. $x^4 + x^2$ **32.** $x^5 + 2x^3$

33. $3x^3 + 48x$ **34.** $6x^3 + 24x$

35. $5x - 20x^3$ **36.** $4x - 36x^3$

37. $4x^2 - 64$ **38.** $9x^2 - 9$

39. $75x^3y - 12xy^3$ **40.** $32x^3y - 18xy^3$

For Problems 41–64, solve each equation.

41. $x^2 = 9$

42. $x^2 = 1$

43. $4 = n^2$

44. $144 = n^2$

45. $9x^2 = 16$

46. $4x^2 = 9$

47. $n^2 - 121 = 0$

48. $n^2 - 81 = 0$

49. $25x^2 = 4$

50. $49x^2 = 36$

51. $3x^2 = 75$

52. $7x^2 = 28$

53. $3x^3 - 48x = 0$

54. $x^3 - x = 0$

55. $n^3 = 16n$

56. $2n^3 = 8n$

57. $5 - 45x^2 = 0$

58. $3 - 12x^2 = 0$

59. $4x^3 - 400x = 0$

60. $2x^3 - 98x = 0$

61. $64x^2 = 81$

62. $81x^2 = 25$

63. $36x^3 = 9x$

64. $64x^3 = 4x$

For Problems 65–76, set up an equation and solve the problem.

65. Forty-nine less than the square of a number equals zero. Find the number.

66. The cube of a number equals nine times the number. Find the number.

67. Suppose that five times the cube of a number equals 80 times the number. Find the number.

68. Ten times the square of a number equals 40. Find the number.

69. The sum of the areas of two squares is 234 square inches. Each side of the larger square is five times the length of a side of the smaller square. Find the length of a side of each square.

70. The difference of the areas of two squares is 75 square feet. Each side of the larger square is twice the length of a side of the smaller square. Find the length of a side of each square.

71. Suppose that the length of a certain rectangle is $2\frac{1}{2}$ times its width, and the area of that same rectangle is 160 square centimeters. Find the length and width of the rectangle.

72. Suppose that the width of a certain rectangle is three-fourths of its length, and the area of that same rectangle is 108 square meters. Find the length and width of the rectangle.

73. The sum of the areas of two circles is 80π square meters. Find the length of a radius of each circle if one of them is twice as long as the other.

74. The area of a triangle is 98 square feet. If one side of the triangle and the altitude to that side are of equal length, find the length.

75. The total surface area of a right circular cylinder is 100π square centimeters. If a radius of the base and the altitude of the cylinder are the same length, find the length of a radius.

76. The total surface area of a right circular cone is 192π square feet. If the slant height of the cone is equal in length to a diameter of the base, find the length of a radius.

THOUGHTS INTO WORDS

77. How do we know that the equation $x^2 + 1 = 0$ has no solutions in the set of real numbers?

78. Why is the following factoring process incomplete?

$16x^2 - 64 = (4x + 8)(4x - 8)$

How could the factoring be done?

79. Consider the following solution:

$$4x^2 - 36 = 0$$
$$4(x^2 - 9) = 0$$
$$4(x + 3)(x - 3) = 0$$
$$4 = 0 \quad \text{or} \quad x + 3 = 0 \quad \text{or} \quad x - 3 = 0$$
$$4 = 0 \quad \text{or} \quad x = -3 \quad \text{or} \quad x = 3$$

The solution set is $\{-3, 3\}$. Is this a correct solution? Do you have any suggestions to offer the person who did this problem?

FURTHER INVESTIGATIONS

The following patterns can be used to factor the sum and difference of two cubes.

$$a^3 + b^3 = (a + b)(a^2 - ab + b^2)$$

$$a^3 - b^3 = (a - b)(a^2 + ab + b^2)$$

Consider these examples.

$$x^3 + 8 = (x)^3 + (2)^3 = (x + 2)(x^2 - 2x + 4)$$

$$x^3 - 1 = (x)^3 - (1)^3 = (x - 1)(x^2 + x + 1)$$

Use the sum and difference of cubes patterns to factor each polynomial.

80. $x^3 + 1$

81. $x^3 - 8$

82. $n^3 - 27$

83. $n^3 + 64$

84. $8x^3 + 27y^3$

85. $27a^3 - 64b^3$

86. $1 - 8x^3$

87. $1 + 27a^3$

88. $x^3 + 8y^3$

89. $8x^3 - y^3$

90. $a^3b^3 - 1$

91. $27x^3 - 8y^3$

92. $8 + n^3$

93. $125x^3 + 8y^3$

94. $27n^3 - 125$

95. $64 + x^3$

6.3 Factoring Trinomials of the Form $x^2 + bx + c$

One of the most common types of factoring used in algebra is the expression of a trinomial as the product of two binomials. In this section we will consider trinomials where the coefficient of the squared term is 1—that is, trinomials of the form $x^2 + bx + c$.

Again, to develop a factoring technique we first look at some multiplication ideas. Consider the product $(x + r)(x + s)$, and use the distributive property to show how each term of the resulting trinomial is formed.

$$(x + r)(x + s) = x(x) + x(s) + r(x) + r(s)$$

$$x^2 \quad + \quad (s + r)x \quad + \quad rs$$

Notice that the coefficient of the middle term is the *sum* of r and s, and the last term is the *product* of r and s. These two relationships are used in the next examples.

EXAMPLE 1

Factor $x^2 + 7x + 12$.

Solution

We need to fill in the blanks with two numbers whose product is 12 and whose sum is 7.

$$x^2 + 7x + 12 = (x + \_\_\_\_\_)(x + \_\_\_\_\_)$$

To assist in finding the numbers, we can set up a table of the factors of 12.

Product	Sum
$1(12) = 12$	$1 + 12 = 13$
$2(6) = 12$	$2 + 6 = 8$
$3(4) = 12$	$3 + 4 = 7$

The bottom line contains the numbers that we need. Thus,

$$x^2 + 7x + 12 = (x + 3)(x + 4)$$ ■

EXAMPLE 2

Factor $x^2 - 11x + 24$.

Solution

To factor $x^2 - 11x + 24$, we want to find two numbers whose product is 24 and whose sum is -11.

Product	Sum
$(-1)(-24) = 24$	$-1 + (-24) = -25$
$(-2)(-12) = 24$	$-2 + (-12) = -14$
$(-3)(-8) = 24$	$-3 + (-8) = -11$
$(-4)(-6) = 24$	$-4 + (-6) = -10$

The third line contains the numbers that we want. Thus

$$x^2 - 11x + 24 = (x - 3)(x - 8)$$ ■

EXAMPLE 3

Factor $x^2 + 3x - 10$.

Solution

To factor $x^2 + 3x - 10$, we want to find two numbers whose product is -10 and whose sum is 3.

Product	Sum
$1(-10) = -10$	$1 + (-10) = -9$
$-1(10) = -10$	$-1 + 10 = 9$
$2(-5) = -10$	$2 + (-5) = -3$
$-2(5) = -10$	$-2 + 5 = 3$

The bottom line is the key line. Thus,

$$x^2 + 3x - 10 = (x + 5)(x - 2)$$ ■

EXAMPLE 4

Factor $x^2 - 2x - 8$.

Solution

We are looking for two numbers whose product is -8 and whose sum is -2.

Product	Sum
$1(-8) = -8$	$1 + (-8) = -7$
$-1(8) = -8$	$-1 + 8 = 7$
$2(-4) = -8$	$2 + (-4) = -2$
$-2(4) = -8$	$-2 + 4 = 2$

The third line has the information we want.

$$x^2 - 2x - 8 = (x - 4)(x + 2)$$

■

The tables in the last four examples illustrate one way of organizing your thoughts for such problems. We show complete tables; that is, for Example 4, we include the bottom line even though the desired numbers are obtained in the third line. If you use such tables, keep in mind that as soon as you get the desired numbers, the table need not be completed any further. Furthermore, you may be able to find the numbers without using a table. The key ideas are the product and sum relationships.

EXAMPLE 5

Factor $x^2 - 13x + 12$.

Solution

Product	Sum
$(-1)(-12) = 12$	$(-1) + (-12) = -13$

We need not complete the table.

$$x^2 - 13x + 12 = (x - 1)(x - 12)$$

■

In the next example, we refer to the concept of absolute value. Recall that the absolute value of any nonzero real number is positive. For example,

$$|4| = 4 \quad \text{and} \quad |-4| = 4$$

EXAMPLE 6

Factor $x^2 - x - 56$.

Solution

Notice that the coefficient of the middle term is -1. Therefore, we are looking for two numbers whose product is -56; because their sum is -1, the absolute value of the negative number must be one larger than the absolute value of the positive number. The numbers are -8 and 7, and we have

$$x^2 - x - 56 = (x - 8)(x + 7)$$

■

EXAMPLE 7

Factor $x^2 + 10x + 12$.

Solution

Product	Sum
$1(12) = 12$	$1 + 12 = 13$
$2(6) = 12$	$2 + 6 = 8$
$3(4) = 12$	$3 + 4 = 7$

Since the table is complete and no two factors of 12 produce a sum of 10, we conclude that

$$x^2 + 10x + 12$$

is not factorable using integers. ■

In a problem such as Example 7, we need to be sure that we have tried all possibilities before we conclude that the trinomial is not factorable.

■ Back to Solving Equations

The property $ab = 0$ if and only if $a = 0$ or $b = 0$ continues to play an important role as we solve equations that involve the factoring ideas of this section. Consider the following examples.

EXAMPLE 8

Solve $x^2 + 8x + 15 = 0$.

Solution

$$x^2 + 8x + 15 = 0$$

$$(x + 3)(x + 5) = 0 \quad \text{Factor the left side}$$

$$x + 3 = 0 \quad \text{or} \quad x + 5 = 0 \quad \text{Use } ab = 0 \text{ if and only if } a = 0 \text{ or } b = 0$$

$$x = -3 \quad \text{or} \quad x = -5$$

The solution set is $\{-5, -3\}$. ■

EXAMPLE 9

Solve $x^2 + 5x - 6 = 0$.

Solution

$$x^2 + 5x - 6 = 0$$
$$(x + 6)(x - 1) = 0$$
$$x + 6 = 0 \quad \text{or} \quad x - 1 = 0$$
$$x = -6 \quad \text{or} \quad x = 1$$

The solution set is $\{-6, 1\}$. ■

EXAMPLE 10

Solve $y^2 - 4y = 45$.

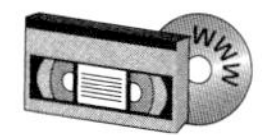

Solution

$$y^2 - 4y = 45$$
$$y^2 - 4y - 45 = 0$$
$$(y - 9)(y + 5) = 0$$
$$y - 9 = 0 \quad \text{or} \quad y + 5 = 0$$
$$y = 9 \quad \text{or} \quad y = -5$$

The solution set is $\{-5, 9\}$. ■

Don't forget that we can always check to be absolutely sure of our solutions. Let's check the solutions for Example 10. If $y = 9$, then $y^2 - 4y = 45$ becomes

$$9^2 - 4(9) \stackrel{?}{=} 45$$
$$81 - 36 \stackrel{?}{=} 45$$
$$45 = 45$$

If $y = -5$, then $y^2 - 4y = 45$ becomes

$$(-5)^2 - 4(-5) \stackrel{?}{=} 45$$
$$25 + 20 \stackrel{?}{=} 45$$
$$45 = 45$$

■ Back to Problem Solving

The more we know about factoring and solving equations, the more easily we can solve word problems.

PROBLEM 1

Find two consecutive integers whose product is 72.

Solution

Let n represent one integer. Then $n + 1$ represents the next integer.

$$n(n + 1) = 72 \quad \text{The product of the two integers is 72}$$

$$n^2 + n = 72$$

$$n^2 + n - 72 = 0$$

$$(n + 9)(n - 8) = 0$$

$$n + 9 = 0 \quad \text{or} \quad n - 8 = 0$$

$$n = -9 \quad \text{or} \quad n = 8$$

If $n = -9$, then $n + 1 = -9 + 1 = -8$. If $n = 8$, then $n + 1 = 8 + 1 = 9$. Thus the consecutive integers are -9 and -8 or 8 and 9. ■

PROBLEM 2

A rectangular plot is 6 meters longer than it is wide. The area of the plot is 16 square meters. Find the length and width of the plot.

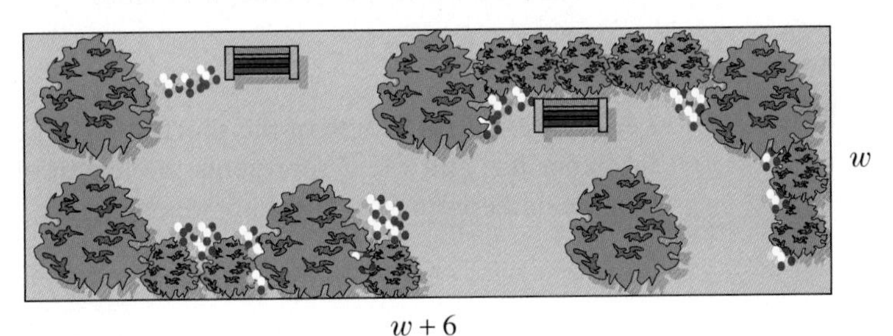

Figure 6.3

Solution

We let w represent the width of the plot, and then $w + 6$ represents the length (see Figure 6.3). Using the area formula $A = lw$, we obtain

$$w(w + 6) = 16$$

$$w^2 + 6w = 16$$

$$w^2 + 6w - 16 = 0$$

$$(w + 8)(w - 2) = 0$$

$$w + 8 = 0 \quad \text{or} \quad w - 2 = 0$$

$$w = -8 \quad \text{or} \quad w = 2$$

The solution of -8 is not possible for the width of a rectangle, so the plot is 2 meters wide and its length $(w + 6)$ is 8 meters. ■

The Pythagorean theorem, an important theorem pertaining to right triangles, can also serve as a guideline for solving certain types of problems. The Pythagorean theorem states that **in any right triangle, the square of the longest side** (called the hypotenuse) **is equal to the sum of the squares of the other two sides** (called legs); see Figure 6.4. We can use this theorem to help solve a problem.

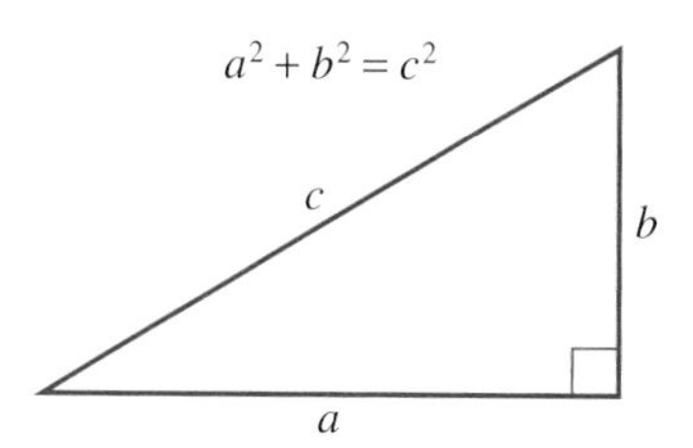

Figure 6.4

PROBLEM 3

Suppose that the lengths of the three sides of a right triangle are consecutive whole numbers. Find the lengths of the three sides.

Solution

Let s represent the length of the shortest leg. Then $s + 1$ represents the length of the other leg, and $s + 2$ represents the length of the hypotenuse. Using the Pythagorean theorem as a guideline, we obtain the following equation:

$$\underbrace{s^2 + (s + 1)^2}_{\text{Sum of squares of two legs}} = \underbrace{(s + 2)^2}_{\text{Square of hypotenuse}}$$

Solving this equation yields

$$s^2 + s^2 + 2s + 1 = s^2 + 4s + 4 \qquad \text{Remember } (a + b)^2 = a^2 + 2ab + b^2$$

$$2s^2 + 2s + 1 = s^2 + 4s + 4$$

$$s^2 + 2s + 1 = 4s + 4 \qquad \text{Add } -s^2 \text{ to both sides}$$

$$s^2 - 2s + 1 = 4$$

$$s^2 - 2s - 3 = 0$$

$$(s - 3)(s + 1) = 0$$

$$s - 3 = 0 \quad \text{or} \quad s + 1 = 0$$

$$s = 3 \quad \text{or} \quad s = -1$$

The solution of -1 is not possible for the length of a side, so the shortest side (s) is of length 3. The other two sides ($s + 1$ and $s + 2$) have lengths of 4 and 5. ■

Problem Set 6.3

For Problems 1–30, factor each trinomial completely. Indicate any that are not factorable using integers.

1. $x^2 + 10x + 24$

2. $x^2 + 9x + 14$

3. $x^2 + 13x + 40$

4. $x^2 + 11x + 24$

5. $x^2 - 11x + 18$

6. $x^2 - 5x + 4$

7. $n^2 - 11n + 28$

8. $n^2 - 7n + 10$

9. $n^2 + 6n - 27$

10. $n^2 + 3n - 18$

11. $n^2 - 6n - 40$

12. $n^2 - 4n - 45$

13. $t^2 + 12t + 24$

14. $t^2 + 20t + 96$

15. $x^2 - 18x + 72$

16. $x^2 - 14x + 32$

17. $x^2 + 5x - 66$

18. $x^2 + 11x - 42$

19. $y^2 - y - 72$

20. $y^2 - y - 30$

21. $x^2 + 21x + 80$

22. $x^2 + 21x + 90$

23. $x^2 + 6x - 72$

24. $x^2 - 8x - 36$

25. $x^2 - 10x - 48$

26. $x^2 - 12x - 64$

27. $x^2 + 3xy - 10y^2$

28. $x^2 - 4xy - 12y^2$

29. $a^2 - 4ab - 32b^2$

30. $a^2 + 3ab - 54b^2$

For Problems 31–50, solve each equation.

31. $x^2 + 10x + 21 = 0$

32. $x^2 + 9x + 20 = 0$

33. $x^2 - 9x + 18 = 0$

34. $x^2 - 9x + 8 = 0$

35. $x^2 - 3x - 10 = 0$

36. $x^2 - x - 12 = 0$

37. $n^2 + 5n - 36 = 0$

38. $n^2 + 3n - 18 = 0$

39. $n^2 - 6n - 40 = 0$

40. $n^2 - 8n - 48 = 0$

41. $t^2 + t - 56 = 0$

42. $t^2 + t - 72 = 0$

43. $x^2 - 16x + 28 = 0$

44. $x^2 - 18x + 45 = 0$

45. $x^2 + 11x = 12$

46. $x^2 + 8x = 20$

47. $x(x - 10) = -16$

48. $x(x - 12) = -35$

49. $-x^2 - 2x + 24 = 0$

50. $-x^2 + 6x + 16 = 0$

For Problems 51–68, set up an equation and solve each problem.

51. Find two consecutive integers whose product is 56.

52. Find two consecutive odd whole numbers whose product is 63.

53. Find two consecutive even whole numbers whose product is 168.

54. One number is 2 larger than another number. The sum of their squares is 100. Find the numbers.

55. Find four consecutive integers such that the product of the two larger integers is 22 less than twice the product of the two smaller integers.

56. Find three consecutive integers such that the product of the two smaller integers is 2 more than ten times the largest integer.

57. One number is 3 smaller than another number. The square of the larger number is 9 larger than ten times the smaller number. Find the numbers.

58. The area of the floor of a rectangular room is 84 square feet. The length of the room is 5 feet more than its width. Find the length and width of the room.

59. Suppose that the width of a certain rectangle is 3 inches less than its length. The area is numerically 6 less than twice the perimeter. Find the length and width of the rectangle.

60. The sum of the areas of a square and a rectangle is 64 square centimeters. The length of the rectangle is 4 centimeters more than a side of the square, and the width of the rectangle is 2 centimeters more than a side of the square. Find the dimensions of the square and the rectangle.

61. The perimeter of a rectangle is 30 centimeters, and the area is 54 square centimeters. Find the length and width of the rectangle. [*Hint*: Let w represent the width; then $15 - w$ represents the length.]

62. The perimeter of a rectangle is 44 inches, and its area is 120 square inches. Find the length and width of the rectangle.

63. An apple orchard contains 84 trees. The number of trees per row is five more than the number of rows. Find the number of rows.

64. A room contains 54 chairs. The number of rows is 3 less than the number of chairs per row. Find the number of rows.

65. Suppose that one leg of a right triangle is 7 feet shorter than the other leg. The hypotenuse is 2 feet longer than the longer leg. Find the lengths of all three sides of the right triangle.

66. Suppose that one leg of a right triangle is 7 meters longer than the other leg. The hypotenuse is 1 meter longer than the longer leg. Find the lengths of all three sides of the right triangle.

67. Suppose that the length of one leg of a right triangle is 2 inches less than the length of the other leg. If the length of the hypotenuse is 10 inches, find the length of each leg.

68. The length of one leg of a right triangle is 3 centimeters more than the length of the other leg. The length of the hypotenuse is 15 centimeters. Find the lengths of the two legs.

THOUGHTS INTO WORDS

69. What does the expression "not factorable using integers" mean to you?

70. Discuss the role that factoring plays in solving equations.

71. Explain how you would solve the equation

$(x - 3)(x + 4) = 0$

and also how you would solve

$(x - 3)(x + 4) = 8.$

FURTHER INVESTIGATIONS

For Problems 72–75, factor each trinomial and assume that all variables appearing as exponents represent positive integers.

72. $x^{2a} + 10x^a + 24$

73. $x^{2a} + 13x^a + 40$

74. $x^{2a} - 2x^a - 8$

75. $x^{2a} + 6x^a - 27$

76. Suppose that we want to factor $n^2 + 26n + 168$ so that we can solve the equation $n^2 + 26n + 168 = 0$. We need to find two positive integers whose product is 168 and whose sum is 26. Since the constant term, 168, is rather large, let's look at it in prime factored form:

$168 = 2 \cdot 2 \cdot 2 \cdot 3 \cdot 7$

Now we can mentally form two numbers by using all of these factors in different combinations. Using two 2s and the 3 in one number and the other 2 and the 7 in another number produces $2 \cdot 2 \cdot 3 = 12$ and $2 \cdot 7 = 14$. Therefore, we can solve the given equation as follows:

$$n^2 + 26n + 168 = 0$$
$$(n + 12)(n + 14) = 0$$
$$n + 12 = 0 \quad \text{or} \quad n + 14 = 0$$
$$n = -12 \quad \text{or} \quad n = -14$$

The solution set is $\{-14, -12\}$.

Solve each of the following equations.

(a) $n^2 + 30n + 216 = 0$

(b) $n^2 + 35n + 294 = 0$

(c) $n^2 - 40n + 384 = 0$

(d) $n^2 - 40n + 375 = 0$

(e) $n^2 + 6n - 432 = 0$

(f) $n^2 - 16n - 512 = 0$

6.4 Factoring Trinomials of the Form $ax^2 + bx + c$

Now let's consider factoring trinomials where the coefficient of the squared term is not 1. We present here an informal trial and error technique that works quite well for certain types of trinomials. This technique simply relies on our knowledge of multiplication of binomials.

EXAMPLE 1

Factor $2x^2 + 7x + 3$.

Solution

By looking at the first term, $2x^2$, and the positive signs of the other two terms, we know that the binomials are of the form

$$(2x + \_\_\_\_)(x + \_\_\_\_)$$

Since the factors of the constant term, 3, are 1 and 3, we have only two possibilities to try:

$$(2x + 3)(x + 1) \qquad \text{or} \qquad (2x + 1)(x + 3)$$

By checking the middle term of both of these products, we find that the second one yields the correct middle term of $7x$. Therefore,

$$2x^2 + 7x + 3 = (2x + 1)(x + 3)$$

■

EXAMPLE 2

Factor $6x^2 - 17x + 5$.

Solution

First, we note that $6x^2$ can be written as $2x \cdot 3x$ or $6x \cdot x$. Second, since the middle term of the trinomial is negative, and the last term is positive, we know that the binomials are of the form

$$(2x - \_\_\_\_)(3x - \_\_\_\_) \qquad \text{or} \qquad (6x - \_\_\_\_)(x - \_\_\_\_)$$

Since the factors of the constant term, 5, are 1 and 5, we have the following possibilities:

$$(2x - 5)(3x - 1) \qquad (2x - 1)(3x - 5)$$

$$(6x - 5)(x - 1) \qquad (6x - 1)(x - 5)$$

By checking the middle term for each of these products, we find that the product $(2x - 5)(3x - 1)$ produces the desired term of $-17x$. Therefore,

$$6x^2 - 17x + 5 = (2x - 5)(3x - 1)$$

■

EXAMPLE 3 Factor $8x^2 - 8x - 30$.

Solution

First, we note that the polynomial $8x^2 - 8x - 30$ has a common factor of 2. Factoring out the common factor gives us $2(4x^2 - 4x - 15)$. Now we need to factor $4x^2 - 4x - 15$.

We note that $4x^2$ can be written as $4x \cdot x$ or $2x \cdot 2x$. The last term, -15, can be written as $(1)(-15)$, $(-1)(15)$, $(3)(-5)$, or $(-3)(5)$. Thus we can generate the possibilities for the binomial factors as follows:

Using 1 and −15	Using −1 and 15
$(4x - 15)(x + 1)$	$(4x - 1)(x + 15)$
$(4x + 1)(x - 15)$	$(4x + 15)(x - 1)$
$(2x + 1)(2x - 15)$	$(2x - 1)(2x + 15)$
Using 3 and −5	**Using −3 and 5**
$(4x + 3)(x - 5)$	$(4x - 3)(x + 5)$
$(4x - 5)(x + 3)$	$(4x + 5)(x - 3)$
✓ $(2x - 5)(2x + 3)$	$(2x + 5)(2x - 3)$

By checking the middle term of each of these products, we find that the product indicated with a check mark produces the desired middle term of $-4x$. Therefore,

$$8x^2 - 8x - 30 = 2(4x^2 - 4x - 15) = 2(2x - 5)(2x + 3)$$ ■

Let's pause for a moment and look back over Examples 1, 2, and 3. Obviously, Example 3 created the most difficulty because we had to consider so many possibilities. We have suggested one possible format for considering the possibilities, but as you practice such problems, you may develop a format of your own that works better for you. Regardless of the format that you use, the key idea is to organize your work so that you consider all possibilities. Let's look at another example.

EXAMPLE 4 Factor $4x^2 + 6x + 9$.

Solution

First, we note that $4x^2$ can be written as $4x \cdot x$ or $2x \cdot 2x$. Second, since the middle term is positive and the last term is positive, we know that the binomials are of the form

$$(4x + \_\_\_\_)(x + \_\_\_\_) \quad \text{or} \quad (2x + \_\_\_\_)(2x + \_\_\_\_)$$

Since 9 can be written as $9 \cdot 1$ or $3 \cdot 3$, we have only the five following possibilities to try:

$(4x + 9)(x + 1)$ $\quad$ $(4x + 1)(x + 9)$

$(4x + 3)(x + 3)$ $\quad$ $(2x + 1)(2x + 9)$

$(2x + 3)(2x + 3)$

When we try all of these possibilities, we find that none of them yields a middle term of $6x$. Therefore, $4x^2 + 6x + 9$ is *not factorable* using integers. ■

Remark: Example 4 illustrates the importance of organizing your work so that you try *all* possibilities before you conclude that a particular trinomial is not factorable.

Another Approach

There is another, more systematic technique that you may wish to use with some trinomials. It is an extension of the method we used in the previous section. Recall that at the beginning of Section 6.3 we looked at the following product:

$$(x + r)(x + s) = x(x) + x(s) + r(x) + r(s)$$
$$= x^2 + \underbrace{(s + r)}_{\text{Sum of } r \text{ and } s}x + \underbrace{rs}_{\text{Product of } r \text{ and } s}$$

Now let's look at this product:

$$(px + r)(qx + s) = px(qx) + px(s) + r(qx) + r(s)$$
$$= (pq)x^2 + (ps + rq)x + rs$$

Notice that the product of the coefficient of the x^2 term, (pq), and the constant term, (rs), is $pqrs$. Likewise, the product of the two coefficients of x, (ps and rq), is also $pqrs$. Therefore, the two coefficients of x must have a sum of $ps + rq$ and a product of $pqrs$. To begin the factoring process we will look for two factors of the product $pqrs$ whose sum is equal to the coefficient of the x term. This may seem a little confusing, but the next few examples illustrate how easy it is to apply.

EXAMPLE 5 Factor $3x^2 + 14x + 8$.

Solution

$$3x^2 + 14x + 8$$

Sum of 14

Product of $3 \cdot 8 = 24$

We need to find two integers whose product is 24 and whose sum is 14. Obviously, 2 and 12 satisfy these conditions. Therefore, we can express the middle term of the trinomial, $14x$, as $2x + 12x$ and proceed as follows:

$$3x^2 + 14x + 8 = 3x^2 + 2x + 12x + 8$$
$$= x(3x + 2) + 4(3x + 2) \quad \text{Factor by grouping}$$
$$= (3x + 2)(x + 4)$$

■

EXAMPLE 6 Factor $16x^2 - 26x + 3$.

Solution

$16x^2 - 26x + 3$ Sum of −26

Product of 16(3) = 48

We need two integers whose product is 48 and whose sum is -26. The integers -2 and -24 satisfy these conditions and allow us to express the middle term, $-26x$, as $-2x - 24x$. Then we can factor as follows:

$$16x^2 - 26x + 3 = 16x^2 - 2x - 24x + 3$$
$$= 2x(8x - 1) - 3(8x - 1) \quad \text{Factor by grouping}$$
$$= (8x - 1)(2x - 3)$$

■

EXAMPLE 7 Factor $6x^2 - 5x - 6$.

Solution

$6x^2 - 5x - 6$ Sum of −5

Product of 6(−6) = −36

We need two integers whose product is -36 and whose sum is -5. Furthermore, since the sum is negative, the absolute value of the negative number must be greater than the absolute value of the positive number. A little searching will determine that the numbers are -9 and 4. Thus we can express the middle term of $-5x$ as $-9x + 4x$ and proceed as follows:

$$6x^2 - 5x - 6 = 6x^2 - 9x + 4x - 6$$
$$= 3x(2x - 3) + 2(2x - 3)$$
$$= (2x - 3)(3x + 2)$$

■

Now that we have shown you two possible techniques for factoring trinomials of the form $ax^2 + bx + c$, the ball is in your court. Practice may not make you perfect at factoring, but it will surely help. We are not promoting one technique over the other; that is an individual choice. Many people find the trial and error technique we presented first very useful if the number of possibilities for the factors is fairly small. However, as the list of possibilities grows, the second technique does have the advantage of being systematic. So perhaps having both techniques at your fingertips is your best bet.

Now We Can Solve More Equations

The ability to factor certain trinomials of the form $ax^2 + bx + c$ provides us with greater equation-solving capabilities. Consider the next examples.

EXAMPLE 8

Solve $3x^2 + 17x + 10 = 0$.

Solution

$$3x^2 + 17x + 10 = 0$$

$$(x + 5)(3x + 2) = 0 \qquad \text{Factoring } 3x^2 + 17x + 10 \text{ as } (x + 5)(3x + 2) \text{ may require some extra work on scratch paper}$$

$$x + 5 = 0 \quad \text{or} \quad 3x + 2 = 0 \qquad ab = 0 \text{ if and only if } a = 0 \text{ or } b = 0$$

$$x = -5 \quad \text{or} \quad 3x = -2$$

$$x = -5 \quad \text{or} \quad x = -\frac{2}{3}$$

The solution set is $\left\{-5, -\frac{2}{3}\right\}$. Check it! ■

EXAMPLE 9

Solve $24x^2 + 2x - 15 = 0$.

Solution

$$24x^2 + 2x - 15 = 0$$

$$(4x - 3)(6x + 5) = 0$$

$$4x - 3 = 0 \quad \text{or} \quad 6x + 5 = 0$$

$$4x = 3 \quad \text{or} \quad 6x = -5$$

$$x = \frac{3}{4} \quad \text{or} \quad x = -\frac{5}{6}$$

The solution set is $\left\{-\frac{5}{6}, \frac{3}{4}\right\}$. ■

Problem Set 6.4

For Problems 1–50, factor each of the trinomials completely. Indicate any that are not factorable using integers.

1. $3x^2 + 7x + 2$
2. $2x^2 + 9x + 4$
3. $6x^2 + 19x + 10$
4. $12x^2 + 19x + 4$
5. $4x^2 - 25x + 6$
6. $5x^2 - 22x + 8$
7. $12x^2 - 31x + 20$
8. $8x^2 - 30x + 7$
9. $5y^2 - 33y - 14$
10. $6y^2 - 4y - 16$
11. $4n^2 + 26n - 48$
12. $4n^2 + 17n - 15$

13. $2x^2 + x + 7$

14. $7x^2 + 19x + 10$

15. $18x^2 + 45x + 7$

16. $10x^2 + x - 5$

17. $7x^2 - 30x + 8$

18. $6x^2 - 17x + 12$

19. $8x^2 + 2x - 21$

20. $9x^2 + 15x - 14$

21. $9t^3 - 15t^2 - 14t$

22. $12t^3 - 20t^2 - 25t$

23. $12y^2 + 79y - 35$

24. $9y^2 + 52y - 12$

25. $6n^2 + 2n - 5$

26. $20n^2 - 27n + 9$

27. $14x^2 + 55x + 21$

28. $15x^2 + 34x + 15$

29. $20x^2 - 31x + 12$

30. $8t^2 - 3t - 4$

31. $16n^2 - 8n - 15$

32. $25n^2 - 20n - 12$

33. $24x^2 - 50x + 25$

34. $24x^2 - 41x + 12$

35. $2x^2 + 25x + 72$

36. $2x^2 + 23x + 56$

37. $21a^2 + a - 2$

38. $14a^2 + 5a - 24$

39. $12a^2 - 31a - 15$

40. $10a^2 - 39a - 4$

41. $12x^2 + 36x + 27$

42. $27x^2 - 36x + 12$

43. $6x^2 - 5xy + y^2$

44. $12x^2 + 13xy + 3y^2$

45. $20x^2 + 7xy - 6y^2$

46. $8x^2 - 6xy - 35y^2$

47. $5x^2 - 32x + 12$

48. $3x^2 - 35x + 50$

49. $8x^2 - 55x - 7$

50. $12x^2 - 67x - 30$

For Problems 51–80, solve each equation.

51. $2x^2 + 13x + 6 = 0$

52. $3x^2 + 16x + 5 = 0$

53. $12x^2 + 11x + 2 = 0$

54. $15x^2 + 56x + 20 = 0$

55. $3x^2 - 25x + 8 = 0$

56. $4x^2 - 31x + 21 = 0$

57. $15n^2 - 41n + 14 = 0$

58. $6n^2 - 31n + 40 = 0$

59. $6t^2 + 37t - 35 = 0$

60. $2t^2 + 15t - 27 = 0$

61. $16y^2 - 18y - 9 = 0$

62. $9y^2 - 15y - 14 = 0$

63. $9x^2 - 6x - 8 = 0$

64. $12n^2 + 28n - 5 = 0$

65. $10x^2 - 29x + 10 = 0$

66. $4x^2 - 16x + 15 = 0$

67. $6x^2 + 19x = -10$

68. $12x^2 + 17x = -6$

69. $16x(x + 1) = 5$

70. $5x(5x + 2) = 8$

71. $35n^2 - 34n - 21 = 0$

72. $18n^2 - 3n - 28 = 0$

73. $4x^2 - 45x + 50 = 0$

74. $7x^2 - 65x + 18 = 0$

75. $7x^2 + 46x - 21 = 0$

76. $2x^2 + 7x - 30 = 0$

77. $12x^2 - 43x - 20 = 0$

78. $14x^2 - 13x - 12 = 0$

79. $18x^2 + 55x - 28 = 0$

80. $24x^2 + 17x - 20 = 0$

THOUGHTS INTO WORDS

81. Explain your thought process when factoring $24x^2 - 17x - 20$.

82. Your friend factors $8x^2 - 32x + 32$ as follows:

$$\begin{aligned} 8x^2 - 32x + 32 &= (4x - 8)(2x - 4) \\ &= 4(x - 2)(2)(x - 2) \\ &= 8(x - 2)(x - 2) \end{aligned}$$

Is she correct? Do you have any suggestions for her?

83. Your friend solves the equation $8x^2 - 32x + 32 = 0$ as follows:

$$8x^2 - 32x + 32 = 0$$

$$(4x - 8)(2x - 4) = 0$$

$$\begin{aligned} 4x - 8 &= 0 \quad \text{or} \quad & 2x - 4 &= 0 \\ 4x &= 8 \quad \text{or} \quad & 2x &= 4 \\ x &= 2 \quad \text{or} \quad & x &= 2 \end{aligned}$$

The solution set is {2}. Is she correct? Do you have any suggestions for her?

FURTHER INVESTIGATIONS

84. Consider the following approach to factoring $20x^2 + 39x + 18$:

$20x^2 + 39x + 18$ — Sum of 39

Product of $20(18) = 360$

We need two integers whose sum is 39 and whose product is 360. To help find these integers, let's prime factor 360.

$360 = 2 \cdot 2 \cdot 2 \cdot 3 \cdot 3 \cdot 5$

Now by grouping these factors in various ways, we find that $2 \cdot 2 \cdot 2 \cdot 3 = 24$, $3 \cdot 5 = 15$, and $24 + 15 = 39$. So the numbers are 15 and 24, and we can express the middle term of the given trinomial, $39x$, as $15x + 24x$. Therefore, we can complete the factoring as follows:

$$\begin{aligned} 20x^2 + 39x + 18 &= 20x^2 + 15x + 24x + 18 \\ &= 5x(4x + 3) + 6(4x + 3) \\ &= (4x + 3)(5x + 6) \end{aligned}$$

Factor each of the following trinomials.

(a) $20x^2 + 41x + 20$

(b) $24x^2 - 79x + 40$

(c) $30x^2 + 23x - 40$

(d) $36x^2 + 65x - 36$

6.5 Factoring, Solving Equations, and Problem Solving

Factoring

Before we summarize our work with factoring techniques, let's look at two more special factoring patterns. These patterns emerge when multiplying binomials. Consider the following examples.

$$(x + 5)^2 = (x + 5)(x + 5) = x^2 + 10x + 25$$

$$(2x + 3)^2 = (2x + 3)(2x + 3) = 4x^2 + 12x + 9$$

$$(4x + 7)^2 = (4x + 7)(4x + 7) = 16x^2 + 56x + 49$$

In general, $(a + b)^2 = (a + b)(a + b) = a^2 + 2ab + b^2$. Also,

$$(x - 6)^2 = (x - 6)(x - 6) = x^2 - 12x + 36$$

$$(3x - 4)^2 = (3x - 4)(3x - 4) = 9x^2 - 24x + 16$$

$$(5x - 2)^2 = (5x - 2)(5x - 2) = 25x^2 - 20x + 4$$

In general, $(a - b)^2 = (a - b)(a - b) = a^2 - 2ab + b^2$. Thus we have the following patterns.

Perfect Square Trinomials

$$a^2 + 2ab + b^2 = (a + b)^2$$

$$a^2 - 2ab + b^2 = (a - b)^2$$

Trinomials of the form $a^2 + 2ab + b^2$ or $a^2 - 2ab + b^2$ are called **perfect square trinomials**. They are easy to recognize because of the nature of their terms. For example, $9x^2 + 30x + 25$ is a perfect square trinomial for these reasons:

1. The first term is a square: $(3x)^2$.

2. The last term is a square: $(5)^2$.

3. The middle term is twice the product of the quantities being squared in the first and last terms: $2(3x)(5)$.

Likewise, $25x^2 - 40xy + 16y^2$ is a perfect square trinomial for these reasons:

1. The first term is a square: $(5x)^2$.

2. The last term is a square: $(4y)^2$.

3. The middle term is twice the product of the quantities being squared in the first and last terms: $2(5x)(4y)$.

Once we know that we have a perfect square trinomial, the factoring process follows immediately from the two basic patterns.

$$9x^2 + 30x + 25 = (3x + 5)^2$$

$$25x^2 - 40xy + 16y^2 = (5x - 4y)^2$$

Here are some additional examples of perfect square trinomials and their factored form.

$$\begin{aligned}
x^2 - 16x + 64 &= (x)^2 - 2(x)(8) + (8)^2 &&= (x - 8)^2 \\
16x^2 - 56x + 49 &= (4x)^2 - 2(4x)(7) + (7)^2 &&= (4x - 7)^2 \\
25x^2 + 20xy + 4y^2 &= (5x)^2 + 2(5x)(2y) + (2y)^2 &&= (5x + 2y)^2 \\
1 + 6y + 9y^2 &= (1)^2 + 2(1)(3y) + (3y)^2 &&= (1 + 3y)^2 \\
4m^2 - 4mn + n^2 &= (2m)^2 - 2(2m)(n) + (n)^2 &&= (2m - n)^2
\end{aligned}$$

You may want to do this step mentally after you feel comfortable with the process

We have considered some basic factoring techniques in this chapter one at a time, but you must be able to apply them as needed in a variety of situations. So, let's first summarize the techniques and then consider some examples.

In this chapter we have discussed these techniques:

1. Factoring by using the distributive property to factor out the greatest common monomial or binomial factor

2. Factoring by grouping

3. Factoring by applying the difference-of-squares pattern

4. Factoring by applying the perfect-square-trinomial pattern

5. Factoring trinomials of the form $x^2 + bx + c$ into the product of two binomials

6. Factoring trinomials of the form $ax^2 + bx + c$ into the product of two binomials

As a general guideline, **always look for a greatest common monomial factor first**, and then proceed with the other factoring techniques.

In each of the following examples we have factored completely whenever possible. Study them carefully, and notice the factoring techniques we used.

1. $2x^2 + 12x + 10 = 2(x^2 + 6x + 5) = 2(x + 1)(x + 5)$

2. $4x^2 + 36 = 4(x^2 + 9)$

Remember that the sum of two squares is not factorable using integers unless there is a common factor

3. $4t^2 + 20t + 25 = (2t + 5)^2$

If you fail to recognize a perfect trinomial square, no harm is done. Simply proceed to factor into the product of two binomials, and then you will recognize that the two binomials are the same

4. $x^2 - 3x - 8$ is not factorable using integers. This becomes obvious from the table.

Product	Sum
$1(-8) = -8$	$1 + (-8) = -7$
$-1(8) = -8$	$-1 + 8 = 7$
$2(-4) = -8$	$2 + (-4) = -2$
$-2(4) = -8$	$-2 + 4 = 2$

No two factors of -8 produce a sum of -3.

5. $6y^2 - 13y - 28 = (2y - 7)(3y + 4)$. We found the binomial factors as follows:

$(y + \_\_\_\_\_)(6y - \_\_\_\_\_)$

or

$(y - \_\_\_\_\_)(6y + \_\_\_\_\_)$

or

$(2y - \_\_\_\_\_)(3y + \_\_\_\_\_)$

or

$(2y + \_\_\_\_\_)(3y - \_\_\_\_\_)$

Factors of 28

$1 \cdot 28$ or $28 \cdot 1$

$2 \cdot 14$ or $14 \cdot 2$

$4 \cdot 7$ or $7 \cdot 4$

6. $32x^2 - 50y^2 = 2(16x^2 - 25y^2) = 2(4x + 5y)(4x - 5y)$

Solving Equations by Factoring

Each time we considered a new factoring technique in this chapter, we used that technique to help solve some equations. It is important that you recognize which technique works for a particular type of equation.

EXAMPLE 1

Solve $x^2 = 25x$.

Solution

$$x^2 = 25x$$

$$x^2 - 25x = 0 \qquad \text{Added } -25x \text{ to both sides}$$

$$x(x - 25) = 0$$

$$x = 0 \quad \text{or} \quad x - 25 = 0$$

$$x = 0 \quad \text{or} \quad x = 25$$

The solution set is {0, 25}. Check it!

EXAMPLE 2

Solve $x^3 - 36x = 0$.

Solution

$$x^3 - 36x = 0$$

$$x(x^2 - 36) = 0$$

$$x(x + 6)(x - 6) = 0$$

$$x = 0 \quad \text{or} \quad x + 6 = 0 \quad \text{or} \quad x - 6 = 0 \qquad \text{If } abc = 0, \text{ then } a = 0 \text{ or } b = 0 \text{ or } c = 0$$

$$x = 0 \quad \text{or} \quad x = -6 \quad \text{or} \quad x = 6$$

The solution set is $\{-6, 0, 6\}$. Does it check?

EXAMPLE 3

Solve $10x^2 - 13x - 3 = 0$.

Solution

$$10x^2 - 13x - 3 = 0$$

$$(5x + 1)(2x - 3) = 0$$

$$5x + 1 = 0 \quad \text{or} \quad 2x - 3 = 0$$

$$5x = -1 \quad \text{or} \quad 2x = 3$$

$$x = -\frac{1}{5} \quad \text{or} \quad x = \frac{3}{2}$$

The solution set is $\left\{-\frac{1}{5}, \frac{3}{2}\right\}$. Does it check?

EXAMPLE 4

Solve $4x^2 - 28x + 49 = 0$.

Solution

$$4x^2 - 28x + 49 = 0$$
$$(2x - 7)^2 = 0$$
$$(2x - 7)(2x - 7) = 0$$
$$2x - 7 = 0 \quad \text{or} \quad 2x - 7 = 0$$
$$2x = 7 \quad \text{or} \quad 2x = 7$$
$$x = \frac{7}{2} \quad \text{or} \quad x = \frac{7}{2}$$

The solution set is $\left\{\frac{7}{2}\right\}$. ■

Pay special attention to the next example. We need to change the form of the original equation before we can apply the property $ab = 0$ if and only if $a = 0$ or $b = 0$. A necessary condition of this property is that an indicated product is set equal to zero.

EXAMPLE 5

Solve $(x + 1)(x + 4) = 40$.

Solution

$$(x + 1)(x + 4) = 40$$
$$x^2 + 5x + 4 = 40$$
$$x^2 + 5x - 36 = 0$$
$$(x + 9)(x - 4) = 0$$
$$x + 9 = 0 \quad \text{or} \quad x - 4 = 0$$
$$x = -9 \quad \text{or} \quad x = 4$$

The solution set is $\{-9, 4\}$. Check it! ■

EXAMPLE 6

Solve $2n^2 + 16n - 40 = 0$.

Solution

$$2n^2 + 16n - 40 = 0$$
$$2(n^2 + 8n - 20) = 0$$
$$n^2 + 8n - 20 = 0 \qquad \text{Multiplied both sides by } \frac{1}{2}$$
$$(n + 10)(n - 2) = 0$$

$$n + 10 = 0 \quad \text{or} \quad n - 2 = 0$$

$$n = -10 \quad \text{or} \quad n = 2$$

The solution set is $\{-10, 2\}$. Does it check? ■

Problem Solving

Throughout this book we highlight the need to *learn a skill, to use that skill to help solve equations*, and then *to use equations to help solve problems*. This approach should be very apparent in this chapter. Our new factoring skills have provided more ways of solving equations, which in turn gives us more power to solve word problems. We conclude the chapter by solving a few more problems.

PROBLEM 1

Find two numbers whose product is 65 if one of the numbers is 3 more than twice the other number.

Solution

Let n represent one of the numbers; then $2n + 3$ represents the other number. Since their product is 65, we can set up and solve the following equation:

$$n(2n + 3) = 65$$

$$2n^2 + 3n - 65 = 0$$

$$(2n + 13)(n - 5) = 0$$

$$2n + 13 = 0 \quad \text{or} \quad n - 5 = 0$$

$$2n = -13 \quad \text{or} \quad n = 5$$

$$n = -\frac{13}{2} \quad \text{or} \quad n = 5$$

If $n = -\frac{13}{2}$, then $2n + 3 = 2\left(-\frac{13}{2}\right) + 3 = -10$. If $n = 5$, then $2n + 3 = 2(5) + 3 = 13$. Thus the numbers are $-\frac{13}{2}$ and -10, or 5 and 13. ■

PROBLEM 2

The area of a triangular sheet of paper is 14 square inches. One side of the triangle is 3 inches longer than the altitude to that side. Find the length of the one side and the length of the altitude to that side.

Solution

Let h represent the altitude to the side. Then $h + 3$ represents the side of the triangle (see Figure 6.5).

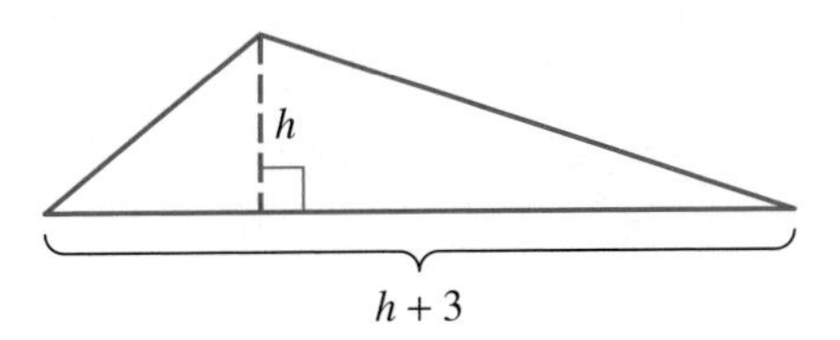

Figure 6.5

Since the formula for finding the area of a triangle is $A = \frac{1}{2}bh$, we have

$$\frac{1}{2}h(h + 3) = 14$$

$$h(h + 3) = 28 \quad \text{Multiplied both sides by 2}$$

$$h^2 + 3h = 28$$

$$h^2 + 3h - 28 = 0$$

$$(h + 7)(h - 4) = 0$$

$$h + 7 = 0 \quad \text{or} \quad h - 4 = 0$$

$$h = -7 \quad \text{or} \quad h = 4$$

The solution of -7 is not reasonable. Thus the altitude is 4 inches and the length of the side to which that altitude is drawn is 7 inches. ■

PROBLEM 3

A strip with a uniform width is shaded along both sides and both ends of a rectangular poster with dimensions 12 inches by 16 inches. How wide is the strip if one-half of the area of the poster is shaded?

Solution

Let x represent the width of the shaded strip of the poster in Figure 6.6. The area of the strip is one-half of the area of the poster; therefore, it is $\frac{1}{2}(12)(16) = 96$ square inches. Furthermore, we can represent the area of the strip around the poster by the words *the area of the poster minus the area of the unshaded portion.*

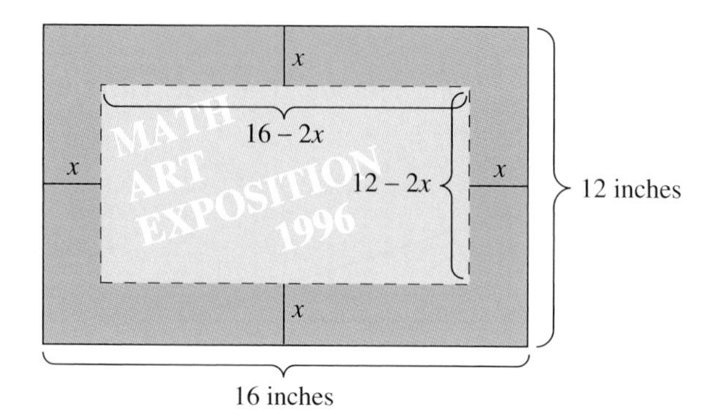

Figure 6.6

Thus we can set up and solve the following equation:

Area of poster − Area of unshaded portion = Area of strip

$$16(12) - (16 - 2x)(12 - 2x) = 96$$

$$192 - (192 - 56x + 4x^2) = 96$$

$$192 - 192 + 56x - 4x^2 = 96$$

$$-4x^2 + 56x - 96 = 0$$

$$x^2 - 14x + 24 = 0$$

$$(x - 12)(x - 2) = 0$$

$$x - 12 = 0 \quad \text{or} \quad x - 2 = 0$$

$$x = 12 \quad \text{or} \quad x = 2$$

Obviously, the strip cannot be 12 inches wide because the total width of the poster is 12 inches. Thus we must disregard the solution of 12 and conclude that the strip is 2 inches wide. ■

Problem Set 6.5

For Problems 1–12, factor each of the perfect square trinomials.

1. $x^2 + 4x + 4$

2. $x^2 + 18x + 81$

3. $x^2 - 10x + 25$

4. $x^2 - 24x + 144$

5. $9n^2 + 12n + 4$

6. $25n^2 + 30n + 9$

7. $16a^2 - 8a + 1$

8. $36a^2 - 84a + 49$

9. $4 + 36x + 81x^2$

10. $1 - 4x + 4x^2$

11. $16x^2 - 24xy + 9y^2$

12. $64x^2 + 16xy + y^2$

For Problems 13–40, factor each polynomial completely. Indicate any that are not factorable using integers.

13. $2x^2 + 17x + 8$

14. $x^2 + 19x$

15. $2x^3 - 72x$

16. $30x^2 - x - 1$

17. $n^2 - 7n - 60$

18. $4n^3 - 100n$

19. $3a^2 - 7a - 4$

20. $a^2 + 7a - 30$

21. $8x^2 + 72$

22. $3y^3 - 36y^2 + 96y$

23. $9x^2 + 30x + 25$

24. $5x^2 - 5x - 6$

25. $15x^2 + 65x + 70$

26. $4x^2 - 20xy + 25y^2$

27. $24x^2 + 2x - 15$

28. $9x^2y - 27xy$

29. $xy + 5y - 8x - 40$

30. $xy - 3y + 9x - 27$

31. $20x^2 + 31xy - 7y^2$

32. $2x^2 - xy - 36y^2$

33. $24x^2 + 18x - 81$

34. $30x^2 + 55x - 50$

35. $12x^2 + 6x + 30$

36. $24x^2 - 8x + 32$

37. $5x^4 - 80$

38. $3x^5 - 3x$

39. $x^2 + 12xy + 36y^2$

40. $4x^2 - 28xy + 49y^2$

For Problems 41–70, solve each equation.

41. $4x^2 - 20x = 0$

42. $-3x^2 - 24x = 0$

43. $x^2 - 9x - 36 = 0$

44. $x^2 + 8x - 20 = 0$

45. $-2x^3 + 8x = 0$

46. $4x^3 - 36x = 0$

47. $6n^2 - 29n - 22 = 0$

48. $30n^2 - n - 1 = 0$

49. $(3n - 1)(4n - 3) = 0$

50. $(2n - 3)(7n + 1) = 0$

51. $(n - 2)(n + 6) = -15$

52. $(n + 3)(n - 7) = -25$

53. $2x^2 = 12x$

54. $-3x^2 = 15x$

55. $t^3 - 2t^2 - 24t = 0$

56. $2t^3 - 16t^2 - 18t = 0$

57. $12 - 40x + 25x^2 = 0$

58. $12 - 7x - 12x^2 = 0$

59. $n^2 - 28n + 192 = 0$

60. $n^2 + 33n + 270 = 0$

61. $(3n + 1)(n + 2) = 12$

62. $(2n + 5)(n + 4) = -1$

63. $x^3 = 6x^2$

64. $x^3 = -4x^2$

65. $9x^2 - 24x + 16 = 0$

66. $25x^2 + 60x + 36 = 0$

67. $x^3 + 10x^2 + 25x = 0$

68. $x^3 - 18x^2 + 81x = 0$

69. $24x^2 + 17x - 20 = 0$

70. $24x^2 + 74x - 35 = 0$

For Problems 71–88, set up an equation and solve each problem.

71. Find two numbers whose product is 15 such that one of the numbers is seven more than four times the other number.

72. Find two numbers whose product is 12 such that one of the numbers is four less than eight times the other number.

73. Find two numbers whose product is -1. One of the numbers is three more than twice the other number.

74. Suppose that the sum of the squares of three consecutive integers is 110. Find the integers.

75. A number is one more than twice another number. The sum of the squares of the two numbers is 97. Find the numbers.

76. A number is one less than three times another number. If the product of the two numbers is 102, find the numbers.

77. In an office building, a room contains 54 chairs. The number of chairs per row is three less than twice the number of rows. Find the number of rows and the number of chairs per row.

78. An apple orchard contains 85 trees. The number of trees in each row is three less than four times the number of rows. Find the number of rows and the number of trees per row.

79. Suppose that the combined area of two squares is 360 square feet. Each side of the larger square is three times as long as a side of the smaller square. How big is each square?

80. The area of a rectangular slab of sidewalk is 45 square feet. Its length is 3 feet more than four times its width. Find the length and width of the slab.

81. The length of a rectangular sheet of paper is 1 centimeter more than twice its width, and the area of the rectangle is 55 square centimeters. Find the length and width of the rectangle.

82. Suppose that the length of a certain rectangle is three times its width. If the length is increased by 2 inches, and the width increased by 1 inch, the newly formed rectangle has an area of 70 square inches. Find the length and width of the original rectangle.

83. The area of a triangle is 51 square inches. One side of the triangle is 1 inch less than three times the length of the altitude to that side. Find the length of that side and the length of the altitude to that side.

84. Suppose that a square and a rectangle have equal areas. Furthermore, suppose that the length of the rectangle is twice the length of a side of the square, and the width of the rectangle is 4 centimeters less than the length of a side of the square. Find the dimensions of both figures.

85. A strip of uniform width is to be cut off of both sides and both ends of a sheet of paper that is 8 inches by 11 inches, in order to reduce the size of the paper to an area of 40 square inches. Find the width of the strip.

86. The sum of the areas of two circles is 100π square centimeters. The length of a radius of the larger circle is 2 centimeters more than the length of a radius of the smaller circle. Find the length of a radius of each circle.

87. The sum of the areas of two circles is 180π square inches. The length of a radius of the smaller circle is 6 inches less than the length of a radius of the larger circle. Find the length of a radius of each circle.

88. A strip of uniform width is shaded along both sides and both ends of a rectangular poster that is 18 inches by 14 inches. How wide is the strip if the unshaded portion of the poster has an area of 165 square inches?

THOUGHTS INTO WORDS

89. When factoring polynomials, why do you think that it is best to look for a greatest common monomial factor first?

90. Explain how you would solve $(4x - 3)(8x + 5) = 0$ and also how you would solve $(4x - 3)(8x + 5) = -9$.

91. Explain how you would solve $(x + 2)(x + 3) = (x + 2)(3x - 1)$. Do you see more than one approach to this problem?

Chapter 6 Summary

(6.1) The distributive property in the form $ab + ac = a(b + c)$ provides the basis for **factoring out a greatest common monomial or binomial factor**.

Rewriting an expression such as $ab + 3a + bc + 3c$ as $a(b + 3) + c(b + 3)$ and then factoring out the common binomial factor of $b + 3$ so that $a(b + 3) + c(b + 3)$ becomes $(b + 3)(a + c)$ is called **factoring by grouping**.

The property $ab = 0$ if and only if $a = 0$ or $b = 0$ provides us with another technique for solving equations.

(6.2) This factoring pattern is called the **difference of two squares**:

$$a^2 - b^2 = (a - b)(a + b)$$

(6.3) The following multiplication pattern provides a technique for factoring trinomials of the form $x^2 + bx + c$:

$$(x + r)(x + s) = x^2 + rs + sx + rs$$
$$= x^2 + (r + s)x + rs$$

Sum of r and s — Product of r and s

(6.4) We presented two different techniques for factoring trinomials of the form

$$ax^2 + bx + c$$

To review these techniques, turn to Section 6.4 and study the examples.

(6.5) As a general guideline for **factoring completely**, always look for a greatest common monomial or binomial factor first, and then proceed with one or more of the following techniques:

1. Apply the difference-of-squares pattern.
2. Apply the perfect-square-trinomial pattern.
3. Factor a trinomial of the form

   $$x^2 + bx + c$$

   into the product of two binomials.
4. Factor a trinomial of the form

   $$ax^2 + bx + c$$

   into the product of two binomials.

Chapter 6 Review Problem Set

For Problems 1–24, factor completely. Indicate any polynomials that are not factorable using integers.

1. $x^2 - 9x + 14$

2. $3x^2 + 21x$

3. $9x^2 - 4$

4. $4x^2 + 8x - 5$

5. $25x^2 - 60x + 36$

6. $n^3 + 13n^2 + 40n$

7. $y^2 + 11y - 12$

8. $3xy^2 + 6x^2y$

9. $x^4 - 1$

10. $18n^2 + 9n - 5$

11. $x^2 + 7x + 24$

12. $4x^2 - 3x - 7$

13. $3n^2 + 3n - 90$

14. $x^3 - xy^2$

15. $2x^2 + 3xy - 2y^2$

16. $4n^2 - 6n - 40$

17. $5x + 5y + ax + ay$

18. $21t^2 - 5t - 4$

19. $2x^3 - 2x$

20. $3x^3 - 108x$

21. $16x^2 + 40x + 25$

22. $xy - 3x - 2y + 6$

23. $15x^2 - 7xy - 2y^2$

24. $6n^4 - 5n^3 + n^2$

For Problems 25–44, solve each equation.

25. $x^2 + 4x - 12 = 0$

26. $x^2 = 11x$

27. $2x^2 + 3x - 20 = 0$

28. $9n^2 + 21n - 8 = 0$

29. $6n^2 = 24$

30. $16y^2 + 40y + 25 = 0$

31. $t^3 - t = 0$

32. $28x^2 + 71x + 18 = 0$

33. $x^2 + 3x - 28 = 0$

34. $(x - 2)(x + 2) = 21$

35. $5n^2 + 27n = 18$

36. $4n^2 + 10n = 14$

37. $2x^3 - 8x = 0$

38. $x^2 - 20x + 96 = 0$

39. $4t^2 + 17t - 15 = 0$

40. $3(x + 2) - x(x + 2) = 0$

41. $(2x - 5)(3x + 7) = 0$

42. $(x + 4)(x - 1) = 50$

43. $-7n - 2n^2 = -15$

44. $-23x + 6x^2 = -20$

Set up an equation and solve each of the following problems.

45. The larger of two numbers is one less than twice the smaller number. The difference of their squares is 33. Find the numbers.

46. The length of a rectangle is 2 centimeters less than five times the width of the rectangle. The area of the rectangle is 16 square centimeters. Find the length and width of the rectangle.

47. Suppose that the combined area of two squares is 104 square inches. Each side of the larger square is five times as long as a side of the smaller square. Find the size of each square.

48. The longer leg of a right triangle is one unit shorter than twice the length of the shorter leg. The hypotenuse is one unit longer than twice the length of the shorter leg. Find the lengths of the three sides of the triangle.

49. The product of two numbers is 26 and one of the numbers is one larger than six times the other number. Find the numbers.

50. Find three consecutive positive odd whole numbers such that the sum of the squares of the two smaller numbers is nine more than the square of the largest number.

51. The number of books per shelf in a bookcase is one less than nine times the number of shelves. If the bookcase contains 140 books, find the number of shelves.

52. The combined area of a square and a rectangle is 225 square yards. The length of the rectangle is eight times the width of the rectangle, and the length of a side of the square is the same as the width of the rectangle. Find the dimensions of the square and the rectangle.

53. Suppose that we want to find two consecutive integers such that the sum of their squares is 613. What are they?

54. If numerically the volume of a cube equals the total surface area of the cube, find the length of an edge of the cube.

55. The combined area of two circles is 53π square meters. The length of a radius of the larger circle is 1 meter more than three times the length of a radius of the smaller circle. Find the length of a radius of each circle.

56. The product of two consecutive odd whole numbers is one less than five times their sum. Find the whole numbers.

57. Sandy has a photograph that is 14 centimeters long and 8 centimeters wide. She wants to reduce the length and width by the same amount so that the area is decreased by 40 square centimeters. By what amount should she reduce the length and width?

58. Suppose that a strip of uniform width is plowed along both sides and both ends of a garden that is 120 feet long and 90 feet wide (see Figure 6.7). How wide is the strip if the garden is half plowed?

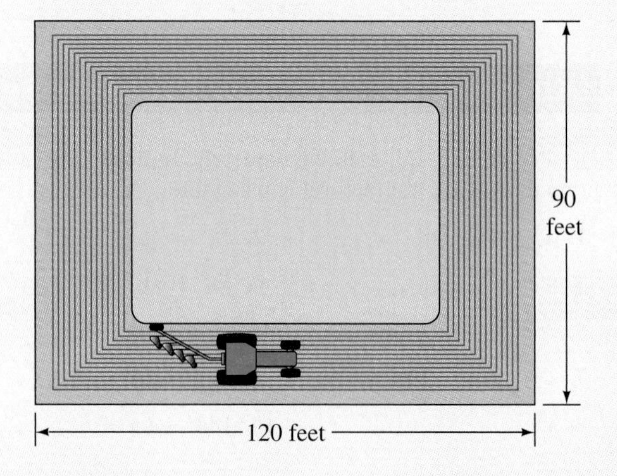

Figure 6.7

Chapter 6 Test

For Problems 1–10, factor each expression completely.

1. $x^2 + 3x - 10$
2. $x^2 - 5x - 24$
3. $2x^3 - 2x$
4. $x^2 + 21x + 108$
5. $18n^2 + 21n + 6$
6. $ax + ay + 2bx + 2by$
7. $4x^2 + 17x - 15$
8. $6x^2 + 24$
9. $30x^3 - 76x^2 + 48x$
10. $28 + 13x - 6x^2$

For Problems 11–21, solve each equation.

11. $7x^2 = 63$
12. $x^2 + 5x - 6 = 0$
13. $4n^2 = 32n$
14. $(3x - 2)(2x + 5) = 0$
15. $(x - 3)(x + 7) = -9$
16. $x^3 + 16x^2 + 48x = 0$
17. $9(x - 5) - x(x - 5) = 0$
18. $3t^2 + 35t = 12$
19. $8 - 10x - 3x^2 = 0$
20. $3x^3 = 75x$
21. $25n^2 - 70n + 49 = 0$

For Problems 22–25, set up an equation and solve each problem.

22. The length of a rectangle is 2 inches less than twice its width. If the area of the rectangle is 112 square inches, find the length of the rectangle.
23. The length of one leg of a right triangle is 4 centimeters more than the length of the other leg. The length of the hypotenuse is 8 centimeters more than the length of the shorter leg. Find the length of the shorter leg.
24. A room contains 112 chairs. The number of chairs per row is five less than three times the number of rows. Find the number of chairs per row.
25. If numerically the volume of a cube equals twice the total surface area, find the length of an edge of the cube.

7

Algebraic Fractions

If Phil can do a certain job in 15 minutes and Mark can do the same job in 10 minutes, then the equation $\frac{m}{15} + \frac{m}{10} = 1$ *can be used to determine how long it would take to do the job if they work together. The equation* $\frac{1}{15} + \frac{1}{10} = \frac{1}{m}$ *could also be used.*

One day, Jeff rode his bicycle 40 miles out into the country. On the way back, he took a different route that was 2 miles longer, and it took him an hour longer to return. If his rate on the way out into the country was 4 miles per hour faster than his rate back, find both rates. We can use the fractional equation $\frac{40}{x} = \frac{42}{x - 4} - 1$ to determine that Jeff rode out into the country at 16 miles per hour and returned at 12 miles per hour.

In Chapter 2 our study of common fractions led naturally to some work with simple algebraic fractions. Then in Chapters 5 and 6, we discussed the basic operations that pertain to polynomials. Now we can use some ideas about polynomials — specifically the factoring techniques — to expand our study of algebraic fractions. This, in turn, gives us more techniques for solving equations, which increases our problem-solving capabilities.

7.1 Simplifying Algebraic Fractions

If the numerator and denominator of a fraction are polynomials, then we call the fraction an **algebraic fraction** or a **rational expression**. Here are some examples of algebraic fractions:

$$\frac{4}{x-2} \qquad \frac{x^2+2x-4}{x^2-9} \qquad \frac{y+x^2}{xy-3} \qquad \frac{x^3+2x^2-3x-4}{x^2-2x-6}$$

Because we must avoid division by zero, no values can be assigned to variables that create a denominator of zero. Thus the fraction $\frac{4}{x-2}$ is meaningful for all real number values of x except for $x = 2$. Rather than making a restriction for each individual fraction, we will simply assume that all denominators represent nonzero real numbers.

Recall that the **fundamental principle of fractions** $\left(\frac{ak}{bk} = \frac{a}{b}\right)$ provides the basis for expressing fractions in reduced (or simplified) form, as the next examples demonstrate.

$$\frac{18}{24} = \frac{3 \cdot \cancel{6}}{4 \cdot \cancel{6}} = \frac{3}{4} \qquad \frac{-42xy}{77y} = -\frac{2 \cdot 3 \cdot \cancel{7} \cdot x \cdot \cancel{y}}{\cancel{7} \cdot 11 \cdot \cancel{y}} = -\frac{6x}{11}$$

$$\frac{15x}{25x} = \frac{3 \cdot \cancel{5} \cdot \cancel{x}}{5 \cdot \cancel{5} \cdot \cancel{x}} = \frac{3}{5} \qquad \frac{28x^2y^2}{-63x^2y^3} = -\frac{4 \cdot \cancel{7} \cdot \cancel{x^2} \cdot \cancel{y^2}}{9 \cdot \cancel{7} \cdot \cancel{x^2} \cdot \underset{y}{\cancel{y^3}}} = -\frac{4}{9y}$$

We can use the factoring techniques from Chapter 6 to factor numerators and/or denominators so that we can apply the fundamental principle of fractions. Several examples should clarify this process.

EXAMPLE 1

Simplify $\frac{x^2+6x}{x^2-36}$.

Solution

$$\frac{x^2+6x}{x^2-36} = \frac{x\cancel{(x+6)}}{(x-6)\cancel{(x+6)}} = \frac{x}{x-6}$$ ■

EXAMPLE 2

Simplify $\frac{a+2}{a^2+4a+4}$.

Solution

$$\frac{a+2}{a^2+4a+4} = \frac{1\cancel{(a+2)}}{(a+2)\cancel{(a+2)}} = \frac{1}{a+2}$$ ■

EXAMPLE 3

Simplify $\frac{x^2 + 4x - 21}{2x^2 + 15x + 7}$.

Solution

$$\frac{x^2 + 4x - 21}{2x^2 + 15x + 7} = \frac{(x - 3)\cancel{(x + 7)}}{(2x + 1)\cancel{(x + 7)}} = \frac{x - 3}{2x + 1}$$

■

EXAMPLE 4

Simplify $\frac{a^2b + ab^2}{ab + b^2}$.

Solution

$$\frac{a^2b + ab^2}{ab + b^2} = \frac{a\cancel{b}\cancel{(a + b)}}{\cancel{b}\cancel{(a + b)}} = a$$

■

EXAMPLE 5

Simplify $\frac{4x^3y - 36xy}{2x^2 - 4x - 30}$.

Solution

$$\frac{4x^3y - 36xy}{2x^2 - 4x - 30} = \frac{4xy(x^2 - 9)}{2(x^2 - 2x - 15)}$$

$$= \frac{\overset{2}{\cancel{4}}xy\cancel{(x + 3)}(x - 3)}{\cancel{2}(x - 5)\cancel{(x + 3)}}$$

$$= \frac{2xy(x - 3)}{x - 5}$$

■

Notice in Example 5 that we left the numerator of the final fraction in factored form. We do this when polynomials other than monomials are involved. Either $\frac{2xy(x - 3)}{x - 5}$ or $\frac{2x^2y - 6xy}{x - 5}$ is an acceptable answer.

Remember that the quotient of any nonzero real number and its opposite is -1. For example, $\frac{7}{-7} = -1$ and $\frac{-9}{9} = -1$. Likewise, the indicated quotient of any polynomial and its opposite is equal to -1. Consider these examples.

$$\frac{x}{-x} = -1 \quad \text{because } x \text{ and } -x \text{ are opposites.}$$

$$\frac{x - y}{y - x} = -1 \quad \text{because } x - y \text{ and } y - x \text{ are opposites.}$$

$$\frac{a^2 - 9}{9 - a^2} = -1 \quad \text{because } a^2 - 9 \text{ and } 9 - a^2 \text{ are opposites.}$$

Use this idea to simplify algebraic fractions in the final examples of this section.

EXAMPLE 6

Simplify $\frac{14 - 7n}{n - 2}$.

Solution

$$\frac{14 - 7n}{n - 2} = \frac{7(2 - n)}{n - 2}$$
$$= 7(-1) \qquad \frac{2 - n}{n - 2} = -1$$
$$= -7$$

■

EXAMPLE 7

Simplify $\frac{x^2 + 4x - 21}{15 - 2x - x^2}$.

Solution

$$\frac{x^2 + 4x - 21}{15 - 2x - x^2} = \frac{(x + 7)(x - 3)}{(5 + x)(3 - x)}$$
$$= \left(\frac{x + 7}{x + 5}\right)(-1) \qquad \frac{x - 3}{3 - x} = -1$$
$$= -\frac{x + 7}{x + 5} \quad \text{or} \quad \frac{-x - 7}{x + 5}$$

■

Problem Set 7.1

For Problems 1–60, simplify each algebraic fraction.

1. $\frac{6x}{14y}$
2. $\frac{8y}{18x}$
3. $\frac{9xy}{24x}$
4. $\frac{12y}{20xy}$
5. $\frac{-15x^2y}{25x}$
6. $\frac{16x^3y^2}{-28x^2y}$
7. $\frac{-36x^4y^3}{-48x^6y^2}$
8. $\frac{-18x^3y}{-36xy^3}$
9. $\frac{12a^2b^5}{-54a^2b^3}$
10. $\frac{-24a^3b^3}{39a^5b^2}$
11. $\frac{32xy^2z^3}{72yz^4}$
12. $\frac{27x^2y^3z^4}{45x^3y^3z}$
13. $\frac{xy}{x^2 - 2x}$
14. $\frac{x^2 + 5x}{xy}$
15. $\frac{8x + 12y}{12}$
16. $\frac{8}{12x - 16y}$
17. $\frac{x^2 + 2x}{x^2 - 7x}$
18. $\frac{x^2 - 6x}{2x^2 + 6x}$
19. $\frac{7 - x}{x - 7}$
20. $\frac{x - 9}{9 - x}$
21. $\frac{15 - 3n}{n - 5}$
22. $\frac{2n^2 - 8n}{4 - n}$
23. $\frac{4x^3 - 4x}{1 - x^2}$
24. $\frac{9 - x^2}{3x^3 - 27x}$
25. $\frac{x^2 - 1}{3x^2 - 3x}$
26. $\frac{5x^2 + 25x}{x^2 - 25}$
27. $\frac{x^2 + xy}{x^2}$
28. $\frac{x^3}{x^3 - x^2y}$
29. $\frac{6x^3 - 15x^2y}{6x^2 + 24xy}$
30. $\frac{6x^2 + 42xy}{16x^3 - 8x^2y}$

31. $\dfrac{n^2 + 2n}{n^2 + 3n + 2}$

32. $\dfrac{n^2 + 9n + 18}{n^2 + 6n}$

33. $\dfrac{2n^2 + 5n - 3}{n^2 - 9}$

34. $\dfrac{3n^2 - 10n - 8}{n^2 - 16}$

35. $\dfrac{2x^2 + 17x + 35}{3x^2 + 19x + 20}$

36. $\dfrac{5x^2 - 32x + 12}{4x^2 - 27x + 18}$

37. $\dfrac{9(x - 1)^2}{12(x - 1)^3}$

38. $\dfrac{18(x + 2)^3}{16(x + 2)^2}$

39. $\dfrac{7x^2 + 61x - 18}{7x^2 + 19x - 6}$

40. $\dfrac{8x^2 - 51x + 18}{8x^2 + 29x - 12}$

41. $\dfrac{10a^2 + a - 3}{15a^2 + 4a - 3}$

42. $\dfrac{6a^2 - 11a - 10}{8a^2 - 22a + 5}$

43. $\dfrac{x^2 + 2xy - 3y^2}{2x^2 - xy - y^2}$

44. $\dfrac{x^2 - 3xy + 2y^2}{x^2 - 4y^2}$

45. $\dfrac{x^2 - 9}{-x^2 - 3x}$

46. $\dfrac{-x^2 - 2x}{x^2 - 4}$

47. $\dfrac{n^2 + 14n + 49}{8n + 56}$

48. $\dfrac{6n - 60}{n^2 - 20n + 100}$

49. $\dfrac{4n^2 - 12n + 9}{2n^2 - n - 3}$

50. $\dfrac{9n^2 + 30n + 25}{3n^2 - n - 10}$

51. $\dfrac{y^2 - 6y - 72}{y^2 - 8y - 84}$

52. $\dfrac{y^2 + 20y + 96}{y^2 + 23y + 120}$

53. $\dfrac{1 - x^2}{x - x^2}$

54. $\dfrac{2x + x^2}{4 - x^2}$

55. $\dfrac{6 - x - 2x^2}{12 + 7x - 10x^2}$

56. $\dfrac{15 + x - 2x^2}{21 - 10x + x^2}$

57. $\dfrac{x^2 + 7x - 18}{12 - 4x - x^2}$

58. $\dfrac{3x - 21}{28 - 4x}$

59. $\dfrac{5x - 40}{80 - 10x}$

60. $\dfrac{x^2 - x - 12}{8 + 2x - x^2}$

THOUGHTS INTO WORDS

61. Explain the role that factoring plays in simplifying algebraic fractions.

62. Which of the following simplification processes are correct? Explain your answer.

$$\frac{2x}{x} = 2 \qquad \frac{x + 2}{x} = 2 \qquad \frac{x(x + 2)}{x} = x + 2$$

FURTHER INVESTIGATIONS

For Problems 63–66, simplify each fraction. You will need to use factoring by grouping.

63. $\dfrac{xy - 3x + 2y - 6}{xy + 5x + 2y + 10}$

64. $\dfrac{xy + 4x - y - 4}{xy + 4x - 4y - 16}$

65. $\dfrac{xy - 6x + y - 6}{xy - 6x + 5y - 30}$

66. $\dfrac{xy - 7x - 5y + 35}{xy - 9x - 5y + 45}$

The link between positive and negative exponents $\left(a^{-n} = \dfrac{1}{a^n}\right)$ along with the property $\dfrac{a^n}{a^m} = a^{n-m}$ can also be used when reducing fractions. Consider this example:

$$\frac{x^3}{x^7} = x^{3-7} = x^{-4} = \frac{1}{x^4}$$

For Problems 67–72, use this approach to express each fraction in reduced form. Give all answers with positive exponents only.

67. $\dfrac{x^3}{x^9}$

68. $\dfrac{x^4}{x^8}$

69. $\dfrac{x^4y^3}{x^7y^5}$

70. $\dfrac{x^5y^2}{x^6y^3}$

71. $\dfrac{28a^2b^3}{-7a^5b^3}$

72. $\dfrac{-44a^3b^4}{4a^3b^6}$

7.2 Multiplying and Dividing Algebraic Fractions

Multiplying Algebraic Fractions

In Chapter 2 we defined the product of two rational numbers as $\frac{a}{b} \cdot \frac{c}{d} = \frac{ac}{bd}$. This definition extends to algebraic fractions in general.

Definition 7.1

If $\frac{A}{B}$ and $\frac{C}{D}$ are rational expressions with $B \neq 0$ and $D \neq 0$, then

$$\frac{A}{B} \cdot \frac{C}{D} = \frac{AC}{BD}$$

In other words, to multiply algebraic fractions we multiply the numerators, multiply the denominators, and **express the product in simplified form**. The following examples illustrate this concept.

1. $\frac{2x}{3y} \cdot \frac{5y}{4x} = \frac{\cancel{2} \cdot 5 \cdot \cancel{x} \cdot \cancel{y}}{3 \cdot \underset{2}{\cancel{4}} \cdot \cancel{x} \cdot \cancel{y}} = \frac{5}{6}$

Notice that we used the commutative property of multiplication to rearrange factors in a more convenient form for recognizing common factors of the numerator and denominator

2. $\frac{4a}{6b} \cdot \frac{8b}{12a^2} = \frac{\cancel{4} \cdot \overset{4}{\cancel{8}} \cdot \cancel{a} \cdot \cancel{b}}{\underset{3}{\cancel{6}} \cdot \underset{3}{\cancel{12}} \cdot \underset{a}{\cancel{a^2}} \cdot \cancel{b}} = \frac{4}{9a}$

3. $\frac{-9x^2}{15xy} \cdot \frac{5y^2}{7x^2y^3} = -\frac{\overset{3}{\cancel{9}} \cdot \cancel{5} \cdot \cancel{x^2} \cdot \cancel{y^2}}{\underset{\cancel{5}}{\cancel{15}} \cdot 7 \cdot \underset{x}{\cancel{x^3}} \cdot \underset{y^2}{\cancel{y^4}}} = -\frac{3}{7xy^2}$

When multiplying algebraic fractions, we sometimes need to factor the numerators and/or denominators so that we can recognize common factors. Consider the next examples.

EXAMPLE 1

Multiply and simplify $\frac{x}{x^2 - 9} \cdot \frac{x + 3}{y}$.

Solution

$$\frac{x}{x^2 - 9} \cdot \frac{x + 3}{y} = \frac{x\cancel{(x + 3)}}{\cancel{(x + 3)}(x - 3)(y)}$$

$$= \frac{x}{y(x - 3)}$$

$\frac{x}{xy - 3y}$ is also an acceptable answer. ■

Remember, when working with algebraic fractions, we are assuming that all denominators represent nonzero real numbers. Therefore, in Example 1, we are claiming that $\frac{x}{x^2 - 9} \cdot \frac{x+3}{y} = \frac{x}{y(x-3)}$ for all real numbers except −3 and 3 for x, and 0 for y.

EXAMPLE 2

Multiply and simplify $\frac{x}{x^2 + 2x} \cdot \frac{x^2 + 10x + 16}{5}$.

Solution

$$\frac{x}{x^2 + 2x} \cdot \frac{x^2 + 10x + 16}{5} = \frac{\cancel{x}(\cancel{x+2})(x+8)}{\cancel{x}(\cancel{x+2})(5)} = \frac{x+8}{5}$$ ■

EXAMPLE 3

Multiply and simplify $\frac{a^2 - 3a}{a + 5} \cdot \frac{a^2 + 3a - 10}{a^2 - 5a + 6}$.

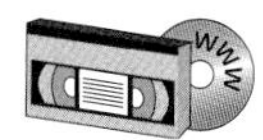

Solution

$$\frac{a^2 - 3a}{a + 5} \cdot \frac{a^2 + 3a - 10}{a^2 - 5a + 6} = \frac{a(\cancel{a-3})(\cancel{a+5})(\cancel{a-2})}{(\cancel{a+5})(\cancel{a-2})(\cancel{a-3})} = a$$ ■

EXAMPLE 4

Multiply and simplify $\frac{6n^2 + 7n - 3}{n + 1} \cdot \frac{n^2 - 1}{2n^2 + 3n}$.

Solution

$$\frac{6n^2 + 7n - 3}{n + 1} \cdot \frac{n^2 - 1}{2n^2 + 3n} = \frac{(\cancel{2n+3})(3n-1)(\cancel{n+1})(n-1)}{(\cancel{n+1})(n)(\cancel{2n+3})}$$

$$= \frac{(3n-1)(n-1)}{n}$$ ■

■ Dividing Algebraic Fractions

Recall that to divide two rational numbers in $\frac{a}{b}$ form, we invert the divisor and multiply. Symbolically we express this as $\frac{a}{b} \div \frac{c}{d} = \frac{a}{b} \cdot \frac{d}{c}$. Furthermore, we call the numbers $\frac{c}{d}$ and $\frac{d}{c}$ *reciprocals* of each other because their product is 1. Thus we can also describe division as to divide by a fraction, multiply by its reciprocal. We define division of algebraic fractions in the same way using the same vocabulary.

Definition 7.2

If $\frac{A}{B}$ and $\frac{C}{D}$ are rational expressions with $B \neq 0$, $D \neq 0$, and $C \neq 0$, then

$$\frac{A}{B} \div \frac{C}{D} = \frac{A}{B} \cdot \frac{D}{C} = \frac{AD}{BC}$$

Consider some examples.

1. $$\frac{4x}{7y} \div \frac{6x^2}{14y^2} = \frac{4x}{7y} \cdot \frac{14y^2}{6x^2} = \frac{\overset{2}{\cancel{4}} \cdot \overset{2}{\cancel{14}} \cdot \cancel{x} \cdot \overset{y}{\cancel{y^2}}}{\cancel{7} \cdot \underset{3}{\cancel{6}} \cdot \underset{x}{\cancel{x^2}} \cdot \cancel{y}} = \frac{4y}{3x}$$

2. $$\frac{-8ab}{9b} \div \frac{18a^3}{15a^2b} = \frac{-8ab}{9b} \cdot \frac{15a^2b}{18a^3} = -\frac{\overset{4}{\cancel{8}} \cdot \overset{5}{\cancel{15}} \cdot \cancel{a^3} \cdot \overset{b}{\cancel{b^2}}}{\underset{3}{\cancel{9}} \cdot \underset{9}{\cancel{18}} \cdot \cancel{a^3} \cdot \cancel{b}} = -\frac{20b}{27}$$

3. $$\frac{x^2y^3}{4ab} \div \frac{5xy^2}{-9a^2b} = \frac{x^2y^3}{4ab} \cdot \frac{-9a^2b}{5xy^2} = -\frac{9 \cdot \overset{x}{\cancel{x^2}} \cdot \overset{y}{\cancel{y^3}} \cdot \overset{a}{\cancel{a^2}} \cdot \cancel{b}}{4 \cdot 5 \cdot \cancel{a} \cdot \cancel{b} \cdot \cancel{x} \cdot \cancel{y^2}} = -\frac{9axy}{20}$$

The key idea when dividing fractions is to *first* convert to an equivalent multiplication problem and then proceed to factor numerator and denominator completely and look for common factors.

EXAMPLE 5

Divide and simplify $\frac{x^2 - 4x}{xy} \div \frac{x^2 - 16}{y^3 + y^2}$.

Solution

$$\frac{x^2 - 4x}{xy} \div \frac{x^2 - 16}{y^3 + y^2} = \frac{x^2 - 4x}{xy} \cdot \frac{y^3 + y^2}{x^2 - 16}$$

$$= \frac{\cancel{x}\cancel{(x - 4)}(\overset{y}{\cancel{y^2}})(y + 1)}{\cancel{x}\cancel{y}(x + 4)\cancel{(x - 4)}}$$

$$= \frac{y(y + 1)}{x + 4}$$

■

EXAMPLE 6

Divide and simplify $\frac{a^2 + 3a - 18}{a^2 + 4} \div \frac{1}{3a^2 + 12}$.

Solution

$$\frac{a^2 + 3a - 18}{a^2 + 4} \div \frac{1}{3a^2 + 12} = \frac{a^2 + 3a - 18}{a^2 + 4} \cdot \frac{3a^2 + 12}{1}$$

$$= \frac{(a + 6)(a - 3)(3)\cancel{(a^2 + 4)}}{\cancel{a^2 + 4}}$$

$$= 3(a + 6)(a - 3)$$

■

EXAMPLE 7

Divide and simplify $\dfrac{2n^2 - 7n - 4}{6n^2 + 7n + 2} \div (n - 4)$.

Solution

$$\frac{2n^2 - 7n - 4}{6n^2 + 7n + 2} \div (n - 4) = \frac{2n^2 - 7n - 4}{6n^2 + 7n + 2} \cdot \frac{1}{n - 4}$$

$$= \frac{\cancel{(2n + 1)}\cancel{(n - 4)}}{\cancel{(2n + 1)}(3n + 2)\cancel{(n - 4)}}$$

$$= \frac{1}{3n + 2}$$

■

In a problem such as Example 7, it may be helpful to write the divisor with a denominator of 1. Thus we can write $n - 4$ as $\dfrac{n - 4}{1}$; its reciprocal then is obviously $\dfrac{1}{n - 4}$.

Problem Set 7.2

For Problems 1–40, perform the indicated multiplications and divisions and express your answers in simplest form.

1. $\dfrac{5}{9} \cdot \dfrac{3}{10}$

2. $\dfrac{7}{8} \cdot \dfrac{12}{14}$

3. $\left(-\dfrac{3}{4}\right)\left(\dfrac{6}{7}\right)$

4. $\left(\dfrac{5}{6}\right)\left(-\dfrac{4}{15}\right)$

5. $\left(\dfrac{17}{9}\right) \div \left(-\dfrac{19}{9}\right)$

6. $\left(-\dfrac{15}{7}\right) \div \left(\dfrac{13}{14}\right)$

7. $\dfrac{8xy}{12y} \cdot \dfrac{6x}{14y}$

8. $\dfrac{9x}{15y} \cdot \dfrac{20xy}{18x}$

9. $\left(-\dfrac{5n^2}{18n}\right)\left(\dfrac{27n}{25}\right)$

10. $\left(\dfrac{4ab}{10}\right)\left(-\dfrac{30a}{22b}\right)$

11. $\dfrac{3a^2}{7} \div \dfrac{6a}{28}$

12. $\dfrac{4x}{11y} \div \dfrac{12x}{33}$

13. $\dfrac{18a^2b^2}{-27a} \div \dfrac{-9a}{5b}$

14. $\dfrac{24ab^2}{25b} \div \dfrac{-12ab}{15a^2}$

15. $24x^3 \div \dfrac{16x}{y}$

16. $14xy^2 \div \dfrac{7y}{9}$

17. $\dfrac{1}{15ab^3} \div \dfrac{-1}{12a}$

18. $\dfrac{-2}{7a^2b^3} \div \dfrac{1}{9ab^4}$

19. $\dfrac{18rs}{34} \div 9r$

20. $\dfrac{8rs}{3} \div 6s$

21. $\dfrac{y}{x + y} \cdot \dfrac{x^2 - y^2}{xy}$

22. $\dfrac{x^2 - 9}{6} \cdot \dfrac{8}{x - 3}$

23. $\dfrac{2x^2 + xy}{xy} \cdot \dfrac{y}{10x + 5y}$

24. $\dfrac{x^2 + y^2}{x - y} \cdot \dfrac{x^2 - xy}{3}$

25. $\dfrac{6ab}{4ab + 4b^2} \div \dfrac{7a - 7b}{a^2 - b^2}$

26. $\dfrac{4ab}{2a^2 - 2ab} \div \dfrac{ab + b}{3a - 3b}$

27. $\dfrac{x^2 + 11x + 30}{x^2 + 4} \cdot \dfrac{5x^2 + 20}{x^2 + 14x + 45}$

28. $\dfrac{x^2 + 15x + 54}{x^2 + 2} \cdot \dfrac{3x^2 + 6}{x^2 + 10x + 9}$

29. $\dfrac{2x^2 - 3xy + y^2}{4x^2y} \div \dfrac{x^2 - y^2}{6x^2y^2}$

30. $\dfrac{2x^2 + xy - y^2}{x^2y} \div \dfrac{5x^2 + 4xy - y^2}{y}$

31. $\dfrac{a + a^2}{15a^2 + 11a + 2} \cdot \dfrac{1 - a}{1 - a^2}$

32. $\dfrac{2a^2 - 11a - 21}{3a^2 + a} \cdot \dfrac{3a^2 - 11a - 4}{2a^2 - 5a - 12}$

33. $\dfrac{2x^2 - 2xy}{x^2 + 4x - 32} \cdot \dfrac{x^2 - 16}{5xy - 5y^2}$

34. $\dfrac{x^3 + 3x^2}{x^2 + 4x + 4} \cdot \dfrac{x^2 - 5x - 14}{x^2 + 3x}$

35. $\dfrac{2x^2 - xy - 3y^2}{(x + y)^2} \div \dfrac{4x^2 - 12xy + 9y^2}{10x - 15y}$

36. $\dfrac{x^2 + 4xy + 4y^2}{x^2} \div \dfrac{x^2 - 4y^2}{x^2 - 2xy}$

37. $\dfrac{(3t - 1)^2}{45t - 15} \div \dfrac{12t^2 + 5t - 3}{20t + 5}$

38. $\dfrac{5t^2 - 3t - 2}{(t - 1)^2} \div \dfrac{5t^2 + 32t + 12}{4t^2 - 3t - 1}$

39. $\dfrac{n^3 - n}{n^2 + 7n + 6} \cdot \dfrac{4n + 24}{n^2 - n}$

40. $\dfrac{2x^2 - 6x - 36}{x^2 + 2x - 48} \cdot \dfrac{x^2 + 5x - 24}{2x^2 - 18}$

For Problems 41–46, perform the indicated operations and express the answers in simplest form. Remember that multiplications and divisions are done in the order that they appear from left to right.

41. $\dfrac{6}{9y} \div \dfrac{30x}{12y^2} \cdot \dfrac{5xy}{4}$

42. $\dfrac{5xy^2}{12y} \cdot \dfrac{18x^2}{15y} \div \dfrac{3}{2xy}$

43. $\dfrac{8x^2}{xy - xy^2} \cdot \dfrac{x - 1}{8x^2 - 8y^2} \div \dfrac{xy}{x + y}$

44. $\dfrac{5x - 20}{x^2 - 9} \cdot \dfrac{x + 3}{x - 4} \div \dfrac{15}{x - 3}$

45. $\dfrac{x^2 + 9x + 18}{x^2 + 3x} \cdot \dfrac{x^2 + 5x}{x^2 - 25} \div \dfrac{x^2 + 8x}{x^2 + 3x - 40}$

46. $\dfrac{4x}{3x + 6y} \cdot \dfrac{5xy}{x^2 - 4} \div \dfrac{10}{x^2 + 4x + 4}$

THOUGHTS INTO WORDS

47. Give a step-by-step description of how to do the following multiplication problem:

$$\frac{x^2 - x}{x^2 - 1} \cdot \frac{x^2 + x - 6}{x^2 + 4x - 12}$$

48. Is $\left(\dfrac{x}{x + 1} \div \dfrac{x - 1}{x}\right) \div \dfrac{1}{x} = \dfrac{x}{x + 1} \div \left(\dfrac{x - 1}{x} \div \dfrac{1}{x}\right)$?
Justify your answer.

49. Explain why the quotient $\dfrac{x - 2}{x + 1} \div \dfrac{x}{x - 1}$ is undefined for $x = -1$, $x = 1$, and $x = 0$ but is defined for $x = 2$.

7.3 Adding and Subtracting Algebraic Fractions

In Chapter 2 we defined addition and subtraction of rational numbers as $\dfrac{a}{b} + \dfrac{c}{b} = \dfrac{a + c}{b}$ and $\dfrac{a}{b} - \dfrac{c}{b} = \dfrac{a - c}{b}$, respectively. These definitions extend to algebraic fractions in general.

Definition 7.3

If $\frac{A}{B}$ and $\frac{C}{B}$ are rational expressions with $B \neq 0$, then

$$\frac{A}{B} + \frac{C}{B} = \frac{A + C}{B} \quad \text{and} \quad \frac{A}{B} - \frac{C}{B} = \frac{A - C}{B}$$

Thus if the denominators of two algebraic fractions are the same, then we can add or subtract the fractions by adding or subtracting the numerators and placing the result over the common denominator. Here are some examples:

$$\frac{5}{x} + \frac{7}{x} = \frac{5 + 7}{x} = \frac{12}{x}$$

$$\frac{8}{xy} - \frac{3}{xy} = \frac{8 - 3}{xy} = \frac{5}{xy}$$

$$\frac{14}{2x + 1} + \frac{15}{2x + 1} = \frac{14 + 15}{2x + 1} = \frac{29}{2x + 1}$$

$$\frac{3}{a - 1} - \frac{4}{a - 1} = \frac{3 - 4}{a - 1} = \frac{-1}{a - 1} \quad \text{or} \quad -\frac{1}{a - 1}$$

In the next examples, notice how we put to use our previous work with simplifying polynomials.

$$\frac{x + 3}{4} + \frac{2x - 3}{4} = \frac{(x + 3) + (2x - 3)}{4} = \frac{3x}{4}$$

$$\frac{x + 5}{7} - \frac{x + 2}{7} = \frac{(x + 5) - (x + 2)}{7} = \frac{x + 5 - x - 2}{7} = \frac{3}{7}$$

$$\frac{3x + 1}{xy} + \frac{2x + 3}{xy} = \frac{(3x + 1) + (2x + 3)}{xy} = \frac{5x + 4}{xy}$$

$$\frac{2(3n + 1)}{n} - \frac{3(n - 1)}{n} = \frac{2(3n + 1) - 3(n - 1)}{n} = \frac{6n + 2 - 3n + 3}{n} = \frac{3n + 5}{n}$$

It may be necessary to simplify the fraction that results from adding or subtracting two fractions.

$$\frac{4x - 3}{8} + \frac{2x + 3}{8} = \frac{(4x - 3) + (2x + 3)}{8} = \frac{6x}{8} = \frac{3x}{4}$$

$$\frac{3n - 1}{12} - \frac{n - 5}{12} = \frac{(3n - 1) - (n - 5)}{12} = \frac{3n - 1 - n + 5}{12}$$

$$= \frac{2n + 4}{12} = \frac{2(n + 2)}{12} = \frac{n + 2}{6}$$

$$\frac{-2x+3}{x^2-4}+\frac{3x-1}{x^2-4}=\frac{(-2x+3)+(3x-1)}{x^2-4}=\frac{x+2}{x^2-4}$$

$$=\frac{\cancel{x+2}}{(\cancel{x+2})(x-2)}$$

$$=\frac{1}{x-2}$$

Recall that to add or subtract rational numbers with different denominators, we first change them to equivalent fractions that have a common denominator. In fact, we found that by using the least common denominator (LCD), our work was easier. Let's carefully review the process because it will also work with algebraic fractions in general.

EXAMPLE 1

Add $\frac{3}{5}+\frac{1}{4}$.

Solution

By inspection, we see that the LCD is 20. Thus we can change both fractions to equivalent fractions that have a denominator of 20.

$$\frac{3}{5}+\frac{1}{4}=\frac{3}{5}\left(\frac{4}{4}\right)+\frac{1}{4}\left(\frac{5}{5}\right)=\frac{12}{20}+\frac{5}{20}=\frac{17}{20}$$

Form of 1 (arrows pointing to $\left(\frac{4}{4}\right)$ and $\left(\frac{5}{5}\right)$)

EXAMPLE 2

Subtract $\frac{5}{18}-\frac{7}{24}$.

Solution

If we cannot find the LCD by inspection, then we can use the prime factorization forms.

$$\left.\begin{aligned}18&=2\cdot3\cdot3\\24&=2\cdot2\cdot2\cdot3\end{aligned}\right\}\longrightarrow \text{LCD}=2\cdot2\cdot2\cdot3\cdot3=72$$

$$\frac{5}{18}-\frac{7}{24}=\frac{5}{18}\left(\frac{4}{4}\right)-\frac{7}{24}\left(\frac{3}{3}\right)=\frac{20}{72}-\frac{21}{72}=-\frac{1}{72}$$

Now let's consider adding and subtracting algebraic fractions with different denominators.

EXAMPLE 3

Add $\frac{x-2}{4} + \frac{3x+1}{3}$.

Solution

By inspection, we see that the LCD is 12.

$$\begin{aligned}\frac{x-2}{4} + \frac{3x+1}{3} &= \left(\frac{x-2}{4}\right)\left(\frac{3}{3}\right) + \left(\frac{3x+1}{3}\right)\left(\frac{4}{4}\right) \\ &= \frac{3(x-2)}{12} + \frac{4(3x+1)}{12} \\ &= \frac{3(x-2) + 4(3x+1)}{12} \\ &= \frac{3x - 6 + 12x + 4}{12} \\ &= \frac{15x - 2}{12}\end{aligned}$$

■

EXAMPLE 4

Subtract $\frac{n-2}{2} - \frac{n-6}{6}$.

Solution

By inspection, we see that the LCD is 6.

$$\begin{aligned}\frac{n-2}{2} - \frac{n-6}{6} &= \left(\frac{n-2}{2}\right)\left(\frac{3}{3}\right) - \frac{n-6}{6} \\ &= \frac{3(n-2)}{6} - \frac{(n-6)}{6} \\ &= \frac{3(n-2) - (n-6)}{6} \\ &= \frac{3n - 6 - n + 6}{6} \\ &= \frac{2n}{6} \\ &= \frac{n}{3}\end{aligned}$$

Don't forget to simplify!

■

It does not create any serious difficulties when the denominators contain variables; our approach remains basically the same.

EXAMPLE 5

Add $\frac{3}{4x} + \frac{7}{3x}$.

Solution

By inspection, we see that the LCD is $12x$.

$$\frac{3}{4x} + \frac{7}{3x} = \frac{3}{4x}\left(\frac{3}{3}\right) + \frac{7}{3x}\left(\frac{4}{4}\right) = \frac{9}{12x} + \frac{28}{12x} = \frac{9 + 28}{12x} = \frac{37}{12x}$$ ■

EXAMPLE 6

Subtract $\frac{11}{12x} - \frac{5}{14x}$.

Solution

$$\left.\begin{aligned} 12x &= 2 \cdot 2 \cdot 3 \cdot x \\ 14x &= 2 \cdot 7 \cdot x \end{aligned}\right\} \longrightarrow \text{LCD} = 2 \cdot 2 \cdot 3 \cdot 7 \cdot x = 84x$$

$$\begin{aligned} \frac{11}{12x} - \frac{5}{14x} &= \frac{11}{12x}\left(\frac{7}{7}\right) - \frac{5}{14x}\left(\frac{6}{6}\right) \\ &= \frac{77}{84x} - \frac{30}{84x} = \frac{77 - 30}{84x} = \frac{47}{84x} \end{aligned}$$ ■

EXAMPLE 7

Add $\frac{2}{y} + \frac{4}{y - 2}$.

Solution

By inspection, we see that the LCD is $y(y - 2)$.

$$\frac{2}{y} + \frac{4}{y - 2} = \frac{2}{y}\underbrace{\left(\frac{y - 2}{y - 2}\right)}_{\text{Form of 1}} + \frac{4}{y - 2}\underbrace{\left(\frac{y}{y}\right)}_{\text{Form of 1}}$$

$$\begin{aligned} &= \frac{2(y - 2)}{y(y - 2)} + \frac{4y}{y(y - 2)} \\ &= \frac{2(y - 2) + 4y}{y(y - 2)} \\ &= \frac{2y - 4 + 4y}{y(y - 2)} = \frac{6y - 4}{y(y - 2)} \end{aligned}$$ ■

Notice the final result in Example 7. The numerator, $6y - 4$, can be factored into $2(3y - 2)$. However, because this produces no common factors with the denominator, the fraction cannot be simplified. Thus the final answer can be left as $\frac{6y - 4}{y(y - 2)}$; it is also acceptable to express it as $\frac{2(3y - 2)}{y(y - 2)}$.

EXAMPLE 8

Subtract $\frac{4}{x+2} - \frac{7}{x+3}$.

Solution

By inspection, we see that the LCD is $(x + 2)(x + 3)$.

$$\begin{aligned}
\frac{4}{x+2} - \frac{7}{x+3} &= \left(\frac{4}{x+2}\right)\left(\frac{x+3}{x+3}\right) - \left(\frac{7}{x+3}\right)\left(\frac{x+2}{x+2}\right) \\
&= \frac{4(x+3)}{(x+2)(x+3)} - \frac{7(x+2)}{(x+3)(x+2)} \\
&= \frac{4(x+3) - 7(x+2)}{(x+2)(x+3)} \\
&= \frac{4x + 12 - 7x - 14}{(x+2)(x+3)} \\
&= \frac{-3x - 2}{(x+2)(x+3)}
\end{aligned}$$

■

Problem Set 7.3

For Problems 1–34, add or subtract as indicated. Be sure to express your answers in simplest form.

1. $\frac{5}{x} + \frac{12}{x}$

2. $\frac{17}{x} - \frac{13}{x}$

3. $\frac{7}{3x} - \frac{5}{3x}$

4. $\frac{4}{5x} + \frac{3}{5x}$

5. $\frac{7}{2n} + \frac{1}{2n}$

6. $\frac{5}{3n} + \frac{4}{3n}$

7. $\frac{9}{4x^2} - \frac{13}{4x^2}$

8. $\frac{12}{5x^2} - \frac{22}{5x^2}$

9. $\frac{x+1}{x} + \frac{3}{x}$

10. $\frac{x-2}{x} + \frac{4}{x}$

11. $\frac{3}{x-1} - \frac{6}{x-1}$

12. $\frac{8}{x+4} - \frac{10}{x+4}$

13. $\frac{x+1}{x} - \frac{1}{x}$

14. $\frac{2x+3}{x} - \frac{3}{x}$

15. $\frac{3t-1}{4} + \frac{2t+3}{4}$

16. $\frac{4t-1}{7} + \frac{8t-5}{7}$

17. $\frac{7a+2}{3} - \frac{4a-6}{3}$

18. $\frac{9a-1}{6} - \frac{4a-2}{6}$

19. $\frac{4n+3}{8} + \frac{6n+5}{8}$

20. $\frac{2n-5}{10} + \frac{6n-1}{10}$

21. $\frac{3n-7}{6} - \frac{9n-1}{6}$

22. $\frac{2n-6}{5} - \frac{7n-1}{5}$

23. $\frac{5x-2}{7x} - \frac{8x+3}{7x}$

24. $\frac{4x+1}{3x} - \frac{2x+5}{3x}$

25. $\frac{3(x+2)}{4x} + \frac{6(x-1)}{4x}$

26. $\frac{4(x-3)}{5x} + \frac{2(x+6)}{5x}$

27. $\frac{6(n-1)}{3n} + \frac{3(n+2)}{3n}$

28. $\frac{2(n-4)}{3n} + \frac{4(n+2)}{3n}$

29. $\frac{2(3x-4)}{7x^2} - \frac{7x-8}{7x^2}$

30. $\dfrac{3(4x-3)}{8x^2} - \dfrac{11x-9}{8x^2}$

31. $\dfrac{a^2}{a+2} - \dfrac{4}{a+2}$

32. $\dfrac{n^2}{n-4} - \dfrac{16}{n-4}$

33. $\dfrac{3x}{(x-6)^2} - \dfrac{18}{(x-6)^2}$

34. $\dfrac{x^2+5x}{(x+1)^2} + \dfrac{4}{(x+1)^2}$

For Problems 35–80, add or subtract as indicated and express your answers in simplest form.

35. $\dfrac{3x}{8} + \dfrac{5x}{4}$

36. $\dfrac{5x}{3} + \dfrac{2x}{9}$

37. $\dfrac{7n}{12} - \dfrac{4n}{3}$

38. $\dfrac{n}{6} - \dfrac{7n}{12}$

39. $\dfrac{y}{6} + \dfrac{3y}{4}$

40. $\dfrac{3y}{4} + \dfrac{7y}{5}$

41. $\dfrac{8x}{3} - \dfrac{3x}{7}$

42. $\dfrac{5y}{6} - \dfrac{3y}{8}$

43. $\dfrac{2x}{6} + \dfrac{3x}{5}$

44. $\dfrac{6x}{9} + \dfrac{7x}{12}$

45. $\dfrac{7n}{8} - \dfrac{3n}{9}$

46. $\dfrac{8n}{10} - \dfrac{7n}{15}$

47. $\dfrac{x+3}{5} + \dfrac{x-4}{2}$

48. $\dfrac{x-2}{5} + \dfrac{x+1}{6}$

49. $\dfrac{x-6}{9} + \dfrac{x+2}{3}$

50. $\dfrac{x-2}{4} + \dfrac{x+4}{8}$

51. $\dfrac{3n-1}{3} + \dfrac{2n+5}{4}$

52. $\dfrac{2n+3}{4} + \dfrac{4n-1}{7}$

53. $\dfrac{4n-3}{6} - \dfrac{3n+5}{18}$

54. $\dfrac{5n-2}{12} - \dfrac{4n+7}{6}$

55. $\dfrac{3x}{4} + \dfrac{x}{6} - \dfrac{5x}{8}$

56. $\dfrac{5x}{2} - \dfrac{3x}{4} - \dfrac{7x}{6}$

57. $\dfrac{x}{5} - \dfrac{3}{10} - \dfrac{7x}{12}$

58. $\dfrac{4x}{3} + \dfrac{5}{9} - \dfrac{11x}{6}$

59. $\dfrac{5}{8x} + \dfrac{1}{6x}$

60. $\dfrac{7}{8x} + \dfrac{5}{12x}$

61. $\dfrac{5}{6y} - \dfrac{7}{9y}$

62. $\dfrac{11}{9y} - \dfrac{8}{15y}$

63. $\dfrac{5}{12x} - \dfrac{11}{16x^2}$

64. $\dfrac{4}{9x} - \dfrac{7}{6x^2}$

65. $\dfrac{3}{2x} - \dfrac{2}{3x} + \dfrac{5}{4x}$

66. $\dfrac{3}{4x} - \dfrac{5}{6x} + \dfrac{10}{9x}$

67. $\dfrac{3}{x-5} + \dfrac{7}{x}$

68. $\dfrac{4}{x-8} + \dfrac{9}{x}$

69. $\dfrac{2}{n-1} - \dfrac{3}{n}$

70. $\dfrac{5}{n+3} - \dfrac{7}{n}$

71. $\dfrac{4}{n} - \dfrac{6}{n+4}$

72. $\dfrac{8}{n} - \dfrac{3}{n-9}$

73. $\dfrac{6}{x} - \dfrac{12}{2x+1}$

74. $\dfrac{2}{x} - \dfrac{6}{3x-2}$

75. $\dfrac{4}{x+4} + \dfrac{6}{x-3}$

76. $\dfrac{7}{x-2} + \dfrac{8}{x+1}$

77. $\dfrac{3}{x-2} - \dfrac{9}{x+1}$

78. $\dfrac{5}{x-1} - \dfrac{4}{x+6}$

79. $\dfrac{3}{2x-1} - \dfrac{4}{3x+1}$

80. $\dfrac{6}{3x-4} - \dfrac{4}{2x+3}$

THOUGHTS INTO WORDS

81. Give a step-by-step description of how to do this addition problem:

$$\frac{3x-1}{6} + \frac{2x+3}{9}$$

82. Why are $\dfrac{3}{x-2}$ and $\dfrac{3}{2-x}$ opposites? What should be the result of adding $\dfrac{3}{x-2}$ and $\dfrac{3}{2-x}$?

83. Suppose that your friend does an addition problem as follows:

$$\frac{5}{8} + \frac{7}{12} = \frac{5(12)+8(7)}{8(12)} = \frac{60+56}{96} = \frac{116}{96} = \frac{29}{24}$$

Is this answer correct? What advice would you offer your friend?

FURTHER INVESTIGATIONS

Consider the addition problem $\frac{9}{x-2}+\frac{4}{2-x}$. Notice that the denominators $x-2$ and $2-x$ are opposites; that is, $-1(2-x)=(x-2)$. In such cases, add the fractions as follows:

$$\frac{9}{x-2}+\frac{4}{2-x}=\frac{9}{x-2}+\frac{4}{2-x}\left(\frac{-1}{-1}\right)$$

$$=\frac{9}{x-2}+\frac{-4}{x-2}$$

$$=\frac{9+(-4)}{x-2}=\frac{5}{x-2}$$

Form of 1

For Problems 84–89, use this approach to help with the additions and subtractions.

84. $\frac{7}{x-1}-\frac{2}{1-x}$

85. $\frac{5}{x-3}+\frac{1}{3-x}$

86. $\frac{x}{x-4}+\frac{4}{4-x}$

87. $\frac{-4}{a-1}+\frac{2}{1-a}$

88. $\frac{1}{x^2-9}+\frac{2}{x+3}-\frac{3}{3-x}$

89. $\frac{n}{2n-1}-\frac{3}{1-2n}$

7.4 More on Addition and Subtraction of Algebraic Fractions

In this section, we expand our work with adding and subtracting rational expressions, and we discuss the process of simplifying complex fractions. Before we begin, however, this seems like an appropriate time to offer a bit of advice regarding your study of algebra. Success in algebra depends on having a good understanding of the concepts as well as being able to perform the various computations. As for the computational work, you should adopt a carefully organized format that shows as many steps as you need in order to minimize the chances of making careless errors. Don't be eager to find shortcuts for certain computations before you have a thorough understanding of the steps involved in the process. This advice is especially appropriate at the beginning of this section.

Study Examples 1–4 very carefully. Note that the same basic procedure is followed in solving each problem:

Step 1 Factor the denominators.

Step 2 Find the LCD.

Step 3 Change each fraction to an equivalent fraction that has the LCD as its denominator.

Step 4 Combine the numerators and place over the LCD.

Step 5 Simplify by performing the addition or subtraction.

Step 6 Look for ways to reduce the resulting fraction.

EXAMPLE 1 Add $\dfrac{3}{x^2 + 2x} + \dfrac{5}{x}$.

Solution

$$\left.\begin{array}{ll}\text{1st denominator:} & x^2 + 2x = x(x + 2) \\ \text{2nd denominator:} & x\end{array}\right\} \longrightarrow \text{LCD is } x(x + 2).$$

$$\frac{3}{x^2 + 2x} + \frac{5}{x} = \frac{3}{x(x + 2)} + \frac{5}{x}\left(\frac{x + 2}{x + 2}\right)$$

This fraction has the LCD as its denominator (arrow to $\frac{3}{x(x+2)}$); Form of 1 (arrow to $\frac{x+2}{x+2}$)

$$= \frac{3}{x(x + 2)} + \frac{5(x + 2)}{x(x + 2)} = \frac{3 + 5(x + 2)}{x(x + 2)}$$

$$= \frac{3 + 5x + 10}{x(x + 2)} = \frac{5x + 13}{x(x + 2)}$$

■

EXAMPLE 2 Subtract $\dfrac{4}{x^2 - 4} - \dfrac{1}{x - 2}$.

Solution

$$\left.\begin{array}{l}x^2 - 4 = (x + 2)(x - 2) \\ x - 2 = x - 2\end{array}\right\} \longrightarrow \text{LCD is } (x + 2)(x - 2).$$

$$\frac{4}{x^2 - 4} - \frac{1}{x - 2} = \frac{4}{(x + 2)(x - 2)} - \left(\frac{1}{x - 2}\right)\left(\frac{x + 2}{x + 2}\right)$$

$$= \frac{4}{(x + 2)(x - 2)} - \frac{1(x + 2)}{(x + 2)(x - 2)}$$

$$= \frac{4 - 1(x + 2)}{(x + 2)(x - 2)} = \frac{4 - x - 2}{(x + 2)(x - 2)}$$

$$= \frac{-x + 2}{(x + 2)(x - 2)}$$

$$= \frac{-1\cancel{(x - 2)}}{(x + 2)\cancel{(x - 2)}} \longrightarrow \text{Note the changing of } -x + 2 \text{ to } -1(x - 2)$$

$$= -\frac{1}{x + 2}$$

■

EXAMPLE 3 Add $\frac{2}{a^2 - 9} + \frac{3}{a^2 + 5a + 6}$.

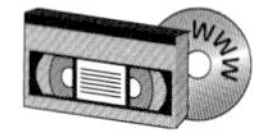

Solution

$$\left.\begin{aligned} a^2 - 9 &= (a + 3)(a - 3) \\ a^2 + 5a + 6 &= (a + 3)(a + 2) \end{aligned}\right\} \longrightarrow \text{LCD is } (a + 3)(a - 3)(a + 2).$$

$$\frac{2}{a^2 - 9} + \frac{3}{a^2 + 5a + 6}$$

$$= \left(\frac{2}{(a + 3)(a - 3)}\right)\left(\frac{a + 2}{a + 2}\right) + \left(\frac{3}{(a + 3)(a + 2)}\right)\left(\frac{a - 3}{a - 3}\right)$$

Form of 1 Form of 1

$$= \frac{2(a + 2)}{(a + 3)(a - 3)(a + 2)} + \frac{3(a - 3)}{(a + 3)(a - 3)(a + 2)}$$

$$= \frac{2(a + 2) + 3(a - 3)}{(a + 3)(a - 3)(a + 2)} = \frac{2a + 4 + 3a - 9}{(a + 3)(a - 3)(a + 2)}$$

$$= \frac{5a - 5}{(a + 3)(a - 3)(a + 2)} \quad \text{or} \quad \frac{5(a - 1)}{(a + 3)(a - 3)(a + 2)}$$ ■

EXAMPLE 4 Perform the indicated operations.

$$\frac{2x}{x^2 - y^2} + \frac{3}{x + y} - \frac{2}{x - y}$$

Solution

$$\left.\begin{aligned} x^2 - y^2 &= (x + y)(x - y) \\ x + y &= x + y \\ x - y &= x - y \end{aligned}\right\} \longrightarrow \text{LCD is } (x + y)(x - y).$$

$$\frac{2x}{x^2 - y^2} + \frac{3}{x + y} - \frac{2}{x - y}$$

$$= \frac{2x}{(x + y)(x - y)} + \left(\frac{3}{x + y}\right)\left(\frac{x - y}{x - y}\right) - \left(\frac{2}{x - y}\right)\left(\frac{x + y}{x + y}\right)$$

$$= \frac{2x}{(x + y)(x - y)} + \frac{3(x - y)}{(x + y)(x - y)} - \frac{2(x + y)}{(x + y)(x - y)}$$

$$= \frac{2x + 3(x - y) - 2(x + y)}{(x + y)(x - y)}$$

$$= \frac{2x + 3x - 3y - 2x - 2y}{(x + y)(x - y)}$$

$$= \frac{3x - 5y}{(x + y)(x - y)}$$ ■

Complex Fractions

Fractional forms that contain fractions in the numerators and/or denominators are called **complex fractions**. Here are some examples of complex fractions:

$$\frac{\frac{2}{3}}{\frac{4}{5}} \qquad \frac{\frac{1}{x}}{\frac{3}{y}} \qquad \frac{\frac{1}{2} + \frac{1}{3}}{\frac{5}{6} - \frac{1}{4}} \qquad \frac{\frac{2}{x} + \frac{2}{y}}{\frac{5}{x} - \frac{1}{y^2}}$$

It is often necessary to *simplify* a complex fraction — that is, to express it as a simple fraction. We will illustrate this process with the next four examples.

EXAMPLE 5

Simplify $\frac{\frac{2}{3}}{\frac{4}{5}}$.

Solution

This type of problem creates no difficulty because it is merely a division problem. Thus,

$$\frac{\frac{2}{3}}{\frac{4}{5}} = \frac{2}{3} \div \frac{4}{5} = \frac{\overset{1}{\cancel{2}}}{3} \cdot \frac{5}{\underset{2}{\cancel{4}}} = \frac{5}{6}$$ ■

EXAMPLE 6

Simplify $\frac{\frac{1}{x}}{\frac{3}{y}}$.

Solution

$$\frac{\frac{1}{x}}{\frac{3}{y}} = \frac{1}{x} \div \frac{3}{y} = \frac{1}{x} \cdot \frac{y}{3} = \frac{y}{3x}$$ ■

EXAMPLE 7 Simplify $\dfrac{\frac{1}{2}+\frac{1}{3}}{\frac{5}{6}-\frac{1}{4}}$.

Let's look at two possible "attack routes" for such a problem.

Solution A

$$\frac{\frac{1}{2}+\frac{1}{3}}{\frac{5}{6}-\frac{1}{4}} = \frac{\frac{3}{6}+\frac{2}{6}}{\frac{10}{12}-\frac{3}{12}} = \frac{\frac{5}{6}}{\frac{7}{12}} = \frac{5}{\cancel{6}_1} \cdot \frac{\cancel{12}^2}{7} = \frac{10}{7}$$

Invert divisor and multiply

Solution B

The least common multiple of all four denominators (2, 3, 6, and 4) is 12. We multiply the entire complex fraction by a form of 1, specifically $\frac{12}{12}$.

$$\frac{\frac{1}{2}+\frac{1}{3}}{\frac{5}{6}-\frac{1}{4}} = \left(\frac{12}{12}\right)\left(\frac{\frac{1}{2}+\frac{1}{3}}{\frac{5}{6}-\frac{1}{4}}\right)$$

$$= \frac{12\left(\frac{1}{2}+\frac{1}{3}\right)}{12\left(\frac{5}{6}-\frac{1}{4}\right)} = \frac{12\left(\frac{1}{2}\right)+12\left(\frac{1}{3}\right)}{12\left(\frac{5}{6}\right)-12\left(\frac{1}{4}\right)}$$

$$= \frac{6+4}{10-3} = \frac{10}{7}$$

■

EXAMPLE 8 Simplify $\dfrac{\frac{2}{x}+\frac{3}{y}}{\frac{5}{x}-\frac{1}{y^2}}$.

Solution A

$$\frac{\frac{2}{x}+\frac{3}{y}}{\frac{5}{x}-\frac{1}{y^2}} = \frac{\frac{2}{x}\left(\frac{y}{y}\right)+\frac{3}{y}\left(\frac{x}{x}\right)}{\frac{5}{x}\left(\frac{y^2}{y^2}\right)-\frac{1}{y^2}\left(\frac{x}{x}\right)} = \frac{\frac{2y}{xy}+\frac{3x}{xy}}{\frac{5y^2}{xy^2}-\frac{x}{xy^2}}$$

$$= \frac{\dfrac{2y + 3x}{xy}}{\dfrac{5y^2 - x}{xy^2}}$$

$$= \frac{2y + 3x}{\cancel{xy}} \cdot \frac{\overset{y}{\cancel{xy^2}}}{5y^2 - x} \quad \text{Invert divisor and multiply}$$

$$= \frac{y(2y + 3x)}{5y^2 - x}$$

Solution B

The least common multiple of all four denominators (x, y, x, and y^2) is xy^2. We multiply the entire complex fraction by a form of 1, specifically, $\dfrac{xy^2}{xy^2}$.

$$\frac{\dfrac{2}{x} + \dfrac{3}{y}}{\dfrac{5}{x} - \dfrac{1}{y^2}} = \left(\frac{xy^2}{xy^2}\right)\left(\frac{\dfrac{2}{x} + \dfrac{3}{y}}{\dfrac{5}{x} - \dfrac{1}{y^2}}\right)$$

$$= \frac{xy^2\left(\dfrac{2}{x} + \dfrac{3}{y}\right)}{xy^2\left(\dfrac{5}{x} - \dfrac{1}{y^2}\right)} = \frac{xy^2\left(\dfrac{2}{x}\right) + xy^2\left(\dfrac{3}{y}\right)}{xy^2\left(\dfrac{5}{x}\right) - xy^2\left(\dfrac{1}{y^2}\right)}$$

$$= \frac{2y^2 + 3xy}{5y^2 - x} \quad \text{or} \quad \frac{y(2y + 3x)}{5y^2 - x}$$

■

Certainly, either approach (Solution A or Solution B) will work for problems such as Examples 7 and 8. You should carefully examine Solution B of each example. This approach works very effectively with algebraic complex fractions where the least common multiple of all the denominators of the simple fractions is easy to find. We can summarize the two methods for simplifying a complex fraction as follows:

1. Simplify the numerator and denominator of the fraction separately. Then divide the simplified numerator by the simplified denominator.

2. Multiply the numerator and denominator of the complex fraction by the least common multiple of all of the denominators that appear in the complex fraction.

EXAMPLE 9

Simplify $\dfrac{\frac{2}{x} - 3}{4 + \frac{5}{y}}$.

Solution

$$\frac{\frac{2}{x} - 3}{4 + \frac{5}{y}} = \left(\frac{xy}{xy}\right)\left(\frac{\frac{2}{x} - 3}{4 + \frac{5}{y}}\right)$$

$$= \frac{(xy)\left(\frac{2}{x}\right) - (xy)(3)}{(xy)(4) + (xy)\left(\frac{5}{y}\right)}$$

$$= \frac{2y - 3xy}{4xy + 5x} \quad \text{or} \quad \frac{y(2 - 3x)}{x(4y + 5)}$$

■

Problem Set 7.4

For Problems 1–40, perform the indicated operations and express answers in simplest form.

1. $\dfrac{4}{x^2 - 4x} + \dfrac{3}{x}$
2. $\dfrac{3}{x^2 + 2x} + \dfrac{7}{x}$
3. $\dfrac{7}{x^2 + 2x} - \dfrac{5}{x}$
4. $\dfrac{9}{x^2 - 5x} - \dfrac{1}{x}$
5. $\dfrac{8}{n} - \dfrac{2}{n^2 - 6n}$
6. $\dfrac{6}{n} - \dfrac{4}{n^2 + 6n}$
7. $\dfrac{4}{n^2 + n} - \dfrac{4}{n}$
8. $\dfrac{8}{n^2 - 2n} + \dfrac{4}{n}$
9. $\dfrac{7}{2x} - \dfrac{x}{x^2 - x}$
10. $\dfrac{3x}{x^2 + 2x} + \dfrac{4}{5x}$
11. $\dfrac{3}{x^2 - 16} + \dfrac{5}{x + 4}$
12. $\dfrac{6}{x^2 - 9} + \dfrac{9}{x - 3}$
13. $\dfrac{8x}{x^2 - 1} - \dfrac{4}{x - 1}$
14. $\dfrac{6x}{x^2 - 4} - \dfrac{3}{x + 2}$
15. $\dfrac{4}{a^2 - 2a} + \dfrac{7}{a^2 + 2a}$
16. $\dfrac{3}{a^2 + 4a} + \dfrac{5}{a^2 - 4a}$
17. $\dfrac{1}{x^2 - 6x} - \dfrac{1}{x^2 + 6x}$
18. $\dfrac{3}{x^2 + 5x} - \dfrac{4}{x^2 - 5x}$
19. $\dfrac{n}{n^2 - 16} - \dfrac{2}{3n + 12}$
20. $\dfrac{n}{n^2 - 25} - \dfrac{2}{3n - 15}$
21. $\dfrac{5x}{6x + 4} + \dfrac{2x}{9x + 6}$
22. $\dfrac{7x}{3x - 12} + \dfrac{3x}{4x - 16}$
23. $\dfrac{x - 1}{5x + 5} - \dfrac{x - 4}{3x + 3}$
24. $\dfrac{2x + 1}{6x + 12} - \dfrac{3x - 4}{8x + 16}$
25. $\dfrac{2}{x^2 + 7x + 12} + \dfrac{3}{x^2 - 9}$
26. $\dfrac{x}{x^2 - 1} + \dfrac{3}{x^2 + 5x + 4}$
27. $\dfrac{x}{x^2 + 6x + 8} - \dfrac{5}{x^2 - 3x - 10}$
28. $\dfrac{x}{x^2 - x - 30} - \dfrac{7}{x^2 - 7x + 6}$
29. $\dfrac{a}{ab + b^2} - \dfrac{b}{a^2 + ab}$
30. $\dfrac{2x}{xy + y^2} - \dfrac{2y}{x^2 + xy}$

31. $\dfrac{3}{x-5} - \dfrac{4}{x^2-25} + \dfrac{5}{x+5}$

32. $\dfrac{2}{x+1} + \dfrac{3}{x^2-1} - \dfrac{5}{x-1}$

33. $\dfrac{10}{x^2-2x} + \dfrac{8}{x^2+2x} - \dfrac{3}{x^2-4}$

34. $\dfrac{1}{x^2+7x} - \dfrac{2}{x^2-7x} - \dfrac{5}{x^2-49}$

35. $\dfrac{3x}{x^2+7x+10} - \dfrac{2}{x+2} + \dfrac{3}{x+5}$

36. $\dfrac{4}{x-3} - \dfrac{3}{x-6} + \dfrac{x}{x^2-9x+18}$

37. $\dfrac{5x}{3x^2+7x-20} - \dfrac{1}{3x-5} - \dfrac{2}{x+4}$

38. $\dfrac{4x}{6x^2+7x+2} - \dfrac{2}{2x+1} - \dfrac{4}{3x+2}$

39. $\dfrac{2}{x+4} - \dfrac{1}{x-3} + \dfrac{2x+1}{x^2+x-12}$

40. $\dfrac{3}{x-5} - \dfrac{4}{x+7} + \dfrac{3x-27}{x^2+2x-35}$

For Problems 41–60, simplify each of the complex fractions.

41. $\dfrac{\frac{1}{2} - \frac{3}{4}}{\frac{1}{6} + \frac{1}{3}}$

42. $\dfrac{\frac{3}{8} + \frac{1}{4}}{\frac{1}{2} + \frac{3}{16}}$

43. $\dfrac{\frac{2}{9} + \frac{1}{3}}{\frac{5}{6} - \frac{2}{3}}$

44. $\dfrac{\frac{7}{8} - \frac{1}{3}}{\frac{1}{6} + \frac{3}{4}}$

45. $\dfrac{3 - \frac{2}{3}}{2 + \frac{1}{4}}$

46. $\dfrac{4 + \frac{3}{5}}{\frac{1}{3} - 2}$

47. $\dfrac{\frac{3}{x}}{\frac{9}{y}}$

48. $\dfrac{\frac{-6}{a}}{\frac{8}{b}}$

49. $\dfrac{\frac{2}{x} + \frac{3}{y}}{\frac{5}{x} - \frac{1}{y}}$

50. $\dfrac{\frac{3}{x^2} - \frac{2}{x}}{\frac{4}{x} - \frac{7}{x^2}}$

51. $\dfrac{\frac{1}{y} - \frac{4}{x^2}}{\frac{7}{x} - \frac{3}{y}}$

52. $\dfrac{\frac{4}{ab} + \frac{2}{b}}{\frac{8}{a} + \frac{1}{b}}$

53. $\dfrac{\frac{6}{x} + 2}{\frac{3}{x} + 4}$

54. $\dfrac{1 - \frac{6}{y}}{3 - \frac{2}{y}}$

55. $\dfrac{\frac{3}{2x^2} - \frac{4}{x}}{\frac{5}{3x} + \frac{7}{x^2}}$

56. $\dfrac{\frac{4}{3x} + \frac{5}{x^2}}{\frac{7}{4x} - \frac{9}{x}}$

57. $\dfrac{\frac{x+2}{4}}{\frac{1}{x} + \frac{3}{2}}$

58. $\dfrac{\frac{3}{x+1} + 2}{-4 + \frac{2}{x+1}}$

59. $\dfrac{\frac{1}{x-1} - 2}{\frac{3}{x-1} + 4}$

60. $\dfrac{\frac{3}{x-2} + \frac{2}{x+2}}{\frac{4}{x+2} - \frac{5}{x-2}}$

For Problems 61–71, answer each question with an algebraic fraction.

61. If by jogging at a constant rate Joan can complete a race in 40 minutes, how much of the course has she completed at the end of m minutes?

62. If Kent can mow the entire lawn in m minutes, what fractional part of the lawn has he mowed at the end of 20 minutes?

63. If Sandy drove k kilometers at a rate of r kilometers per hour, how long did it take her to make the trip?

64. If Roy traveled m miles in h hours, what was his rate in miles per hour?

65. If l liters of gasoline cost d dollars, what is the price per liter?

66. If p pounds of candy cost c cents, what is the price per pound?

67. Suppose that the product of two numbers is 34, and one of the numbers is n. What is the other number?

68. If a cold water faucet, when opened, can fill a tank in 3 hours, how much of the tank is filled at the end of h hours? (See Figure 7.1.)

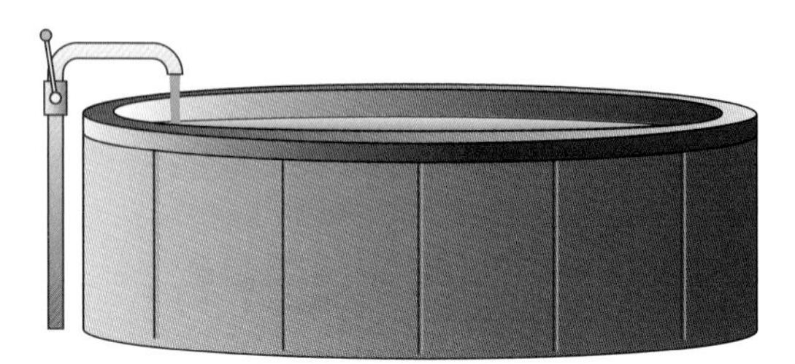

Figure 7.1

69. If the area of a rectangle is 47 square inches, and the length is l inches, what is the width of the rectangle?

70. If the area of a rectangle is 56 square centimeters, and the width is w centimeters, what is the length of the rectangle?

71. If the area of a triangle is 48 square feet and the length of one side is b feet, what is the length of the altitude to that side?

THOUGHTS INTO WORDS

72. Which of the two techniques presented in the text would you use to simplify $\dfrac{\dfrac{1}{4}+\dfrac{1}{3}}{\dfrac{3}{4}-\dfrac{1}{6}}$? Which technique would you use to simplify $\dfrac{\dfrac{3}{8}-\dfrac{5}{7}}{\dfrac{7}{9}+\dfrac{6}{25}}$? Explain your choice for each problem.

FURTHER INVESTIGATIONS

For Problems 73–76, simplify each complex fraction.

73. $1-\dfrac{n}{1-\dfrac{1}{n}}$

74. $2-\dfrac{3n}{1+\dfrac{4}{n}}$

75. $\dfrac{3x}{4-\dfrac{2}{x}}-1$

76. $\dfrac{5x}{3+\dfrac{1}{x}}+2$

7.5 Fractional Equations and Problem Solving

We will consider two basic types of fractional equations in this text. One type has only constants as denominators, and the other type has some variables in the denominators. In Chapter 3 we considered fractional equations that had only constants in the denominators. Let's review our approach to these equations because we will be using that same basic technique to solve any fractional equations.

EXAMPLE 1 Solve $\frac{x-2}{3} + \frac{x+1}{4} = \frac{1}{6}$.

Solution

$$\frac{x-2}{3} + \frac{x+1}{4} = \frac{1}{6}$$

$$12\left(\frac{x-2}{3} + \frac{x+1}{4}\right) = 12\left(\frac{1}{6}\right)$$ Multiply both sides by 12, the LCD of all three denominators

$$12\left(\frac{x-2}{3}\right) + 12\left(\frac{x+1}{4}\right) = 12\left(\frac{1}{6}\right)$$

$$4(x-2) + 3(x+1) = 2$$

$$4x - 8 + 3x + 3 = 2$$

$$7x - 5 = 2$$

$$7x = 7$$

$$x = 1$$

The solution set is {1}. (Check it!) ■

If an equation contains a variable in one or more denominators, then we proceed in essentially the same way except that we must avoid any value of the variable that makes a denominator zero. Consider the next example.

EXAMPLE 2 Solve $\frac{3}{x} + \frac{1}{2} = \frac{5}{x}$.

Solution

First, we need to realize that x *cannot equal zero.* Then we can proceed in the usual way.

$$\frac{3}{x} + \frac{1}{2} = \frac{5}{x}$$

$$2x\left(\frac{3}{x} + \frac{1}{2}\right) = 2x\left(\frac{5}{x}\right)$$ Multiply both sides by $2x$, the LCD of all denominators

$$6 + x = 10$$

$$x = 4$$

✔ **Check**

$$\frac{3}{x} + \frac{1}{2} = \frac{5}{x} \quad \text{becomes} \quad \frac{3}{4} + \frac{1}{2} \stackrel{?}{=} \frac{5}{4} \quad \text{when } x = 4$$

$$\frac{3}{4} + \frac{2}{4} \stackrel{?}{=} \frac{5}{4}$$

$$\frac{5}{4} = \frac{5}{4}$$

The solution set is {4}. ■

EXAMPLE 3 Solve $\frac{5}{x+2} = \frac{2}{x-1}$.

Solution

Because neither denominator can be zero, we know that $x \neq -2$ and $x \neq 1$.

$$\frac{5}{x+2} = \frac{2}{x-1}$$

$$(x+2)(x-1)\left(\frac{5}{x+2}\right) = (x+2)(x-1)\left(\frac{2}{x-1}\right) \quad \text{Multiply both sides by } (x+2)(x-1) \text{, the LCD}$$

$$5(x-1) = 2(x+2)$$

$$5x - 5 = 2x + 4$$

$$3x = 9$$

$$x = 3$$

Because the only restrictions are $x \neq -2$ and $x \neq 1$, the solution set is $\{3\}$. (Check it!) ■

EXAMPLE 4 Solve $\frac{2}{x-2} + 2 = \frac{x}{x-2}$.

Solution

No denominator can be zero, so $x \neq 2$.

$$\frac{2}{x-2} + 2 = \frac{x}{x-2}$$

$$(x-2)\left(\frac{2}{x-2} + 2\right) = (x-2)\left(\frac{x}{x-2}\right) \quad \text{Multiply by } x - 2 \text{, the LCD}$$

$$2 + 2(x-2) = x$$

$$2 + 2x - 4 = x$$

$$2x - 2 = x$$

$$x = 2$$

Two cannot be a solution because it will produce a denominator of zero. There is no solution to the given equation; the solution set is $\varnothing$. ■

Example 4 illustrates the importance of recognizing the restrictions that must be placed on possible values of a variable. We will indicate such restrictions at the beginning of our solution.

EXAMPLE 5 Solve $\frac{125-n}{n} = 4 + \frac{10}{n}$.

Solution

$$\frac{125-n}{n} = 4 + \frac{10}{n}, \quad n \neq 0 \quad \text{Note the necessary restriction}$$

$$n\left(\frac{125-n}{n}\right) = n\left(4 + \frac{10}{n}\right) \quad \text{Multiply both sides by } n$$

$$125 - n = 4n + 10$$
$$115 = 5n$$
$$23 = n$$

The only restriction is $n \neq 0$, and the solution set is $\{23\}$. ■

■ Back to Problem Solving

We are now ready to solve more problems, specifically those that translate into fractional equations.

PROBLEM 1

One number is 10 larger than another number. The indicated quotient of the smaller number divided by the larger number reduces to $\frac{3}{5}$. Find the numbers.

Solution

We let n represent the smaller number. Then $n + 10$ represents the larger number. The second sentence in the statement of the problem translates into the following equation:

$$\frac{n}{n+10} = \frac{3}{5}, \qquad n \neq -10$$

$$5n = 3(n + 10) \qquad \text{Cross products are equal}$$
$$5n = 3n + 30$$
$$2n = 30$$
$$n = 15$$

If n is 15, then $n + 10$ is 25. Thus the numbers are 15 and 25. To check, consider the quotient of the smaller number divided by the larger number.

$$\frac{15}{25} = \frac{3 \cdot \cancel{5}}{5 \cdot \cancel{5}} = \frac{3}{5}$$ ■

PROBLEM 2

One angle of a triangle has a measure of 40°, and the measures of the other two angles are in the ratio of 5 to 2. Find the measures of the other two angles.

Solution

The sum of the measures of the other two angles is $180° - 40° = 140°$. Let y represent the measure of one angle. Then $140 - y$ represents the measure of the other angle.

$$\frac{y}{140-y} = \frac{5}{2}, \qquad y \neq 140$$

$$2y = 5(140 - y) \qquad \text{Cross products are equal}$$
$$2y = 700 - 5y$$

$$7y = 700$$
$$y = 100$$

If $y = 100$, then $140 - y = 40$. Therefore the measures of the other two angles of the triangle are 100° and 40°. ■

In Chapter 4, we solved some uniform motion problems in which the formula $d = rt$ played an important role. Let's consider another one of those problems; keep in mind that we can also write the formula $d = rt$ as $\frac{d}{r} = t$ or $\frac{d}{t} = r$.

PROBLEM 3

Wendy rides her bicycle 30 miles in the same time that it takes Kim to ride her bicycle 20 miles. If Wendy rides 5 miles per hour faster than Kim, find the rate of each.

Solution

Let r represent Kim's rate. Then $r + 5$ represents Wendy's rate. Let's record the information of this problem in a table.

	Distance	Rate	Time = $\frac{\text{Distance}}{\text{Rate}}$
Kim	20	r	$\frac{20}{r}$
Wendy	30	$r + 5$	$\frac{30}{r+5}$

We can use the fact that their times are equal as a guideline.

Kim's Time = Wendy's Time

$$\frac{\text{Distance Kim rides}}{\text{Rate Kim rides}} = \frac{\text{Distance Wendy rides}}{\text{Rate Wendy rides}}$$

$$\frac{20}{r} = \frac{30}{r+5}, \qquad r \neq 0 \text{ and } r \neq -5$$

$$20(r + 5) = 30r$$
$$20r + 100 = 30r$$
$$100 = 10r$$
$$10 = r$$

Therefore, Kim rides at 10 miles per hour, and Wendy rides at $10 + 5 = 15$ miles per hour. ■

Problem Set 7.5

For Problems 1–40, solve each of the equations.

1. $\frac{x}{2} + \frac{x}{3} = 10$

2. $\frac{x}{8} - \frac{x}{6} = -1$

3. $\frac{x}{6} - \frac{4x}{3} = \frac{1}{9}$

4. $\frac{3x}{4} + \frac{x}{5} = \frac{3}{10}$

5. $\frac{n}{2} + \frac{n-1}{6} = \frac{5}{2}$

6. $\frac{n+2}{7} + \frac{n}{3} = \frac{12}{7}$

7. $\frac{t-3}{4} + \frac{t+1}{9} = -1$

8. $\frac{t-2}{4} - \frac{t+3}{7} = 1$

9. $\frac{2x+3}{3} + \frac{3x-4}{4} = \frac{17}{4}$

10. $\frac{3x-1}{4} + \frac{2x-3}{5} = -2$

11. $\frac{x-4}{8} - \frac{x+5}{4} = 3$

12. $\frac{x+6}{9} - \frac{x-2}{5} = \frac{7}{15}$

13. $\frac{3x+2}{5} - \frac{2x-1}{6} = \frac{2}{15}$

14. $\frac{4x-1}{3} - \frac{2x+5}{8} = \frac{1}{6}$

15. $\frac{1}{x} + \frac{2}{3} = \frac{7}{6}$

16. $\frac{2}{x} + \frac{1}{4} = \frac{13}{20}$

17. $\frac{5}{3n} - \frac{1}{9} = \frac{1}{n}$

18. $\frac{9}{n} - \frac{1}{4} = \frac{7}{n}$

19. $\frac{1}{2x} + 3 = \frac{4}{3x}$

20. $\frac{2}{3x} + 1 = \frac{5}{4x}$

21. $\frac{4}{5t} - 1 = \frac{3}{2t}$

22. $\frac{1}{6t} - 2 = \frac{7}{8t}$

23. $\frac{-5}{4h} + \frac{7}{6h} = \frac{1}{4}$

24. $\frac{3}{h} + \frac{5}{2h} = 1$

25. $\frac{90-n}{n} = 10 + \frac{2}{n}$

26. $\frac{51-n}{n} = 7 + \frac{3}{n}$

27. $\frac{n}{49-n} = 3 + \frac{1}{49-n}$

28. $\frac{n}{57-n} = 10 + \frac{2}{57-n}$

29. $\frac{x}{x+3} - 2 = \frac{-3}{x+3}$

30. $\frac{4}{x-2} = \frac{5}{x+6}$

31. $\frac{7}{x+3} = \frac{5}{x-9}$

32. $\frac{x}{x-4} - 2 = \frac{4}{x-4}$

33. $\frac{x}{x+2} + 3 = \frac{1}{x+2}$

34. $\frac{x}{x-5} - 4 = \frac{3}{x-5}$

35. $-1 - \frac{5}{x-2} = \frac{3}{x-2}$

36. $\frac{x-1}{x} - 2 = \frac{3}{2}$

37. $1 + \frac{n+1}{2n} = \frac{3}{4}$

38. $\frac{3}{n-1} + 4 = \frac{2}{n-1}$

39. $\frac{h}{2} - \frac{h}{4} + \frac{h}{3} = 1$

40. $\frac{h}{4} + \frac{h}{5} - \frac{h}{6} = 1$

For Problems 41–52, set up an equation and solve each problem.

41. The numerator of a fraction is 8 less than the denominator. The fraction in its simplest form is $\frac{5}{6}$. Find the fraction.

42. One number is 12 larger than another number. The indicated quotient of the smaller number divided by the larger reduces to $\frac{2}{3}$. Find the numbers.

43. What number must be added to the numerator and denominator of $\frac{2}{5}$ to produce a fraction equivalent to $\frac{4}{5}$?

44. What number must be subtracted from the numerator and denominator of $\frac{29}{31}$ to produce a fraction equivalent to $\frac{11}{12}$?

45. One angle of a triangle has a measure of 60°, and the measures of the other two angles are in the ratio of 2 to 3. Find the measures of the other two angles.

46. The measure of angle A of a triangle is 20° more than the measure of angle B. The measures of the angles are in a ratio of 3 to 4. Find the measure of each angle.

47. The ratio of the measures of the complement of an angle to its supplement is 1 to 4. Find the measure of the angle.

48. One angle of a triangle has a measure of 45°, and the measures of the other two angles are in the ratio of 2 to 1. Find the measures of the other two angles.

49. It took Heidi 3 hours and 20 minutes longer to ride her bicycle 125 miles than it took Abby to ride 75 miles. If they both rode at the same rate, find this rate.

50. Two trains left a depot traveling in opposite directions at the same rate. One train traveled 338 miles in 2 hours more time than it took the other train to travel 234 miles. Find the rate of the trains.

51. Kent drives his Mazda 270 miles in the same time that Dave drives his Datsun 250 miles. If Kent averages 4 miles per hour faster than Dave, find their rates.

52. An airplane travels 2050 miles in the same time that a car travels 260 miles. If the rate of the plane is 358 miles per hour faster than the rate of the car, find the rate of each.

THOUGHTS INTO WORDS

53. **(a)** Explain how to do the addition problem

$$\frac{3}{x+2} + \frac{5}{x-1}.$$

(b) Explain how to solve the equation

$$\frac{3}{x+2} + \frac{5}{x-1} = 0.$$

54. How can you tell by inspection that $\frac{x}{x-4} = \frac{4}{x-4}$ has no solution?

55. How would you help someone solve the equation $\frac{1}{x} + \frac{2}{x} = \frac{3}{x}$?

FURTHER INVESTIGATIONS

For Problems 56–59, solve each of the equations.

56. $\frac{3}{2n} + \frac{1}{n} = \frac{5}{3n}$

57. $\frac{1}{2n} + \frac{4}{n} = \frac{9}{2n}$

58. $\frac{n+1}{2} + \frac{n}{3} = \frac{1}{2}$

59. $\frac{1}{n+2} + \frac{2}{n+3} = \frac{3n+7}{(n+2)(n+3)}$

7.6 More Fractional Equations and Problem Solving

Let's begin this section by considering a few more fractional equations. We will continue to solve them using the same basic techniques as in the preceding section.

EXAMPLE 1

Solve $\frac{10}{8x-2} - \frac{6}{4x-1} = \frac{1}{9}$.

Solution

$$\frac{10}{8x-2} - \frac{6}{4x-1} = \frac{1}{9}, \qquad x \neq \frac{1}{4}$$

Do you see why x cannot equal $\frac{1}{4}$?

$$\frac{10}{2(4x-1)} - \frac{6}{4x-1} = \frac{1}{9}$$

Factor the first denominator

$$18(4x-1)\left(\frac{10}{2(4x-1)} - \frac{6}{4x-1}\right) = 18(4x-1)\left(\frac{1}{9}\right)$$

Multiply both sides by $18(4x - 1)$, the LCD

$$9(10) - 18(6) = 2(4x-1)$$

$$90 - 108 = 8x - 2$$

$$-18 = 8x - 2$$

$$-16 = 8x$$

$$-2 = x$$

Be sure that the solution -2 checks!

The solution set is $\{-2\}$. ■

Remark: In the second step of the solution for Example 1, you may choose to reduce $\frac{10}{2(4x-1)}$ to $\frac{5}{4x-1}$. Then the left side, $\frac{5}{4x-1} - \frac{6}{4x-1}$, simplifies to $\frac{-1}{4x-1}$. This forms the proportion $\frac{-1}{4x-1} = \frac{1}{9}$, which can be solved easily using the *cross multiplication method.*

EXAMPLE 2

Solve $\frac{2n}{n+3} + \frac{5n}{n^2-9} = 2$.

Solution

$$\frac{2n}{n+3} + \frac{5n}{n^2-9} = 2, \qquad n \neq -3 \text{ and } n \neq 3$$

$$\frac{2n}{n+3} + \frac{5n}{(n+3)(n-3)} = 2$$

$$(n+3)(n-3)\left(\frac{2n}{n+3}+\frac{5n}{(n+3)(n-3)}\right)=(n+3)(n-3)(2)$$

$$2n(n-3)+5n=2(n^2-9)$$

$$2n^2-6n+5n=2n^2-18$$

$$-6n+5n=-18 \quad \text{Add } -2n^2 \text{ to both sides}$$

$$-n=-18$$

$$n=18$$

The solution set is $\{18\}$. ■

EXAMPLE 3

Solve $n+\frac{1}{n}=\frac{10}{3}$.

Solution

$$n+\frac{1}{n}=\frac{10}{3}, \qquad n\neq 0$$

$$3n\left(n+\frac{1}{n}\right)=3n\left(\frac{10}{3}\right)$$

$$3n^2+3=10n$$

$$3n^2-10n+3=0$$

$$(3n-1)(n-3)=0$$

Remember when we used the factoring techniques to help solve equations of this type in Chapter 6?

$$3n-1=0 \quad \text{or} \quad n-3=0$$

$$3n=1 \quad \text{or} \quad n=3$$

$$n=\frac{1}{3}$$

The solution set is $\left\{\frac{1}{3}, 3\right\}$. ■

■ Problem Solving

Recall that $\frac{2}{3}$ and $\frac{3}{2}$ are called multiplicative inverses, or *reciprocals*, of each other because their product is 1. In general, the reciprocal of any nonzero real number n is the number $\frac{1}{n}$. Let's use this idea to solve a problem.

PROBLEM 1

The sum of a number and its reciprocal is $\frac{26}{5}$. Find the number.

Solution

We let n represent the number. Then $\frac{1}{n}$ represents its reciprocal.

$$\text{Number} + \text{Its reciprocal} = \frac{26}{5}$$

$$n + \frac{1}{n} = \frac{26}{5}, \quad n \neq 0$$

$$5n\left(n + \frac{1}{n}\right) = 5n\left(\frac{26}{5}\right) \quad \text{Multiply both sides by } 5n\text{, the LCD}$$

$$5n^2 + 5 = 26n$$

$$5n^2 - 26n + 5 = 0$$

$$(5n - 1)(n - 5) = 0$$

$$5n - 1 = 0 \quad \text{or} \quad n - 5 = 0$$

$$5n = 1 \quad \text{or} \quad n = 5$$

$$n = \frac{1}{5}$$

If the number is $\frac{1}{5}$, its reciprocal is $\frac{1}{\frac{1}{5}} = 5$. If the number is 5, its reciprocal is $\frac{1}{5}$. ■

Now let's consider another uniform motion problem, which is a slight variation of those we studied in the previous section. Again, keep in mind that we always use the distance–rate–time relationships in these problems.

PROBLEM 2

To travel 60 miles, it takes Sue, riding a moped, 2 hours less than it takes LeAnn, riding a bicycle, to travel 50 miles (see Figure 7.2). Sue travels 10 miles per hour faster than LeAnn. Find the times and rates of both girls.

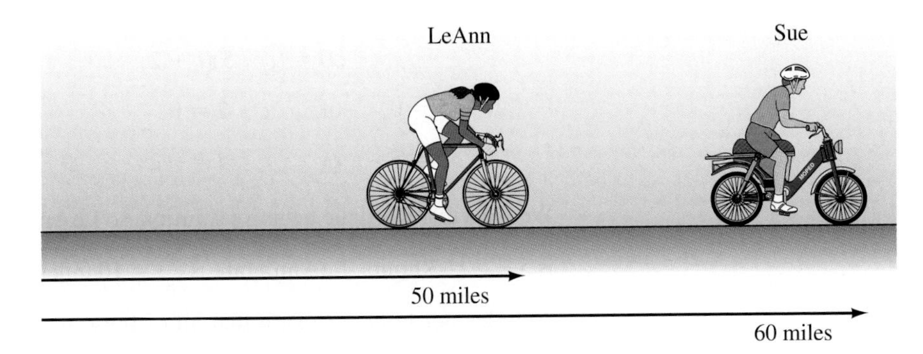

Figure 7.2

Solution

We let t represent LeAnn's time. Then $t - 2$ represents Sue's time. We can record the information from the problem in the table.

	Distance	Time	Rate $\left(r = \frac{d}{t}\right)$
LeAnn	50	t	$\frac{50}{t}$
Sue	60	$t - 2$	$\frac{60}{t - 2}$

We use the fact that Sue travels 10 miles per hour faster than LeAnn as a guideline to set up an equation.

Sue's Rate = LeAnn's Rate + 10

$$\frac{60}{t - 2} = \frac{50}{t} + 10, \qquad t \neq 2 \text{ and } t \neq 0$$

Solving this equation yields

$$\frac{60}{t - 2} = \frac{50}{t} + 10$$

$$t(t - 2)\left(\frac{60}{t - 2}\right) = t(t - 2)\left(\frac{50}{t} + 10\right)$$

$$60t = 50(t - 2) + 10t(t - 2)$$

$$60t = 50t - 100 + 10t^2 - 20t$$

$$0 = 10t^2 - 30t - 100$$

$$0 = t^2 - 3t - 10$$

$$0 = (t - 5)(t + 2)$$

$$t - 5 = 0 \quad \text{or} \quad t + 2 = 0$$

$$t = 5 \quad \text{or} \quad t = -2$$

We must disregard the negative solution, so LeAnn's time is 5 hours, and Sue's time is $5 - 2 = 3$ hours. LeAnn's rate is $\frac{50}{5} = 10$ miles per hour, and Sue's rate is $\frac{60}{3} =$ 20 miles per hour. (Be sure that all of these results check back into the original problem!) ■

There is another class of problems that we commonly refer to as work problems, or sometimes as *rate-time* problems. For example, if a certain machine produces 120 items in 10 minutes, then we say that it is producing at a rate of $\frac{120}{10} = 12$ items per minute. Likewise, if a person can do a certain job in 5 hours, then that person is working at a rate of $\frac{1}{5}$ of the job per hour. In general, if Q is the quantity of something done in t units of time, then the rate r is given by $r = \frac{Q}{t}$. The rate is stated in terms of *so much quantity per unit of time.* The uniform-motion problems we discussed earlier are a special kind of rate-time problem where the *quantity* is distance. The use of tables to organize information, as we illustrated with the uniform-motion problems, is a convenient aid for some rate-time problems. Let's consider some problems.

PROBLEM 3

Printing press A can produce 35 fliers per minute, and press B can produce 50 fliers per minute. Printing press A is set up and starts a job, and then 15 minutes later printing press B is started, and both presses continue printing until 2225 fliers are produced. How long would printing press B be used?

Solution

We let m represent the number of minutes that printing press B is used. Then $m + 15$ represents the number of minutes that press A is used. The information in the problem can be organized in a table.

	Rate	Time	Quantity = Rate × Time
Press A	35	$m + 15$	$35(m + 15)$
Press B	50	m	$50\,m$

Since the total quantity (total number of fliers) is 2225 fliers, we can set up and solve the following equation:

$$35(m + 15) + 50m = 2225$$
$$35m + 525 + 50m = 2225$$
$$85m = 1700$$
$$m = 20$$

Therefore, printing press B must be used for 20 minutes. ■

PROBLEM 4

Bill can mow a lawn in 45 minutes, and Jennifer can mow the same lawn in 30 minutes. How long would it take the two of them working together to mow the lawn? (See Figure 7.3.)

Figure 7.3

Remark: Before you look at the solution of this problem, *estimate* the answer. Remember that Jennifer can mow the lawn by herself in 30 minutes.

Solution

Bill's rate is $\frac{1}{45}$ of the lawn per minute, and Jennifer's rate is $\frac{1}{30}$ of the lawn per minute. If we let m represent the number of minutes that they work together, then $\frac{1}{m}$ represents the rate when working together. Therefore, since the sum of the individual rates must equal the rate working together, we can set up and solve the following equation:

$$\frac{1}{30} + \frac{1}{45} = \frac{1}{m}, \qquad m \neq 0$$

$$90m\left(\frac{1}{30} + \frac{1}{45}\right) = 90m\left(\frac{1}{m}\right) \qquad \text{Multiply both sides by } 90m\text{, the LCD}$$

$$3m + 2m = 90$$

$$5m = 90$$

$$m = 18$$

It should take them 18 minutes to mow the lawn when working together. (How close was your estimate?) ■

PROBLEM 5

It takes Amy twice as long to deliver papers as it does Nancy. How long would it take each girl by herself if they can deliver the papers together in 40 minutes?

Solution

We let m represent the number of minutes that it takes Nancy by herself. Then $2m$ represents Amy's time by herself. Therefore, Nancy's rate is $\frac{1}{m}$, and Amy's rate is $\frac{1}{2m}$. Since the combined rate is $\frac{1}{40}$, we can set up and solve the following equation:

$$\underset{\text{Nancy's rate}}{\frac{1}{m}} + \underset{\text{Amy's rate}}{\frac{1}{2m}} = \underset{\text{Combined rate}}{\frac{1}{40}}, \quad m \neq 0$$

$$40m\left(\frac{1}{m} + \frac{1}{2m}\right) = 40m\left(\frac{1}{40}\right)$$

$$40 + 20 = m$$

$$60 = m$$

Therefore, Nancy can deliver the papers by herself in 60 minutes, and Amy can deliver them by herself in $2(60) = 120$ minutes. ■

One final example of this section outlines another approach that some people find meaningful for work problems. This approach represents the fractional parts of a job. For example, if a person can do a certain job in 7 hours, then at the end of 3 hours, that person has finished $\frac{3}{7}$ of the job. (Again, we assume a constant rate of work.) At the end of 5 hours, $\frac{5}{7}$ of the job has been done—in general, at the end of h hours, $\frac{h}{7}$ of the job has been completed. Let's use this idea to solve a work problem.

PROBLEM 6

It takes Pat 12 hours to install a wood floor. After he had been working for 3 hours, he was joined by his brother Mike, and together they finished the floor in 5 hours. How long would it take Mike to install the floor by himself?

Solution

Let h represent the number of hours that it would take Mike to install the floor by himself.

The fractional part of the job that Pat does equals his working rate times his time. Because it takes Pat 12 hours to do the entire floor, his working rate is $\frac{1}{12}$.

He works for 8 hours (3 hours before Mike and then 5 hours with Mike). Therefore, Pat's part of the job is $\frac{1}{12}(8) = \frac{8}{12}$. The fractional part of the job that Mike does equals his working rate times his time. Because h represents Mike's time to install the floor, his working rate is $\frac{1}{h}$. He works for 5 hours. Therefore, Mike's part of the job is $\frac{1}{h}(5) = \frac{5}{h}$. Adding the two fractional parts together results in 1 entire job being done. Let's also show this information in chart form and set up our guideline. Then we can set up and solve the equation.

	Time to do entire job	**Working rate**	**Time working**	**Fractional part of the job done**
Pat	12	$\frac{1}{12}$	8	$\frac{8}{12}$
Mike	h	$\frac{1}{h}$	5	$\frac{5}{h}$

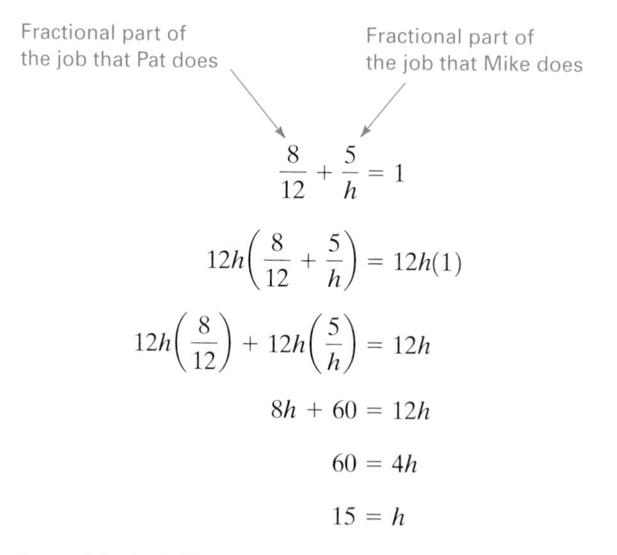

$$\frac{8}{12} + \frac{5}{h} = 1$$

$$12h\left(\frac{8}{12} + \frac{5}{h}\right) = 12h(1)$$

$$12h\left(\frac{8}{12}\right) + 12h\left(\frac{5}{h}\right) = 12h$$

$$8h + 60 = 12h$$

$$60 = 4h$$

$$15 = h$$

It would take Mike 15 hours to install the floor by himself. ■

We emphasize a point made earlier. Don't become discouraged if solving word problems is still giving you trouble. The development of problem-solving skills is a long-term objective. If you continue to work hard and give it your best shot, you will gradually become more and more confident in your approach to solving problems. Don't be afraid to try some different approaches on your own. Our problem-solving suggestions simply provide a framework for you to build on.

Problem Set 7.6

For Problems 1–32, solve each equation.

1. $\frac{4}{x} + \frac{7}{6} = \frac{1}{x} + \frac{2}{3x}$

2. $\frac{2}{3x} - \frac{9}{x} = -\frac{25}{9}$

3. $\frac{3}{2x + 2} + \frac{4}{x + 1} = \frac{11}{12}$

4. $\frac{5}{2x - 6} + \frac{1}{x - 3} = \frac{7}{2}$

5. $\frac{5}{2n - 10} - \frac{3}{n - 5} = 1$

6. $\frac{7}{3x + 6} - \frac{2}{x + 2} = 2$

7. $\frac{3}{2t} - \frac{5}{t} = \frac{7}{5t} + 1$

8. $\frac{2}{3t} + \frac{3}{4t} = 1 - \frac{5}{2t}$

9. $\frac{x}{x - 2} + \frac{4}{x + 2} = 1$

10. $\frac{2x}{x + 1} - \frac{3}{x - 1} = 2$

11. $\frac{x}{x - 4} - \frac{2x}{x + 4} = -1$

12. $\frac{2x}{x + 2} + \frac{x}{x - 2} = 3$

13. $\frac{3n}{n + 3} - \frac{n}{n - 3} = 2$

14. $\frac{4n}{n - 5} - \frac{2n}{n + 5} = 2$

15. $\frac{3}{t^2 - 4} + \frac{5}{t + 2} = \frac{2}{t - 2}$

16. $\frac{t}{2t - 8} + \frac{16}{t^2 - 16} = \frac{1}{2}$

17. $\frac{4}{x - 1} - \frac{2x - 3}{x^2 - 1} = \frac{6}{x + 1}$

18. $\frac{3x - 1}{x^2 - 9} + \frac{4}{x + 3} = \frac{5}{x - 3}$

19. $8 + \frac{5}{y^2 + 2y} = \frac{3}{y + 2}$

20. $2 + \frac{4}{y - 1} = \frac{4}{y^2 - y}$

21. $n + \frac{1}{n} = \frac{17}{4}$

22. $n + \frac{3}{n} = 4$

23. $\frac{15}{4n} + \frac{15}{4(n + 4)} = 1$

24. $\frac{10}{7x} + \frac{10}{7(x + 3)} = 1$

25. $x - \frac{5x}{x - 2} = \frac{-10}{x - 2}$

26. $\frac{x + 1}{x - 3} - \frac{3}{x} = \frac{12}{x^2 - 3x}$

27. $\frac{t}{4t - 4} + \frac{5}{t^2 - 1} = \frac{1}{4}$

28. $\frac{x}{3x - 6} + \frac{4}{x^2 - 4} = \frac{1}{3}$

29. $\frac{3}{n - 5} + \frac{4}{n + 7} = \frac{2n + 11}{n^2 + 2n - 35}$

30. $\frac{2}{n + 3} + \frac{3}{n - 4} = \frac{2n - 1}{n^2 - n - 12}$

31. $\frac{a}{a + 2} + \frac{3}{a + 4} = \frac{14}{a^2 + 6a + 8}$

32. $3 + \frac{6}{t - 3} = \frac{6}{t^2 - 3t}$

For Problems 33–50, set up an equation and solve the problem.

33. The sum of a number and twice its reciprocal is $\frac{9}{2}$. Find the number.

34. The sum of a number and three times its reciprocal is 4. Find the number.

35. A number is $\frac{21}{10}$ larger than its reciprocal. Find the number.

36. Suppose that the reciprocal of a number subtracted from the number yields $\frac{5}{6}$. Find the number.

37. Suppose that Celia rides her bicycle 60 miles in 2 hours less time than it takes Tom to ride his bicycle 85 miles. If Celia rides 3 miles per hour faster than Tom, find their respective rates.

38. To travel 300 miles, it takes a freight train 2 hours longer than it takes an express train to travel 280 miles. The rate of the express train is 20 miles per hour faster than the rate of the freight train. Find the rates of both trains.

39. One day, Jeff rides his bicycle out into the country 40 miles (see Figure 7.4). On the way back, he takes a different route that is 2 miles longer and it takes him an hour longer to return. If his rate on the way out into the country is 4 miles per hour faster than his rate back, find both rates.

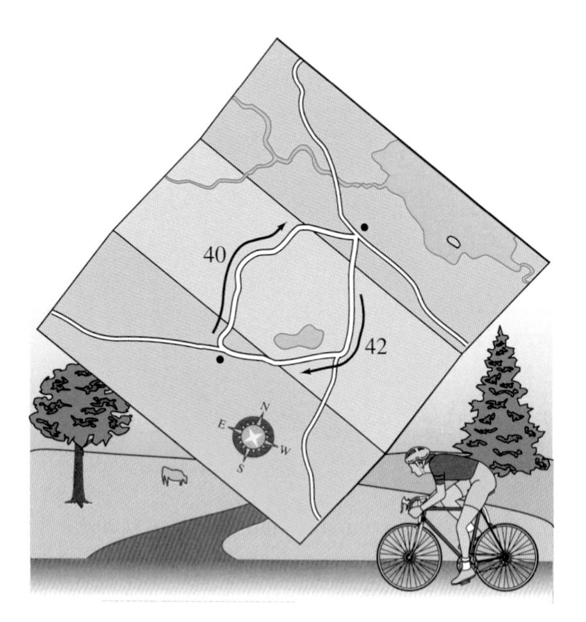

Figure 7.4

40. Rita jogs for 8 miles and then walks an additional 12 miles. She jogs at a rate twice her walking rate and she covers the entire distance of 20 miles in 4 hours. Find the rate she jogs and the rate she walks.

41. A water tank can be filled by an inlet pipe in 5 minutes. A drain pipe will empty the tank in 6 minutes. If by mistake the drain is left open as the tank is being filled, how long will it take before the tank overflows?

42. Betty can do a job in 10 minutes. Doug can do the same job in 15 minutes. If they work together, how long will it take them to complete the job?

43. It takes Barry twice as long to deliver papers as it does Mike. How long would it take each if they can deliver the papers together in 40 minutes?

44. Working together, Cindy and Sharon can address envelopes in 12 minutes. Cindy could do the addressing by herself in 20 minutes. How long would it take Sharon to address the envelopes by herself?

45. Mark can overhaul an engine in 20 hours, and Phil can do the same job by himself in 30 hours. If they both work together for a time, and then Mark finishes the job by himself in 5 hours, how long did they work together?

46. Working together, Pam and Laura can complete a job in $1\frac{1}{2}$ hours. When working alone, it takes Laura 4 hours longer than Pam to do the job. How long does it take each of them working alone?

47. A copy center has two copiers. Copier A can produce copies at a rate of 40 pages per minute, and copier B does 30 pages per minute. How long will copier B need to run if copier A has been copying by itself for 6 minutes, and then both copier A and B are used until 520 copies are made?

48. It takes two pipes 3 hours to fill a water tank. Pipe B can fill the tank alone in 8 hours more than it takes pipe A to fill the tank alone. How long would it take each pipe to fill the tank by itself?

49. In a survivor competition, the Pachena tribe can shuck 300 oysters in 10 minutes less time than it takes the Tchaika tribe. If the Pachena tribe shucks oysters at a rate of 5 oysters per minute faster than the Tchaika tribe, find the rate of each tribe.

50. Machine A can wrap 600 pieces of candy in 5 minutes less time than it takes machine B to wrap 600 pieces of candy. If the rate of machine A is 20 candies per minute faster than machine B, find the rate of each machine.

A selection of additional word problems is in Appendix B. All Appendix problems that are referenced as (7.5) or (7.6) are appropriate for your practice.

THOUGHTS INTO WORDS

51. Write a paragraph or two summarizing the new ideas about problem solving that you have acquired thus far in this course.

FURTHER INVESTIGATIONS

For Problems 52–54, solve each equation.

52. $\dfrac{3x - 1}{x^2 - 9} + \dfrac{4}{x + 3} = \dfrac{7}{x - 3}$

53. $\dfrac{x - 2}{x^2 - 1} + \dfrac{3}{x + 1} = \dfrac{-5}{x - 1}$

54. $\dfrac{7x - 12}{x^2 - 16} - \dfrac{5}{x + 4} = \dfrac{2}{x - 4}$

Chapter 7 Summary

(7.1) The fundamental principle of fractions

$$\left(\frac{ak}{bk} = \frac{a}{b}\right)$$

provides the basis for simplifying algebraic fractions.

(7.2) To **multiply** algebraic fractions, multiply the numerators, multiply the denominators, and express the product in simplified form.

To **divide** algebraic fractions, invert the divisor and multiply.

(7.3) Addition and **subtraction** of algebraic fractions are based on the following definitions.

$$\frac{a}{b} + \frac{c}{b} = \frac{a+c}{b} \qquad \text{Addition}$$

$$\frac{a}{b} - \frac{c}{b} = \frac{a-c}{b} \qquad \text{Subtraction}$$

(7.4) Use the following procedure when adding and subtracting fractions:

1. Find the least common denominator.
2. Change each fraction to an equivalent fraction that has the LCD as its denominator.
3. Add or subtract the numerators and place this result over the LCD.
4. Look for possibilities to simplify the final fraction.

Fractional forms that contain fractions in the numerators and/or denominators are called **complex fractions.**

(7.5) and (7.6) To solve a fractional equation, it is often convenient to begin by multiplying both sides of the equation by the LCD, which clears the equation of all fractions.

Chapter 7 Review Problem Set

For Problems 1–4, simplify each algebraic fraction.

1. $\dfrac{56x^3y}{72xy^3}$

2. $\dfrac{x^2 - 9x}{x^2 - 6x - 27}$

3. $\dfrac{3n^2 - n - 10}{n^2 - 4}$

4. $\dfrac{16a^2 + 24a + 9}{20a^2 + 7a - 6}$

For Problems 5–15, perform the indicated operations and express your answers in simplest form.

5. $\dfrac{7x^2y^2}{12y^3} \cdot \dfrac{18y}{28x}$

6. $\dfrac{x^2y}{x^2 + 2x} \cdot \dfrac{x^2 - x - 6}{y}$

7. $\dfrac{n^2 - 2n - 24}{n^2 + 11n + 28} \div \dfrac{n^3 - 6n^2}{n^2 - 49}$

8. $\dfrac{4a^2 + 4a + 1}{(a + 6)^2} \div \dfrac{6a^2 - 5a - 4}{3a^2 + 14a - 24}$

9. $\dfrac{3x + 4}{5} + \dfrac{2x - 7}{4}$

10. $\dfrac{7}{3x} + \dfrac{5}{4x} - \dfrac{2}{8x^2}$

11. $\dfrac{7}{n} + \dfrac{3}{n - 1}$

12. $\dfrac{2}{a - 4} - \dfrac{3}{a - 2}$

13. $\dfrac{2x}{x^2 - 3x} - \dfrac{3}{4x}$

14. $\dfrac{2}{x^2 + 7x + 10} + \dfrac{3}{x^2 - 25}$

15. $\dfrac{5x}{x^2 - 4x - 21} - \dfrac{3}{x - 7} + \dfrac{4}{x + 3}$

For Problems 16 and 17, simplify each complex fraction.

16. $\dfrac{\dfrac{3}{x} - \dfrac{4}{y^2}}{\dfrac{4}{y} + \dfrac{5}{x}}$

17. $\dfrac{\dfrac{2}{x} - 1}{3 + \dfrac{5}{y}}$

For Problems 18–29, solve each equation.

18. $\dfrac{2x - 1}{3} + \dfrac{3x - 2}{4} = \dfrac{5}{6}$

19. $\frac{5}{3x} - 2 = \frac{7}{2x} + \frac{1}{5x}$

20. $\frac{67 - x}{x} = 6 + \frac{4}{x}$

21. $\frac{5}{2n + 3} = \frac{6}{3n - 2}$

22. $\frac{x}{x - 3} + \frac{5}{x + 3} = 1$

23. $n + \frac{1}{n} = 2$

24. $\frac{n - 1}{n^2 + 8n - 9} - \frac{n}{n + 9} = 4$

25. $\frac{6}{7x} - \frac{1}{6} = \frac{5}{6x}$

26. $n + \frac{1}{n} = \frac{5}{2}$

27. $\frac{n}{5} = \frac{10}{n - 5}$

28. $\frac{-1}{2x - 5} + \frac{2x - 4}{4x^2 - 25} = \frac{5}{6x + 15}$

29. $1 + \frac{1}{n - 1} = \frac{1}{n^2 - n}$

For Problems 30–35, set up an equation and solve each problem.

30. It takes Nancy three times as long to complete a task as it does Becky. How long would it take each of them to complete the task if working together they can do it in 2 hours?

31. The sum of a number and twice its reciprocal is 3. Find the number.

32. The denominator of a fraction is twice the numerator. If 4 is added to the numerator and 18 to the denominator, a fraction that is equivalent to $\frac{4}{9}$ is produced. Find the original fraction.

33. Lanette can ride her moped 44 miles in the same time that Todd rides his bicycle 30 miles. If Lanette rides 7 miles per hour faster than Todd, find their rates.

34. Jim rode his bicycle 36 miles in 4 hours. For the first 20 miles, he rode at a constant rate, and then for the last 16 miles, he reduced his rate by 2 miles per hour. Find his rate for the last 16 miles.

35. An inlet pipe can fill a tank in 10 minutes. A drain can empty the tank in 12 minutes. If the tank is empty and both the inlet pipe and drain are open, how long will it be before the tank overflows?

Chapter 7 Test

For Problems 1–4, simplify each algebraic fraction.

1. $\dfrac{72x^4y^5}{81x^2y^4}$

2. $\dfrac{x^2 + 6x}{x^2 - 36}$

3. $\dfrac{2n^2 - 7n - 4}{3n^2 - 8n - 16}$

4. $\dfrac{2x^3 + 7x^2 - 15x}{x^3 - 25x}$

For Problems 5–14, perform the indicated operations and express answers in simplest form.

5. $\left(\dfrac{8x^2y}{7x}\right)\left(\dfrac{21xy^3}{12y^2}\right)$

6. $\dfrac{x^2 - 49}{x^2 + 7x} \div \dfrac{x^2 - 4x - 21}{x^2 - 2x}$

7. $\dfrac{x^2 - 5x - 36}{x^2 - 15x + 54} \cdot \dfrac{x^2 - 2x - 24}{x^2 + 7x}$

8. $\dfrac{3x - 1}{6} - \dfrac{2x - 3}{8}$

9. $\dfrac{n + 2}{3} - \dfrac{n - 1}{5} + \dfrac{n - 6}{6}$

10. $\dfrac{3}{2x} - \dfrac{5}{6} + \dfrac{7}{9x}$

11. $\dfrac{6}{n} - \dfrac{4}{n - 1}$

12. $\dfrac{2x}{x^2 + 6x} - \dfrac{3}{4x}$

13. $\dfrac{9}{x^2 + 4x - 32} + \dfrac{5}{x + 8}$

14. $\dfrac{-3}{6x^2 - 7x - 20} - \dfrac{5}{3x^2 - 14x - 24}$

For Problems 15–22, solve each of the equations.

15. $\dfrac{x + 3}{5} - \dfrac{x - 2}{6} = \dfrac{23}{30}$

16. $\dfrac{5}{8x} - 2 = \dfrac{3}{x}$

17. $n + \dfrac{4}{n} = \dfrac{13}{3}$

18. $\dfrac{x}{8} = \dfrac{6}{x - 2}$

19. $\dfrac{x}{x - 1} + \dfrac{2}{x + 1} = \dfrac{8}{3}$

20. $\dfrac{3}{2x + 1} = \dfrac{5}{3x - 6}$

21. $\dfrac{4}{n^2 - n} - \dfrac{3}{n - 1} = -1$

22. $\dfrac{3n - 1}{3} + \dfrac{2n + 5}{4} = \dfrac{4n - 6}{9}$

For Problems 23–25, set up an equation and solve the problem.

23. The sum of a number and twice its reciprocal is $3\frac{2}{3}$. Find the number.

24. Wendy can ride her bicycle 42 miles in the same time that it takes Betty to ride her bicycle 36 miles. Wendy rides 2 miles per hour faster than Betty. Find Wendy's rate.

25. Garth can mow a lawn in 20 minutes, and Alex can mow the same lawn in 30 minutes. How long would it take the two of them working together to mow the lawn?

Chapters 1–7 Cumulative Review Problem Set

For Problems 1–8, evaluate each algebraic expression for the given values of the variables. First you may want to simplify the expression or change its form by factoring.

1. $3x - 2xy - 7x + 5xy$ for $x = \frac{1}{2}$ and $y = 3$

2. $7(a - b) - 3(a - b) - (a - b)$ for $a = -3$ and $b = -5$

3. $\frac{xy + yz}{y}$ for $x = \frac{2}{3}, y = \frac{5}{6}$, and $z = \frac{3}{4}$

4. $ab + b^2$ for $a = 0.4$ and $b = 0.6$

5. $x^2 - y^2$ for $x = -6$ and $y = 4$

6. $x^2 + 5x - 36$ for $x = -9$

7. $\frac{x^2 + 2x}{x^2 + 5x + 6}$ for $x = -6$

8. $\frac{x^2 + 3x - 10}{x^2 - 9x + 14}$ for $x = 4$

For Problems 9–16, evaluate each of the expressions.

9. 3^{-3}

10. $\left(\frac{2}{3}\right)^{-1}$

11. $\left(\frac{1}{2} + \frac{1}{3}\right)^0$

12. $\left(\frac{1}{3} + \frac{1}{4}\right)^{-1}$

13. -4^{-2}

14. $\left(\frac{2}{3}\right)^{-2}$

15. $\frac{1}{\left(\frac{2}{5}\right)^{-2}}$

16. $(-3)^{-3}$

For Problems 17–32, perform the indicated operations and express your answers in simplest form.

17. $\frac{7}{5x} + \frac{2}{x} - \frac{3}{2x}$

18. $\frac{4x}{5y} \div \frac{12x^2}{10y^2}$

19. $\frac{4}{x - 6} + \frac{3}{x + 4}$

20. $\frac{2}{x^2 - 4x} - \frac{3}{x^2}$

21. $\frac{x^2 - 8x}{x^2 - x - 56} \cdot \frac{x^2 - 49}{3xy}$

22. $\frac{5}{x^2 - x - 12} - \frac{3}{x - 4}$

23. $(-5x^2y)(7x^3y^4)$

24. $(9ab^3)^2$

25. $(-3n^2)(5n^2 + 6n - 2)$

26. $(5x - 1)(3x + 4)$

27. $(2x + 5)^2$

28. $(x + 2)(2x^2 - 3x - 1)$

29. $(x^2 - x - 1)(x^2 + 2x - 3)$

30. $(-2x - 1)(3x - 7)$

31. $\frac{24x^2y^3 - 48x^4y^5}{8xy^2}$

32. $(28x^2 - 19x - 20) \div (4x - 5)$

For Problems 33–42, factor each polynomial completely.

33. $3x^3 + 15x^2 + 27x$

34. $x^2 - 100$

35. $5x^2 - 22x + 8$

36. $8x^2 - 22x - 63$

37. $n^2 + 25n + 144$

38. $nx + ny - 2x - 2y$

39. $3x^3 - 3x$

40. $2x^3 - 6x^2 - 108x$

41. $36x^2 - 60x + 25$

42. $3x^2 - 5xy - 2y^2$

For Problems 43–57, solve each of the equations.

43. $3(x - 2) - 2(x + 6) = -2(x + 1)$

44. $x^2 = -11x$

45. $0.2x - 3(x - 0.4) = 1$

46. $\frac{3n - 1}{4} = \frac{5n + 2}{7}$

47. $5n^2 - 5 = 0$

48. $x^2 + 5x - 6 = 0$

49. $n + \frac{4}{n} = 4$

50. $\frac{2x + 1}{2} + \frac{3x - 4}{3} = 1$

51. $2(x - 1) - x(x - 1) = 0$

52. $\frac{3}{2x} - 1 = \frac{5}{3x} + 2$

53. $6t^2 + 19t - 7 = 0$

54. $(2x - 1)(x - 8) = 0$

55. $(x + 1)(x + 6) = 24$

56. $\frac{x}{x - 2} - \frac{7}{x + 1} = 1$

57. $\frac{1}{n} - \frac{2}{n - 1} = \frac{3}{n}$

For Problems 58–67, set up an equation or an inequality to help solve each problem.

58. One leg of a right triangle is 2 inches longer than the other leg. The hypotenuse is 4 inches longer than the shorter leg. Find the lengths of the three sides of the right triangle.

59. Twenty percent of what number is 15?

60. How many milliliters of a 65% solution of hydrochloric acid must be added to 40 milliliters of a 30% solution of hydrochloric acid to obtain a 55% hydrochloric acid solution?

61. The material for a landscaping border 28 feet long was bent into the shape of a rectangle. The length of the rectangle was 2 feet more than the width. Find the dimensions of the rectangle.

62. Two motorcyclists leave Daytona Beach at the same time and travel in opposite directions. If one travels at 55 miles per hour and the other travels at 65 miles per hour, how long will it take for them to be 300 miles apart?

63. Find the length of an altitude of a trapezoid with bases of 10 centimeters and 22 centimeters and an area of 120 square centimeters.

64. If a car uses 16 gallons of gasoline for a 352-mile trip, at the same rate of consumption, how many gallons will it use on a 594-mile trip?

65. Swati had scores of 89, 92, 87, and 90 on her first four history exams. What score must she get on the fifth exam to have an average of 90 or higher for the five exams?

66. Two less than three times a certain number is less than 10. Find all positive integers that satisfy this relationship.

67. One more than four times a certain number is greater than 15. Find all real numbers that satisfy this relationship.

For additional word problems see Appendix B. All Appendix problems with references to chapters 3–7 would be appropriate.

8

Coordinate Geometry and Linear Systems

Using the concept of slope, we can set up and solve the proportion $\frac{30}{100} = \frac{y}{5280}$ *to determine how much vertical change a highway with a 30% grade has in a horizontal distance of 1 mile.*

René Descartes, a French mathematician of the 17th century, transformed geometric problems into an algebraic setting so that he could use the tools of algebra to solve those problems. This interface of algebraic and geometric ideas is the foundation of a branch of mathematics called analytic geometry; today more commonly called **coordinate geometry**.

We started to make this connection between algebra and geometry in Chapter 3 when graphing the solution sets of inequalities in one variable. For example, the solution set for $3x + 1 \geq 4$ is $\{x|x \geq 1\}$, and a geometric picture of this solution set is shown in Figure 8.1.

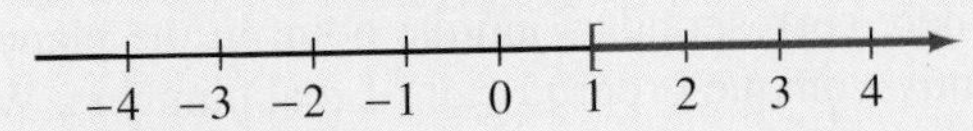

Figure 8.1

In this chapter we will associate pairs of real numbers with points in a geometric plane. This will provide the basis for obtaining pictures of algebraic equations and inequalities in two variables. Finally, we will work with systems of equations that will provide us with even more problem-solving power.

8.1 Cartesian Coordinate System

In Section 2.3 we introduced the real number line (Figure 8.2) as the result of setting up a one-to-one correspondence between the set of real numbers and the points on a line. Recall that the number associated with a point on the line is called the coordinate of that point.

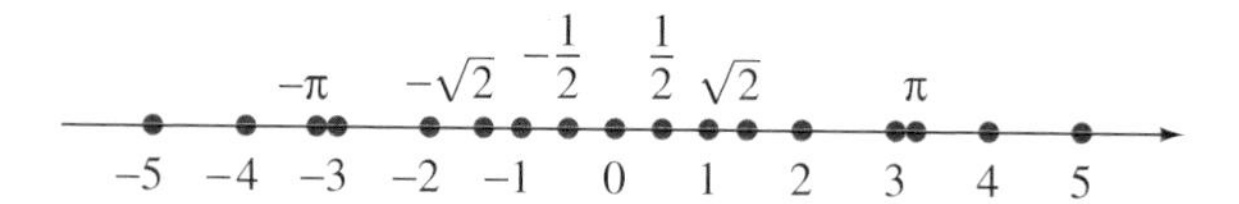

Figure 8.2

Now let's consider two lines (one vertical and one horizontal) that are perpendicular to each other at the point we associate with zero on both lines (Figure 8.3). We refer to these number lines as the horizontal and vertical **axes** or together as the **coordinate axes**; they partition the plane into four parts called **quadrants**. The quadrants are numbered counterclockwise from I to IV as indicated in Figure 8.3. The point of intersection of the two axes is called the **origin**.

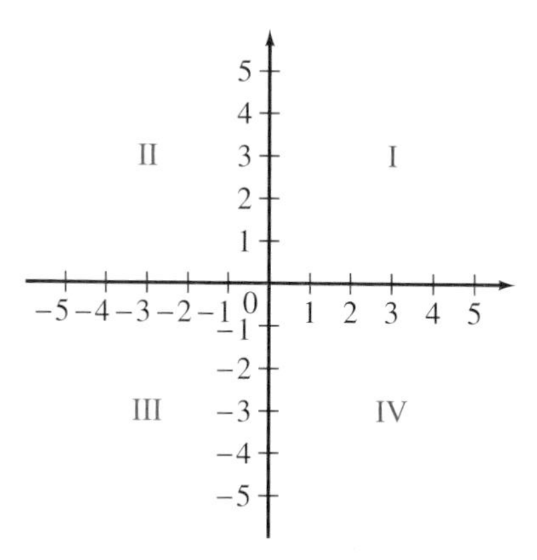

Figure 8.3

It is now possible to set up a one-to-one correspondence between ordered pairs of real numbers and the points in a plane. To each ordered pair of real numbers there corresponds a unique point in the plane, and to each point there corresponds a unique ordered pair of real numbers. We have indicated a part of this correspondence in Figure 8.4. The ordered pair $(3, 1)$ corresponds to point A and denotes that the point A is located 3 units to the right of and 1 unit up from the origin. (The ordered pair $(0, 0)$ corresponds to the origin.) The ordered pair $(-2, 4)$ corresponds to point B and denotes that point B is located 2 units to the left of and 4 units up from the origin. Make sure that you agree with all of the other points plotted in Figure 8.4.

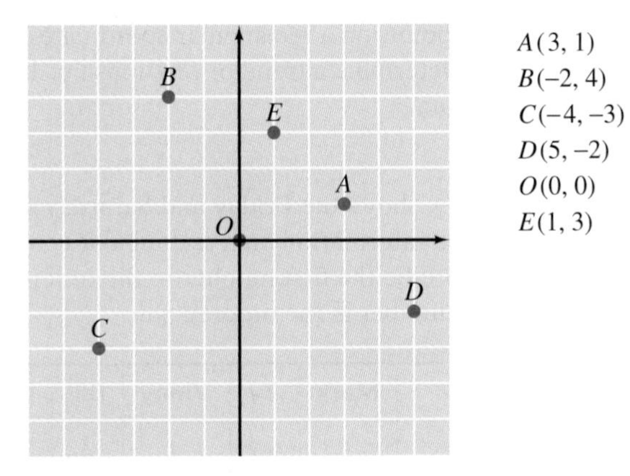

Figure 8.4

Remark: The notation $(-2, 4)$ was used earlier in this text to indicate an interval of the real number line. Now we are using the same notation to indicate an ordered pair of real numbers. This double meaning should not be confusing because the context of the material will always indicate which meaning of the notation is being used. Throughout this chapter, we will be using the ordered-pair interpretation.

In general we refer to the real numbers a and b in an ordered pair (a, b) associated with a point as the **coordinates of the point**. The first number, a, called the **abscissa**, is the directed distance of the point from the vertical axis measured parallel to the horizontal axis. The second number, b, called the **ordinate**, is the directed distance of the point from the horizontal axis measured parallel to the vertical axis (see Figure 8.5(a)). Thus in the first quadrant, all points have a positive abscissa and a positive ordinate. In the second quadrant, all points have a negative abscissa and a positive ordinate. We have indicated the signs in all four quadrants in Figure 8.5(b). This system of associating points with ordered pairs of real numbers is called the **Cartesian coordinate system** or the **rectangular coordinate system**.

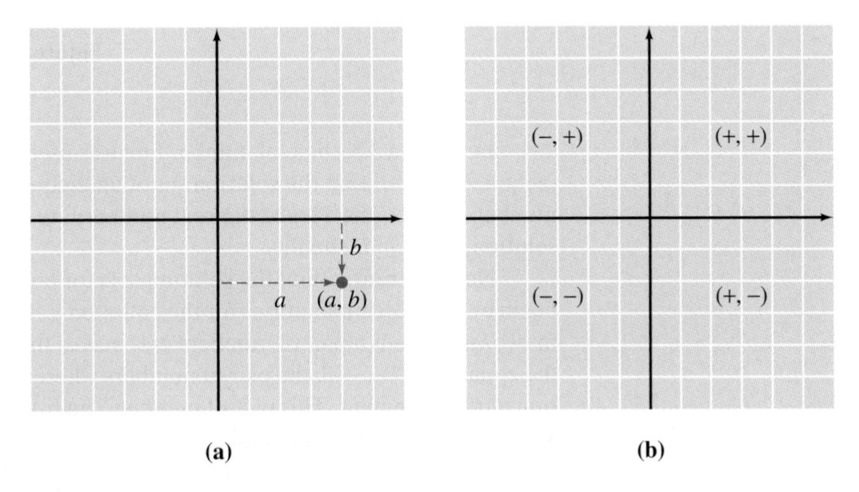

Figure 8.5

Plotting points on a rectangular coordinate system can be helpful when analyzing data to determine a trend or relationship. The following example shows the plot of some data.

EXAMPLE 1

The chart below shows the Friday and Saturday scores of golfers in terms of par. Plot the charted information on a rectangular coordinate system. For each golfer, let Friday's score be the first number in the ordered pair, and let Saturday's score be the second number in the ordered pair.

	Mark	Ty	Vinay	Bill	Herb	Rod
Friday's score	1	−2	−1	4	−3	0
Saturday's score	3	−2	0	7	−4	1

Solution

The ordered pairs are as follows:

Mark $(1, 3)$ Ty $(-2, -2)$

Vinay $(-1, 0)$ Bill $(4, 7)$

Herb $(-3, -4)$ Rod $(0, 1)$

The points are plotted on the rectangular coordinate system in Figure 8.6. In the study of statistics, this graph of the charted data would be called a scatterplot. For this plot, the points appear to approximate a straight-line path, which suggests that there is a linear correlation between Friday's score and Saturday's score.

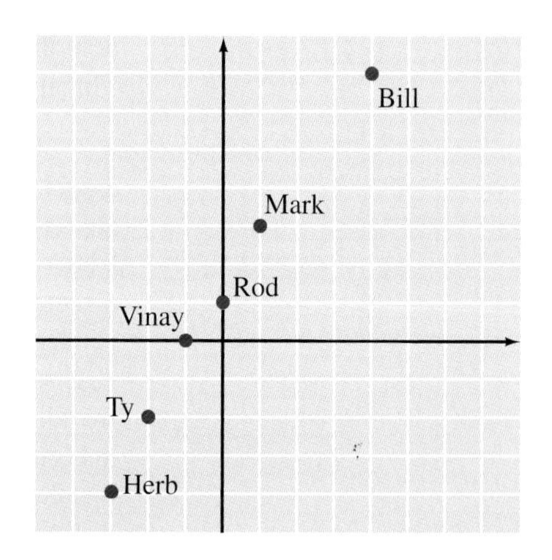

Figure 8.6 ■

In Section 3.5, the real number line was used to display the solution set of an inequality. Recall that the solution set of an inequality such as $x > 3$ has an infinite number of solutions. Hence the real number line is an effective way to display the solution set of an inequality.

Now we want to find the solution sets for equations with two variables. Let's begin by considering solutions for the equation $y = x + 3$. A solution of an equation in two variables is a pair of numbers that makes the equation a true statement. For the equation $y = x + 3$, the pair of numbers where $x = 4$ and $y = 7$ makes the equation a true statement. This pair of numbers can be written as an ordered pair $(4, 7)$. When using the variables x and y, we agree that the first number of an ordered pair is the value for x, and the second number is a value for y. Likewise $(2, 5)$ is a solution for $y = x + 3$ because $5 = 2 + 3$. We can find an infinite number of ordered pairs that satisfy the equation $y = x + 3$ by choosing a value for x and determining the corresponding value for y that satisfies the equation. The table below lists some of the solutions for the equation $y = x + 3$.

Choose x	**Determine y from $y = x + 3$**	**Solution for $y = x + 3$**
0	3	(0, 3)
1	4	(1, 4)
3	6	(3, 6)
5	8	(5, 8)
−1	2	(−1, 2)
−3	0	(−3, 0)
−5	−2	(−5, −2)

Because the number of solutions for the equation $y = x + 3$ is infinite, we do not have a convenient way to list the solution set. This is similar to inequalities where the solution set is infinite. For inequalities we used a real number line to display the solution set. Now we will use the rectangular coordinate system to display the solution set of an equation in two variables.

On a rectangular coordinate system where we label the horizontal axis as the x axis and the vertical axis as the y axis, we can locate the point associated with each ordered pair of numbers in the table. These points are shown in Figure 8.7(a). These points are only some of the infinite solutions of the equation $y = x + 3$. The straight line in Figure 8.7(b) that connects the points represents all the solutions of the equation and is called the graph of the equation $y = x + 3$.

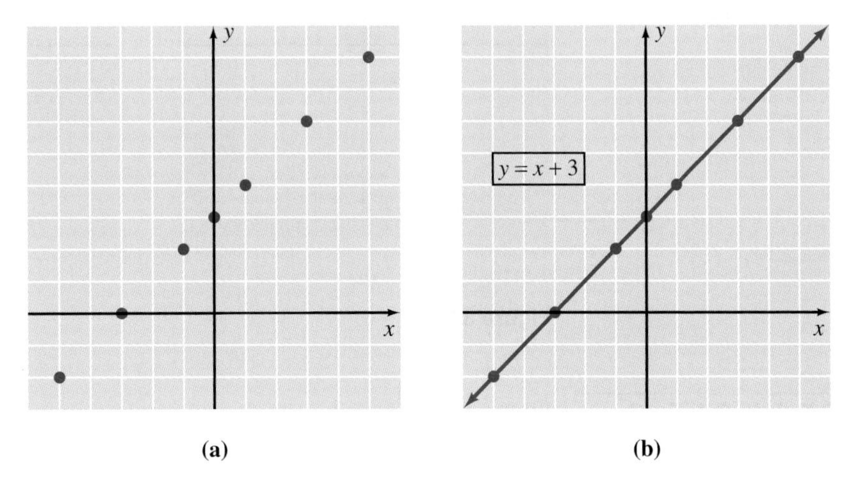

Figure 8.7

The next examples further illustrate the process of graphing equations.

EXAMPLE 2 Graph $y = 2x + 1$.

Solution

First we set up a table of some of the solutions for the equation $y = 2x + 1$. You can choose any value for the variable x and find the corresponding y value. The values we chose for x are in the table and include positive integers, zero, and negative integers.

x	y	Solutions (x, y)
0	1	(0, 1)
1	3	(1, 3)
2	5	(2, 5)
−1	−1	(−1, −1)
−2	−3	(−2, −3)
−3	−5	(−3, −5)

From the table we plot the points associated with the ordered pairs as shown in Figure 8.8(a). Connecting these points with a straight line produces the graph of the equation $y = 2x + 1$ as shown in Figure 8.8(b).

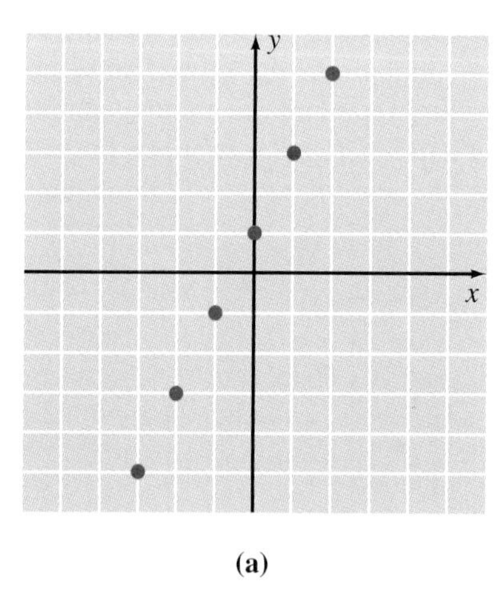

(a)

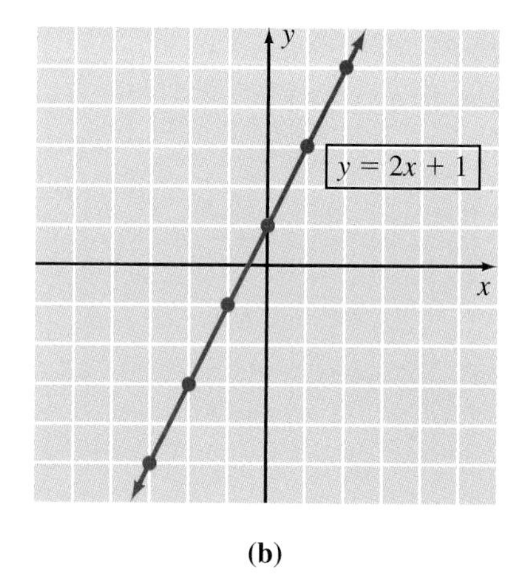

(b)

Figure 8.8

EXAMPLE 3 Graph $y = -\frac{1}{2}x + 3$.

Solution

First we set up a table of some of the solutions for the equation $y = -\frac{1}{2}x + 3$. You can choose any value for the variable x and find the corresponding y value; however the values we chose for x are numbers that are divisible by 2. This is not necessary but does produce integer values for y.

x	y	Solutions (x, y)
0	3	(0, 3)
2	2	(2, 2)
4	1	(4, 1)
−2	4	(−2, 4)
−4	5	(−4, 5)

From the table we plot the points associated with the ordered pairs as shown in Figure 8.9(a). Connecting these points with a straight line produces the graph of the equation $y = -\frac{1}{2}x + 3$ as shown in Figure 8.9(b).

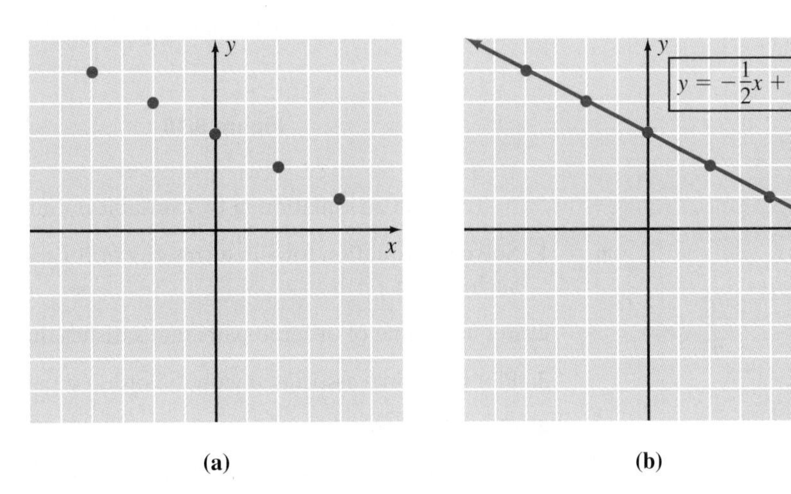

Figure 8.9

EXAMPLE 4

Graph $6x + 3y = 9$

Solution

First, let's change the form of the equation to make it easier to find solutions of the equation $6x + 3y = 9$. We can solve either for x in terms of y or for y in terms of x. Typically the equation is solved for y in terms of x so that's what we show here.

$$6x + 3y = 9$$

$$3y = -6x + 9$$

$$y = \frac{-6x}{3} + \frac{9}{3}$$

$$y = -2x + 3$$

Now we can set up a table of values. Plotting these points and connecting them produces Figure 8.10.

x	y
0	3
1	1
2	−1
−1	5

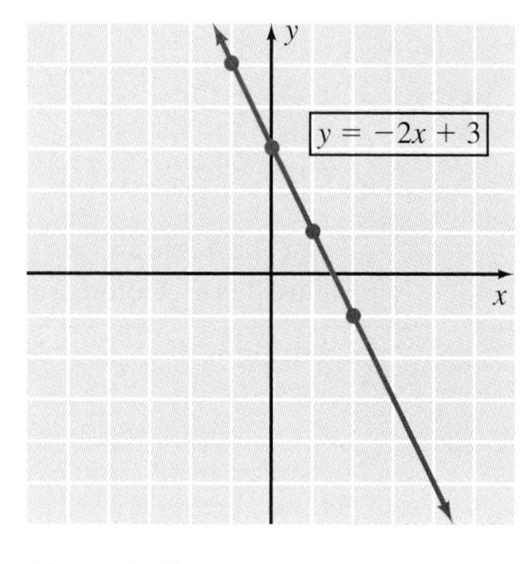

Figure 8.10

■

To graph an equation in two variables, x and y, keep these steps in mind:

1. Solve the equation for y in terms of x or for x in terms of y, if it is not already in such a form.
2. Set up a table of ordered pairs that satisfies the equation.
3. Plot the points associated with the ordered pairs.
4. Connect the points.

We conclude this section with two more examples that illustrate step 1.

EXAMPLE 5

Solve $4x + 9y = 12$ for y.

Solution

$$4x + 9y = 12$$

$$9y = -4x + 12 \quad \text{Subtracted } 4x \text{ from both sides}$$

$$y = -\frac{4}{9}x + \frac{12}{9} \quad \text{Divided both sides by 9}$$

$$y = -\frac{4}{9}x + \frac{4}{3}$$

■

EXAMPLE 6 Solve $2x - 6y = 3$ for y.

Solution

$$2x - 6y = 3$$

$$-6y = -2x + 3 \quad \text{Subtracted } 2x \text{ from each side}$$

$$y = \frac{-2}{-6}x + \frac{3}{-6} \quad \text{Divided both sides by } -6$$

$$y = \frac{1}{3}x - \frac{1}{2} \quad \text{Reduce the fractions}$$

■

Problem Set 8.1

For Problems 1–10, plot the points on a rectangular coordinate system, and determine if the points fall in a straight line.

1. $(1, 4), (-2, 1), (-4, -1), (-3, 0)$
2. $(0, 2), (3, -1), (5, -3), (-3, 5)$
3. $(0, 3), (-2, 7), (3, -6)$
4. $(1, 4), (-1, 6), (6, -3)$
5. $(3, -1), (-3, -3), (0, -2)$
6. $(0, 1), (2, 0), (-2, 2)$
7. $(-4, 0), (-2, -1), (6, -5)$
8. $(1, 3), (-2, -6), (0, 0)$
9. $(-2, 4), (2, -4), (1, 0)$
10. $(2, 2), (-3, -6), (-1, -1)$
11. Maria, a biology student, designed an experiment to test the effects of changing the amount of light and the amount of water given to selected plants. In the experiment, the amounts of water and light given to the plant were randomly changed. The chart shows the amount of light and water above or below the normal amount given to the plant for six days. Plot the charted information on a rectangular coordinate system. Let the change in light be the first number in the ordered pair, and let the change in water be the second number in the ordered pair.

	Mon.	Tue.	Wed.	Thu.	Fri.	Sat.
Change in amount of light	1	−2	−1	4	−3	0
Change in amount of water	−3	4	−1	0	−5	1

12. Chase is studying the monthly percent changes in the stock price for two different companies. Using the data in the table below, plot the points for each month. Let the percent change for XM Inc. be the first number of the ordered pair, and let the percent change for Icom be the second number in the ordered pair.

	Jan.	Feb.	Mar.	Apr.	May	Jun.
XM Inc.	1	−2	−1	4	−3	0
Icom	−3	4	2	−5	−1	1

For Problems 13–20, (a) determine which of the ordered pairs satisfy the given equation, and then (b) graph the equation by plotting the points that satisfied the equation.

13. $y = -x + 4$; $(1, 3), (0, 4), (2, -1), (-2, 6), (-1, 5)$
14. $y = -2x + 3$; $(1, 1), (1.5, 0), (3, -1), (0, 3), (-1, 5)$
15. $y = 2x - 3$; $(1, 2), (-3, 0), (0, -3), (2, 1), (-1, -5)$
16. $y = x + 4$; $(-1, 3), (2, -1), (-2, 2), (-1, 5), (0, 4)$

17. $y = \frac{2}{3}x$; $(1, 2), (-3, -2), (0, 0), (3, 6), (-6, -2)$

18. $y = \frac{1}{2}x$; $(1, 2), (2, 1), \left(0, \frac{1}{2}\right), (-2, 1), \left(-1, -\frac{1}{2}\right)$

19. $y = \frac{2}{5}x - \frac{1}{5}$; $\left(1, \frac{1}{5}\right), \left(-3, \frac{4}{5}\right), \left(-1, -\frac{3}{5}\right), \left(2, \frac{2}{5}\right), (-2, -1)$

20. $y = -\frac{2}{3}x + \frac{1}{3}$; $\left(1, \frac{1}{3}\right), (0, 1), (5, -3), \left(\frac{1}{2}, 0\right), (-1, 0)$

For Problems 21–30, solve the given equation for the variable indicated.

21. $3x + 7y = 13$ for y

22. $5x + 9y = 17$ for y

23. $x - 3y = 9$ for x

24. $2x - 7y = 5$ for x

25. $-x + 5y = 14$ for y

26. $-2x - y = 9$ for y

27. $-3x + y = 7$ for x

28. $-x - y = 9$ for x

29. $-2x + 3y = -5$ for y

30. $3x - 4y = -7$ for y

For Problems 31–50, graph each of the equations.

31. $y = x + 1$

32. $y = x + 4$

33. $y = x - 2$

34. $y = -x - 1$

35. $y = 3x - 4$

36. $y = -2x - 1$

37. $y = -4x + 4$

38. $y = 2x - 6$

39. $y = \frac{1}{2}x + 3$

40. $y = \frac{1}{2}x - 2$

41. $x + 2y = 4$

42. $x + 3y = 6$

43. $2x - 5y = 10$

44. $5x - 2y = 10$

45. $y = -\frac{1}{3}x + 2$

46. $y = -\frac{2}{3}x - 3$

47. $y = x$

48. $y = -x$

49. $y = -3x + 2$

50. $3x - y = 4$

THOUGHTS INTO WORDS

51. How would you convince someone that there are infinitely many ordered pairs of real numbers that satisfy the equation $x + y = 9$?

52. Explain why no points of the graph of the equation $y = x$ will be in the second quadrant.

FURTHER INVESTIGATIONS

Not all graphs of equations are straight lines. Some equations produce graphs that are smooth curves. For example, the graph of $y = x^2$ will produce a smooth curve that is called a parabola. To graph $y = x^2$, let's proceed as in the previous problems and plot points associated with solutions to the equation $y = x^2$, and connect the points with a smooth curve.

x	y	Solutions (x, y)
0	0	(0, 0)
1	1	(1, 1)
2	4	(2, 4)
3	9	(3, 9)
−1	1	(−1, 1)
−2	4	(−2, 4)
−3	9	(−3, 9)

Then we plot the points associated with the solutions as shown in Figure 8.11(a). Finally, we connect the points with the smooth curve shown in Figure 8.11(b).

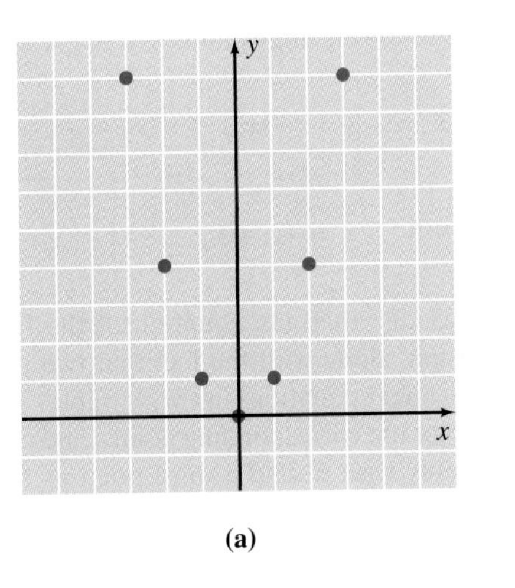

(a)

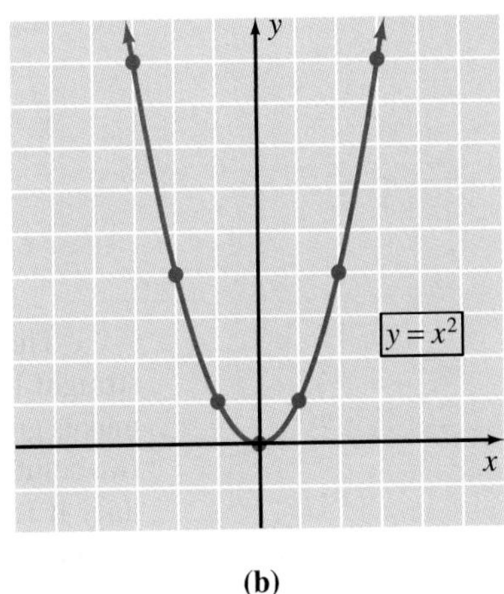

(b)

Figure 8.11

For Problems 53–62, graph each of the equations. Be sure to find a sufficient number of solutions so that the graph of the equation can be determined.

53. $y = x^2 + 2$

54. $y = x^2 - 3$

55. $y = -x^2$

56. $y = (x - 2)^2$

57. $y = x^3$

58. $y = x^3 - 4$

59. $y = (x - 3)^3$

60. $y = -x^3$

61. $y = x^4$

62. $y = -x^4$

8.2 Graphing Linear Equations and Inequalities

In the preceding section we graphed equations whose graphs were straight lines. In general any equation of the form $Ax + By = C$, where A, B, and C are constants (A and B not both zero) and x and y are variables, is a *linear equation in two variables*, and its graph is a straight line.

We should clarify two points about this description of a linear equation in two variables. First, the choice of x and y for variables is arbitrary. We could use any two letters to represent the variables. An equation such as $3m + 2n = 7$ can be considered a linear equation in two variables. So that we are not constantly changing the labeling of the coordinate axes when graphing equations, however, it is much easier to use the same two variables in all equations. Thus we will go along with convention and use x and y as our variables. Second, the statement "any equation of the form $Ax + By = C$" technically means "any equation of the form $Ax + By = C$ or *equivalent* to the form." For example, the equation $y = x + 3$, which has a straight line graph, is equivalent to $-x + y = 3$.

All of the following are examples of linear equations in two variables.

$$y = x + 3 \qquad y = -3x + 2 \qquad y = \frac{2}{5}x + 1 \qquad y = 2x$$

$$3x - 2y = 6 \qquad x - 4y = 5 \qquad 5x - y = 10 \qquad y = \frac{2x + 4}{3}$$

The knowledge that any equation of the form $Ax + By = C$ produces a straight line graph, along with the fact that two points determine a straight line, makes graphing linear equations in two variables a simple process. We merely find two solutions, plot the corresponding points, and connect the points with a straight line. It is probably wise to find a third point as a check point. Let's consider an example.

EXAMPLE 1

Graph $2x - 3y = 6$.

Solution

We recognize that equation $2x - 3y = 6$ is a linear equation in two variables, and therefore its graph will be a straight line. All that is necessary is to find two solutions and connect the points with a straight line. We will, however, also find a third solution to serve as a check point.

Let $x = 0$; then $2(0) - 3y = 6$

$$-3y = 6$$

$$y = -2 \qquad \text{Thus } (0, -2) \text{ is a solution}$$

Let $y = 0$; then $2x - 3(0) = 6$

$$2x = 6$$

$$x = 3 \qquad \text{Thus } (3, 0) \text{ is a solution}$$

Let $x = -3$; then $2(-3) - 3y = 6$

$$-6 - 3y = 6$$

$$-3y = 12$$

$$y = -4 \qquad \text{Thus } (-3, -4) \text{ is a solution}$$

We can plot the points associated with these three solutions and connect them with a straight line to produce the graph of $2x - 3y = 6$ in Figure 8.12.

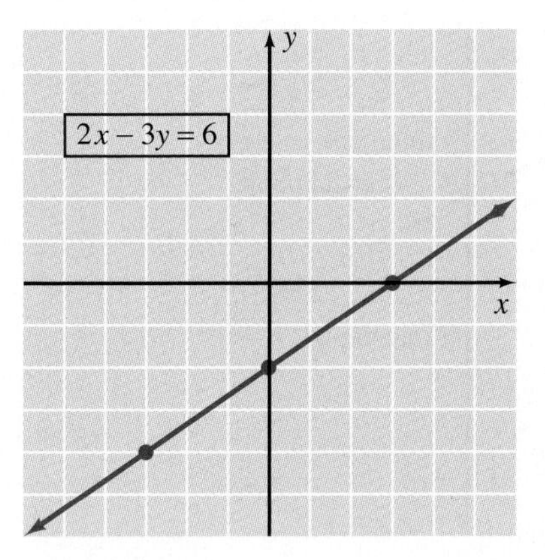

Figure 8.12

Let us briefly review our approach to Example 1. Notice that we did not begin the solution by solving either for y in terms of x or for x in terms of y. The reason for this is that we know the graph is a straight line and therefore there is no need for an extensive table of values. Thus there is no real benefit to changing the form of the original equation. The first two solutions indicate where the line intersects the coordinate axes. The ordinate of the point $(0, -2)$ is called the ***y* intercept**, and the abscissa of the point $(3, 0)$ is called the ***x* intercept** of this graph. That is, the graph of the equation $2x - 3y = 6$ has a y intercept of -2 and an x intercept of 3. In general, the intercepts are often easy to find. You can let $x = 0$ and solve for y to find the y intercept, and let $y = 0$ and solve for x to find the x intercept. The third solution, $(-3, -4)$, serves as a check point. If $(-3, -4)$ had not been on the line determined by the two intercepts, then we would have known that we had committed an error.

EXAMPLE 2

Graph $x + 2y = 4$.

Solution

Without showing all of our work, we present the following table indicating the intercepts and a check point.

x	y	
0	2	Intercepts
4	0	
2	1	Check point

We plot the points $(0, 2)$, $(4, 0)$, and $(2, 1)$ and connect them with a straight line to produce the graph in Figure 8.13.

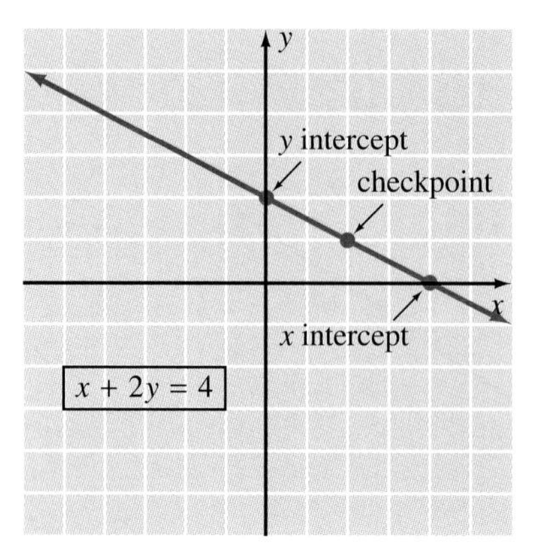

Figure 8.13

■

EXAMPLE 3

Graph $2x + 3y = 7$.

Solution

The intercepts and a check point are given in the table. Finding intercepts may involve fractions, but the computation is usually easy. We plot the points from the table and show the graph of $2x + 3y = 7$ in Figure 8.14.

x	y	
0	$\frac{7}{3}$	Intercepts
$\frac{7}{2}$	0	
2	1	Check point

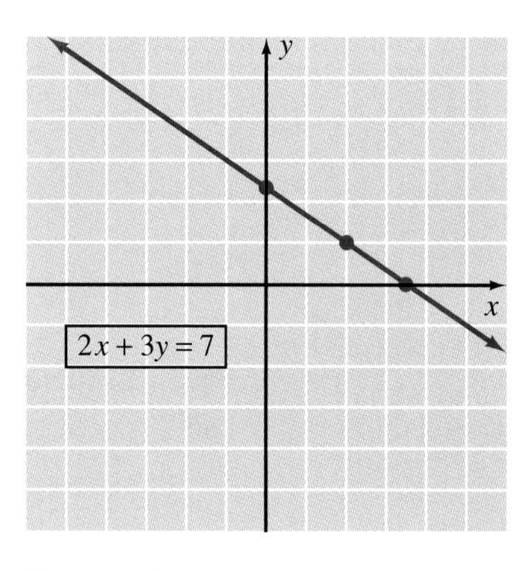

Figure 8.14

■

EXAMPLE 4

Graph $y = 2x$.

Solution

Notice that (0, 0) is a solution; thus, this line intersects both axes at the origin. Since both the x intercept and the y intercept are determined by the origin, (0, 0), we need another point to graph the line. Then a third point should be found to serve as a check point. These results are summarized in the table. The graph of $y = 2x$ is shown in Figure 8.15.

x	y	
0	0	Intercept
2	4	Additional point
−1	−2	Check point

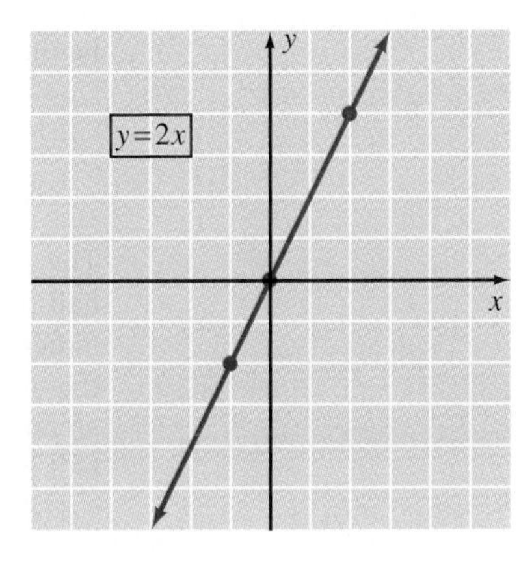

Figure 8.15

EXAMPLE 5 Graph $x = 3$.

Solution

Since we are considering linear equations in *two variables*, the equation $x = 3$ is equivalent to $x + 0(y) = 3$. Now we can see that any value of y can be used, but the x value must always be 3. Therefore, some of the solutions are (3, 0), (3, 1), (3, 2), (3, −1), and (3, −2). The graph of all of the solutions is the vertical line indicated in Figure 8.16.

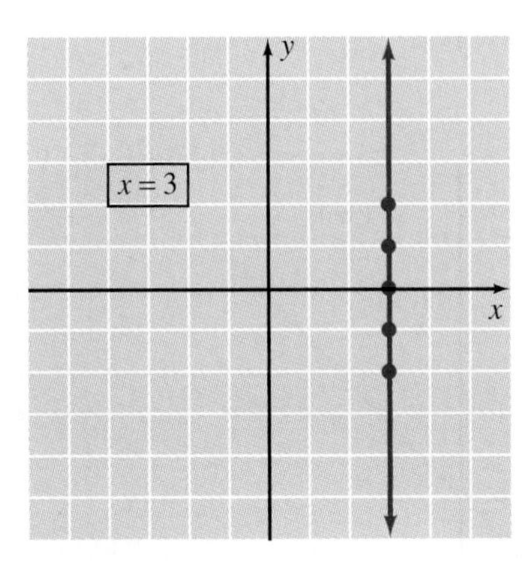

Figure 8.16

Graphing Linear Inequalities

Linear inequalities in two variables are of the form $Ax + By > C$ or $Ax + By < C$, where A, B, and C are real numbers. (Combined linear equality and inequality statements are of the form $Ax + By \geq C$ or $Ax + By \leq C$.) Graphing linear inequalities is almost as easy as graphing linear equations. Our discussion leads to a simple step-by-step process. First we consider the following equation and related inequalities:

$$x - y = 2$$

$$x - y > 2$$

$$x - y < 2$$

The graph of $x - y = 2$ is shown in Figure 8.17. The line divides the plane into two half-planes, one above the line and one below the line. In Figure 8.18(a) we have indicated the coordinates for several points above the line. Note that for each point, the ordered pair of real numbers satisfies the inequality $x - y < 2$. This is true for *all points* in the half-plane above the line. Therefore, the graph of $x - y < 2$ is the half-plane above the line, indicated by the shaded region in Figure 8.18(b). We use a dashed line to indicate that points on the line do not satisfy $x - y < 2$.

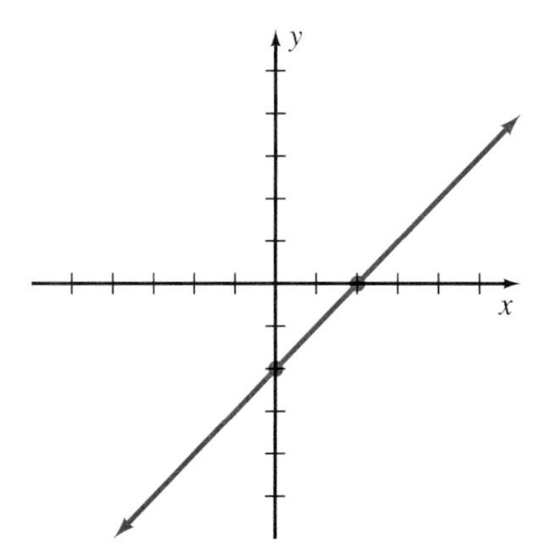

Figure 8.17

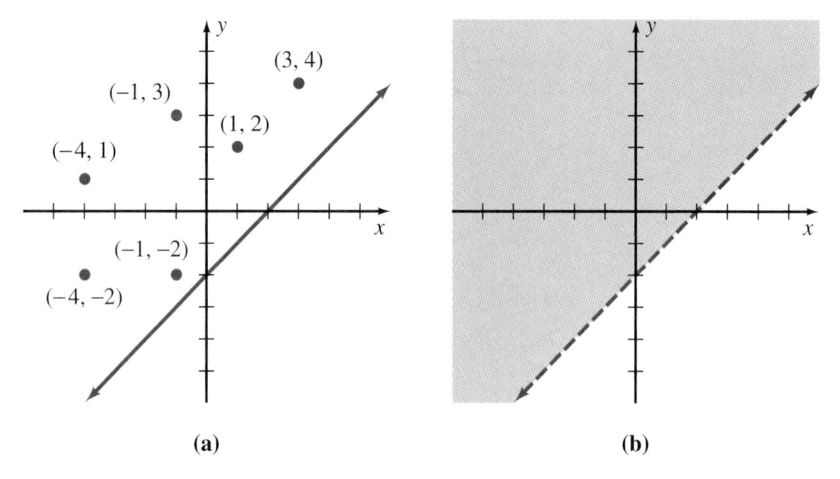

Figure 8.18

In Figure 8.19(a), we have indicated the coordinates of several points below the line $x - y = 2$. Note that for each point, the ordered pair of real numbers satisfies the inequality $x - y > 2$. This is true for *all points* in the half-plane below the line. Therefore, the graph of $x - y > 2$ is the half-plane below the line, indicated by the shaded region in Figure 8.19(b).

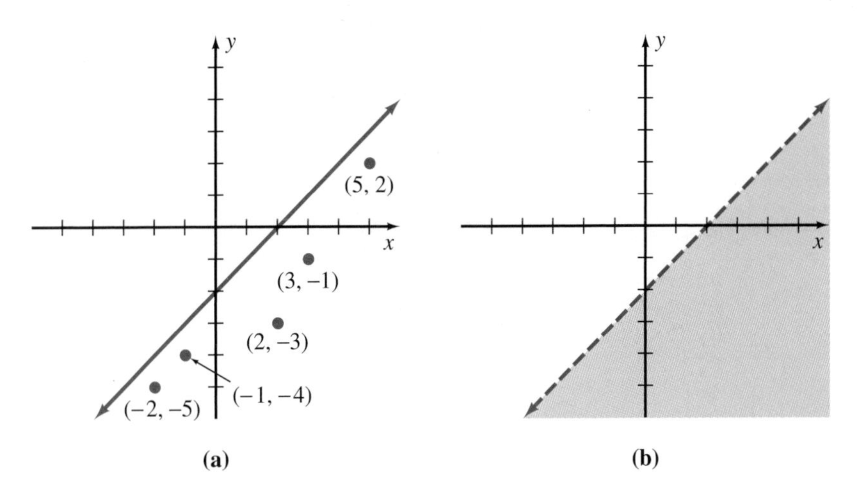

Figure 8.19

Based on this discussion, we suggest the following steps for graphing linear inequalities:

1. Graph the corresponding equality. Use a solid line if equality is included in the given statement and a dashed line if equality is not included.

2. Choose a *test point* not on the line, and substitute its coordinates into the inequality statement. (The origin is a convenient point to use if it is not on the line.)

3. The graph of the given inequality is
   (a) the half-plane that contains the test point if the inequality is satisfied by the coordinates of the point, or
   (b) the half-plane that does not contain the test point if the inequality is not satisfied by the coordinates of the point.

We can apply these steps to some examples.

EXAMPLE 6 Graph $2x + y > 4$.

Solution

Step 1 Graph $2x + y = 4$ as a dashed line because equality is not included in the given statement $2x + y > 4$; see Figure 8.20(a).

Step 2 Choose the origin as a test point, and substitute its coordinates into the inequality.

$$2x + y > 4 \quad \text{becomes } 2(0) + 0 > 4$$

which is a false statement.

Step 3 Since the test point does not satisfy the given inequality, the graph is the half-plane that does not contain the test point. Thus the graph of $2x + y > 4$ is the half-plane above the line, indicated in Figure 8.20(b).

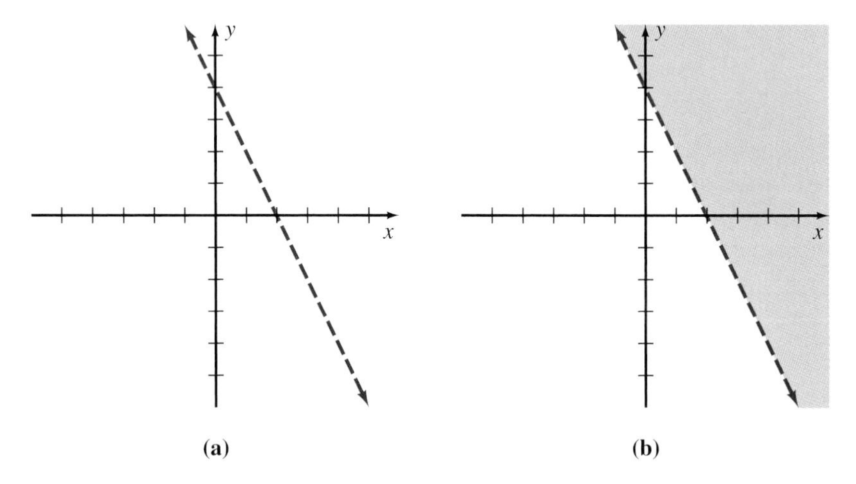

Figure 8.20

EXAMPLE 7

Graph $y \leq 2x$.

Solution

Step 1 Graph $y = 2x$ as a solid line, because equality is included in the given statement [see Figure 8.21(a)].

Step 2 Since the origin is on the line, we need to choose another point as a test point. Let's use (3, 2).

$$y \leq 2x \quad \text{becomes } 2 \leq 2(3)$$

which is a true statement.

Step 3 Since the test point satisfies the given inequality, the graph is the half-plane that contains the test point. Thus the graph of $y \leq 2x$ is the line along with the half-plane below the line indicated in Figure 8.21(b).

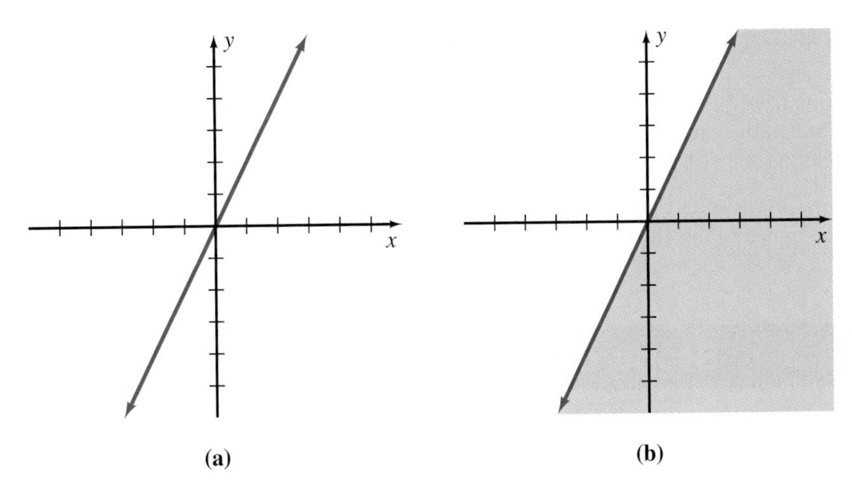

(a) (b)

Figure 8.21

■

Problem Set 8.2

For Problems 1–36, graph each linear equation.

1. $x + y = 2$
2. $x + y = 4$
3. $x - y = 3$
4. $x - y = 1$
5. $x - y = -4$
6. $-x + y = 5$
7. $x + 2y = 2$
8. $x + 3y = 5$
9. $3x - y = 6$
10. $2x - y = -4$
11. $3x - 2y = 6$
12. $2x - 3y = 4$
13. $x - y = 0$
14. $x + y = 0$
15. $y = 3x$
16. $y = -2x$
17. $x = -2$
18. $y = 3$
19. $y = 0$
20. $x = 0$
21. $y = -2x - 1$
22. $y = 3x - 4$
23. $y = \frac{1}{2}x + 1$
24. $y = \frac{2}{3}x - 2$
25. $y = -\frac{1}{3}x - 2$
26. $y = -\frac{3}{4}x - 1$
27. $4x + 5y = -10$
28. $3x + 5y = -9$
29. $-2x + y = -4$
30. $-3x + y = -5$
31. $3x - 4y = 7$
32. $4x - 3y = 10$
33. $y + 4x = 0$
34. $y - 5x = 0$
35. $x = 2y$
36. $x = -3y$

For Problems 37–56, graph each linear inequality.

37. $x + y > 1$
38. $2x + y > 4$
39. $3x + 2y < 6$
40. $x + 3y < 3$
41. $2x - y \geq 4$
42. $x - 2y \geq 2$
43. $4x - 3y \leq 12$
44. $3x - 4y \leq 12$
45. $y > -x$
46. $y < x$
47. $2x - y \geq 0$
48. $3x - y \leq 0$
49. $-x + 2y < -2$
50. $-2x + y > -2$
51. $y \leq \frac{1}{2}x - 2$
52. $y \geq -\frac{1}{2}x + 1$
53. $y \geq -x + 4$
54. $y \leq -x - 3$
55. $3x + 4y > -12$
56. $4x + 3y > -12$

THOUGHTS INTO WORDS

57. Your friend is having trouble understanding why the graph of the equation $y = 3$ is a horizontal line that contains the point $(0, 3)$. What can you do to help him?

58. How do we know that the graph of $y = -4x$ is a straight line that contains the origin?

59. Do all graphs of linear equations have x intercepts? Explain your answer.

60. How do we know that the graphs of $x - y = 4$ and $-x + y = -4$ are the same line?

8.3 Slope of a Line

In Figure 8.22, note that the line associated with $4x - y = 4$ is *steeper* than the line associated with $2x - 3y = 6$. Mathematically, we use the concept of **slope** to discuss the steepness of lines. The slope of a line is the ratio of the vertical change to the horizontal change as we move from one point on a line to another point. We indicate this in Figure 8.23 with the points P_1 and P_2.

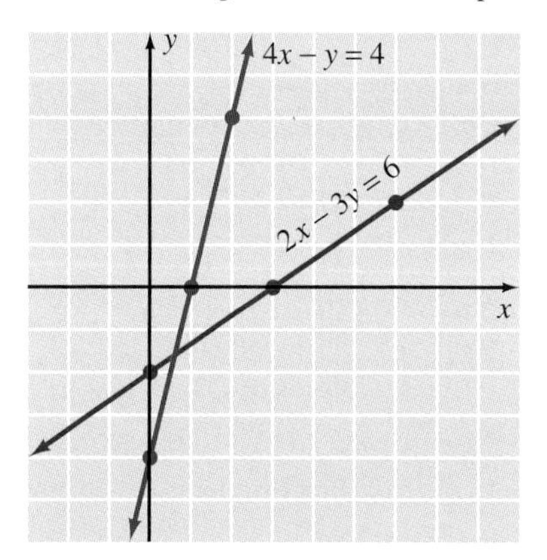

Figure 8.22

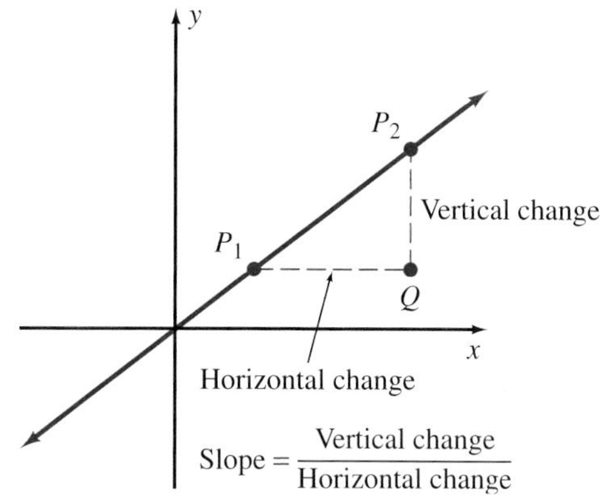

Figure 8.23

We can give a precise definition for slope by considering the coordinates of the points P_1, P_2, and Q in Figure 8.24. Since P_1 and P_2 represent any two points on the line, we assign the coordinates (x_1, y_1) to P_1 and (x_2, y_2) to P_2. The point Q is the same distance from the y axis as P_2 and the same distance from the x axis as P_1. Thus we assign the

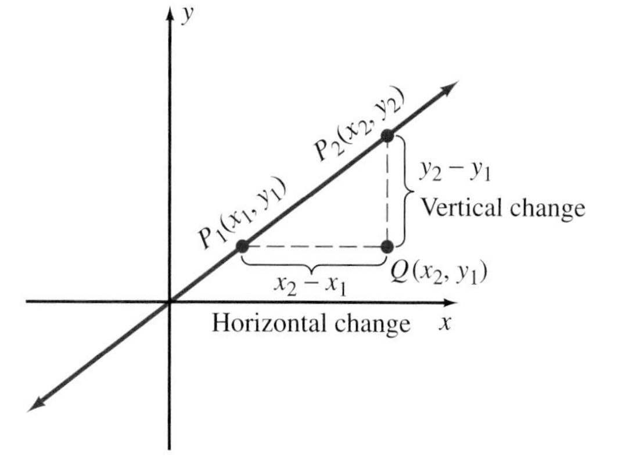

Figure 8.24

coordinates (x_2, y_1) to Q (see Figure 8.24). It should now be apparent that the vertical change is $y_2 - y_1$, and the horizontal change is $x_2 - x_1$. Thus we have the following definition for slope.

Definition 8.1

If points P_1 and P_2 with coordinates (x_1, y_1) and (x_2, y_2), respectively, are any two different points on a line, then the slope of the line (denoted by m) is

$$m = \frac{y_2 - y_1}{x_2 - x_1}, \qquad x_1 \neq x_2$$

Using Definition 8.1, we can easily determine the slope of a line if we know the coordinates of two points on the line.

EXAMPLE 1

Find the slope of the line determined by each of the following pairs of points.

(a) (2, 1) and (4, 6)

(b) (3, 2) and (−4, 5)

(c) (−4, −3) and (−1, −3)

Solution

(a) Let (2, 1) be P_1 and (4, 6) be P_2 as in Figure 8.25; then we have

$$m = \frac{y_2 - y_1}{x_2 - x_1} = \frac{6 - 1}{4 - 2} = \frac{5}{2}$$

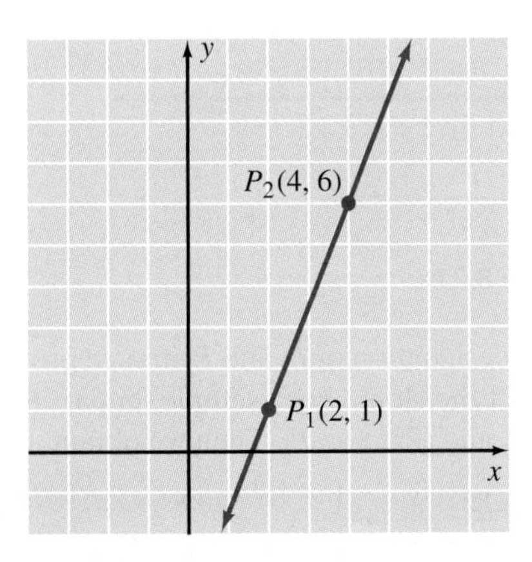

Figure 8.25

(b) Let (3, 2) be P_1 and (−4, 5) be P_2 as in Figure 8.26.

$$m = \frac{y_2 - y_1}{x_2 - x_1} = \frac{5 - 2}{-4 - 3} = \frac{3}{-7} = -\frac{3}{7}$$

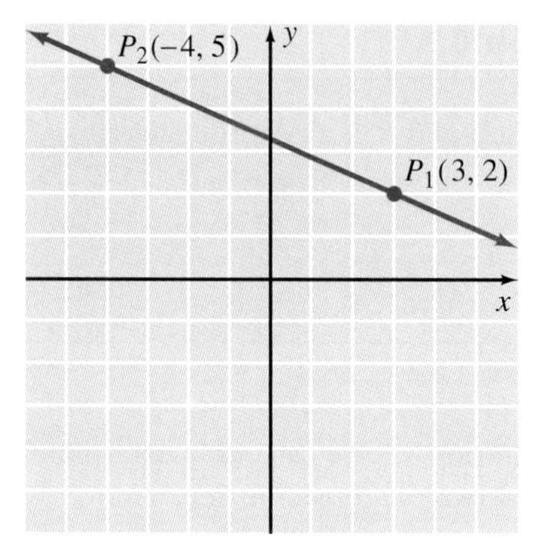

Figure 8.26

(c) Let (−4, −3) be P_1 and (−1, −3) be P_2 as in Figure 8.27.

$$m = \frac{y_2 - y_1}{x_2 - x_1} = \frac{-3 - (-3)}{-1 - (-4)} = \frac{0}{3} = 0$$

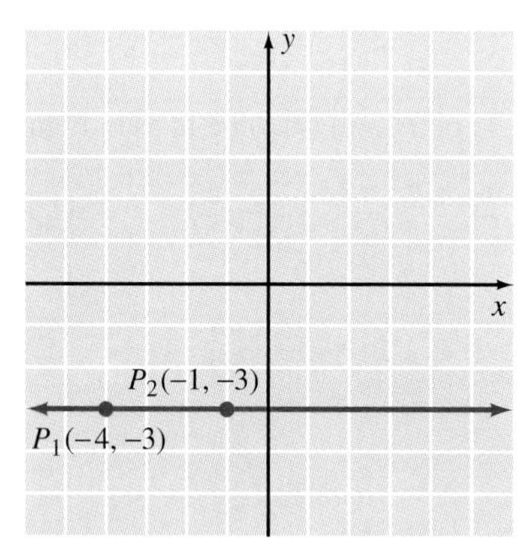

Figure 8.27 ■

The designation of P_1 and P_2 in such problems is arbitrary and does not affect the value of the slope. For example, in part (a) of Example 1 suppose that we let (4, 6) be P_1 and (2, 1) be P_2. Then we obtain

$$m = \frac{y_2 - y_1}{x_2 - x_1} = \frac{1 - 6}{2 - 4} = \frac{-5}{-2} = \frac{5}{2}$$

The parts of Example 1 illustrate the three basic possibilities for slope; that is, the slope of a line can be *positive, negative*, or *zero*. A line that has a positive slope rises as we move from left to right, as in part (a). A line that has a negative slope falls as we move from left to right, as in part (b). A horizontal line, as in part (c), has a slope of 0. Finally, we need to realize that **the concept of slope is undefined for vertical lines**. This is because, for any vertical line, the change in x as we move from one point to another is zero. Thus the ratio $\frac{y_2 - y_1}{x_2 - x_1}$ will have a denominator of zero and be undefined. So in Definition 8.1, the restriction $x_1 \neq x_2$ is made.

EXAMPLE 2

Find the slope of the line determined by the equation $3x + 4y = 12$.

Solution

Since we can use any two points on the line to determine the slope of the line, let's find the intercepts.

$$\text{If } x = 0, \text{ then } 3(0) + 4y = 12$$
$$4y = 12$$
$$y = 3 \qquad \text{Thus } (0, 3) \text{ is on the line}$$

$$\text{If } y = 0, \text{ then } 3x + 4(0) = 12$$
$$3x = 12$$
$$x = 4 \qquad \text{Thus } (4, 0) \text{ is on the line}$$

Using (0, 3) as P_1 and (4, 0) as P_2, we have

$$m = \frac{y_2 - y_1}{x_2 - x_1} = \frac{0 - 3}{4 - 0} = \frac{-3}{4} = -\frac{3}{4}$$

■

We need to emphasize one final idea pertaining to the concept of slope. The slope of a line is a **ratio** of vertical change to horizontal change. A slope of $\frac{3}{4}$ means that for every 3 units of vertical change, there is a corresponding 4 units of horizontal change. So starting at some point on the line, we could move to other points on the line as follows:

$\frac{3}{4} = \frac{6}{8}$ by moving 6 units *up* and 8 units to the *right*

$\frac{3}{4} = \frac{15}{20}$ by moving 15 units *up* and 20 units to the *right*

$\frac{3}{4} = \frac{\frac{3}{2}}{2}$ by moving $1\frac{1}{2}$ units *up* and 2 units to the *right*

$\frac{3}{4} = \frac{-3}{-4}$ by moving 3 units *down* and 4 units to the *left*

Likewise, a slope of $-\frac{5}{6}$ indicates that starting at some point on the line, we could move to other points on the line as follows:

$-\frac{5}{6} = \frac{-5}{6}$ by moving 5 units *down* and 6 units to the *right*

$-\frac{5}{6} = \frac{5}{-6}$ by moving 5 units *up* and 6 units to the *left*

$-\frac{5}{6} = \frac{-10}{12}$ by moving 10 units *down* and 12 units to the *right*

$-\frac{5}{6} = \frac{15}{-18}$ by moving 15 units *up* and 18 units to the *left*

EXAMPLE 3

Graph the line that passes through the point $(0, -2)$ and has a slope of $\frac{1}{3}$.

Solution

To begin, plot the point $(0, -2)$. Furthermore, because the slope $= \frac{\text{vertical change}}{\text{horizontal change}} = \frac{1}{3}$, we can locate another point on the line by starting from the point $(0, -2)$ and moving 1 unit up and 3 units to the right to obtain the point $(3, -1)$. Because two points determine a line, we can draw the line (Figure 8.28).

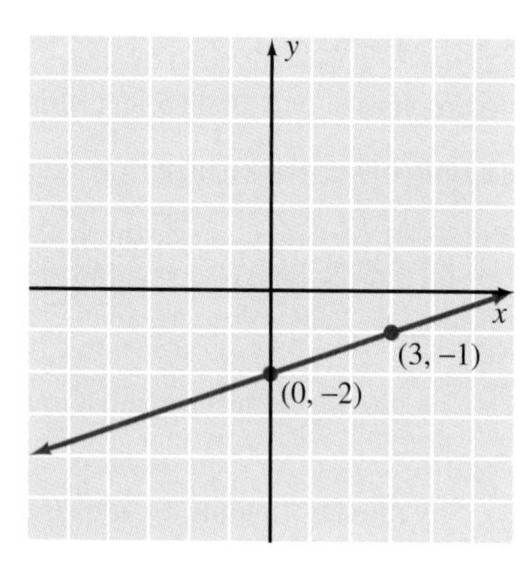

Figure 8.28

Remark: Because $m = \frac{1}{3} = \frac{-1}{-3}$, we can locate another point by moving 1 unit down and 3 units to the left from the point $(0, -2)$. ■

EXAMPLE 4 Graph the line that passes through the point $(1, 3)$ and has a slope of -2.

Solution

To graph the line, plot the point $(1, 3)$. We know that $m = -2 = \frac{-2}{1}$. Furthermore, because the slope $= \frac{\text{vertical change}}{\text{horizontal change}} = \frac{-2}{1}$, we can locate another point on the line by starting from the point $(1, 3)$ and moving 2 units down and 1 unit to the right to obtain the point $(2, 1)$. Because two points determine a line, we can draw the line (Figure 8.29).

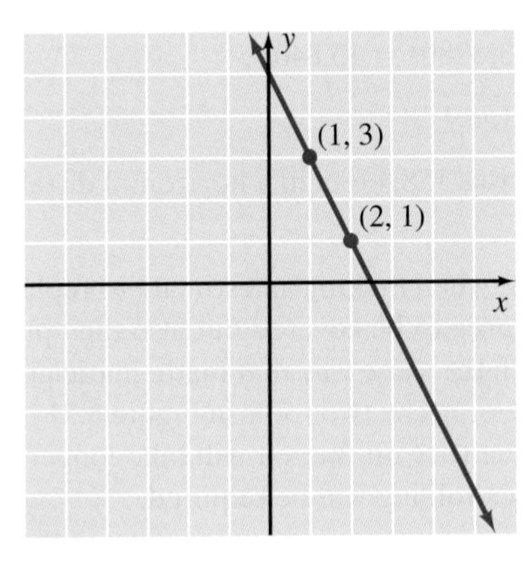

Figure 8.29

Remark: Because $m = -2 = \frac{-2}{1} = \frac{2}{-1}$ we can locate another point by moving 2 units up and 1 unit to the left from the point $(1, 3)$. ■

■ Applications of Slope

The concept of slope has many real-world applications even though the word "slope" is often not used. For example, the highway in Figure 8.30 is said to have a "grade" of 17%. This means that for every horizontal distance of 100 feet, the highway rises or drops 17 feet. In other words, the absolute value of the slope of the highway is $\frac{17}{100}$.

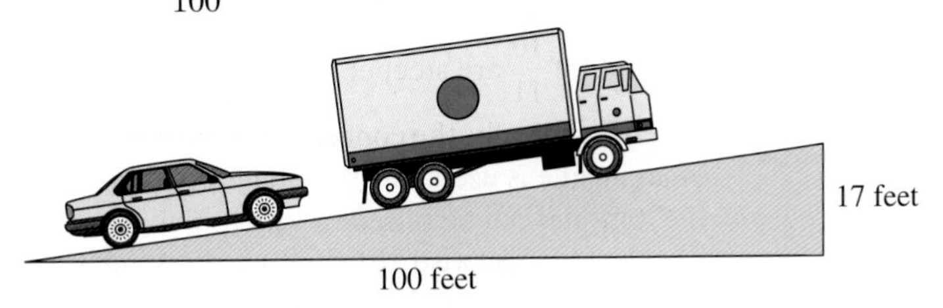

Figure 8.30

PROBLEM 1

A certain highway has a 3% grade. How many feet does it rise in a horizontal distance of 1 mile?

Solution

A 3% grade means a slope of $\frac{3}{100}$. Therefore, if we let y represent the unknown vertical distance and use the fact that 1 mile = 5280 feet, we can set up and solve the following proportion:

$$\frac{3}{100} = \frac{y}{5280}$$

$$100y = 3(5280) = 15{,}840$$

$$y = 158.4$$

The highway rises 158.4 feet in a horizontal distance of 1 mile. ■

A roofer, when making an estimate to replace a roof, is concerned about not only the total area to be covered but also the "pitch" of the roof. (Contractors do not define pitch the same way that mathematicians define slope, but both terms refer to "steepness.") The two roofs in Figure 8.31 might require the same number of shingles, but the roof on the left will take longer to complete because the pitch is so great that scaffolding will be required.

Figure 8.31

The concept of slope is also used in the construction of flights of stairs. The terms "rise" and "run" are commonly used, and the steepness (slope) of the stairs can be expressed as the ratio of rise to run. In Figure 8.32, the stairs on the left with the ratio of $\frac{10}{11}$ are steeper than the stairs on the right, which have a ratio of $\frac{7}{11}$.

Technically, the concept of slope is involved in most situations where the idea of an incline is used. Hospital beds are constructed so that both the head-end and the foot-end can be raised or lowered; that is, the slope of either end of the bed can be changed. Likewise, treadmills are designed so that the incline (slope) of the platform can be raised or lowered as desired. Perhaps you can think of several other applications of the concept of slope.

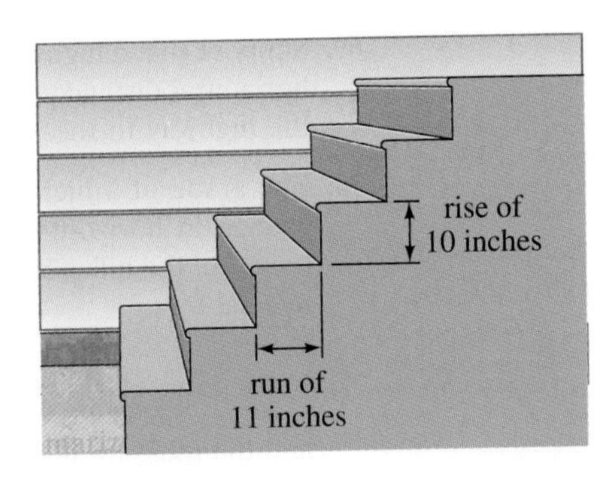

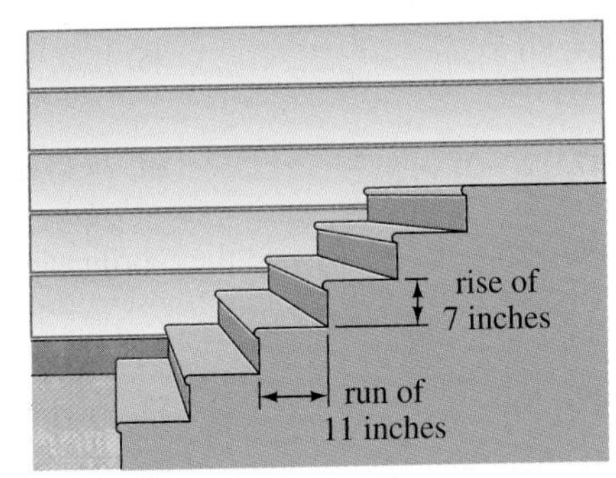

Figure 8.32

Problem Set 8.3

For Problems 1–20, find the slope of the line determined by each pair of points.

1. $(7, 5), (3, 2)$
2. $(9, 10), (6, 2)$
3. $(-1, 3), (-6, -4)$
4. $(-2, 5), (-7, -1)$
5. $(2, 8), (7, 2)$
6. $(3, 9), (8, 4)$
7. $(-2, 5), (1, -5)$
8. $(-3, 4), (2, -6)$
9. $(4, -1), (-4, -7)$
10. $(5, -3), (-5, -9)$
11. $(3, -4), (2, -4)$
12. $(-3, -6), (5, -6)$
13. $(-6, -1), (-2, -7)$
14. $(-8, -3), (-2, -11)$
15. $(-2, 4), (-2, -6)$
16. $(-4, -5), (-4, 9)$
17. $(-1, 10), (-9, 2)$
18. $(-2, 12), (-10, 2)$
19. $(a, b), (c, d)$
20. $(a, 0), (0, b)$
21. Find y if the line through the points $(7, 8)$ and $(2, y)$ has a slope of $\frac{4}{5}$.
22. Find y if the line through the points $(12, 14)$ and $(3, y)$ has a slope of $\frac{4}{3}$.
23. Find x if the line through the points $(-2, -4)$ and $(x, 2)$ has a slope of $-\frac{3}{2}$.
24. Find x if the line through the points $(6, -4)$ and $(x, 6)$ has a slope of $-\frac{5}{4}$.

For Problems 25–32, you are given one point on a line and the slope of the line. Find the coordinates of three other points on the line.

25. $(3, 2), m = \frac{2}{3}$
26. $(4, 1), m = \frac{5}{6}$
27. $(-2, -4), m = \frac{1}{2}$
28. $(-6, -2), m = \frac{2}{5}$
29. $(-3, 4), m = -\frac{3}{4}$
30. $(-2, 6), m = -\frac{3}{7}$
31. $(4, -5), m = -2$
32. $(6, -2), m = 4$

For Problems 33–40, sketch the line determined by each pair of points and decide whether the slope of the line is positive, negative, or zero.

33. $(2, 8), (7, 1)$
34. $(1, -2), (7, -8)$
35. $(-1, 3), (-6, -2)$
36. $(7, 3), (4, -6)$
37. $(-2, 4), (6, 4)$
38. $(-3, -4), (5, -4)$
39. $(-3, 5), (2, -7)$
40. $(-1, -1), (1, -9)$

For Problems 41–48, graph the line that passes through the given point and has the given slope.

41. $(3, 1), m = \frac{2}{3}$
42. $(-1, 0), m = \frac{3}{4}$
43. $(-2, 3), m = -1$
44. $(1, -4), m = -3$

45. $(0, 5), m = -\frac{1}{4}$

46. $(-3, 4), m = -\frac{3}{2}$

47. $(2, -2), m = \frac{3}{2}$

48. $(3, -4), m = \frac{5}{2}$

For Problems 49–68, find the coordinates of two points on the given line, and then use those coordinates to find the slope of the line.

49. $3x + 2y = 6$

50. $4x + 3y = 12$

51. $5x - 4y = 20$

52. $7x - 3y = 21$

53. $x + 5y = 6$

54. $2x + y = 4$

55. $2x - y = -7$

56. $x - 4y = -6$

57. $y = 3$

58. $x = 6$

59. $-2x + 5y = 9$

60. $-3x - 7y = 10$

61. $6x - 5y = -30$

62. $7x - 6y = -42$

63. $y = -3x - 1$

64. $y = -2x + 5$

65. $y = 4x$

66. $y = 6x$

67. $y = \frac{2}{3}x - \frac{1}{2}$

68. $y = -\frac{3}{4}x + \frac{1}{5}$

69. Suppose that a highway rises a distance of 135 feet in a horizontal distance of 2640 feet. Express the grade of the highway to the nearest tenth of a percent.

70. The grade of a highway up a hill is 27%. How much change in horizontal distance is there if the vertical height of the hill is 550 feet? Express the answer to the nearest foot.

71. If the ratio of rise to run is to be $\frac{3}{5}$ for some stairs, and the measure of the rise is 19 centimeters, find the measure of the run to the nearest centimeter.

72. If the ratio of rise to run is to be $\frac{2}{3}$ for some stairs and the measure of the run is 28 centimeters, find the measure of the rise to the nearest centimeter.

73. A county ordinance requires a $2\frac{1}{4}\%$ "fall" for a sewage pipe from the house to the main pipe at the street. How much vertical drop must there be for a horizontal distance of 45 feet? Express the answer to the nearest tenth of a foot.

THOUGHTS INTO WORDS

74. How would you explain the concept of slope to someone who was absent from class the day it was discussed?

75. If one line has a slope of $\frac{2}{3}$, and another line has a slope of 2, which line is steeper? Explain your answer.

76. Why do we say that the slope of a vertical line is undefined?

77. Suppose that a line has a slope of $\frac{3}{4}$ and contains the point $(5, 2)$. Are the points $(-3, -4)$ and $(14, 9)$ also on the line? Explain your answer.

8.4 Writing Equations of Lines

There are two basic types of problems in analytic or coordinate geometry:

1. Given an algebraic equation, find its geometric graph.
2. Given a set of conditions pertaining to a geometric figure, determine its algebraic equation.

We discussed problems of type 1 in the first two sections of this chapter. Now we want to consider a few problems of type 2 that deal with straight lines. In other words, given certain facts about a line, we need to be able to write its algebraic equation.

EXAMPLE 1

Find the equation of the line that has a slope of $\frac{3}{4}$ and contains the point (1, 2).

Solution

First, we draw the line as indicated in Figure 8.33. Since the slope is $\frac{3}{4}$, we can find a second point by moving 3 units up and 4 units to the right of the given point (1, 2). [The point (5, 5) merely helps to draw the line; it will not be used in analyzing the problem.] Now we choose a point (x, y) that represents any point on the line other than the given point (1, 2). The slope determined by (1, 2) and (x, y) is $\frac{3}{4}$. Thus

$$\frac{y-2}{x-1} = \frac{3}{4}$$

$$3(x-1) = 4(y-2)$$

$$3x - 3 = 4y - 8$$

$$3x - 4y = -5$$

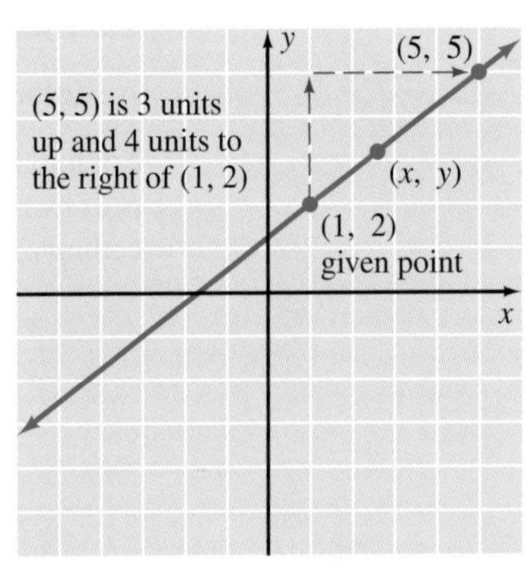

Figure 8.33

■

EXAMPLE 2

Find the equation of the line that contains (3, 4) and (−2, 5).

Solution

First, we draw the line determined by the two given points in Figure 8.34. Since we know two points, we can find the slope.

$$m = \frac{y_2 - y_1}{x_2 - x_1} = \frac{5-4}{-2-3} = \frac{1}{-5} = -\frac{1}{5}$$

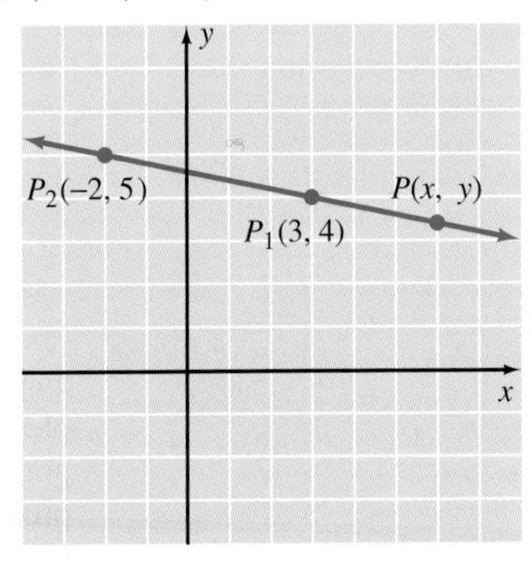

Figure 8.34

Now we can use the same approach as in Example 1. We form an equation using a variable point (x, y), one of the two given points (we choose P_1), and the slope of $-\frac{1}{5}$.

$$\frac{y-4}{x-3} = \frac{1}{-5} \qquad -\frac{1}{5} = \frac{1}{-5}$$

$$x - 3 = -5y + 20$$

$$x + 5y = 23$$

■

EXAMPLE 3 Find the equation of the line that has a slope of $\frac{1}{4}$ and a y intercept of 2.

Solution

A y intercept of 2 means that the point $(0, 2)$ is on the line. Since the slope is $\frac{1}{4}$, we can find another point by moving 1 unit up and 4 units to the right of $(0, 2)$. The line is drawn in Figure 8.35. We choose variable point (x, y) and proceed as in the preceding examples.

$$\frac{y-2}{x-0} = \frac{1}{4}$$

$$x = 4y - 8$$

$$x - 4y = -8$$

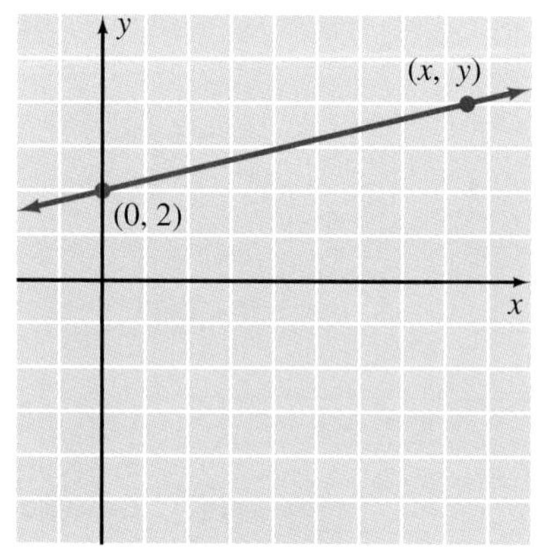

Figure 8.35

■

It may be helpful for you to pause for a moment and look back over Examples 1, 2, and 3. Notice that we used the same basic approach in all three examples; that is, we chose a variable point (x, y) and used it, along with another known point, to determine the equation of the line. You should also recognize that the approach we take in these examples can be generalized to produce some special forms of equations for straight lines.

Point-Slope Form

EXAMPLE 4

Find the equation of the line that has a slope of m and contains the point (x_1, y_1).

Solution

We choose (x, y) to represent any other point on the line in Figure 8.36. The slope of the line given by

$$m = \frac{y - y_1}{x - x_1}$$

enables us to obtain

$$y - y_1 = m(x - x_1)$$

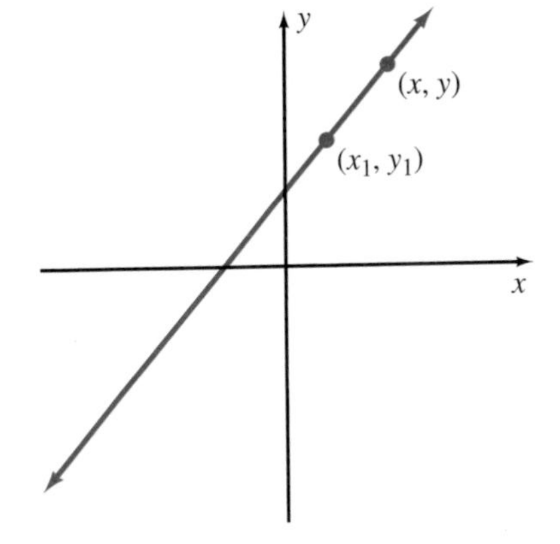

Figure 8.36

We refer to the equation

$$y - y_1 = m(x - x_1)$$

as the **point-slope form** of the equation of a straight line. Instead of the approach we used in Example 1, we could use the point-slope form to write the equation of a line with a given slope that contains a given point—as the next example illustrates.

EXAMPLE 5

Write the equation of the line that has a slope of $\frac{3}{5}$ and contains the point $(2, -4)$.

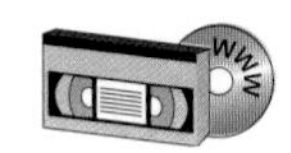

Solution

Substituting $\frac{3}{5}$ for m and $(2, -4)$ for (x_1, y_1) in the point-slope form, we obtain

$$y - y_1 = m(x - x_1)$$

$$y - (-4) = \frac{3}{5}(x - 2)$$

$$y + 4 = \frac{3}{5}(x - 2)$$

$$5(y + 4) = 3(x - 2) \qquad \text{Multiply both sides by 5}$$

$$5y + 20 = 3x - 6$$

$$26 = 3x - 5y$$

■

■ Slope-Intercept Form

Now consider the equation of a line that has a slope of m and a y intercept of b (see Figure 8.37). A y intercept of b means that the line contains the point $(0, b)$; therefore, we can use the point-slope form.

$$y - y_1 = m(x - x_1)$$

$$y - b = m(x - 0) \qquad y_1 = b \text{ and } x_1 = 0$$

$$y - b = mx$$

$$y = mx + b$$

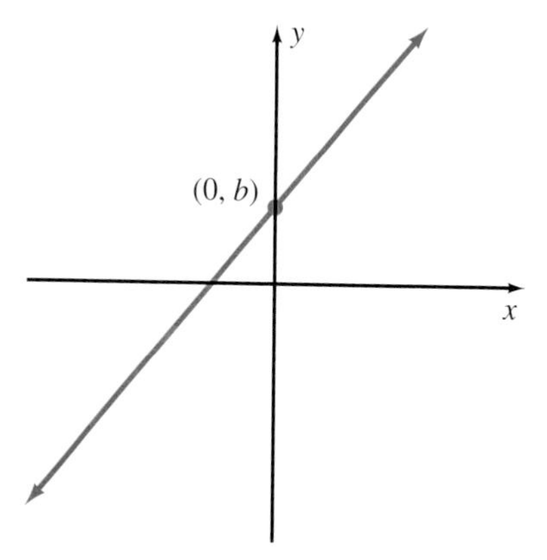

Figure 8.37

The equation

$$y = mx + b$$

is called the **slope-intercept form** of the equation of a straight line. We use it for three primary purposes, as the next three examples illustrate.

EXAMPLE 6

Find the equation of the line that has a slope of $\frac{1}{4}$ and a y intercept of 2.

Solution

This is a restatement of Example 3, but this time we will use the slope-intercept form of a line ($y = mx + b$) to write its equation. From the statement of the problem we know that $m = \frac{1}{4}$ and $b = 2$. Thus substituting these values for m and b into $y = mx + b$, we obtain

$$y = mx + b$$

$$y = \frac{1}{4}x + 2$$

$$4y = x + 8$$

$$x - 4y = -8 \quad \text{Same result as in Example 3}$$

■

Remark: It is acceptable to leave answers in slope-intercept form. We did not do that in Example 6 because we wanted to show that it was the same result as in Example 3.

EXAMPLE 7

Find the slope of the line with the equation $2x + 3y = 4$.

Solution

We can solve the equation for y in terms of x, and then compare it to the slope-intercept form to determine its slope.

$$2x + 3y = 4$$

$$3y = -2x + 4$$

$$y = -\frac{2}{3}x + \frac{4}{3}$$

Compare this result to $y = mx + b$, and you see that the slope of the line is $-\frac{2}{3}$. Furthermore, the y intercept is $\frac{4}{3}$. ■

EXAMPLE 8

Graph the line determined by the equation $y = \frac{2}{3}x - 1$.

Solution

Comparing the given equation to the general slope-intercept form, we see that the slope of the line is $\frac{2}{3}$, and the y intercept is -1. Because the y intercept is -1, we can plot the point $(0, -1)$. Then because the slope is $\frac{2}{3}$, let's move 3 units to the right and 2 units up from $(0, -1)$ to locate the point $(3, 1)$. The two points $(0, -1)$ and $(3, 1)$ determine the line in Figure 8.38.

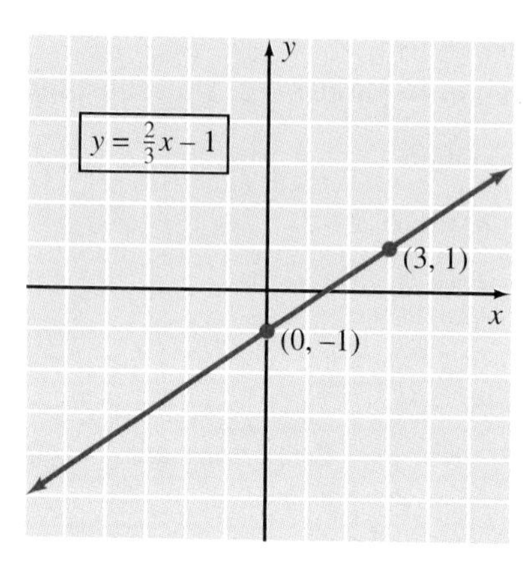

Figure 8.38

In general, **if the equation of a nonvertical line is written in slope-intercept form, the coefficient of x is the slope of the line, and the constant term is the y intercept**. (Remember that the concept of slope is not defined for a vertical line.) Let's consider a few more examples.

EXAMPLE 9

Find the slope and y intercept of each of the following lines and graph the lines.

(a) $5x - 4y = 12$ **(b)** $-y = 3x - 4$ **(c)** $y = 2$

Solution

(a) We change $5x - 4y = 12$ to slope-intercept form to get

$$5x - 4y = 12$$
$$-4y = -5x + 12$$
$$4y = 5x - 12$$
$$y = \frac{5}{4}x - 3$$

The slope of the line is $\frac{5}{4}$ (the coefficient of x) and the y intercept is -3 (the constant term). To graph the line, we plot the y intercept, -3. Then, because the slope is $\frac{5}{4}$, we can determine a second point, $(4, 2)$, by moving 5 units up and 4 units to the right from the y intercept. The graph is shown in Figure 8.39.

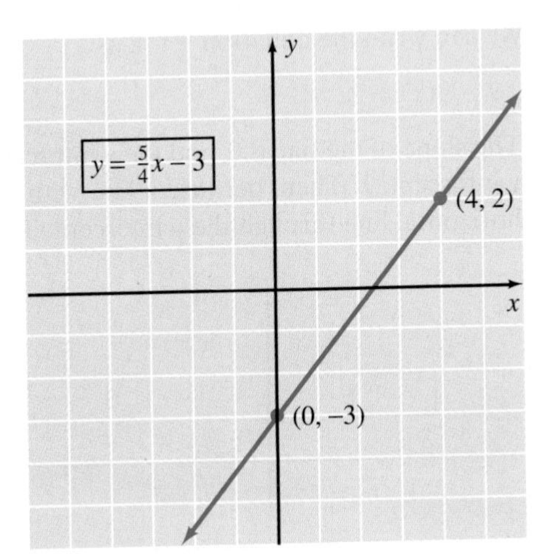

Figure 8.39

(b) We multiply both sides of the given equation by -1 to change it to slope-intercept form.

$$-y = 3x - 4$$
$$y = -3x + 4$$

The slope of the line is -3, and the y intercept is 4. To graph the line, we plot the y intercept, 4. Then, because the slope is $-3 = \frac{-3}{1}$, we can find a second point, (1, 1), by moving 3 units down and 1 unit to the right from the y intercept. The graph is shown in Figure 8.40.

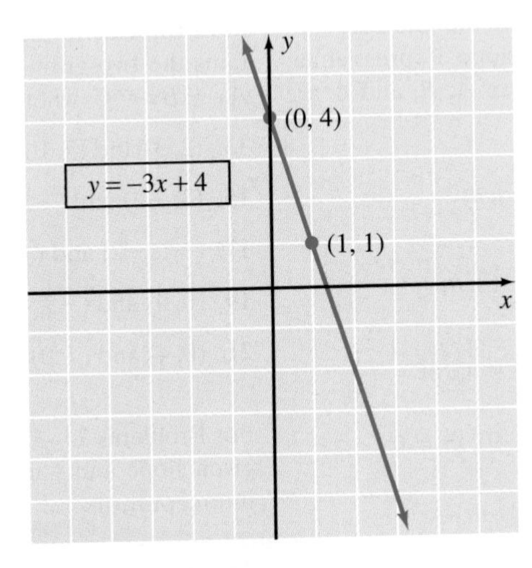

Figure 8.40

(c) We can write the equation $y = 2$ as

$$y = 0(x) + 2$$

The slope of the line is 0, and the y intercept is 2. To graph the line, we plot the y intercept, 2. Then, because a line with a slope of 0 is horizontal, we draw a horizontal line through the y intercept. The graph is shown in Figure 8.41.

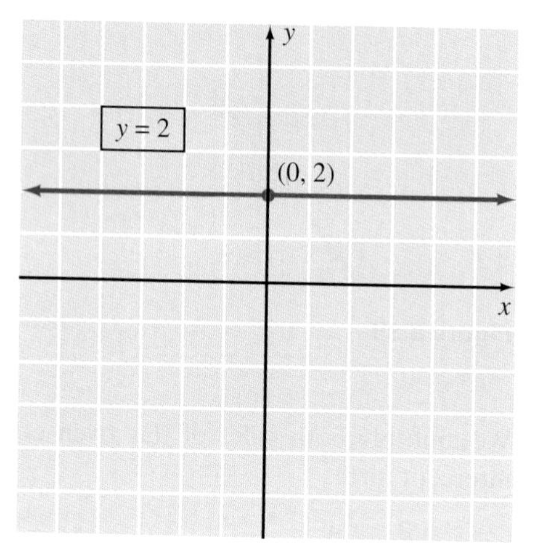

Figure 8.41

Problem Set 8.4

For Problems 1–12, find the equation of the line that contains the given point and has the given slope. Express equations in the form $Ax + By = C$, where A, B, and C are integers.

1. $(2, 3),\ m = \frac{2}{3}$

2. $(5, 2),\ m = \frac{3}{7}$

3. $(-3, -5),\ m = \frac{1}{2}$

4. $(5, -6),\ m = \frac{3}{5}$

5. $(-4, 8),\ m = -\frac{1}{3}$

6. $(-2, -4),\ m = -\frac{5}{6}$

7. $(3, -7),\ m = 0$

8. $(-3, 9),\ m = 0$

9. $(0, 0),\ m = -\frac{4}{9}$

10. $(0, 0),\ m = \frac{5}{11}$

11. $(-6, -2),\ m = 3$

12. $(2, -10),\ m = -2$

For Problems 13–22, find the equation of the line that contains the two given points. Express equations in the form $Ax + By = C$, where A, B, and C are integers.

13. $(2, 3)$ and $(7, 10)$

14. $(1, 4)$ and $(9, 10)$

15. $(3, -2)$ and $(-1, 4)$

16. $(-2, 8)$ and $(4, -2)$

17. $(-1, -2)$ and $(-6, -7)$

18. $(-8, -7)$ and $(-3, -1)$

19. $(0, 0)$ and $(-3, -5)$

20. $(5, -8)$ and $(0, 0)$

21. $(0, 4)$ and $(7, 0)$

22. $(-2, 0)$ and $(0, -9)$

For Problems 23–32, find the equation of the line with the given slope and y intercept. Leave your answers in slope-intercept form.

23. $m = \frac{3}{5}$ and $b = 2$

24. $m = \frac{5}{9}$ and $b = 4$

25. $m = 2$ and $b = -1$

26. $m = 4$ and $b = -3$

27. $m = -\frac{1}{6}$ and $b = -4$

28. $m = -\frac{5}{7}$ and $b = -1$

29. $m = -1$ and $b = \frac{5}{2}$

30. $m = -2$ and $b = \frac{7}{3}$

31. $m = -\frac{5}{9}$ and $b = -\frac{1}{2}$

32. $m = -\frac{7}{12}$ and $b = -\frac{2}{3}$

For Problems 33–44, determine the slope and y intercept of the line represented by the given equation, and graph the line.

33. $y = -2x - 5$

34. $y = \frac{2}{3}x + 4$

35. $3x - 5y = 15$

36. $7x + 5y = 35$

37. $-4x + 9y = 18$

38. $-6x + 7y = -14$

39. $-y = -\frac{3}{4}x + 4$

40. $5x - 2y = 0$

41. $-2x - 11y = 11$

42. $-y = \frac{2}{3}x + \frac{11}{2}$

43. $9x + 7y = 0$

44. $-5x - 13y = 26$

THOUGHTS INTO WORDS

45. Explain the importance of the slope-intercept form ($y = mx + b$) of the equation of a line.

46. What does it mean to say that two points "determine" a line?

47. How would you describe coordinate geometry to a group of elementary algebra students?

48. How can you tell by inspection that $y = 2x - 4$ and $y = -3x - 1$ are not parallel lines?

FURTHER INVESTIGATIONS

The following two properties pertain to parallel and perpendicular lines.

(a) Two nonvertical lines are parallel if and only if their slopes are equal.

(b) Two nonvertical lines are perpendicular if and only if the product of their slopes is -1.

For Problems 49–54, determine whether each pair of lines is (a) parallel, (b) perpendicular, or (c) intersecting lines that are not perpendicular.

49. $\begin{pmatrix} 5x - 2y = 6 \\ 2x + 5y = 9 \end{pmatrix}$

50. $\begin{pmatrix} 2x - y = 9 \\ 6x - 3y = 5 \end{pmatrix}$

51. $\begin{pmatrix} 4x - 3y = 12 \\ 3x - 4y = 12 \end{pmatrix}$

52. $\begin{pmatrix} 9x + 2y = 18 \\ 2x - 9y = 13 \end{pmatrix}$

53. $\begin{pmatrix} x - 3y = 7 \\ 5x - 15y = 9 \end{pmatrix}$

54. $\begin{pmatrix} 7x - 2y = 14 \\ 7x + 2y = 5 \end{pmatrix}$

55. Write the equation of the line that contains the point (4, 3) and is parallel to the line $2x - 3y = 6$.

56. Write the equation of the line that contains the point $(-1, 3)$ and is perpendicular to the line $3x - y = 4$.

8.5 Solving Linear Systems by Graphing

Suppose we graph $x - 2y = 4$ and $x + 2y = 8$ on the same set of axes, as shown in Figure 8.42. The ordered pair (6, 1), which is associated with the point of intersection of the two lines, satisfies both equations. That is to say, (6, 1) is the solution for $x - 2y = 4$ and $x + 2y = 8$. To check this, we can substitute 6 for x and 1 for y in both equations.

$$x - 2y = 4 \quad \text{becomes } 6 - 2(1) = 4$$

$$x + 2y = 8 \quad \text{becomes } 6 + 2(1) = 8$$

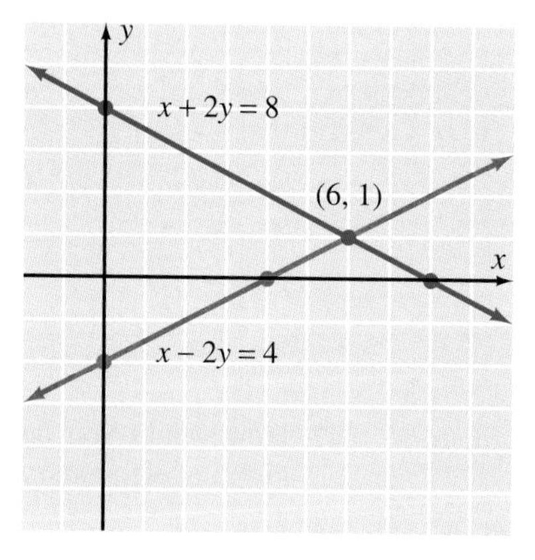

Figure 8.42

Thus we say that $\{(6, 1)\}$ is the solution set of the system

$$\begin{pmatrix} x - 2y = 4 \\ x + 2y = 8 \end{pmatrix}$$

Two or more linear equations in two variables considered together are called a **system of linear equations**. Here are three systems of linear equations:

$$\begin{pmatrix} x - 2y = 4 \\ x + 2y = 8 \end{pmatrix} \quad \begin{pmatrix} 5x - 3y = 9 \\ 3x + 7y = 12 \end{pmatrix} \quad \begin{pmatrix} 4x - y = 5 \\ 2x + y = 9 \\ 7x - 2y = 13 \end{pmatrix}$$

To **solve a system of linear equations** means to find all of the ordered pairs that are solutions of all of the equations in the system. There are several techniques for solving systems of linear equations. We will use three of them in this chapter—a graphing method in this section and two other methods in the following sections.

To solve a system of linear equations by **graphing**, we proceed as in the opening discussion of this section. We graph the equations on the same set of axes, and then the ordered pairs associated with any points of intersection are the solutions to the system. Let's consider another example.

EXAMPLE 1 Solve the system $\begin{pmatrix} x + y = 5 \\ x - 2y = -4 \end{pmatrix}$.

Solution

We can find the intercepts and a check point for each of the lines.

$x + y = 5$

x	y	
0	5	Intercepts
5	0	
2	3	Check point

$x - 2y = -4$

x	y	
0	5	Intercepts
−4	0	
−2	1	Check point

Figure 8.43 shows the graphs of the two equations. It appears that (2, 3) is the solution of the system. To check it we can substitute 2 for x and 3 for y in both equations.

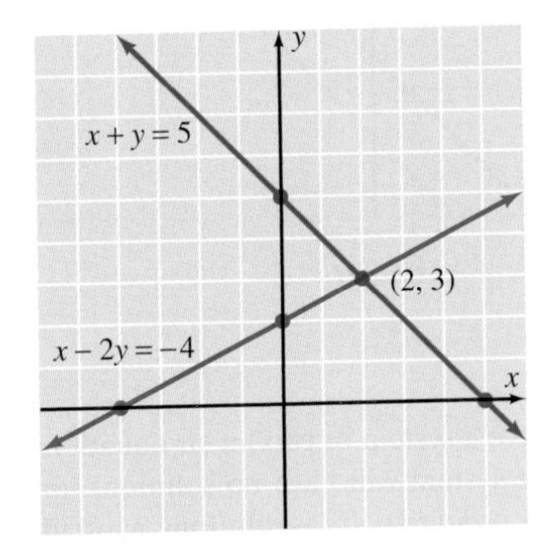

Figure 8.43

$x + y$ becomes $2 + 3 = 5$ A true statement

$x - 2y = -4$ becomes $2 - 2(3) = -4$ A true statement

Therefore, $\{(2, 3)\}$ is the solution set. ■

It should be evident that solving systems of equations by graphing requires accurate graphs. In fact, unless the solutions are integers it is really quite difficult to obtain exact solutions from a graph. For this reason the systems in this section have integer solutions. Furthermore, checking a solution takes on additional significance when the graphing approach is used. By checking you can be absolutely sure that you are *reading* the correct solution from the graph.

Figure 8.44 shows the three possible cases for the graph of a system of two linear equations in two variables.

Case I The graphs of the two equations are two lines intersecting in one point. There is one solution, and we call the system a **consistent system**.

Case II The graphs of the two equations are parallel lines. There is no solution, and we call the system an **inconsistent system**.

Case III The graphs of the two equations are the same line. There are infinitely many solutions to the system. Any pair of real numbers that satisfies one of the equations also satisfies the other equation, and we say the equations are **dependent**.

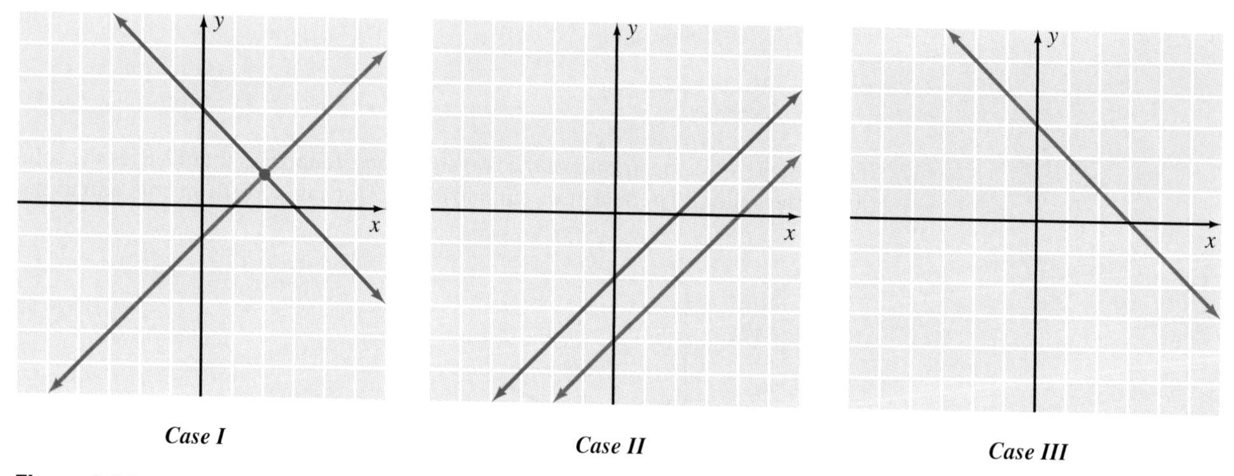

Case I *Case II* *Case III*

Figure 8.44

Thus as we solve a system of two linear equations in two variables, we know what to expect. The system will have no solutions, one ordered pair as a solution, or infinitely many ordered pairs as solutions. Most of the systems that we will be working with in this text have one solution.

An example of case I was given in Example 1 (Figure 8.43). The next two examples illustrate the other cases.

EXAMPLE 2

Solve the system $\begin{pmatrix} 2x + 3y = 6 \\ 2x + 3y = 12 \end{pmatrix}$.

Solution

$2x + 3y = 6$

x	y
0	2
3	0
−3	4

$2x + 3y = 12$

x	y
0	4
6	0
3	2

Figure 8.45 shows the graph of the system. Since the lines are parallel, there is no solution to the system. The solution set is $\varnothing$.

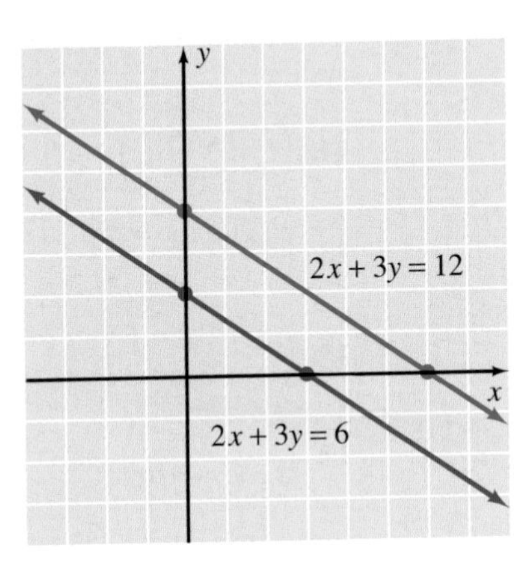

Figure 8.45

■

EXAMPLE 3

Solve the system $\begin{pmatrix} x + y = 3 \\ 2x + 2y = 6 \end{pmatrix}$.

Solution

$x + y = 3$

x	y
0	3
3	0
1	2

$2x + 2y = 6$

x	y
0	3
3	0
1	2

Figure 8.46 shows the graph of this system. Since the graphs of both equations are the same line, there are infinitely many solutions to the system. Any ordered pair of real numbers that satisfies one equation also satisfies the other equation.

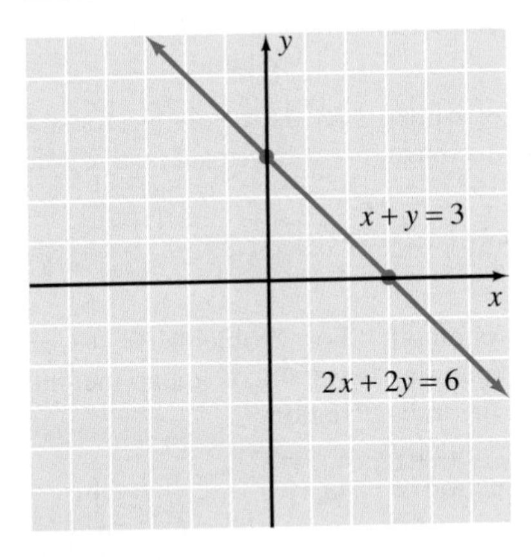

Figure 8.46

■

Problem Set 8.5

For Problems 1–10, decide whether the given ordered pair is a solution of the given system of equations.

1. $\begin{pmatrix} 5x + y = 9 \\ 3x - 2y = 4 \end{pmatrix}$, $(1, 4)$

2. $\begin{pmatrix} 3x - 2y = 8 \\ 2x + y = 3 \end{pmatrix}$, $(2, -1)$

3. $\begin{pmatrix} x - 3y = 17 \\ 2x + 5y = -21 \end{pmatrix}$, $(2, -5)$

4. $\begin{pmatrix} 2x + 7y = -5 \\ 4x - y = 6 \end{pmatrix}$, $(1, -1)$

5. $\begin{pmatrix} y = 2x \\ 3x - 4y = 5 \end{pmatrix}$, $(-1, -2)$

6. $\begin{pmatrix} -x + 2y = 8 \\ x - 2y = 8 \end{pmatrix}$, $(-2, 3)$

7. $\begin{pmatrix} 6x - 5y = 5 \\ 3x + 4y = -4 \end{pmatrix}$, $(0, -1)$

8. $\begin{pmatrix} y = 3x - 1 \\ 5x - 2y = -1 \end{pmatrix}$, $(3, 8)$

9. $\begin{pmatrix} -3x - y = 4 \\ -2x + 3y = -23 \end{pmatrix}$, $(4, -5)$

10. $\begin{pmatrix} 5x + 4y = -15 \\ 2x - 7y = -6 \end{pmatrix}$, $(-3, 0)$

For Problems 11–30, use the graphing method to solve each system.

11. $\begin{pmatrix} x + y = 1 \\ x - y = 3 \end{pmatrix}$

12. $\begin{pmatrix} x - y = 2 \\ x + y = -4 \end{pmatrix}$

13. $\begin{pmatrix} x + 2y = 4 \\ 2x - y = 3 \end{pmatrix}$

14. $\begin{pmatrix} 2x - y = -8 \\ x + y = 2 \end{pmatrix}$

15. $\begin{pmatrix} x + 3y = 6 \\ x + 3y = 3 \end{pmatrix}$

16. $\begin{pmatrix} y = -2x \\ y - 3x = 0 \end{pmatrix}$

17. $\begin{pmatrix} x + y = 0 \\ x - y = 0 \end{pmatrix}$

18. $\begin{pmatrix} 3x - y = 3 \\ 3x - y = -3 \end{pmatrix}$

19. $\begin{pmatrix} 3x - 2y = 5 \\ 2x + 5y = -3 \end{pmatrix}$

20. $\begin{pmatrix} 2x + 3y = 1 \\ 4x - 3y = -7 \end{pmatrix}$

21. $\begin{pmatrix} y = -2x + 3 \\ 6x + 3y = 9 \end{pmatrix}$

22. $\begin{pmatrix} y = 2x + 5 \\ x + 3y = -6 \end{pmatrix}$

23. $\begin{pmatrix} y = 5x - 2 \\ 4x + 3y = 13 \end{pmatrix}$

24. $\begin{pmatrix} y = x - 2 \\ 2x - 2y = 4 \end{pmatrix}$

25. $\begin{pmatrix} y = 4 - 2x \\ y = 7 - 3x \end{pmatrix}$

26. $\begin{pmatrix} y = 3x + 4 \\ y = 5x + 8 \end{pmatrix}$

27. $\begin{pmatrix} y = 2x \\ 3x - 2y = -2 \end{pmatrix}$

28. $\begin{pmatrix} y = 3x \\ 4x - 3y = 5 \end{pmatrix}$

29. $\begin{pmatrix} 7x - 2y = -8 \\ x = -2 \end{pmatrix}$

30. $\begin{pmatrix} 3x + 8y = -1 \\ y = -2 \end{pmatrix}$

THOUGHTS INTO WORDS

31. Discuss the strengths and weaknesses of solving a system of linear equations by graphing.

32. Determine a system of two linear equations for which the solution is (5, 7). Do any other systems have the same solution set? If so, find at least one more system.

33. Is it possible for a system of two linear equations to have exactly two solutions? Defend your answer.

FURTHER INVESTIGATIONS

The graphing approach works very well to solve a system of linear inequalities. The solution set of a system of linear inequalities is the intersection of the solution sets of the individual inequalities. Therefore, we can graph each inequality and then shade the region that represents the intersection of the solution sets of each inequality.

For Problems 34–43, indicate the solution set for each system of linear inequalities by shading the appropriate region.

34. $\begin{pmatrix} 2x + 3y > 6 \\ x - y < 2 \end{pmatrix}$

35. $\begin{pmatrix} x - 2y < 4 \\ 3x + y > 3 \end{pmatrix}$

36. $\begin{pmatrix} x - 3y \geq 3 \\ 3x + \ \ y \leq 3 \end{pmatrix}$

37. $\begin{pmatrix} 4x + 3y \leq 12 \\ 4x - \ \ y \geq \ \ 4 \end{pmatrix}$

38. $\begin{pmatrix} y \geq 2x \\ y < \ \ x \end{pmatrix}$

39. $\begin{pmatrix} y \leq -x \\ y > -3x \end{pmatrix}$

40. $\begin{pmatrix} y < -x + 1 \\ y > -x - 1 \end{pmatrix}$

41. $\begin{pmatrix} y > x - 2 \\ y < x + 3 \end{pmatrix}$

42. $\begin{pmatrix} y < \frac{1}{2}x + 2 \\ y < \frac{1}{2}x - 1 \end{pmatrix}$

43. $\begin{pmatrix} y > -\frac{1}{2}x - 2 \\ y > -\frac{1}{2}x + 1 \end{pmatrix}$

8.6 Elimination-by-Addition Method

We have used the addition property of equality (if $a = b$, then $a + c = b + c$) to help solve equations that contain one variable. An extension of the addition property forms the basis for another method of solving systems of linear equations. Property 8.1 states that when we add two equations, the resulting equation will be equivalent to the original ones. We can use this property to help solve a system of equations.

Property 8.1

For all real numbers a, b, c, and d, if $a = b$ and $c = d$, then

$$a + c = b + d$$

EXAMPLE 1

Solve $\begin{pmatrix} x + y = 12 \\ x - y = \ \ 2 \end{pmatrix}$.

Solution

$$\begin{aligned} x + y &= 12 \\ x - y &= \ \ 2 \\ \hline 2x \qquad &= 14 \end{aligned}$$ This is the result of adding the two equations

Solving this new equation in one variable, we obtain

$$2x = 14$$

$$\boxed{x = 7}$$

Now we can substitute the value of 7 for x in one of the original equations. Thus

$$x + y = 12$$

$$7 + y = 12$$

$$\boxed{y = 5}$$

To check, we see that

$$x + y = 12 \quad \text{and} \quad x - y = 2$$

$$7 + 5 = 12 \quad \text{and} \quad 7 - 5 = 2$$

Thus, the solution set of the system is $\{(7, 5)\}$. ■

Note in Example 1 that by adding the two original equations, we obtained a simple equation that contains only one variable. Adding equations to eliminate a variable is the key idea behind the **elimination-by-addition method** for solving systems of linear equations. The next example further illustrates this point.

EXAMPLE 2

Solve $\begin{pmatrix} 2x + 3y = -26 \\ 4x - 3y = 2 \end{pmatrix}$.

Solution

$$\begin{aligned} 2x + 3y &= -26 \\ 4x - 3y &= 2 \\ \hline 6x \quad &= -24 \quad \text{Added the two equations} \\ x &= -4 \end{aligned}$$

We substitute -4 for x in one of the two original equations.

$$\begin{aligned} 2x + 3y &= -26 \\ 2(-4) + 3y &= -26 \\ -8 + 3y &= -26 \\ 3y &= -18 \\ y &= -6 \end{aligned}$$

The solution set is $\{(-4, -6)\}$. ■

It may be necessary to change the form of one, or perhaps both, of the original equations before adding them. The next example demonstrates this idea.

EXAMPLE 3

Solve $\begin{pmatrix} y = x - 21 \\ x + y = -3 \end{pmatrix}$.

Solution

$y = x - 21$ — Subtract x from both sides → $-x + y = -21$

$x + y = -3$ — Leave alone → $x + y = -3$

$$\begin{aligned} -x + y &= -21 \\ x + y &= -3 \\ \hline 2y &= -24 \\ y &= -12 \end{aligned}$$

We substitute -12 for y in one of the original equations.

$$x + y = -3$$
$$x + (-12) = -3$$
$$\boxed{x = 9}$$

The solution set is $\{(9, -12)\}$. ■

Frequently, the multiplication property of equality needs to be applied first so that adding the equations will eliminate a variable, as the next example illustrates.

EXAMPLE 4

Solve $\begin{pmatrix} 2x + 5y = 29 \\ 3x - y = 1 \end{pmatrix}$.

Solution

Notice that adding the equations as they are would not eliminate a variable. However, we observe that multiplying the bottom equation by 5 and then adding this newly formed, but equivalent, equation to the top equation will eliminate the y variable terms.

$$\begin{array}{lcr} 2x + 5y = 29 & \xrightarrow{\text{Leave alone}} & 2x + 5y = 29 \\ 3x - y = 1 & \xrightarrow{\text{Multiply both sides by 5}} & 15x - 5y = 5 \\ & & \hline 17x = 34 \\ & & \boxed{x = 2} \end{array}$$

We substitute 2 for x in one of the original equations.

$$3x - y = 1$$
$$3(2) - y = 1$$
$$6 - y = 1$$
$$-y = -5$$
$$\boxed{y = 5}$$

The solution set is $\{(2, 5)\}$. ■

Notice in these problems that after finding the value of one of the variables, we substitute this number into one of the *original* equations to find the value of the other variable. It doesn't matter which of the two original equations you use, so pick the easiest one to solve.

Sometimes we need to apply the multiplication property to both equations. Let's look at an example of this type.

EXAMPLE 5

Solve $\begin{pmatrix} 2x + 3y = 4 \\ 9x - 2y = -13 \end{pmatrix}$.

Solution A

If we want to eliminate the x variable terms, then we want their coefficients to be equal in number (absolute value) and opposite in sign. To eliminate the x variable terms we can multiply the top equation by 9, and multiply the bottom equation by -2.

$$\begin{array}{lcr} 2x + 3y = 4 & \xrightarrow{\text{Multiply both sides by 9}} & 18x + 27y = 36 \\ 9x - 2y = -13 & \xrightarrow{\text{Multiply both sides by } -2} & -18x + 4y = 26 \\ & & \hline 31y = 62 \\ & & y = 2 \end{array}$$

We substitute 2 for y in one of the original equations.

$$\begin{aligned} 2x + 3y &= 4 \\ 2x + 3(2) &= 4 \\ 2x + 6 &= 4 \\ 2x &= -2 \\ x &= -1 \end{aligned}$$

The solution set is $\{(-1, 2)\}$.

Solution B

If we want to eliminate the y variable terms, then we want their coefficients to be equal in number (absolute value) and opposite in sign. To eliminate the y variable terms we can multiply the top equation by 2, and multiply the bottom equation by 3.

$$\begin{array}{lcr} 2x + 3y = 4 & \xrightarrow{\text{Multiply both sides by 2}} & 4x + 6y = 8 \\ 9x - 2y = -13 & \xrightarrow{\text{Multiply both sides by 3}} & 27x - 6y = -39 \\ & & \hline 31x = -31 \\ & & x = -1 \end{array}$$

We substitute -1 for x in one of the original equations.

$$\begin{aligned} 2x + 3y &= 4 \\ 2(-1) + 3y &= 4 \\ -2 + 3y &= 4 \\ 3y &= 6 \\ y &= 2 \end{aligned}$$

The solution set is $\{(-1, 2)\}$. ■

Look carefully at Solutions A and B for Example 5. Especially notice the first steps, where we applied the multiplication property of equality to the two equations. In Solution A we multiplied by numbers so that adding the resulting equations eliminated the x variable. In Solution B we multiplied so that the y variable was eliminated when we added the resulting equations. Either approach will work; pick the one that involves the easiest computation.

EXAMPLE 6

Solve $\begin{pmatrix} 3x - 4y = 7 \\ 5x + 3y = 9 \end{pmatrix}$.

Solution

Let's eliminate the y variable terms by multiplying the top equation by 3 and the bottom equation by 4.

$$3x - 4y = 7 \quad \xrightarrow{\text{Multiply both sides by 3}} \quad 9x - 12y = 21$$

$$5x + 3y = 9 \quad \xrightarrow{\text{Multiply both sides by 4}} \quad 20x + 12y = 36$$

$$29x = 57$$

$$x = \frac{57}{29}$$

Since substituting $\frac{57}{29}$ for x in one of the original equations will produce some messy calculations, let's solve for y by eliminating the x variable terms.

$$3x - 4y = 7 \quad \xrightarrow{\text{Multiply both sides by } -5} \quad -15x + 20y = -35$$

$$5x + 3y = 9 \quad \xrightarrow{\text{Multiply both sides by 3}} \quad 15x + 9y = 27$$

$$29y = -8$$

$$y = -\frac{8}{29}$$

The solution set is $\left\{\left(\frac{57}{29}, -\frac{8}{29}\right)\right\}$. ■

Problem Solving

Many word problems that we solved earlier in this text using one equation in one variable can also be solved using a system of two linear equations in two variables. In fact, many times you may find that it seems quite natural to use two variables. It may also seem more meaningful at times to use variables other than x and y. Let's consider two examples.

PROBLEM 1

The cost of 3 tennis balls and 2 golf balls is \$7. Furthermore, the cost of 6 tennis balls and 3 golf balls is \$12. Find the cost of 1 tennis ball and the cost of 1 golf ball.

Solution

We can use t to represent the cost of one tennis ball and g the cost of one golf ball. The problem translates into the following system of equations.

$3t + 2g = 7$ The cost of 3 tennis balls and 2 golf balls is \$7

$6t + 3g = 12$ The cost of 6 tennis balls and 3 golf balls is \$12

Solving this system by the elimination-by-addition method, we obtain

$3t + 2g = 7$ Multiply both sides by -2 → $-6t - 4g = -14$

$6t + 3g = 12$ Leave alone → $6t + 3g = 12$

$$-g = -2$$

$$g = 2$$

We substitute 2 for g in one of the original equations.

$$3t + 2g = 7$$

$$3t + 2(2) = 7$$

$$3t + 4 = 7$$

$$3t = 3$$

$$t = 1$$

The cost of a tennis ball is \$1 and the cost of a golf ball is \$2. ■

PROBLEM 2

Niki invested \$5000, part of it at 4% interest and the rest at 6%. Her total interest earned for a year was \$265. How much did she invest at each rate?

Solution

We let x represent the amount invested at 4%, and we let y represent the amount invested at 6%. The problem translates into the following system:

$x + y = 5000$ Niki invested \$5000

$0.04x + 0.06y = 265$ Her total interest earned for a year was \$265

Multiplying the second equation by 100 produces $4x + 6y = 26{,}500$. Then we have the following equivalent system to solve:

$$\begin{pmatrix} x + y = 5000 \\ 4x + 6y = 26{,}500 \end{pmatrix}$$

Using the elimination-by-addition method, we can proceed as follows:

$$\begin{array}{rcl} x + y = 5000 & \xrightarrow{\text{Multiply both sides by } -4} & -4x - 4y = -20{,}000 \\ 4x + 6y = 26{,}500 & \xrightarrow{\text{Leave alone}} & 4x + 6y = 26{,}500 \\ & & \hline 2y = 6500 \\ & & y = 3250 \end{array}$$

Substituting 3250 for y in $x + y = 5000$ yields

$$x + y = 5000$$

$$x + 3250 = 5000$$

$$x = 1750$$

Therefore, we know that Niki invested \$1750 at 4% interest and \$3250 at 6%. ■

Before you tackle the word problems in this next problem set, it might be helpful to review the problem-solving suggestions we offered in Section 3.3. Those suggestions continue to apply here except that now you have the flexibility of using two equations and two unknowns. Don't forget that to check a word problem you need to see whether your answers satisfy the conditions stated in the original problem.

Problem Set 8.6

For Problems 1–24, solve each system using the elimination-by-addition method.

1. $\begin{pmatrix} x + y = 14 \\ x - y = -2 \end{pmatrix}$

2. $\begin{pmatrix} x - y = -14 \\ x + y = 6 \end{pmatrix}$

3. $\begin{pmatrix} x + 4y = -21 \\ 3x - 4y = 1 \end{pmatrix}$

4. $\begin{pmatrix} -3x + 2y = -21 \\ 3x - 7y = 36 \end{pmatrix}$

5. $\begin{pmatrix} y = 6 - x \\ x - y = -18 \end{pmatrix}$

6. $\begin{pmatrix} x + y = -10 \\ x = y + 6 \end{pmatrix}$

7. $\begin{pmatrix} 5x + y = 23 \\ 3x - 2y = 19 \end{pmatrix}$

8. $\begin{pmatrix} 4x - 5y = -36 \\ x + 2y = 30 \end{pmatrix}$

9. $\begin{pmatrix} x + 2y = 5 \\ 3x - 2y = 6 \end{pmatrix}$

10. $\begin{pmatrix} 2x - y = 3 \\ -2x + 5y = 7 \end{pmatrix}$

11. $\begin{pmatrix} y = -x \\ 2x - y = -2 \end{pmatrix}$

12. $\begin{pmatrix} 3x = 2y \\ 8x + 20y = 19 \end{pmatrix}$

13. $\begin{pmatrix} 4x + 5y = 9 \\ 5x - 6y = -50 \end{pmatrix}$

14. $\begin{pmatrix} 2x + 7y = 46 \\ 3x - 4y = -18 \end{pmatrix}$

15. $\begin{pmatrix} 9x - 7y = 29 \\ 5x - 3y = 17 \end{pmatrix}$

16. $\begin{pmatrix} 7x + 5y = -6 \\ 4x + 3y = -4 \end{pmatrix}$

17. $\begin{pmatrix} 6x + 5y = -6 \\ 8x - 3y = 21 \end{pmatrix}$

18. $\begin{pmatrix} 3x - 8y = -23 \\ 7x + 4y = -48 \end{pmatrix}$

19. $\begin{pmatrix} 2x - 7y = -1 \\ 9x + 4y = -2 \end{pmatrix}$

20. $\begin{pmatrix} 6x - 2y = -3 \\ 5x - 9y = 1 \end{pmatrix}$

21. $\begin{pmatrix} x + y = 750 \\ 0.07x + 0.08y = 57.5 \end{pmatrix}$

22. $\begin{pmatrix} x + y = 700 \\ 0.06x + 0.09y = 54 \end{pmatrix}$

23. $\begin{pmatrix} 0.09x + 0.11y = 31 \\ y = x + 100 \end{pmatrix}$

24. $\begin{pmatrix} 0.08x + 0.1y = 56 \\ y = 2x \end{pmatrix}$

For Problems 25–40, solve each problem by setting up and solving a system of two linear equations in two variables.

25. The sum of two numbers is 30, and their difference is 12. Find the numbers.

26. The sum of two numbers is 20. If twice the smaller is subtracted from the larger, the result is 2. Find the numbers.

27. The difference of two numbers is 7. If three times the smaller is subtracted from twice the larger, the result is 6. Find the numbers.

28. The difference of two numbers is 17. If the larger is increased by three times the smaller, the result is 37. Find the numbers.

29. One number is twice another number. The sum of three times the smaller and five times the larger is 78. Find the numbers.

30. One number is three times another number. The sum of four times the smaller and seven times the larger is 175. Find the numbers.

31. Three lemons and 2 apples cost \$2.35. Two lemons and 3 apples cost \$2.40. Find the cost of 1 lemon and the cost of 1 apple.

32. Chip used two sites and downloaded 19 songs for his MP3 player for a total cost of \$20.00. The songs from Muchmore Music cost \$1.00 each, and the songs from iTunes cost \$1.20 each. How many songs did Chip download from each service?

33. Larry has \$1.45 in change in his pocket, consisting of dimes and quarters. If he has a total of ten coins, how many of each kind does he have?

34. Suppose that Cindy has 90 cents in change in her purse, consisting of nickels and dimes. If she has a total of 13 coins, how many of each kind does she have in her purse?

35. Suppose that a library buys a total of 35 books, which cost \$546. The hardcover books cost \$22 each, and the soft cover books cost \$6 per book. How many books of each type did the library buy?

36. Some punch that contains 10% grapefruit juice is mixed with some punch that contains 5% grapefruit juice to produce 10 quarts of punch that is 8% grapefruit juice. How many quarts of 10% and 5% grapefruit juice must be used?

37. A 10% salt solution is to be mixed with a 15% salt solution to produce 10 gallons of a 13% salt solution. How many gallons of the 10% solution and how many gallons of the 15% solution will be needed?

38. The income from a student production was \$101,000. The price of a student ticket was \$13, and nonstudent tickets were \$25 each. Five thousand people attended the production. How many tickets of each kind were sold?

39. Sidney invested \$13,000, part of it at 5% and the rest at 6%. If his total yearly interest was \$730, how much did he invest at each rate?

40. Heather invested \$1100, part of it at 8% and the rest at 9%. Her total interest earned for a year was \$95. How much did she invest at each rate?

THOUGHTS INTO WORDS

41. Explain how you would solve the system $\begin{pmatrix} 2x - 3y = 5 \\ 4x + 7y = 9 \end{pmatrix}$ using the elimination-by-addition method.

42. Give a general description of how to apply the elimination-by-addition method.

FURTHER INVESTIGATIONS

For Problems 43–46, solve each system using the elimination-by-addition method.

43. $\begin{pmatrix} \frac{1}{2}x - \frac{1}{3}y = -2 \\ \frac{3}{2}x + \frac{2}{3}y = 34 \end{pmatrix}$

44. $\begin{pmatrix} x - 2y = 0 \\ 3x + 5y = 0 \end{pmatrix}$

45. $\begin{pmatrix} x - 4y = 6 \\ 2x - 8y = 3 \end{pmatrix}$

46. $\begin{pmatrix} y = 1 - 2x \\ 6x + 3y = 3 \end{pmatrix}$

8.7 Substitution Method

A third method of solving systems of equations is called the **substitution method**. Like the addition method, it produces exact solutions and can be used on any system of linear equations; however, some systems lend themselves more to the substitution method than others. We will consider a few examples to demonstrate the use of the substitution method.

EXAMPLE 1

Solve $\begin{pmatrix} y = x + 10 \\ x + y = 14 \end{pmatrix}$.

Solution

Because the first equation states that y equals $x + 10$, we can substitute $x + 10$ for y in the second equation.

$$x + y = 14 \xrightarrow{\text{Substitute } x + 10 \text{ for } y} x + (x + 10) = 14$$

Now we have an equation with one variable that can be solved in the usual way.

$$x + (x + 10) = 14$$
$$2x + 10 = 14$$
$$2x = 4$$
$$x = 2$$

Substituting 2 for x in one of the original equations, we can find the value of y.

$$y = x + 10$$
$$y = 2 + 10$$
$$y = 12$$

The solution set is $\{(2, 12)\}$. ■

EXAMPLE 2

Solve $\begin{pmatrix} 3x + 5y = -7 \\ x = 2y + 5 \end{pmatrix}$.

Solution

Because the second equation states that x equals $2y + 5$, we can substitute $2y + 5$ for x in the first equation.

$$3x + 5y = -7 \xrightarrow{\text{Substitute } 2y + 5 \text{ for } x} 3(2y + 5) + 5y = -7$$

Solving this equation, we have

$$3(2y + 5) + 5y = -7$$
$$6y + 15 + 5y = -7$$

$$11y + 15 = -7$$
$$11y = -22$$
$$y = -2$$

Substituting -2 for y in one of the two original equations produces

$$x = 2y + 5$$
$$x = 2(-2) + 5$$
$$x = -4 + 5$$
$$x = 1$$

The solution set is $\{(1, -2)\}$. ■

Note that the key idea behind the substitution method is the elimination of a variable, but the elimination is done by a substitution rather than by addition of the equations. The substitution method is especially convenient to use when at least one of the equations is of the form *y equals* or *x equals*. In Example 1 the first equation is of the form *y equals*, and in Example 2 the second equation is of the form *x equals*. Let's consider another example using the substitution method.

EXAMPLE 3

Solve $\begin{pmatrix} 2x + 3y = -30 \\ y = \frac{2}{3}x - 6 \end{pmatrix}$.

Solution

The second equation allows us to substitute $\frac{2}{3}x - 6$ for y in the first equation.

$$2x + 3y = -30 \xrightarrow{\text{Substitute } \frac{2}{3}x - 6 \text{ for } y} 2x + 3\left(\frac{2}{3}x - 6\right) = -30$$

Solving this equation produces

$$2x + 3\left(\frac{2}{3}x - 6\right) = -30$$
$$2x + 2x - 18 = -30$$
$$4x - 18 = -30$$
$$4x = -12$$
$$x = -3$$

Now we can substitute -3 for x in one of the original equations.

$$y = \frac{2}{3}x - 6$$
$$y = \frac{2}{3}(-3) - 6$$

$$y = -2 - 6$$

$$\boxed{y = -8}$$

The solution set is $\{(-3, -8)\}$. ■

It may be necessary to change the form of one of the equations before we make a substitution. The following examples clarify this point.

EXAMPLE 4

Solve $\begin{pmatrix} 4x - 5y = 55 \\ x + y = -2 \end{pmatrix}$.

Solution

We can easily change the form of the second equation to make it ready for the substitution method.

$$x + y = -2$$

$$y = -2 - x \quad \text{Added } -x \text{ to both sides}$$

Now, we can substitute $-2 - x$ for y in the first equation.

$$4x - 5y = 55 \xrightarrow{\text{Substitute } -2 - x \text{ for } y} 4x - 5(-2 - x) = 55$$

Solving this equation, we obtain

$$4x - 5(-2 - x) = 55$$

$$4x + 10 + 5x = 55$$

$$9x + 10 = 55$$

$$9x = 45$$

$$\boxed{x = 5}$$

Substituting 5 for x in one of the *original* equations produces

$$x + y = -2$$

$$5 + y = -2$$

$$\boxed{y = -7}$$

The solution set is $\{(5, -7)\}$. ■

In Example 4, we could have started by changing the form of the first equation to make it ready for substitution. However, you should be able to look ahead and see that this would produce a fractional form to substitute. We were able to avoid any messy calculations with fractions by changing the form of the second equation instead of the first. Sometimes when using the substitution method, you cannot avoid fractional forms. The next example is a case in point.

EXAMPLE 5

Solve $\begin{pmatrix} 3x + 2y = 8 \\ 2x - 3y = -38 \end{pmatrix}$.

Solution

Looking ahead, we see that changing the form of either equation will produce a fractional form. Therefore, we will merely pick the first equation and solve for y.

$$3x + 2y = 8$$

$$2y = 8 - 3x \quad \text{Added } -3x \text{ to both sides}$$

$$y = \frac{8 - 3x}{2} \quad \text{Multiplied both sides by } \frac{1}{2}$$

Now we can substitute $\frac{8 - 3x}{2}$ for y in the second equation and determine the value of x.

$$2x - 3y = -38$$

$$2x - 3\left(\frac{8 - 3x}{2}\right) = -38$$

$$2x - \frac{24 - 9x}{2} = -38$$

$$4x - 24 + 9x = -76 \quad \text{Multiplied both sides by 2}$$

$$13x - 24 = -76$$

$$13x = -52$$

$$x = -4$$

Substituting -4 for x in one of the original equations, we have

$$3x + 2y = 8$$

$$3(-4) + 2y = 8$$

$$-12 + 2y = 8$$

$$2y = 20$$

$$y = 10$$

The solution set is $\{(-4, 10)\}$. ■

■ Which Method to Use

We have now studied three methods of solving systems of linear equations—the graphing method, the elimination-by-addition method, and the substitution method. As we indicated earlier, the graphing method is quite restrictive and works well only when the solutions are integers or when we need only approximate answers. Both the elimination-by-addition method and the substitution method can be used to obtain exact solutions for any system of linear equations in two variables. The method you choose may depend upon the original form of the equations. Next we consider two examples to illustrate this point.

EXAMPLE 6

Solve $\begin{pmatrix} 7x - 5y = -52 \\ y = 3x - 4 \end{pmatrix}$.

Solution

Because the second equation indicates that we can substitute $3x - 4$ for y, this system lends itself to the substitution method.

$$7x - 5y = -52 \xrightarrow{\text{Substitute } 3x - 4 \text{ for } y} 7x - 5(3x - 4) = -52$$

Solving this equation, we obtain

$$7x - 5(3x - 4) = -52$$
$$7x - 15x + 20 = -52$$
$$-8x + 20 = -52$$
$$-8x = -72$$
$$x = 9$$

Substituting 9 for x in one of the original equations produces

$$y = 3x - 4$$
$$y = 3(9) - 4$$
$$y = 27 - 4$$
$$y = 23$$

The solution set is $\{(9, 23)\}$. ■

EXAMPLE 7

Solve $\begin{pmatrix} 10x + 7y = 19 \\ 2x - 6y = -11 \end{pmatrix}$.

Solution

Because changing the form of either of the two equations in preparation for the substitution method would produce a fractional form, we are probably better off to use the elimination-by-addition method. Furthermore, we should notice that the coefficients of x lend themselves to this method.

$$10x + 7y = 19 \xrightarrow{\text{Leave alone}} 10x + 7y = 19$$
$$2x - 6y = -11 \xrightarrow{\text{Multiply by } -5} -10x + 30y = 55$$
$$37y = 74$$
$$y = 2$$

Substituting 2 for y in the first equation of the given system produces

$$10x + 7y = 19$$
$$10x + 7(2) = 19$$
$$10x + 14 = 19$$
$$10x = 5$$
$$x = \frac{5}{10} = \frac{1}{2}$$

The solution set is $\left\{\left(\frac{1}{2}, 2\right)\right\}$. ■

In Section 8.5, we explained that you can tell by graphing the equations whether the system has no solutions, one solution, or infinitely many solutions. That is, the two lines may be parallel (no solutions), they may intersect in one point (one solution), or they may coincide (infinitely many solutions). From a practical viewpoint, the systems that have one solution deserve most of our attention. However, we do need to be able to deal with the other situations as they arise. The next two examples demonstrate what occurs when we hit a "no solution" or "infinitely many solutions" situation when we are using either the elimination-by-addition or substitution method.

EXAMPLE 8

Solve the system $\begin{pmatrix} y = 2x - 1 \\ 6x - 3y = 7 \end{pmatrix}$.

Solution

Because the first equation indicates that we can substitute $2x - 1$ for y, this system lends itself to the substitution method.

$$6x - 3y = 7 \quad \xrightarrow{\text{Substitute } 2x - 1 \text{ for } y} \quad 6x - 3(2x - 1) = 7$$

Now we solve this equation.

$$6x - 6x + 3 = 7$$
$$0 + 3 = 7$$
$$3 = 7$$

The false numerical statement, $3 = 7$, implies that the system has no solutions. Thus the solution set is $\emptyset$. (You may want to graph the two lines to verify this conclusion!) ■

EXAMPLE 9

Solve the system $\begin{pmatrix} 2x - 3y = 4 \\ 10x - 15y = 20 \end{pmatrix}$.

Solution

We use the elimination-by-addition method.

$$\begin{array}{rcr} 2x - 3y = 4 & \xrightarrow{\text{Multiply both sides by } -5} & -10x + 15y = -20 \\ 10x - 15y = 20 & \xrightarrow{\text{Leave alone}} & 10x - 15y = 20 \\ \hline & & 0 + 0 = 0 \end{array}$$

The true numerical statement, $0 + 0 = 0$, implies that the system has infinitely many solutions. Any ordered pair that satisfies one of the equations also satisfies the other equation. ■

Problem Solving

We will conclude this section with three word problems.

PROBLEM 1

The length of a rectangle is 1 centimeter less than three times the width. The perimeter of the rectangle is 94 centimeters. Find the length and width of the rectangle.

Solution

We let w represent the width of the rectangle, and we let l represent the length of the rectangle (see Figure 8.47). The problem translates into the following system of equations:

$l = 3w - 1$ The length of the rectangle is 1 centimeter less than three times the width

$2l + 2w = 94$ The perimeter of the rectangle is 94 centimeters

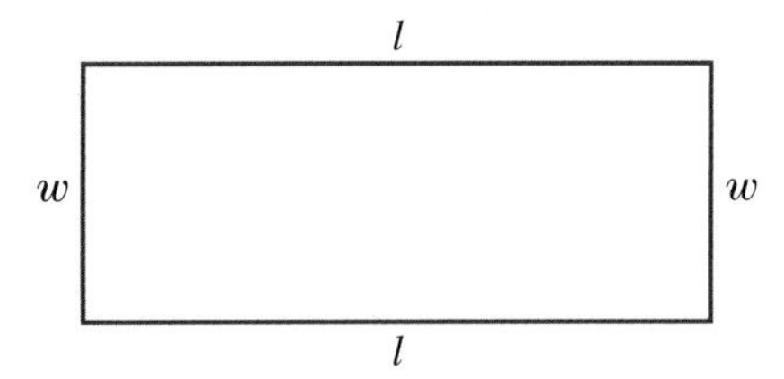

Figure 8.47

Multiplying both sides of the second equation by one-half produces the equivalent equation $l + w = 47$, so we have the following system to solve:

$$\begin{pmatrix} l = 3w - 1 \\ l + w = 47 \end{pmatrix}$$

The first equation indicates that we can substitute $3w - 1$ for l in the second equation.

$$l + w = 47 \xrightarrow{\text{Substitute } 3w - 1 \text{ for } l} 3w - 1 + w = 47$$

Solving this equation yields

$$3w - 1 + w = 47$$

$$4w - 1 = 47$$

$$4w = 48$$

$$w = 12$$

Substituting 12 for w in one of the original equations produces

$$l = 3w - 1$$

$$l = 3(12) - 1$$

$$l = 36 - 1$$

$$l = 35$$

The rectangle is 12 centimeters wide and 35 centimeters long. ■

PROBLEM 2

The proceeds from selling hamburgers and hot dogs at the baseball game were \$575.50. The price of a hot dog was \$2.50, and the price of a hamburger was \$3.00. If a total of 213 hot dogs and hamburgers were sold, how many of each kind were sold?

Solution

Let x equal the number of hot dogs sold, and let y equal the number of hamburgers sold. The problem translates into this system:

The number sold → $x + y = 213$

The proceeds from the sales → $2.50x + 3.00y = 575.50$

$$\begin{pmatrix} x + y = 213 \\ 2.50x + 3.00y = 575.50 \end{pmatrix}$$

Let's begin by solving the first equation for y.

$$x + y = 213$$

$$y = 213 - x$$

Now we will substitute $213 - x$ for y in the second equation and solve for x.

$$2.50x + 3.00(213 - x) = 575.50$$

$$2.50x + 639.00 - 3.00x = 575.50$$

$$-0.5x + 639.00 = 575.50$$

$$-0.5x = -63.50$$

$$x = 127$$

Therefore, there were 127 hot dogs sold and $213 - 127 = 86$ hamburgers sold. ■

Problem Set 8.7

For Problems 1–26, solve each system using the substitution method.

1. $\begin{pmatrix} y = 2x - 1 \\ x + y = 14 \end{pmatrix}$

2. $\begin{pmatrix} y = 3x + 4 \\ x + y = 52 \end{pmatrix}$

3. $\begin{pmatrix} x - y = -14 \\ y = -3x - 2 \end{pmatrix}$

4. $\begin{pmatrix} x - y = -23 \\ y = -2x + 5 \end{pmatrix}$

5. $\begin{pmatrix} 4x - 3y = -6 \\ y = -2x + 7 \end{pmatrix}$

6. $\begin{pmatrix} 8x - y = -8 \\ y = 4x + 5 \end{pmatrix}$

7. $\begin{pmatrix} x + y = 1 \\ 3x + 6y = 7 \end{pmatrix}$

8. $\begin{pmatrix} 2x - 4y = -9 \\ x + y = 3 \end{pmatrix}$

9. $\begin{pmatrix} 2x - y = 12 \\ x = \frac{3}{4}y \end{pmatrix}$

10. $\begin{pmatrix} 4x - 5y = 6 \\ y = \frac{2}{3}x \end{pmatrix}$

11. $\begin{pmatrix} y = \frac{3}{2}x \\ 6x - 5y = 15 \end{pmatrix}$

12. $\begin{pmatrix} x = \frac{3}{5}y \\ 4x - 3y = 12 \end{pmatrix}$

13. $\begin{pmatrix} 4y - 1 = x \\ 2x - 8y = 3 \end{pmatrix}$

14. $\begin{pmatrix} y = 5x - 2 \\ y = 2x + 7 \end{pmatrix}$

15. $\begin{pmatrix} 7x + 2y = -2 \\ 6x + 5y = 18 \end{pmatrix}$

16. $\begin{pmatrix} 2x + 3y = 31 \\ 3x + 5y = -20 \end{pmatrix}$

17. $\begin{pmatrix} 8x - 3y = -9 \\ x + 5y = -71 \end{pmatrix}$

18. $\begin{pmatrix} 9x - 2y = -18 \\ 4x - y = -7 \end{pmatrix}$

19. $\begin{pmatrix} 4x - 6y = 1 \\ 2x + 3y = 4 \end{pmatrix}$

20. $\begin{pmatrix} 2x + 3y = 1 \\ 4x + 6y = 2 \end{pmatrix}$

21. $\begin{pmatrix} 5x + 7y = 3 \\ 3x - 2y = 0 \end{pmatrix}$

22. $\begin{pmatrix} 7x - 3y = 4 \\ 2x + 5y = 0 \end{pmatrix}$

23. $\begin{pmatrix} 0.05x + 0.07y = 33 \\ y = x + 300 \end{pmatrix}$

24. $\begin{pmatrix} 0.06x + 0.08y = 15 \\ y = x + 100 \end{pmatrix}$

25. $\begin{pmatrix} x + y = 13 \\ 0.05x + 0.1y = 1.15 \end{pmatrix}$

26. $\begin{pmatrix} x + y = 17 \\ 0.1x + 0.25y = 3.2 \end{pmatrix}$

For Problems 27–46, solve each system using either the elimination-by-addition or the substitution method, whichever seems more appropriate to you.

27. $\begin{pmatrix} 5x - 4y = 14 \\ 7x + 3y = -32 \end{pmatrix}$

28. $\begin{pmatrix} 2x + 3y = 13 \\ 3x - 5y = -28 \end{pmatrix}$

29. $\begin{pmatrix} 2x + 9y = 6 \\ y = -x \end{pmatrix}$

30. $\begin{pmatrix} y = 3x + 2 \\ 4x - 3y = -21 \end{pmatrix}$

31. $\begin{pmatrix} x + y = 22 \\ 0.6x + 0.5y = 12 \end{pmatrix}$

32. $\begin{pmatrix} x + y = 13 \\ 0.4x + 0.5y = 6 \end{pmatrix}$

33. $\begin{pmatrix} 4x - y = 0 \\ 7x + 2y = 9 \end{pmatrix}$

34. $\begin{pmatrix} x - 2y = 0 \\ 5x = 8y + 12 \end{pmatrix}$

35. $\begin{pmatrix} 2x + y = 1 \\ 6x - 7y = -57 \end{pmatrix}$

36. $\begin{pmatrix} 7x - 9y = 11 \\ x - 3y = 3 \end{pmatrix}$

37. $\begin{pmatrix} 6x - y = -1 \\ 10x + 2y = 13 \end{pmatrix}$

38. $\begin{pmatrix} x + 4y = -5 \\ 6x - 5y = -1 \end{pmatrix}$

39. $\begin{pmatrix} 4x + 8y = 20 \\ x + 2y = 5 \end{pmatrix}$

40. $\begin{pmatrix} x = 5y - 5 \\ 2x - 10y = 2 \end{pmatrix}$

41. $\begin{pmatrix} 3x - 8y = -5 \\ x = 2y \end{pmatrix}$

42. $\begin{pmatrix} x + y = -3 \\ 5x + 6y = -22 \end{pmatrix}$

43. $\begin{pmatrix} 5y - 2x = -4 \\ 10y = 3x + 4 \end{pmatrix}$

44. $\begin{pmatrix} 3x - 7 = 2y + 15 \\ 3x = 6y + 18 \end{pmatrix}$

45. $\begin{pmatrix} x = -y - 1 \\ 6x - 5y = 4 \end{pmatrix}$

46. $\begin{pmatrix} 4x + 7y = 1 \\ y = 2x + 3 \end{pmatrix}$

For Problems 47–60, solve each problem by setting up and solving a system of two linear equations in two variables. Use either the elimination-by-addition or the substitution method to solve the systems.

47. Suppose that the sum of two numbers is 46, and the difference of the numbers is 22. Find the numbers.

48. The sum of two numbers is 52. The larger number is 2 more than four times the smaller number. Find the numbers.

49. Suppose that a youth hostel rents double rooms at \$35 per day and single rooms at \$25 per day. If a total of 50 rooms was rented one day for \$1600, how many of each kind were rented?

50. The total receipts from ticket sales at a concert amounted to \$15,600. Student tickets were sold at \$24 each and nonstudent tickets at \$36 each. The number

of student tickets sold was five times the number of nonstudent tickets sold. How many student tickets and how many nonstudent tickets were sold?

51. In a class of 50 students, the number of women is 2 more than five times the number of men. How many women are there in the class?

52. In a recent survey, 1000 registered voters were asked about their political preferences. The number of men in the survey was 5 less than one-half the number of women. Find the number of men in the survey.

53. The perimeter of a rectangle is 94 inches. The length of the rectangle is 7 inches more than the width. Find the dimensions of the rectangle.

54. Two angles are supplementary, and the measure of one of them is 20° less than three times the measure of the other angle. Find the measure of each angle.

55. A deposit slip listed \$700 in cash to be deposited. There were 100 bills, some of them five-dollar bills and the remainder ten-dollar bills. How many bills of each denomination were deposited?

56. Suppose that Larry has a number of dimes and quarters totaling \$12.05. The number of quarters is 5 more than twice the number of dimes. How many coins of each kind does he have?

57. Suppose that Sue has three times as many nickels as pennies in her collection. Together her pennies and nickels have a value of \$4.80. How many pennies and how many nickels does she have?

58. Tina invested some money at 8% interest and some money at 9%. She invested \$250 more at 9% than she invested at 8%. Her total yearly interest from the two investments was \$48. How much did Tina invest at each rate?

59. One solution contains 40% alcohol, and a second solution contains 70% alcohol. How many liters of each solution should be mixed to make 30 liters that contain 50% alcohol?

60. One solution contains 30% alcohol, and a second solution contains 70% alcohol. How many liters of each solution should be mixed to make 10 liters containing 40% alcohol

61. The length of a rectangle is 3 meters less than four times the width. The perimeter of the rectangle is 74 meters. Find the length and width of the rectangle.

THOUGHTS INTO WORDS

62. Explain how you would solve the system $\begin{pmatrix} 5x - 4y = 10 \\ 3x - y = 6 \end{pmatrix}$ using the substitution method.

63. How do you decide whether to solve a system of linear equations by using the elimination-by-addition method or the substitution method?

64. What do you see as the strengths and weaknesses of the elimination-by-addition method and the substitution method?

FURTHER INVESTIGATIONS

For Problems 65–68, use the substitution method to solve each system.

65. $\begin{pmatrix} y = \frac{2}{3}x - \frac{1}{2} \\ \frac{3}{2}x - y = -\frac{1}{3} \end{pmatrix}$

66. $\begin{pmatrix} 5x - 3y = 0 \\ 4x + 7y = 0 \end{pmatrix}$

67. $\begin{pmatrix} y = 2x - 1 \\ 6x - 3y = 3 \end{pmatrix}$

68. $\begin{pmatrix} x = -4y + 5 \\ 2x + 8y = -1 \end{pmatrix}$

Chapter 8 Summary

(8.1) The **Cartesian** or **rectangular coordinate system** involves a one-to-one correspondence between ordered pairs of real numbers and the points of a plane. The system provides the basis for a study of coordinate geometry, which links algebra and geometry and deals with two basic kinds of problems:

1. Given an algebraic equation, find its geometric graph.
2. Given a set of conditions pertaining to a geometric figure, find its algebraic equation.

(8.2) Any equation of the form $Ax + By = C$, where A, B, and C are constants (A and B not both zero) and x and y are variables, is a **linear equation in two variables** and its graph is a **straight line**.

To **graph** a linear equation, we can find two solutions (the intercepts are usually easy to determine), plot the corresponding points, and connect the points with a straight line.

(8.3) If points P_1 and P_2 with coordinates (x_1, y_1) and (x_2, y_2), respectively, are any two points on a line, then the **slope** of the line (denoted by m) is given by

$$m = \frac{y_2 - y_1}{x_2 - x_1}, \qquad x_1 \neq x_2$$

The slope of a line is a **ratio** of vertical change to horizontal change. The slope of a line can be negative, positive, or zero. The concept of slope is not defined for vertical lines.

(8.4) You should review Examples 1, 2, and 3 of this section to pull together a general approach to writing equations of lines given certain conditions. The equation

$$y - y_1 = m(x - x_1)$$

is called the **point-slope form** of the equation of a straight line. The equation

$$y = mx + b$$

is called the **slope-intercept form** of the equation of a straight line. If the equation of a nonvertical line is written in slope-intercept form, then the coefficient of x is the slope of the line, and the constant term is the y intercept.

(8.5) Solving a system of two linear equations by graphing produces one of the following three possibilities:

1. The graphs of the two equations are two intersecting lines, which indicates **one solution** for the system. The system is called a **consistent system.**
2. The graphs of the two equations are two parallel lines, which indicates **no solution** for the system, and the system is called an **inconsistent system.**
3. The graphs of the two equations are the same line, which indicates **infinitely many solutions** for the system. We refer to the equations as a set of **dependent** equations.

Linear inequalities in two variables are of the form

$$Ax + By > C \qquad \text{or} \qquad Ax + By < C$$

To **graph a linear inequality**, we suggest the following steps.

1. First graph the corresponding equality. Use a solid line if equality is included in the original statement and a dashed line if equality is not included.
2. Choose a test point not on the line, and substitute its coordinates into the inequality.
3. The graph of the original inequality is
   (a) the half-plane that contains the test point if the inequality is satisfied by that point, or
   (b) the half-plane that does not contain the test point if the inequality is not satisfied by the point.

(8.6) The **elimination-by-addition method** for solving a system of two linear equations relies on the property

$$\text{if } a = b \text{ and } c = d, \quad \text{then } a + c = b + d$$

Equations are added to eliminate a variable.

(8.7) The **substitution method** for solving a system of two linear equations involves solving one of the two equations for one variable and **substituting** this expression into the other equation.

To review the problem of choosing a method for solving a particular system of two linear equations, return to Section 8.7 and study Examples 6 and 7 one more time.

Chapter 8 Review Problem Set

For Problems 1–10, graph each of the equations.

1. $2x - 5y = 10$ **2.** $y = -\frac{1}{3}x + 1$

3. $y = -2x$ **4.** $3x + 4y = 12$

5. $2x - 3y = 0$ **6.** $2x + y = 2$

7. $x - y = 4$ **8.** $x + 2y = -2$

9. $y = \frac{2}{3}x - 1$ **10.** $y = 3x$

For Problems 11–16, determine the slope and y intercept and graph the line.

11. $2x - 5y = 10$ **12.** $y = -\frac{1}{3}x + 1$

13. $x + 2y = 2$ **14.** $3x + y = -2$

15. $2x - y = 4$ **16.** $3x - 4y = 12$

17. Find the slope of the line determined by the points $(3, -4)$ and $(-2, 5)$.

18. Find the slope of the line $5x - 6y = 30$.

19. Write the equation of the line that has a slope of $-\frac{5}{7}$ and contains the point $(2, -3)$.

20. Write the equation of the line that contains the points $(2, 5)$ and $(-1, -3)$.

21. Write the equation of the line that has a slope of $\frac{2}{9}$ and a y intercept of -1.

22. Write the equation of the line that contains the point $(2, 4)$ and is perpendicular to the x axis.

23. Solve the system $\begin{pmatrix} 2x + y = 4 \\ x - y = 5 \end{pmatrix}$ by using the graphing method.

For Problems 24–35, solve each system by using either the elimination-by-addition method or the substitution method.

24. $\begin{pmatrix} 2x - y = 1 \\ 3x - 2y = -5 \end{pmatrix}$ **25.** $\begin{pmatrix} 2x + 5y = 7 \\ x = -3y + 1 \end{pmatrix}$

26. $\begin{pmatrix} 3x + 2y = 7 \\ 4x - 5y = 3 \end{pmatrix}$ **27.** $\begin{pmatrix} 9x + 2y = 140 \\ x + 5y = 135 \end{pmatrix}$

28. $\begin{pmatrix} \frac{1}{2}x + \frac{1}{4}y = -5 \\ \frac{2}{3}x - \frac{1}{2}y = 0 \end{pmatrix}$

29. $\begin{pmatrix} x + y = 1000 \\ 0.07x + 0.09y = 82 \end{pmatrix}$

30. $\begin{pmatrix} y = 5x + 2 \\ 10x - 2y = 1 \end{pmatrix}$ **31.** $\begin{pmatrix} 5x - 7y = 9 \\ y = 3x - 2 \end{pmatrix}$

32. $\begin{pmatrix} 10t + u = 6u \\ t + u = 12 \end{pmatrix}$

33. $\begin{pmatrix} t = 2u \\ 10t + u - 36 = 10u + t \end{pmatrix}$

34. $\begin{pmatrix} u = 2t + 1 \\ 10t + u + 10u + t = 110 \end{pmatrix}$

35. $\begin{pmatrix} y = -\frac{2}{3}x \\ \frac{1}{3}x - y = -9 \end{pmatrix}$

For Problems 36–39, graph each of the inequalities.

36. $y > \frac{2}{3}x - 1$ **37.** $x - 2y \leq 4$

38. $y \leq -2x$ **39.** $3x + 2y > -6$

Solve each of the following problems by setting up and solving a system of two linear equations in two variables.

40. The sum of two numbers is 113. The larger number is 1 less than twice the smaller number. Find the numbers.

41. Last year Mark invested a certain amount of money at 6% annual interest and \$500 more than that amount at 8%. He received \$390.00 in interest. How much did he invest at each rate?

42. Cindy has 43 coins consisting of nickels and dimes. The total value of the coins is \$3.40. How many coins of each kind does she have?

43. The length of a rectangle is 1 inch more than three times the width. If the perimeter of the rectangle is 50 inches, find the length and width.

44. The width of a rectangle is 5 inches less than the length. If the perimeter of the rectangle is 38 inches, find the length and width

45. Alex has 32 coins consisting of quarters and dimes. The total value of the coins is $4.85. How many coins of each kind does he have?

46. Two angles are complementary, and one of them is 6° less than twice the other one. Find the measure of each angle.

47. Two angles are supplementary, and the larger angle is 20° less than three times the smaller angle. Find the measure of each angle.

48. Four cheeseburgers and five milkshakes cost a total of $25.50. Two milkshakes cost $1.75 more than one cheeseburger. Find the cost of a cheeseburger and also find the cost of a milkshake.

49. Three bottles of orange juice and two bottles of water cost $6.75. On the other hand, two bottles of juice and three bottles of water cost $6.15. Find the cost per bottle of each.

Chapter 8 Test

For Problems 1–4, determine the slope and y intercept, and graph each equation.

1. $5x + 3y = 15$

2. $-2x + y = -4$

3. $y = -\frac{1}{2}x - 2$

4. $3x + y = 0$

5. Find x if the line through the points $(4, 7)$ and $(x, 13)$ has a slope of $\frac{3}{2}$.

6. Find y if the line through the points $(1, y)$ and $(6, 5)$ has a slope of $-\frac{3}{5}$.

7. If a line has a slope of $\frac{1}{4}$ and passes through the point $(3, 5)$, find the coordinates of two other points on the line.

8. If a line has a slope of -3 and passes through the point $(2, 1)$, find the coordinates of two other points on the line.

9. Suppose that a highway rises a distance of 85 feet over a horizontal distance of 1850 feet. Express the grade of the highway to the nearest tenth of a percent.

10. Find the x intercept of the graph of $y = 4x + 8$.

For Problems 11–13, express each equation in $Ax + By = C$ form, where A, B, and C are integers.

11. Determine the equation of the line that has a slope of $-\frac{3}{5}$ and a y intercept of 4.

12. Determine the equation of the line containing the point $(4, -2)$ and having a slope of $\frac{4}{9}$.

13. Determine the equation of the line that contains the points $(4, 6)$ and $(-2, -3)$.

14. Solve the system $\begin{pmatrix} 3x - 2y = -4 \\ 2x + 3y = 19 \end{pmatrix}$ by graphing.

15. Solve the system $\begin{pmatrix} x - 3y = -9 \\ 4x + 7y = 40 \end{pmatrix}$ using the elimination-by-addition method.

16. Solve the system $\begin{pmatrix} 5x + y = -14 \\ 6x - 7y = -66 \end{pmatrix}$ using the substitution method.

17. Solve the system $\begin{pmatrix} 2x - 7y = 26 \\ 3x + 2y = -11 \end{pmatrix}$.

18. Solve the system $\begin{pmatrix} 8x + 5y = -6 \\ 4x - y = 18 \end{pmatrix}$.

For Problems 19–23, graph each equation or inequality.

19. $5x + 3y = 15$

20. $y = \frac{2}{3}x$

21. $y = 2x - 3$

22. $y \geq 2x - 4$

23. $x + 3y < -3$

For Problems 24 and 25, solve each problem by setting up and solving a system of two linear equations in two variables.

24. Three reams of paper and 4 notebooks cost \$19.63. Four reams of paper and 1 notebook cost \$16.25. Find the cost of each item.

25. The length of a rectangle is 1 inch less than twice the width of the rectangle. If the perimeter of the rectangle is 40 inches, find the length of the rectangle.

9

Roots and Radicals

Suppose that a car is traveling at 65 miles per hour on a highway during a rainstorm. Suddenly something darts across the highway, and the driver hits the brake pedal. How far will the car skid on the wet pavement?

Suppose that a car is traveling at 65 miles per hour on a highway during a rainstorm. Suddenly, something darts across the highway, and the driver hits the brake pedal. How far will the car skid on the wet pavement? We can use the formula $S = \sqrt{30Df}$, where S represents the speed of the car, D the length of skid marks, and f a coefficient of friction, to determine that the car will skid approximately 400 feet.

In Section 2.3 we used $\sqrt{2}$ and $\sqrt{3}$ as examples of irrational numbers. Irrational numbers in decimal form are nonrepeating decimals. For example, $\sqrt{2} = 1.414213562373\ldots$, where the three dots indicate that the decimal expansion continues indefinitely. In Chapter 2, we stated that we would return to the irrationals in Chapter 9. The time has come for us to extend our skills relative to the set of irrational numbers.

9.1 Roots and Radicals

To **square a number** means to raise it to the second power — that is, to use the number as a factor twice.

$4^2 = 4 \cdot 4 = 16$ Read "four squared equals sixteen"

$10^2 = 10 \cdot 10 = 100$

$\left(\frac{1}{2}\right)^2 = \frac{1}{2} \cdot \frac{1}{2} = \frac{1}{4}$

$(-3)^2 = (-3)(-3) = 9$

A **square root of a number** is one of its two equal factors. Thus 4 is a square root of 16 because $4 \cdot 4 = 16$. Likewise, -4 is also a square root of 16 because $(-4)(-4) = 16$. In general, a is a square root of b if $a^2 = b$. The following generalizations are a direct consequence of the previous statement.

1. Every positive real number has two square roots; one is positive and the other is negative. They are opposites of each other.
2. Negative real numbers have no real number square roots because any nonzero real number is positive when squared.
3. The square root of 0 is 0.

The symbol $\sqrt{}$, called a **radical sign**, is used to designate the nonnegative square root. The number under the radical sign is called the **radicand**. The entire expression (such as $\sqrt{16}$) is called a **radical**.

$\sqrt{16} = 4$ $\sqrt{16}$ indicates the nonnegative or **principal square root** of 16

$-\sqrt{16} = -4$ $-\sqrt{16}$ indicates the negative square root of 16

$\sqrt{0} = 0$ Zero has only one square root. Technically, we could write $-\sqrt{0} = -0 = 0$

$\sqrt{-4}$ is not a real number.

$-\sqrt{-4}$ is not a real number.

In general, the following definition is useful.

Definition 9.1

If $a \geq 0$ and $b \geq 0$, then $\sqrt{b} = a$ if and only if $a^2 = b$; a is called the **principal square root of b**.

If a is a number that is the square of an integer, then $\sqrt{a}$ and $-\sqrt{a}$ are rational numbers. For example, $\sqrt{1}$, $\sqrt{4}$, and $\sqrt{25}$ are the rational numbers 1, 2, and 5, respectively. The numbers 1, 4, and 25 are called **perfect squares** because each represents the square of some integer. The following chart contains the squares of the whole numbers from 1 through 20, inclusive. You should know these values so that you can immediately recognize such square roots as $\sqrt{81} = 9$, $\sqrt{144} = 12$, $\sqrt{289} = 17$, and so on from the list. Furthermore, perfect squares of multiples of 10 are easy to recognize. For example, because $30^2 = 900$, we know that $\sqrt{900} = 30$.

$1^2 = 1$	$8^2 = 64$	$15^2 = 225$
$2^2 = 4$	$9^2 = 81$	$16^2 = 256$
$3^2 = 9$	$10^2 = 100$	$17^2 = 289$
$4^2 = 16$	$11^2 = 121$	$18^2 = 324$
$5^2 = 25$	$12^2 = 144$	$19^2 = 361$
$6^2 = 36$	$13^2 = 169$	$20^2 = 400$
$7^2 = 49$	$14^2 = 196$	

Knowing this listing of perfect squares can also help you with square roots of some fractions. Consider the next examples.

$$\sqrt{\frac{16}{25}} = \frac{4}{5} \quad \text{because} \quad \left(\frac{4}{5}\right)^2 = \frac{16}{25}$$

$$\sqrt{\frac{36}{49}} = \frac{6}{7} \quad \text{because} \quad \left(\frac{6}{7}\right)^2 = \frac{36}{49}$$

$$\sqrt{0.09} = 0.3 \quad \text{because } (0.3)^2 = 0.09$$

If a is a positive integer that is *not* the square of an integer, then $\sqrt{a}$ and $-\sqrt{a}$ are irrational numbers. For example, $\sqrt{2}$, $-\sqrt{2}$, $\sqrt{23}$, $\sqrt{31}$, $\sqrt{52}$, and $-\sqrt{75}$ are irrational numbers. Remember that irrational numbers have nonrepeating, nonterminating decimal representations. For example, $\sqrt{2} = 1.414213562373 \ldots$, where the decimal never repeats a block of digits.

For practical purposes, we often need to use a rational approximation of an irrational number. The calculator is a very useful tool for finding such approximations. Be sure that you can use your calculator to find the following approximate square roots. Each approximate square root has been rounded to the nearest thousandth. If you don't have a calculator available, turn to Appendix A and use the table of square roots.

$$\sqrt{19} = 4.359 \qquad \sqrt{38} = 6.164 \qquad \sqrt{72} = 8.485 \qquad \sqrt{93} = 9.644$$

To **cube a number** means to raise it to the third power—that is, to use the number as a factor three times.

$2^3 = 2 \cdot 2 \cdot 2 = 8$ Read "two cubed equals eight"

$4^3 = 4 \cdot 4 \cdot 4 = 64$

$$\left(\frac{2}{3}\right)^3 = \frac{2}{3} \cdot \frac{2}{3} \cdot \frac{2}{3} = \frac{8}{27}$$

$$(-2)^3 = (-2)(-2)(-2) = -8$$

A **cube root of a number** is one of its three equal factors. Thus 2 is a cube root of 8 because $2 \cdot 2 \cdot 2 = 8$. (In fact, 2 is the only real number that is a cube root of 8.) Furthermore, -2 is a cube root of -8 because $(-2)(-2)(-2) = -8$. (In fact, -2 is the only real number that is a cube root of -8.)

In general, a is a cube root of b if $a^3 = b$. The following generalizations are a direct consequence of the previous statement.

1. Every positive real number has one positive real number cube root.

2. Every negative real number has one negative real number cube root.

3. The cube root of 0 is 0.

Remark: Technically, every nonzero real number has three cube roots, but only one of them is a real number. The other two roots are classified as complex numbers. We are restricting our work at this time to the set of real numbers.

The symbol $\sqrt[3]{}$ designates the cube root of a number. Thus we can write

$$\sqrt[3]{8} = 2 \qquad \sqrt[3]{\frac{1}{27}} = \frac{1}{3}$$

$$\sqrt[3]{-8} = -2 \qquad \sqrt[3]{-\frac{1}{27}} = -\frac{1}{3}$$

In general, the following definition is useful.

Definition 9.2

$\sqrt[3]{b} = a$ if and only if $a^3 = b$.

In Definition 9.2, if $b \geq 0$ then $a \geq 0$, whereas if $b < 0$ then $a < 0$. The number a is called **the principal cube root of *b*** or simply **the cube root of *b***. In the radical $\sqrt[3]{b}$, the 3 is called the **index** of the radical. When working with square roots, we commonly omit writing the index; so we write $\sqrt{b}$ instead of $\sqrt[2]{b}$. The concept of root can be extended to fourth roots, fifth roots, sixth roots, and, in general nth roots. However, in this text we will restrict our work to square roots and cube roots until Section 11.3.

You should become familiar with the following perfect cubes so that you can recognize their roots without a calculator or a table.

$2^3 = 8 \qquad 5^3 = 125 \qquad 8^3 = 512$

$3^3 = 27 \qquad 6^3 = 216 \qquad 9^3 = 729$

$4^3 = 64 \qquad 7^3 = 343 \qquad 10^3 = 1000$

For example, you should recognize that $\sqrt[3]{343} = 7$.

Adding and Subtracting Radical Expressions

Recall our use of the distributive property as the basis for combining similar terms. Here are three examples:

$$3x + 2x = (3 + 2)x = 5x$$
$$7y - 4y = (7 - 4)y = 3y$$
$$9a^2 + 5a^2 = (9 + 5)a^2 = 14a^2$$

In a like manner, we can often simplify expressions that contain radicals by using the distributive property.

$$5\sqrt{2} + 7\sqrt{2} = (5 + 7)\sqrt{2} = 12\sqrt{2}$$
$$8\sqrt{5} - 2\sqrt{5} = (8 - 2)\sqrt{5} = 6\sqrt{5}$$
$$4\sqrt{7} + 6\sqrt{7} + 3\sqrt{11} - \sqrt{11} = (4 + 6)\sqrt{7} + (3 - 1)\sqrt{11} = 10\sqrt{7} + 2\sqrt{11}$$
$$6\sqrt[3]{7} + 4\sqrt[3]{7} - 2\sqrt[3]{7} = (6 + 4 - 2)\sqrt[3]{7} = 8\sqrt[3]{7}$$

Note that if we want **to simplify when adding or subtracting radical expressions, the radicals must have the same radicand and index**. Also note the form we use to indicate multiplication when a radical is involved. For example, $5 \cdot \sqrt{2}$ is written as $5\sqrt{2}$.

Now suppose that we need to evaluate $5\sqrt{2} - \sqrt{2} + 4\sqrt{2} - 2\sqrt{2}$ to the nearest tenth. We can either evaluate the expression as it stands or first simplify it by combining radicals and then evaluate that result. Let's use the latter approach. (It would probably be a good idea for you to do it both ways for checking purposes.)

$$\begin{aligned} 5\sqrt{2} - \sqrt{2} + 4\sqrt{2} - 2\sqrt{2} &= (5 - 1 + 4 - 2)\sqrt{2} \\ &= 6\sqrt{2} \\ &= 8.5 \quad \text{to the nearest tenth} \end{aligned}$$

EXAMPLE 1

Find a rational approximation, to the nearest tenth, for

$$7\sqrt{3} + 9\sqrt{5} + 2\sqrt{3} - 3\sqrt{5} + 13\sqrt{3}$$

Solution

First, we simplify the given expression and then evaluate that result.

$$\begin{aligned} 7\sqrt{3} + 9\sqrt{5} + 2\sqrt{3} - 3\sqrt{5} + 13\sqrt{3} &= (7 + 2 + 13)\sqrt{3} + (9 - 3)\sqrt{5} \\ &= 22\sqrt{3} + 6\sqrt{5} \\ &= 51.5 \quad \text{to the nearest tenth} \end{aligned}$$

■

Applications of Radicals

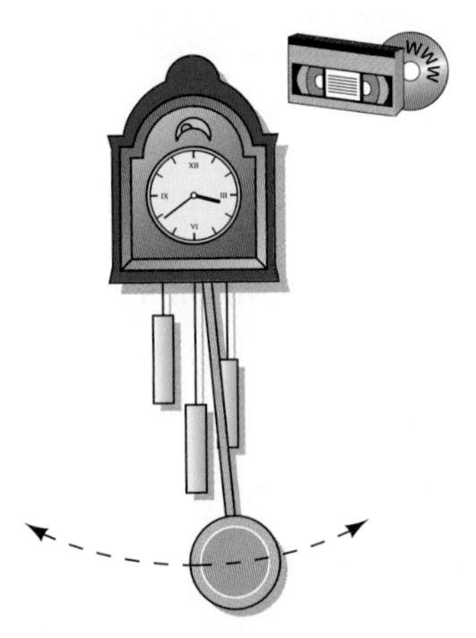

Figure 9.1

Many real-world applications involve radical expressions. For example, the *period* of a pendulum is the time it takes to swing from one side to the other and back (see Figure 9.1). A formula for the period is

$$T = 2\pi\sqrt{\frac{L}{32}}$$

where T is the period of the pendulum expressed in seconds, and L is the length of the pendulum in feet.

PROBLEM 1 Find the period, to the nearest tenth of a second, of a pendulum 2.5 feet long.

Solution

We use 3.14 as an approximation for π and substitute 2.5 for L in the formula.

$$T = 2\pi\sqrt{\frac{L}{32}}$$

$$= 2(3.14)\sqrt{\frac{2.5}{32}}$$

$$= 1.8 \quad \text{to the nearest tenth}$$

The period is approximately 1.8 seconds. ■

Police use the formula $S = \sqrt{30Df}$ to estimate a car's speed based on the length of skid marks (see Figure 9.2). In this formula, S represents the car's speed in miles per hour, D the length of skid marks measured in feet, and f a coefficient of friction. For a particular situation, the coefficient of friction is a constant that depends on the type and condition of the road surface.

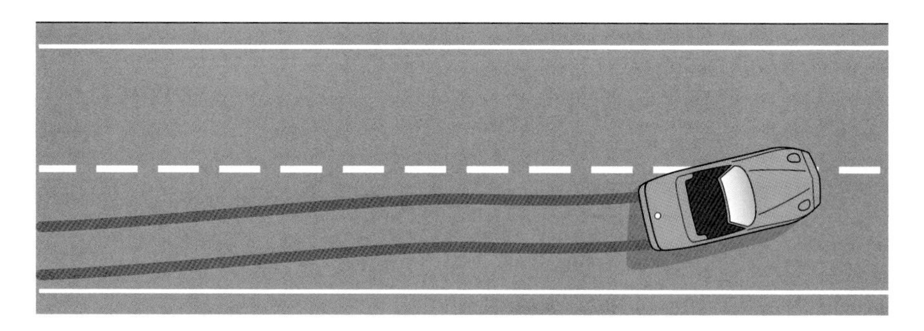

Figure 9.2

PROBLEM 2 Using 0.40 as a coefficient of friction, find how fast a car was moving if it skidded 225 feet. Express the answer to the nearest mile per hour.

Solution

We substitute 0.40 for f and 225 for D in the formula $S = \sqrt{30Df}$.

$$S = \sqrt{30(225)(0.40)} = 52 \quad \text{to the nearest whole number}$$

The car was traveling at approximately 52 miles per hour. ■

Problem Set 9.1

For Problems 1–26, evaluate each radical without using a calculator or a table.

1. $\sqrt{49}$
2. $\sqrt{100}$
3. $-\sqrt{64}$
4. $-\sqrt{36}$
5. $\sqrt{121}$
6. $\sqrt{144}$
7. $\sqrt{3600}$
8. $\sqrt{2500}$
9. $-\sqrt{1600}$
10. $-\sqrt{900}$
11. $\sqrt{6400}$
12. $\sqrt{400}$
13. $\sqrt{324}$
14. $-\sqrt{361}$
15. $\sqrt{\frac{25}{9}}$
16. $\sqrt{\frac{1}{225}}$
17. $\sqrt{0.16}$
18. $\sqrt{0.0121}$
19. $\sqrt[3]{27}$
20. $\sqrt[3]{\frac{1}{8}}$
21. $-\sqrt[3]{-8}$
22. $\sqrt[3]{\frac{125}{8}}$
23. $-\sqrt[3]{729}$
24. $\sqrt[3]{-125}$
25. $\sqrt[3]{-216}$
26. $\sqrt[3]{64}$

For Problems 27–40, use a calculator to evaluate each radical.

27. $\sqrt{576}$
28. $\sqrt{7569}$
29. $\sqrt{2304}$
30. $\sqrt{9801}$
31. $\sqrt{784}$
32. $\sqrt{1849}$
33. $\sqrt{4225}$
34. $\sqrt{2704}$
35. $\sqrt{3364}$
36. $\sqrt{1444}$
37. $\sqrt[3]{3375}$
38. $\sqrt[3]{1728}$
39. $\sqrt[3]{9261}$
40. $\sqrt[3]{5832}$

For Problems 41–48, use a calculator to find a rational approximation of each square root. Express your answers to the nearest hundredth.

41. $\sqrt{19}$
42. $\sqrt{34}$
43. $\sqrt{50}$
44. $\sqrt{66}$
45. $\sqrt{75}$
46. $\sqrt{90}$
47. $\sqrt{95}$
48. $\sqrt{98}$

For Problems 49–58, use a calculator to find a whole number approximation for each expression.

49. $\sqrt{4325}$
50. $\sqrt{7500}$
51. $\sqrt{1175}$
52. $\sqrt{1700}$
53. $\sqrt{9501}$
54. $\sqrt{8050}$
55. $\sqrt[3]{7814}$
56. $\sqrt[3]{1456}$
57. $\sqrt{1000}$
58. $\sqrt{3400}$

For Problems 59–70, simplify each expression by using the distributive property.

59. $7\sqrt{2} + 14\sqrt{2}$
60. $9\sqrt{3} + 4\sqrt{3}$
61. $17\sqrt{7} - 9\sqrt{7}$
62. $19\sqrt{5} - 8\sqrt{5}$
63. $4\sqrt[3]{2} + 7\sqrt[3]{2}$
64. $8\sqrt[3]{6} - 10\sqrt[3]{6}$
65. $9\sqrt[3]{7} + 2\sqrt[3]{5} - 6\sqrt[3]{7}$
66. $8\sqrt{7} + 13\sqrt{7} - 9\sqrt{7}$
67. $8\sqrt{2} - 4\sqrt{3} - 9\sqrt{2} + 6\sqrt{3}$
68. $7\sqrt{5} - 9\sqrt{6} + 14\sqrt{5} - 2\sqrt{6}$
69. $6\sqrt{7} + 5\sqrt{10} - 8\sqrt{10} - 4\sqrt{7} - 11\sqrt{7} + \sqrt{10}$
70. $\sqrt{3} - \sqrt{5} + 4\sqrt{5} - 3\sqrt{3} - 9\sqrt{5} - 16\sqrt{3}$

For Problems 71–84, find a rational approximation, to the nearest tenth, for each radical expression.

71. $9\sqrt{3} + \sqrt{3}$
72. $6\sqrt{2} + 14\sqrt{2}$
73. $9\sqrt{5} - 3\sqrt{5}$
74. $18\sqrt{6} - 12\sqrt{6}$
75. $14\sqrt{2} - 15\sqrt{2}$
76. $7\sqrt{3} - 12\sqrt{3}$
77. $8\sqrt{7} - 4\sqrt{7} + 6\sqrt{7}$
78. $9\sqrt{5} - 2\sqrt{5} + \sqrt{5}$
79. $4\sqrt{3} - 2\sqrt{2}$
80. $3\sqrt{2} + \sqrt{3} - \sqrt{5}$
81. $9\sqrt{6} - 3\sqrt{5} + 2\sqrt{6} - 7\sqrt{5} - \sqrt{6}$
82. $8\sqrt{7} - 2\sqrt{10} + 4\sqrt{7} - 3\sqrt{10} - 7\sqrt{7} + 4\sqrt{10}$
83. $4\sqrt{11} - 5\sqrt{11} - 7\sqrt{11} + 2\sqrt{11} - 3\sqrt{11}$
84. $14\sqrt{13} - 17\sqrt{13} + 3\sqrt{13} - 4\sqrt{13} - 5\sqrt{13}$
85. Use the formula from Problem 1 to find the periods of pendulums that have lengths of 2 feet, 3.5 feet, and 4 feet. Express the answers to the nearest tenth of a second.
86. Use the formula from Problem 2 with a coefficient of friction of 0.35 to find the speeds of cars that leave skid marks of lengths 150 feet, 200 feet, and 275 feet. Express the answers to the nearest mile per hour.

87. The time T, measured in seconds, that it takes for an object to fall d feet (neglecting air resistance) is given by the formula $T = \sqrt{\frac{d}{16}}$. Find the times that it takes objects to fall 75 feet, 125 feet, and 5280 feet. Express the answers to the nearest tenth of a second.

THOUGHTS INTO WORDS

88. Why is $\sqrt{-4}$ not a real number?

89. How could you find a whole number approximation for $\sqrt{1450}$ if you did not have a calculator available?

9.2 Simplifying Radicals

Note the following facts that pertain to square roots:

$$\sqrt{4 \cdot 9} = \sqrt{36} = 6$$

$$\sqrt{4}\sqrt{9} = 2 \cdot 3 = 6$$

Thus we observe that $\sqrt{4 \cdot 9} = \sqrt{4}\sqrt{9}$. This illustrates a general property.

Property 9.1

For any nonnegative real numbers a and b,

$$\sqrt{ab} = \sqrt{a}\sqrt{b}$$

In other words, we say that **the square root of a product is equal to the product of the square roots**.

Property 9.1 and the definition of square root provide the basis for expressing radical expressions in simplest radical form. For now, simplest radical form means that *the radicand contains no factors other than 1 that are perfect squares*. We present some examples to illustrate this meaning of simplest radical form.

1. $\sqrt{8} = \sqrt{4 \cdot 2} = \sqrt{4}\sqrt{2} = 2\sqrt{2}$

4 is a perfect square; $\sqrt{4} = 2$

2. $\sqrt{45} = \sqrt{9 \cdot 5} = \sqrt{9}\sqrt{5} = 3\sqrt{5}$

9 is a perfect square; $\sqrt{9} = 3$

3. $\sqrt{48} = \sqrt{16 \cdot 3} = \sqrt{16}\sqrt{3} = 4\sqrt{3}$

16 is a perfect square; $\sqrt{16} = 4$

The first step in each example is to express the radicand of the given radical as the product of two factors, at least one of which is a perfect square other than 1. Also observe the radicands of the final radicals. In each case, the radicand *cannot* be expressed as the product of two factors, at least one of which is a perfect square other than 1. We say that the final radicals, $2\sqrt{2}$, $3\sqrt{5}$, and $4\sqrt{3}$, are in simplest radical form.

You may vary the steps somewhat in changing to simplest radical form, but the final result should be the same. Consider another sequence of steps to change $\sqrt{48}$ to simplest form.

$$\sqrt{48} = \sqrt{4 \cdot 12} = \sqrt{4}\sqrt{12} = 2\sqrt{12} = 2\sqrt{4 \cdot 3} = 2\sqrt{4}\sqrt{3} = 2 \cdot 2\sqrt{3} = 4\sqrt{3}$$

4 is a perfect square; This is not in simplest form; 4 is a perfect square; Same result as in example 3

Another variation of the technique for changing radicals to simplest form is to prime-factor the radicand and then to look for perfect squares in exponential form. We will redo the previous examples.

4. $\sqrt{8} = \sqrt{2 \cdot 2 \cdot 2} = \sqrt{2^2 \cdot 2} = \sqrt{2^2}\sqrt{2} = 2\sqrt{2}$

Prime factors of 8; 2^2 is a perfect square

5. $\sqrt{45} = \sqrt{3 \cdot 3 \cdot 5} = \sqrt{3^2 \cdot 5} = \sqrt{3^2}\sqrt{5} = 3\sqrt{5}$

Prime factors of 45; 3^2 is a perfect square

6. $\sqrt{48} = \sqrt{2 \cdot 2 \cdot 2 \cdot 2 \cdot 3} = \sqrt{2^4 \cdot 3} = \sqrt{2^4}\sqrt{3} = 2^2\sqrt{3} = 4\sqrt{3}$

Prime factors of 48; 2^4 is a perfect square; $\sqrt{2^4} = 2^2$ because $2^2 \cdot 2^2 = 2^4$

The next examples further illustrate the process of changing to simplest radical form. Only the major steps are shown, so be sure that you can fill in the details.

7. $\sqrt{56} = \sqrt{4}\sqrt{14} = 2\sqrt{14}$

8. $\sqrt{75} = \sqrt{25}\sqrt{3} = 5\sqrt{3}$

9. $\sqrt{108} = \sqrt{2 \cdot 2 \cdot 3 \cdot 3 \cdot 3} = \sqrt{2^2 \cdot 3^2}\sqrt{3} = 6\sqrt{3}$

10. $5\sqrt{12} = 5\sqrt{4}\sqrt{3} = 5 \cdot 2 \cdot \sqrt{3} = 10\sqrt{3}$

We can extend Property 9.1 to apply to cube roots.

Property 9.2

For any real numbers a and b,

$$\sqrt[3]{ab} = \sqrt[3]{a}\sqrt[3]{b}$$

Now, using Property 9.2, we can simplify radicals involving cube roots. Here it is helpful to recognize the perfect cubes that we listed in the previous section.

11. $\sqrt[3]{24} = \sqrt[3]{8}\sqrt[3]{3} = 2\sqrt[3]{3}$

Perfect cube

12. $\sqrt[3]{108} = \sqrt[3]{27}\sqrt[3]{4} = 3\sqrt[3]{4}$

Perfect cube

13. $\sqrt[3]{375} = \sqrt[3]{125}\sqrt[3]{3} = 5\sqrt[3]{3}$

Perfect cube

Radicals That Contain Variables

Before we discuss the process of simplifying radicals that contain variables, there is one technicality that we should call to your attention. Let's look at some examples to clarify the point. Consider the radical $\sqrt{x^2}$.

Let $x = 3$; then $\sqrt{x^2} = \sqrt{3^2} = \sqrt{9} = 3$.

Let $x = -3$; then $\sqrt{x^2} = \sqrt{(-3)^2} = \sqrt{9} = 3$.

Thus if $x \geq 0$, then $\sqrt{x^2} = x$, *but* if $x < 0$, then $\sqrt{x^2} = -x$. Using the concept of absolute value, we can state that for all real numbers, $\sqrt{x^2} = |x|$.

Now consider the radical $\sqrt{x^3}$. Because x^3 is negative when x is negative, we need to restrict x to the nonnegative reals when working with $\sqrt{x^3}$. Thus we can write: If $x \geq 0$, then $\sqrt{x^3} = \sqrt{x^2}\sqrt{x} = x\sqrt{x}$, and no absolute value sign is necessary. Finally, let's consider the radical $\sqrt[3]{x^3}$.

Let $x = 2$; then $\sqrt[3]{x^3} = \sqrt[3]{2^3} = \sqrt[3]{8} = 2$.

Let $x = -2$; then $\sqrt[3]{x^3} = \sqrt[3]{(-2)^3} = \sqrt[3]{-8} = -2$.

Thus it is correct to write: $\sqrt[3]{x^3} = x$ for all real numbers, and again no absolute value sign is necessary.

The previous discussion indicates that technically every radical expression involving variables in the radicand needs to be analyzed individually as to the necessary restrictions imposed on the variables. However, to avoid considering such restrictions on a problem-to-problem basis we shall merely assume that all variables represent positive real numbers.

14. $\sqrt{x^2y} = \sqrt{x^2}\sqrt{y} = x\sqrt{y}$

15. $\sqrt{4x^3} = \sqrt{4x^2}\sqrt{x} = 2x\sqrt{x}$

$4x^2$ is a perfect square because $(2x)(2x) = 4x^2$

16. $\sqrt{8xy^3} = \sqrt{4y^2}\sqrt{2xy} = 2y\sqrt{2xy}$

17. $\sqrt{27x^5y^3} = \sqrt{9x^4y^2}\sqrt{3xy} = 3x^2y\sqrt{3xy}$

$\sqrt{9x^4y^2} = 3x^2y$

18. $\sqrt[3]{40x^4y^5} = \sqrt[3]{8x^3y^3}\sqrt[3]{5xy^2} = 2xy\sqrt[3]{5xy^2}$

Perfect cube

If the numerical coefficient of the radicand is quite large, you may want to look at it in prime factored form. The next example demonstrates this idea.

19. $\sqrt{180a^6b^3} = \sqrt{2 \cdot 2 \cdot 3 \cdot 3 \cdot 5 \cdot a^6 \cdot b^2 \cdot b}$

$= \sqrt{36a^6b^2}\sqrt{5b}$

$= 6a^3b\sqrt{5b}$

When simplifying expressions that contain radicals, we must often first change the radicals to simplest form, and then apply the distributive property.

EXAMPLE 1

Simplify $5\sqrt{8} + 3\sqrt{2}$.

Solution

$$\begin{aligned} 5\sqrt{8} + 3\sqrt{2} &= 5\sqrt{4}\sqrt{2} + 3\sqrt{2} \\ &= 5 \cdot 2 \cdot \sqrt{2} + 3\sqrt{2} \\ &= 10\sqrt{2} + 3\sqrt{2} \\ &= (10 + 3)\sqrt{2} \\ &= 13\sqrt{2} \end{aligned}$$

■

EXAMPLE 2

Simplify $2\sqrt{27} - 5\sqrt{48} + 4\sqrt{3}$.

Solution

$$\begin{aligned} 2\sqrt{27} - 5\sqrt{48} + 4\sqrt{3} &= 2\sqrt{9}\sqrt{3} - 5\sqrt{16}\sqrt{3} + 4\sqrt{3} \\ &= 2(3)\sqrt{3} - 5(4)\sqrt{3} + 4\sqrt{3} \\ &= 6\sqrt{3} - 20\sqrt{3} + 4\sqrt{3} \\ &= (6 - 20 + 4)\sqrt{3} \\ &= -10\sqrt{3} \end{aligned}$$

■

EXAMPLE 3

Simplify $\frac{1}{4}\sqrt{45} + \frac{1}{3}\sqrt{20}$.

Solution

$$\begin{aligned} \frac{1}{4}\sqrt{45} + \frac{1}{3}\sqrt{20} &= \frac{1}{4}\sqrt{9}\sqrt{5} + \frac{1}{3}\sqrt{4}\sqrt{5} \\ &= \frac{1}{4}(3)\sqrt{5} + \frac{1}{3}(2)\sqrt{5} \\ &= \frac{3}{4}\sqrt{5} + \frac{2}{3}\sqrt{5} \\ &= \left(\frac{3}{4} + \frac{2}{3}\right)\sqrt{5} = \left(\frac{9}{12} + \frac{8}{12}\right)\sqrt{5} \\ &= \frac{17}{12}\sqrt{5} \end{aligned}$$

■

EXAMPLE 4

Simplify $5\sqrt[3]{24} + 7\sqrt[3]{375}$.

Solution

$$\begin{aligned} 5\sqrt[3]{24} + 7\sqrt[3]{375} &= 5\sqrt[3]{8}\sqrt[3]{3} + 7\sqrt[3]{125}\sqrt[3]{3} \\ &= 5(2)\sqrt[3]{3} + 7(5)\sqrt[3]{3} \\ &= 10\sqrt[3]{3} + 35\sqrt[3]{3} \\ &= 45\sqrt[3]{3} \end{aligned}$$

■

Problem Set 9.2

For Problems 1–30, change each radical to simplest radical form.

1. $\sqrt{24}$
2. $\sqrt{54}$
3. $\sqrt{18}$
4. $\sqrt{50}$
5. $\sqrt{27}$
6. $\sqrt{12}$
7. $\sqrt{40}$
8. $\sqrt{90}$
9. $\sqrt[3]{-54}$
10. $\sqrt[3]{32}$
11. $\sqrt{80}$
12. $\sqrt{125}$
13. $\sqrt{117}$
14. $\sqrt{126}$
15. $4\sqrt{72}$
16. $8\sqrt{98}$
17. $3\sqrt[3]{40}$
18. $4\sqrt[3]{375}$
19. $-5\sqrt{20}$
20. $-6\sqrt{45}$
21. $-8\sqrt{96}$
22. $-4\sqrt{54}$
23. $\frac{3}{2}\sqrt{8}$
24. $\frac{5}{2}\sqrt{32}$
25. $\frac{3}{4}\sqrt{12}$
26. $\frac{4}{5}\sqrt{27}$
27. $-\frac{2}{3}\sqrt{45}$
28. $-\frac{3}{5}\sqrt{125}$
29. $-\frac{1}{4}\sqrt[3]{32}$
30. $\frac{2}{3}\sqrt[3]{81}$

For Problems 31–52, express each radical in simplest radical form. All variables represent nonnegative real numbers.

31. $\sqrt{x^2y^3}$
32. $\sqrt{xy^4}$
33. $\sqrt{2x^2y}$
34. $\sqrt{3x^2y^2}$
35. $\sqrt{8x^2}$
36. $\sqrt{24x^3}$
37. $\sqrt{27a^3b}$
38. $\sqrt{45a^2b^4}$
39. $\sqrt[3]{64x^4y^2}$
40. $\sqrt[3]{56x^3y^5}$
41. $\sqrt{63x^4y^2}$
42. $\sqrt{28x^3y}$
43. $3\sqrt{48x^2}$
44. $5\sqrt{12x^2y^2}$
45. $-6\sqrt{72x^7}$
46. $-8\sqrt{80y^9}$
47. $\frac{2}{9}\sqrt{54xy}$
48. $\frac{4}{3}\sqrt{20xy}$
49. $\frac{1}{8}\sqrt[3]{250x^4}$
50. $\frac{1}{3}\sqrt[3]{81x^3}$
51. $-\frac{2}{3}\sqrt{169a^8}$
52. $-\frac{2}{7}\sqrt{196a^{10}}$

For Problems 53–66, simplify each expression.

53. $7\sqrt{32} + 5\sqrt{2}$
54. $6\sqrt{48} + 5\sqrt{3}$
55. $4\sqrt{45} - 9\sqrt{5}$
56. $7\sqrt{24} - 12\sqrt{6}$
57. $2\sqrt[3]{54} + 6\sqrt[3]{16}$
58. $2\sqrt[3]{24} - 4\sqrt[3]{81}$
59. $4\sqrt{63} - 7\sqrt{28}$
60. $2\sqrt{40} - 7\sqrt{90}$
61. $5\sqrt{12} + 3\sqrt{27} - 2\sqrt{75}$
62. $4\sqrt{18} - 6\sqrt{50} - 3\sqrt{72}$
63. $\frac{1}{2}\sqrt{20} + \frac{2}{3}\sqrt{45} - \frac{1}{4}\sqrt{80}$
64. $\frac{1}{3}\sqrt{12} - \frac{3}{2}\sqrt{48} + \frac{3}{4}\sqrt{108}$
65. $3\sqrt{8} - 5\sqrt{20} - 7\sqrt{18} - 9\sqrt{125}$
66. $5\sqrt{27} - 3\sqrt{24} + 8\sqrt{54} - 7\sqrt{75}$

THOUGHTS INTO WORDS

67. Explain how you would help someone express $5\sqrt{72}$ in simplest radical form.

68. Explain your thought process when expressing $\sqrt{153}$ in simplest radical form.

FURTHER INVESTIGATIONS

69. Express each of the following in simplest radical form. The divisibility rules given in Problem Set 1.2 should be of some help.

    (a) $\sqrt{162}$ (b) $\sqrt{279}$

    (c) $\sqrt{275}$ (d) $\sqrt{212}$

70. Use a calculator to evaluate each expression in Problems 53–66. Then evaluate the simplified expression that you obtained when doing those problems. Your two results for each problem should be the same.

71. Sometimes we can reach a fairly good estimate of a radical expression by using whole number approxima-

tions. For example, $5\sqrt{35} + 7\sqrt{50}$ is approximately $5(6) + 7(7) = 79$. Using a calculator, we find that $5\sqrt{35} + 7\sqrt{50} = 79.1$ to the nearest tenth. In this case our whole number estimate is very good. For each of the following, first make a whole number estimate, and then use a calculator to see how well you estimated.

(a) $3\sqrt{10} - 4\sqrt{24} + 6\sqrt{65}$

(b) $9\sqrt{27} + 5\sqrt{37} - 3\sqrt{80}$

(c) $12\sqrt{5} + 13\sqrt{18} + 9\sqrt{47}$

(d) $3\sqrt{98} - 4\sqrt{83} - 7\sqrt{120}$

(e) $4\sqrt{170} + 2\sqrt{198} + 5\sqrt{227}$

(f) $-3\sqrt{256} - 6\sqrt{287} + 11\sqrt{321}$

72. Evaluate $\sqrt{x^2}$ for $x = 5, x = 4, x = -3, x = 9, x = -8$, and $x = -11$. For which values of x does $\sqrt{x^2} = x$? For which values of x does $\sqrt{x^2} = -x$? For which values of x does $\sqrt{x^2} = |x|$?

9.3 More on Simplifying Radicals

Another property of roots is motivated by the following examples.

$$\sqrt{\frac{36}{9}} = \sqrt{4} = 2 \quad \text{and} \quad \frac{\sqrt{36}}{\sqrt{9}} = \frac{6}{3} = 2$$

Thus we see that $\sqrt{\frac{36}{9}} = \frac{\sqrt{36}}{\sqrt{9}}$.

$$\sqrt[3]{\frac{64}{8}} = \sqrt[3]{8} = 2 \quad \text{and} \quad \frac{\sqrt[3]{64}}{\sqrt[3]{8}} = \frac{4}{2} = 2$$

Thus we see that $\sqrt[3]{\frac{64}{8}} = \frac{\sqrt[3]{64}}{\sqrt[3]{8}}$.

We can state the following general property.

Property 9.3

For any nonnegative real numbers a and b, $b \neq 0$, $\sqrt{\frac{a}{b}} = \frac{\sqrt{a}}{\sqrt{b}}$.

For any real numbers a and b, $b \neq 0$, $\sqrt[3]{\frac{a}{b}} = \frac{\sqrt[3]{a}}{\sqrt[3]{b}}$.

To evaluate a radical such as $\sqrt{\frac{25}{4}}$, in which the numerator and denominator of the fractional radicand are perfect squares, you may use Property 9.3 or merely rely on the definition of square root.

$$\sqrt{\frac{25}{4}} = \frac{\sqrt{25}}{\sqrt{4}} = \frac{5}{2}$$

or

$$\sqrt{\frac{25}{4}} = \frac{5}{2} \quad \text{because} \quad \frac{5}{2} \cdot \frac{5}{2} = \frac{25}{4}$$

Sometimes it is easier to do the indicated division first and then find the square root, as you can see in the next example.

$$\sqrt{\frac{324}{9}} = \sqrt{36} = 6$$

Now we can extend our concept of simplest radical form. An algebraic expression that contains a radical is said to be in **simplest radical form** if the following conditions are satisfied:

1. No fraction appears within a radical sign. $\left(\sqrt{\frac{2}{3}} \text{ violates this condition.}\right)$
2. No radical appears in the denominator. $\left(\frac{5}{\sqrt{8}} \text{ violates this condition.}\right)$
3. No radicand when expressed in prime factored form contains a factor raised to a power greater than or equal to the index. ($\sqrt{8} = \sqrt{2^3}$ violates this condition.)

The next examples show how to simplify expressions that do not meet these three stated conditions.

EXAMPLE 1

Simplify $\sqrt{\frac{13}{4}}$.

Solution

$$\sqrt{\frac{13}{4}} = \frac{\sqrt{13}}{\sqrt{4}} = \frac{\sqrt{13}}{2}$$

■

EXAMPLE 2

Simplify $\sqrt[3]{\frac{16}{27}}$.

Solution

$$\sqrt[3]{\frac{16}{27}} = \frac{\sqrt[3]{16}}{\sqrt[3]{27}} = \frac{\sqrt[3]{16}}{3} = \frac{\sqrt[3]{8}\sqrt[3]{2}}{3} = \frac{2\sqrt[3]{2}}{3}$$

↑ Don't stop here. The radical in the numerator can be simplified

■

EXAMPLE 3

Simplify $\frac{\sqrt{12}}{\sqrt{16}}$.

Solution A

$$\frac{\sqrt{12}}{\sqrt{16}} = \frac{\sqrt{12}}{4} = \frac{\sqrt{4}\sqrt{3}}{4} = \frac{2\sqrt{3}}{4} = \frac{\sqrt{3}}{2}$$

Solution B

$$\frac{\sqrt{12}}{\sqrt{16}} = \sqrt{\frac{12}{16}} = \sqrt{\frac{3}{4}} = \frac{\sqrt{3}}{\sqrt{4}} = \frac{\sqrt{3}}{2}$$

Reduce the fraction

The two approaches to Example 3 illustrate the need to **think first and then push the pencil**. You may find one approach easier than another.

Now let's consider an example where neither the numerator nor the denominator of the radicand is a perfect square. Keep in mind that an expression is not simplified if there is a radical in the denominator.

EXAMPLE 4

Simplify $\sqrt{\frac{2}{3}}$.

Solution A

Form of 1

$$\sqrt{\frac{2}{3}} = \frac{\sqrt{2}}{\sqrt{3}} = \frac{\sqrt{2}}{\sqrt{3}} \cdot \frac{\sqrt{3}}{\sqrt{3}} = \frac{\sqrt{6}}{3}$$

Solution B

$$\sqrt{\frac{2}{3}} = \sqrt{\frac{2}{3} \cdot \frac{3}{3}} = \sqrt{\frac{6}{9}} = \frac{\sqrt{6}}{\sqrt{9}} = \frac{\sqrt{6}}{3}$$

We refer to the process we used to simplify the radical in Example 4 as **rationalizing the denominator**. Notice that the denominator becomes a rational number. There is more than one way to rationalize the denominator, as the next example shows.

EXAMPLE 5

Simplify $\frac{\sqrt{5}}{\sqrt{8}}$.

Solution A

$$\frac{\sqrt{5}}{\sqrt{8}} = \frac{\sqrt{5}}{\sqrt{8}} \cdot \frac{\sqrt{8}}{\sqrt{8}} = \frac{\sqrt{40}}{8} = \frac{\sqrt{4}\sqrt{10}}{8} = \frac{2\sqrt{10}}{8} = \frac{\sqrt{10}}{4}$$

Solution B

$$\frac{\sqrt{5}}{\sqrt{8}} = \frac{\sqrt{5}}{\sqrt{8}} \cdot \frac{\sqrt{2}}{\sqrt{2}} = \frac{\sqrt{10}}{\sqrt{16}} = \frac{\sqrt{10}}{4}$$

Solution C

$$\frac{\sqrt{5}}{\sqrt{8}} = \frac{\sqrt{5}}{\sqrt{4}\sqrt{2}} = \frac{\sqrt{5}}{2\sqrt{2}} = \frac{\sqrt{5}}{2\sqrt{2}} \cdot \frac{\sqrt{2}}{\sqrt{2}} = \frac{\sqrt{10}}{4}$$ ■

Study the following examples, and check that the answers are in simplest radical form according to the three conditions we listed on page 399.

EXAMPLE 6

Simplify each of these expressions.

(a) $\frac{3}{\sqrt{x}}$ **(b)** $\sqrt{\frac{2x}{3y}}$ **(c)** $\frac{3\sqrt{5}}{\sqrt{6}}$ **(d)** $\sqrt{\frac{4x^2}{9y}}$

Solution

(a) $\frac{3}{\sqrt{x}} = \frac{3}{\sqrt{x}} \cdot \frac{\sqrt{x}}{\sqrt{x}} = \frac{3\sqrt{x}}{x}$

(b) $\sqrt{\frac{2x}{3y}} = \frac{\sqrt{2x}}{\sqrt{3y}} = \frac{\sqrt{2x}}{\sqrt{3y}} \cdot \frac{\sqrt{3y}}{\sqrt{3y}} = \frac{\sqrt{6xy}}{3y}$

(c) $\frac{3\sqrt{5}}{\sqrt{6}} = \frac{3\sqrt{5}}{\sqrt{6}} \cdot \frac{\sqrt{6}}{\sqrt{6}} = \frac{3\sqrt{30}}{6} = \frac{\sqrt{30}}{2}$

(d) $\sqrt{\frac{4x^2}{9y}} = \frac{\sqrt{4x^2}}{\sqrt{9y}} = \frac{2x}{\sqrt{9}\sqrt{y}} = \frac{2x}{3\sqrt{y}} = \frac{2x}{3\sqrt{y}} \cdot \frac{\sqrt{y}}{\sqrt{y}} = \frac{2x\sqrt{y}}{3y}$ ■

Let's return again to the idea of simplifying expressions that contain radicals. Sometimes it may appear as if no simplifying can be done; however, after the individual radicals have been changed to simplest form, the distributive property may apply.

EXAMPLE 7

Simplify $5\sqrt{2} + \frac{3}{\sqrt{2}}$.

Solution

$$\begin{aligned}5\sqrt{2} + \frac{3}{\sqrt{2}} &= 5\sqrt{2} + \frac{3}{\sqrt{2}} \cdot \frac{\sqrt{2}}{\sqrt{2}} = 5\sqrt{2} + \frac{3\sqrt{2}}{2} \\ &= \left(5 + \frac{3}{2}\right)\sqrt{2} = \left(\frac{10}{2} + \frac{3}{2}\right)\sqrt{2} \\ &= \frac{13}{2}\sqrt{2} \quad \text{or} \quad \frac{13\sqrt{2}}{2}\end{aligned}$$ ■

EXAMPLE 8 Simplify $\sqrt{\frac{3}{2}} + \sqrt{24}$.

Solution

$$\sqrt{\frac{3}{2}} + \sqrt{24} = \frac{\sqrt{3}}{\sqrt{2}} + \sqrt{24} = \frac{\sqrt{3}}{\sqrt{2}} \cdot \frac{\sqrt{2}}{\sqrt{2}} + \sqrt{4}\sqrt{6}$$

$$= \frac{\sqrt{6}}{2} + 2\sqrt{6}$$

$$= \left(\frac{1}{2} + 2\right)\sqrt{6}$$

$$= \left(\frac{1}{2} + \frac{4}{2}\right)\sqrt{6}$$

$$= \frac{5}{2}\sqrt{6}$$

■

EXAMPLE 9 Simplify $3\sqrt{\frac{7}{4}} - 14\sqrt{\frac{1}{7}} + 5\sqrt{28}$.

Solution

$$3\sqrt{\frac{7}{4}} - 14\sqrt{\frac{1}{7}} + 5\sqrt{28} = \frac{3\sqrt{7}}{\sqrt{4}} - \frac{14\sqrt{1}}{\sqrt{7}} + 5\sqrt{4}\sqrt{7}$$

$$= \frac{3\sqrt{7}}{2} - \frac{14}{\sqrt{7}} \cdot \frac{\sqrt{7}}{\sqrt{7}} + 10\sqrt{7}$$

$$= \frac{3\sqrt{7}}{2} - \frac{14\sqrt{7}}{7} + 10\sqrt{7}$$

$$= \frac{3\sqrt{7}}{2} - 2\sqrt{7} + 10\sqrt{7}$$

$$= \left(\frac{3}{2} - 2 + 10\right)\sqrt{7}$$

$$= \left(\frac{3}{2} - \frac{4}{2} + \frac{20}{2}\right)\sqrt{7} = \frac{19}{2}\sqrt{7}$$

■

Problem Set 9.3

For Problems 1–10, evaluate each radical.

1. $\sqrt{\frac{16}{25}}$
2. $\sqrt{\frac{4}{49}}$
3. $-\sqrt{\frac{81}{9}}$
4. $-\sqrt{\frac{64}{16}}$
5. $\sqrt{\frac{1}{64}}$
6. $\sqrt{\frac{100}{121}}$
7. $\sqrt[3]{\frac{125}{64}}$
8. $\sqrt[3]{\frac{-27}{8}}$
9. $-\sqrt{\frac{25}{256}}$
10. $-\sqrt{\frac{289}{225}}$

For Problems 11–40, change each radical to simplest radical form.

11. $\sqrt{\frac{19}{25}}$
12. $\sqrt{\frac{17}{4}}$
13. $\sqrt{\frac{8}{49}}$
14. $\sqrt{\frac{24}{25}}$
15. $\frac{\sqrt[3]{375}}{\sqrt[3]{216}}$
16. $\frac{\sqrt[3]{16}}{\sqrt[3]{54}}$
17. $\frac{\sqrt{12}}{\sqrt{36}}$
18. $\frac{\sqrt{20}}{\sqrt{64}}$
19. $\sqrt{\frac{3}{2}}$
20. $\sqrt{\frac{2}{5}}$
21. $\sqrt{\frac{5}{8}}$
22. $\sqrt{\frac{7}{12}}$
23. $\frac{\sqrt{56}}{\sqrt{8}}$
24. $\frac{\sqrt{55}}{\sqrt{11}}$
25. $\frac{\sqrt{63}}{\sqrt{7}}$
26. $\frac{\sqrt{96}}{\sqrt{6}}$
27. $\frac{\sqrt{5}}{\sqrt{18}}$
28. $\frac{\sqrt{3}}{\sqrt{32}}$
29. $\frac{\sqrt{4}}{\sqrt{27}}$
30. $\frac{\sqrt{9}}{\sqrt{48}}$
31. $\sqrt{\frac{1}{24}}$
32. $\sqrt{\frac{1}{12}}$
33. $\frac{2\sqrt{3}}{\sqrt{5}}$
34. $\frac{3\sqrt{2}}{\sqrt{6}}$
35. $\frac{4\sqrt{2}}{3\sqrt{3}}$
36. $\frac{2\sqrt{5}}{7\sqrt{8}}$
37. $\frac{3\sqrt{7}}{4\sqrt{12}}$
38. $\frac{6\sqrt{12}}{5\sqrt{24}}$
39. $\sqrt{4\frac{1}{9}}$
40. $\sqrt{3\frac{1}{4}}$

For Problems 41–60, change each radical to simplest radical form. All variables represent positive real numbers.

41. $\frac{3}{\sqrt{x}}$
42. $\frac{2}{\sqrt{xy}}$
43. $\frac{5}{\sqrt{2x}}$
44. $\frac{7}{\sqrt{3y}}$
45. $\sqrt{\frac{3}{x}}$
46. $\sqrt{\frac{8}{x}}$
47. $\sqrt{\frac{12}{x^2}}$
48. $\sqrt{\frac{27}{4y^2}}$
49. $\frac{\sqrt{2x}}{\sqrt{5y}}$
50. $\frac{\sqrt{3y}}{\sqrt{32x}}$
51. $\frac{\sqrt{5x}}{\sqrt{27y}}$
52. $\frac{\sqrt{3x^2}}{\sqrt{5y^2}}$
53. $\frac{\sqrt{2x^3}}{\sqrt{8y}}$
54. $\frac{\sqrt{5x^2}}{\sqrt{45y^3}}$
55. $\sqrt{\frac{9}{x^3}}$
56. $\sqrt{\frac{25}{y^5}}$
57. $\frac{4}{\sqrt{x^7}}$
58. $\frac{14}{\sqrt{x^5}}$
59. $\frac{3\sqrt{x}}{2\sqrt{y^3}}$
60. $\frac{5\sqrt{x}}{7\sqrt{xy}}$

For Problems 61–72, simplify each expression.

61. $7\sqrt{3} + \sqrt{\frac{1}{3}}$
62. $-3\sqrt{2} + \sqrt{\frac{1}{2}}$
63. $4\sqrt{10} - \sqrt{\frac{2}{5}}$
64. $8\sqrt{5} - 3\sqrt{\frac{1}{5}}$
65. $-2\sqrt{5} - 5\sqrt{\frac{1}{5}}$
66. $6\sqrt{7} + 4\sqrt{\frac{1}{7}}$
67. $-3\sqrt{6} - \frac{5\sqrt{2}}{\sqrt{3}}$
68. $4\sqrt{8} - \frac{6}{\sqrt{2}}$
69. $4\sqrt{12} + \frac{3}{\sqrt{3}} - 5\sqrt{27}$
70. $3\sqrt{8} - 4\sqrt{18} - \frac{6}{\sqrt{2}}$
71. $\frac{9\sqrt{5}}{\sqrt{3}} - 6\sqrt{60} + \frac{10\sqrt{3}}{\sqrt{5}}$
72. $-2\sqrt{3} - 3\sqrt{48} + \frac{3}{\sqrt{3}}$

THOUGHTS INTO WORDS

73. Your friend simplifies $\sqrt{\frac{6}{8}}$ as follows:

$$\sqrt{\frac{6}{8}} = \frac{\sqrt{6}}{\sqrt{8}} = \frac{\sqrt{6}}{\sqrt{8}} \cdot \frac{\sqrt{8}}{\sqrt{8}} = \frac{\sqrt{48}}{8} = \frac{\sqrt{16}\sqrt{3}}{8}$$
$$= \frac{4\sqrt{3}}{8} = \frac{\sqrt{3}}{2}$$

Could you show him a much shorter way to simplify this expression?

74. Is the expression $3\sqrt{2} + \sqrt{50}$ in simplest radical form? Why or why not?

FURTHER INVESTIGATIONS

To rationalize the denominator of algebraic expressions that have cube roots, you must multiply by a factor in the form of 1 that makes the radicand in the denominator a perfect cube. Consider the following example and refer to a list of perfect cubes if you do not have them memorized.

Simplify $\sqrt[3]{\frac{5}{2}}$.

$$\sqrt[3]{\frac{5}{2}} = \frac{\sqrt[3]{5}}{\sqrt[3]{2}} = \frac{\sqrt[3]{5}}{\sqrt[3]{2}} \cdot \frac{\sqrt[3]{4}}{\sqrt[3]{4}} = \frac{\sqrt[3]{20}}{\sqrt[3]{8}} = \frac{\sqrt[3]{20}}{2}$$

Form of 1 (arrow pointing to $\frac{\sqrt[3]{4}}{\sqrt[3]{4}}$)

For Problems 75–80, change each radical to simplest form.

75. $\frac{\sqrt[3]{7}}{\sqrt[3]{3}}$ **76.** $\frac{\sqrt[3]{6}}{\sqrt[3]{4}}$ **77.** $\sqrt[3]{\frac{8}{25}}$

78. $\sqrt[3]{\frac{2}{9}}$ **79.** $\sqrt[3]{\frac{7}{32}}$ **80.** $\sqrt[3]{\frac{9}{10}}$

9.4 Products and Quotients Involving Radicals

We use Property 9.1 ($\sqrt{ab} = \sqrt{a}\sqrt{b}$) and Property 9.2 ($\sqrt[3]{ab} = \sqrt[3]{a}\sqrt[3]{b}$) to multiply radical expressions, and in some cases, to simplify the resulting radical. The following examples illustrate several types of multiplication problems that involve radicals.

EXAMPLE 1

Multiply and simplify where possible.

(a) $\sqrt{3}\sqrt{12}$ **(b)** $\sqrt{3}\sqrt{15}$ **(c)** $\sqrt{7}\sqrt{8}$

(d) $\sqrt[3]{4}\sqrt[3]{6}$ **(e)** $(3\sqrt{2})(4\sqrt{3})$ **(f)** $(2\sqrt[3]{9})(4\sqrt[3]{6})$

Solution

(a) $\sqrt{3}\sqrt{12} = \sqrt{36} = 6$

(b) $\sqrt{3}\sqrt{15} = \sqrt{45} = \sqrt{9}\sqrt{5} = 3\sqrt{5}$

(c) $\sqrt{7}\sqrt{8} = \sqrt{56} = \sqrt{4}\sqrt{14} = 2\sqrt{14}$

(d) $\sqrt[3]{4}\sqrt[3]{6} = \sqrt[3]{24} = \sqrt[3]{8}\sqrt[3]{3} = 2\sqrt[3]{3}$

(e) $(3\sqrt{2})(4\sqrt{3}) = 3 \cdot 4 \cdot \sqrt{2} \cdot \sqrt{3} = 12\sqrt{6}$

(f) $(2\sqrt[3]{9})(4\sqrt[3]{6}) = 2 \cdot 4 \cdot \sqrt[3]{9} \cdot \sqrt[3]{6}$

$$= 8\sqrt[3]{54}$$
$$= 8\sqrt[3]{27}\sqrt[3]{2}$$
$$= 8 \cdot 3 \cdot \sqrt[3]{2}$$
$$= 24\sqrt[3]{2}$$

■

Recall how we use the distributive property when we find the product of a monomial and a polynomial. For example, $2x(3x + 4) = 2x(3x) + 2x(4) = 6x^2 + 8x$. Likewise, the distributive property and Properties 9.1 and 9.2 provide the basis for finding certain special products involving radicals. The next examples demonstrate this idea.

EXAMPLE 2

Multiply and simplify where possible.

(a) $\sqrt{2}(\sqrt{3} + \sqrt{5})$ **(b)** $\sqrt{3}(\sqrt{12} - \sqrt{6})$

(c) $\sqrt{8}(\sqrt{2} - 3)$ **(d)** $\sqrt{x}(\sqrt{x} + \sqrt{y})$

(e) $\sqrt[3]{2}(\sqrt[3]{4} + \sqrt[3]{10})$

Solution

(a) $\sqrt{2}(\sqrt{3} + \sqrt{5}) = \sqrt{2}\sqrt{3} + \sqrt{2}\sqrt{5} = \sqrt{6} + \sqrt{10}$

(b) $\sqrt{3}(\sqrt{12} - \sqrt{6}) = \sqrt{3}\sqrt{12} - \sqrt{3}\sqrt{6}$

$$= \sqrt{36} - \sqrt{18}$$
$$= 6 - \sqrt{9}\sqrt{2}$$
$$= 6 - 3\sqrt{2}$$

(c) $\sqrt{8}(\sqrt{2} - 3) = \sqrt{8}\sqrt{2} - (\sqrt{8})(3)$

$$= \sqrt{16} - 3\sqrt{8}$$
$$= 4 - 3\sqrt{4}\sqrt{2}$$
$$= 4 - 6\sqrt{2}$$

(d) $\sqrt{x}(\sqrt{x} + \sqrt{y}) = \sqrt{x}\sqrt{x} + \sqrt{x}\sqrt{y}$

$$= \sqrt{x^2} + \sqrt{xy}$$
$$= x + \sqrt{xy}$$

(e) $\sqrt[3]{2}(\sqrt[3]{4} + \sqrt[3]{10}) = \sqrt[3]{8} + \sqrt[3]{20} = 2 + \sqrt[3]{20}$

■

The distributive property plays a central role when we find the product of two binomials. For example, $(x + 2)(x + 3) = x(x + 3) + 2(x + 3) = x^2 + 3x + 2x + 6 = x^2 + 5x + 6$. We can find the product of two binomial expressions involving radicals in a similar fashion.

EXAMPLE 3

Multiply and simplify.

(a) $(\sqrt{3} + \sqrt{5})(\sqrt{2} + \sqrt{6})$ **(b)** $(\sqrt{7} - 3)(\sqrt{7} + 6)$

Solution

(a)
$$\begin{aligned}(\sqrt{3} + \sqrt{5})(\sqrt{2} + \sqrt{6}) &= \sqrt{3}(\sqrt{2} + \sqrt{6}) + \sqrt{5}(\sqrt{2} + \sqrt{6})\\ &= \sqrt{3}\sqrt{2} + \sqrt{3}\sqrt{6} + \sqrt{5}\sqrt{2} + \sqrt{5}\sqrt{6}\\ &= \sqrt{6} + \sqrt{18} + \sqrt{10} + \sqrt{30}\\ &= \sqrt{6} + 3\sqrt{2} + \sqrt{10} + \sqrt{30}\end{aligned}$$

(b)
$$\begin{aligned}(\sqrt{7} - 3)(\sqrt{7} + 6) &= \sqrt{7}(\sqrt{7} + 6) - 3(\sqrt{7} + 6)\\ &= \sqrt{7}\sqrt{7} + 6\sqrt{7} - 3\sqrt{7} - 18\\ &= 7 + 6\sqrt{7} - 3\sqrt{7} - 18\\ &= -11 + 3\sqrt{7}\end{aligned}$$

■

If the binomials are of the form $(a + b)(a - b)$, then we can use the multiplication pattern $(a + b)(a - b) = a^2 - b^2$.

EXAMPLE 4

Multiply and simplify.

(a) $(\sqrt{6} + 2)(\sqrt{6} - 2)$ **(b)** $(3 - \sqrt{5})(3 + \sqrt{5})$

(c) $(\sqrt{8} + \sqrt{5})(\sqrt{8} - \sqrt{5})$

Solution

(a) $(\sqrt{6} + 2)(\sqrt{6} - 2) = (\sqrt{6})^2 - 2^2 = 6 - 4 = 2$

(b) $(3 - \sqrt{5})(3 + \sqrt{5}) = 3^2 - (\sqrt{5})^2 = 9 - 5 = 4$

(c) $(\sqrt{8} + \sqrt{5})(\sqrt{8} - \sqrt{5}) = (\sqrt{8})^2 - (\sqrt{5})^2 = 8 - 5 = 3$ ■

Note that in each part of Example 4, the final product contains no radicals. This happens whenever we multiply expressions such as $\sqrt{a} + \sqrt{b}$ and $\sqrt{a} - \sqrt{b}$, where a and b are rational numbers.

$$(\sqrt{a} + \sqrt{b})(\sqrt{a} - \sqrt{b}) = (\sqrt{a})^2 - (\sqrt{b})^2 = a - b$$

Expressions such as $\sqrt{8} + \sqrt{5}$ and $\sqrt{8} - \sqrt{5}$ are called **conjugates** of each other. Likewise, $\sqrt{6} + 2$ and $\sqrt{6} - 2$ are conjugates, as are $3 - \sqrt{5}$ and $3 + \sqrt{5}$. Now let's see how we can use conjugates to rationalize denominators.

EXAMPLE 2

Solve $\sqrt{x} = -3$.

Solution

$$\begin{aligned} \sqrt{x} &= -3 \\ (\sqrt{x})^2 &= (-3)^2 \quad \text{Square both sides} \\ x &= 9 \end{aligned}$$

Because $\sqrt{9} \neq -3$, 9 is not a solution and the solution set is $\varnothing$. ■

In general, squaring both sides of an equation produces an equation that has all of the solutions of the original equation, but it may also have some extra solutions that do not satisfy the original equation. (Such extra solutions are called **extraneous solutions** or **roots**.) Therefore, when using the "squaring" property (Property 9.4), you *must* check each potential solution in the original equation.

We now consider some examples to demonstrate different situations that arise when solving radical equations.

EXAMPLE 3

Solve $\sqrt{2x + 1} = 5$.

Solution

$$\begin{aligned} \sqrt{2x + 1} &= 5 \\ (\sqrt{2x + 1})^2 &= 5^2 \quad \text{Square both sides} \\ 2x + 1 &= 25 \\ 2x &= 24 \\ x &= 12 \end{aligned}$$

✔ **Check**

$$\begin{aligned} \sqrt{2x + 1} &= 5 \\ \sqrt{2(12) + 1} &\stackrel{?}{=} 5 \\ \sqrt{24 + 1} &\stackrel{?}{=} 5 \\ \sqrt{25} &\stackrel{?}{=} 5 \\ 5 &= 5 \end{aligned}$$

The solution set is $\{12\}$. ■

EXAMPLE 4 Solve $\sqrt{3x + 4} = -4$.

Solution

$$\sqrt{3x + 4} = -4$$

$$(\sqrt{3x + 4})^2 = (-4)^2 \quad \text{Square both sides}$$

$$3x + 4 = 16$$

$$3x = 12$$

$$x = 4$$

Check

$$\sqrt{3x + 4} = -4$$

$$\sqrt{3(4) + 4} \stackrel{?}{=} -4$$

$$\sqrt{16} \stackrel{?}{=} -4$$

$$4 \neq -4$$

Since 4 does not check (4 is an extraneous root), the equation $\sqrt{3x + 4} = -4$ has no real number solutions. The solution set is $\varnothing$. ■

EXAMPLE 5 Solve $3\sqrt{2y + 1} = 5$.

Solution

$$3\sqrt{2y + 1} = 5$$

$$\sqrt{2y + 1} = \frac{5}{3} \quad \text{Divided both sides by 3}$$

$$(\sqrt{2y + 1})^2 = \left(\frac{5}{3}\right)^2$$

$$2y + 1 = \frac{25}{9}$$

$$2y = \frac{25}{9} - 1$$

$$2y = \frac{25}{9} - \frac{9}{9}$$

$$2y = \frac{16}{9}$$

$$\frac{1}{2}(2y) = \frac{1}{2}\left(\frac{16}{9}\right) \quad \text{Multiplied both sides by } \frac{1}{2}$$

$$y = \frac{8}{9}$$

✔ **Check**

$$3\sqrt{2y + 1} = 5$$
$$3\sqrt{2\left(\frac{8}{9}\right) + 1} \stackrel{?}{=} 5$$
$$3\sqrt{\frac{16}{9} + \frac{9}{9}} \stackrel{?}{=} 5$$
$$3\sqrt{\frac{25}{9}} \stackrel{?}{=} 5$$
$$3\left(\frac{5}{3}\right) \stackrel{?}{=} 5$$
$$5 = 5$$

The solution set is $\left\{\frac{8}{9}\right\}$. ■

EXAMPLE 6 Solve $\sqrt{5s - 2} = \sqrt{2s + 19}$.

Solution

$$\sqrt{5s - 2} = \sqrt{2s + 19}$$
$$(\sqrt{5s - 2})^2 = (\sqrt{2s + 19})^2 \quad \text{Square both sides}$$
$$5s - 2 = 2s + 19$$
$$3s = 21$$
$$s = 7$$

✔ **Check**

$$\sqrt{5s - 2} = \sqrt{2s + 19}$$
$$\sqrt{5(7) - 2} \stackrel{?}{=} \sqrt{2(7) + 19}$$
$$\sqrt{33} = \sqrt{33}$$

The solution set is $\{7\}$. ■

EXAMPLE 7 Solve $\sqrt{2y - 4} = y - 2$.

Solution

$$\sqrt{2y - 4} = y - 2$$
$$(\sqrt{2y - 4})^2 = (y - 2)^2 \quad \text{Square both sides}$$
$$2y - 4 = y^2 - 4y + 4$$

$$0 = y^2 - 6y + 8$$

$$0 = (y - 4)(y - 2) \quad \text{Factor the right side}$$

$$y - 4 = 0 \quad \text{or} \quad y - 2 = 0 \quad \text{Remember the property: } ab = 0 \text{ if and only if } a = 0 \text{ or } b = 0$$

$$y = 4 \quad \text{or} \quad y = 2$$

Check

$$\sqrt{2y - 4} = y - 2 \qquad \sqrt{2y - 4} = y - 2$$

$$\sqrt{2(4) - 4} \stackrel{?}{=} 4 - 2 \qquad \sqrt{2(2) - 4} \stackrel{?}{=} 2 - 2$$

$$\sqrt{4} \stackrel{?}{=} 2 \quad \text{or} \quad \sqrt{0} \stackrel{?}{=} 0$$

$$2 = 2 \qquad 0 = 0$$

The solution set is $\{2, 4\}$. ■

EXAMPLE 8 Solve $\sqrt{x} + 6 = x$.

Solution

$$\sqrt{x} + 6 = x$$

$$\sqrt{x} = x - 6 \quad \text{We added } -6 \text{ to both sides so that the term with the radical is alone on one side of the equation}$$

$$(\sqrt{x})^2 = (x - 6)^2$$

$$x = x^2 - 12x + 36$$

$$0 = x^2 - 13x + 36$$

$$0 = (x - 4)(x - 9)$$

$$x - 4 = 0 \quad \text{or} \quad x - 9 = 0 \quad \text{Apply } ab = 0 \text{ if and only if } a = 0 \text{ or } b = 0$$

$$x = 4 \quad \text{or} \quad x = 9$$

Check

$$\sqrt{x} + 6 = x \qquad \sqrt{x} + 6 = x$$

$$\sqrt{4} + 6 \stackrel{?}{=} 4 \qquad \sqrt{9} + 6 \stackrel{?}{=} 9$$

$$2 + 6 \stackrel{?}{=} 4 \quad \text{or} \quad 3 + 6 \stackrel{?}{=} 9$$

$$8 \neq 4 \qquad 9 = 9$$

The solution set is $\{9\}$. ■

Note in Example 8 that we changed the form of the original equation, $\sqrt{x} + 6 = x$, to $\sqrt{x} = x - 6$ before we squared both sides. Squaring both sides of $\sqrt{x} + 6 = x$ produces $x + 12\sqrt{x} + 36 = x^2$, a more complex equation that still contains a radical. So again, it pays to think ahead a few steps before carrying out the details of the problem.

■ Another Look at Applications

In Section 9.1 we used the formula $S = \sqrt{30Df}$ to approximate how fast a car was traveling based on the length of skid marks. (Remember that S represents the speed of the car in miles per hour, D the length of skid marks measured in feet, and f a coefficient of friction.) This same formula can be used to estimate the lengths of skid marks that are produced by cars traveling at different rates on various types of road surfaces. To use the formula for this purpose, we change the form of the equation by solving for D.

$$\sqrt{30Df} = S$$

$$30Df = S^2 \quad \text{The result of squaring both sides of the original equation}$$

$$D = \frac{S^2}{30f} \quad \text{D, S, and f are positive numbers, so this final equation and the original one are equivalent}$$

PROBLEM 1

Suppose that for a particular road surface the coefficient of friction is 0.35. How far will a car traveling at 60 miles per hour skid when the brakes are applied?

Solution

We substitute 0.35 for f and 60 for S in the formula $D = \frac{S^2}{30f}$.

$$D = \frac{60^2}{30(0.35)} = 343 \quad \text{to the nearest whole number}$$

The car will skid approximately 343 feet. ■

Remark: Pause for a moment and think about the result in Example 8. The coefficient of friction of 0.35 applies to a wet concrete road surface. Note that a car traveling at 60 miles per hour will skid farther than the length of a football field.

Problem Set 9.5

For Problems 1–40, solve each equation. Be sure to check all potential solutions in the original equation.

1. $\sqrt{x} = 7$

2. $\sqrt{x} = 12$

3. $\sqrt{2x} = 6$

4. $\sqrt{3x} = 9$

5. $\sqrt{3x} = -6$

6. $\sqrt{2x} = -8$

7. $\sqrt{4x} = 3$

8. $\sqrt{5x} = 4$

9. $3\sqrt{x} = 2$

10. $4\sqrt{x} = 3$

11. $\sqrt{2n - 3} = 5$

12. $\sqrt{3n + 1} = 7$

13. $\sqrt{5y + 2} = -1$

14. $\sqrt{4n - 3} - 4 = 0$

15. $\sqrt{6x - 5} - 3 = 0$

16. $\sqrt{5x + 3} = -4$

17. $5\sqrt{x} = 30$

18. $6\sqrt{x} = 42$

19. $\sqrt{3a - 2} = \sqrt{2a + 4}$

20. $\sqrt{4a + 3} = \sqrt{5a - 4}$

21. $\sqrt{7x - 3} = \sqrt{4x + 3}$

22. $\sqrt{8x - 6} = \sqrt{4x + 11}$

23. $2\sqrt{y} + 1 = 5$

24. $3\sqrt{y} - 2 = 4$

25. $\sqrt{x + 3} = x + 3$

26. $\sqrt{x + 7} = x + 7$

27. $\sqrt{-2x + 28} = x - 2$

28. $\sqrt{-2x} = x + 4$

29. $\sqrt{3n - 4} = \sqrt{n}$

30. $\sqrt{5n - 1} = \sqrt{2n}$

31. $\sqrt{3x} = x - 6$

32. $2\sqrt{x} = x - 3$

33. $4\sqrt{x} + 5 = x$

34. $\sqrt{-x} - 6 = x$

35. $\sqrt{x^2 + 27} = x + 3$

36. $\sqrt{x^2 - 35} = x - 5$

37. $\sqrt{x^2 + 2x + 3} = x + 2$

38. $\sqrt{x^2 + x + 4} = x + 3$

39. $\sqrt{8x} - 2 = x$

40. $\sqrt{2x - 4} = x - 6$

41. Go to the formula given in the solution to Problem 1; use a coefficient of friction of 0.95. How far will a car skid at 40 miles per hour? At 55 miles per hour? At 65 miles per hour? Express the answers to the nearest foot.

42. Solve the formula $T = 2\pi\sqrt{\frac{L}{32}}$ for L. (Remember that in this formula, which was used in Section 9.1, T represents the period of a pendulum expressed in seconds, and L represents the length of the pendulum in feet.)

43. In Problem 42 you should have obtained the equation $L = \frac{8T^2}{\pi^2}$. What is the length of a pendulum that has a period of 2 seconds? Of 2.5 seconds? Of 3 seconds? Use 3.14 as an approximation for π and express the answers to the nearest tenth of a foot.

THOUGHTS INTO WORDS

44. Explain in your own words why possible solutions for radical equations *must* be checked.

45. Your friend attempts to solve the equation

$$3 + 2\sqrt{x} = x$$

as follows:

$$(3 + 2\sqrt{x})^2 = x^2$$

$$9 + 12\sqrt{x} + 4x = x^2$$

At this step, she stops and doesn't know how to proceed. What help can you give her?

FURTHER INVESTIGATIONS

For Problems 46–49, solve each of the equations.

46. $\sqrt{x + 1} = 5 - \sqrt{x - 4}$

47. $\sqrt{x - 2} = \sqrt{x + 7} - 1$

48. $\sqrt{2n - 1} + \sqrt{n - 5} = 3$

49. $\sqrt{2n + 1} - \sqrt{n - 3} = 2$

50. Suppose that we solve the equation $x^2 = a^2$ for x as follows:

$$x^2 - a^2 = 0$$

$$(x + a)(x - a) = 0$$

$$x + a = 0 \quad \text{or} \quad x - a = 0$$

$$x = -a \quad \text{or} \quad x = a$$

What is the significance of this result?

To solve an equation with cube roots, we raise each side of the equation to the third power. Consider the following example.

$$\sqrt[3]{x - 4} = 5$$

$$(\sqrt[3]{x - 4})^3 = 5^3$$

$$x - 4 = 125$$

$$x = 129$$

For Problems 51–55, solve each equation.

51. $\sqrt[3]{x + 7} = 4$

52. $\sqrt[3]{x - 12} = -2$

53. $\sqrt[3]{2x - 4} = 6$

54. $\sqrt[3]{4x - 5} = 3$

55. $\sqrt[3]{4x - 8} = 10$

Chapter 9 Summary

(9.1) The number a is a **square root** of b if $a^2 = b$. The symbol $\sqrt{\ }$ is called a **radical sign** and indicates the **nonnegative** or **principal square root**. The number under the radical sign is called the **radicand**. An entire expression, such as $\sqrt{37}$, is called a **radical**. The ***n*th root** of b is written $\sqrt[n]{b}$, and n is called the **index**.

The distributive property provides the basis for combining similar terms in radical expressions. Using the distributive property, the expression $5\sqrt{2} + 7\sqrt{2}$ simplifies to $(5 + 7)\sqrt{2} = 12\sqrt{2}$. Remember that radical expressions are similar terms when they have the same radicand and index.

(9.2) The properties

$$\sqrt{ab} = \sqrt{a}\sqrt{b} \quad \text{and} \quad \sqrt[3]{ab} = \sqrt[3]{a}\sqrt[3]{b}$$

provide the basis for changing radicals to **simplest radical form**.

(9.3) Use the properties

$$\sqrt{\frac{a}{b}} = \frac{\sqrt{a}}{\sqrt{b}} \quad \text{and} \quad \sqrt[3]{\frac{a}{b}} = \frac{\sqrt[3]{a}}{\sqrt[3]{b}}$$

when **rationalizing the denominator**.

An algebraic expression that contains radicals is said to be in **simplest radical form** if it satisfies the following conditions:

1. No fraction appears within a radical sign. $\left(\sqrt{\frac{3}{5}}\right.$ violates this condition.$\left.\right)$
2. No radical appears in the denominator. $\left(\frac{3}{\sqrt{10}}\right.$ violates this condition.$\left.\right)$
3. No radicand when expressed in prime factored form contains a factor raised to a power equal to or greater than the index. ($\sqrt{8} = \sqrt{2^3}$ violates this condition.)

(9.4) Use the properties $\sqrt{ab} = \sqrt{a}\sqrt{b}$ and $\sqrt[3]{ab} = \sqrt[3]{a}\sqrt[3]{b}$ to multiply radicals and to simplify the resulting radical when appropriate.

(9.5) Use the property of equality, if $a = b$, then $a^2 = b^2$, to solve equations that contain radicals.

In general, squaring both sides of an equation produces an equation that has all of the solutions of the original equation, but it may also have some extra solutions that do not satisfy the original equation. Therefore, when using the "squaring" property, you must check each potential solution in the original equation.

Chapter 9 Review Problem Set

For Problems 1–6, evaluate each expression without using a table or a calculator.

1. $\sqrt{64}$ **2.** $-\sqrt{49}$ **3.** $\sqrt{1600}$

4. $\sqrt{\frac{81}{25}}$ **5.** $-\sqrt{\frac{4}{9}}$ **6.** $\sqrt{\frac{49}{36}}$

For Problems 7–20, change each number to simplest radical form.

7. $\sqrt{20}$ **8.** $\sqrt{32}$ **9.** $5\sqrt{8}$

10. $\sqrt{80}$ **11.** $2\sqrt[3]{-125}$ **12.** $\frac{\sqrt[3]{40}}{\sqrt[3]{8}}$

13. $\frac{\sqrt{36}}{\sqrt{7}}$ **14.** $\sqrt{\frac{7}{8}}$ **15.** $\sqrt{\frac{8}{24}}$

16. $\frac{3\sqrt{2}}{\sqrt{5}}$ **17.** $\frac{4\sqrt{3}}{\sqrt{12}}$ **18.** $\frac{5\sqrt{2}}{2\sqrt{3}}$

19. $\frac{-3\sqrt{2}}{\sqrt{27}}$ **20.** $\frac{4\sqrt{6}}{3\sqrt{12}}$

For Problems 21–24, use the fact that $\sqrt{3} = 1.73$, to the nearest hundredth, to help evaluate each of the following. Express your final answers to the nearest tenth.

21. $\sqrt{27}$ **22.** $\dfrac{2}{\sqrt{3}}$

23. $3\sqrt{12} + \sqrt{48}$ **24.** $2\sqrt{27} - 2\sqrt{75}$

For Problems 25–36, change each expression to simplest radical form. All variables represent positive real numbers.

25. $\sqrt{12a^2b^3}$ **26.** $\sqrt{50xy^4}$

27. $\sqrt{48x^3y^2}$ **28.** $\sqrt[3]{125a^2b}$

29. $\dfrac{4}{3}\sqrt{27xy^2}$ **30.** $\dfrac{3}{4}(\sqrt[3]{24x^3})$

31. $\dfrac{\sqrt{2x}}{\sqrt{5y}}$ **32.** $\dfrac{\sqrt{72x}}{\sqrt{16y}}$ **33.** $\sqrt{\dfrac{4}{x}}$

34. $\sqrt{\dfrac{2x^3}{9}}$ **35.** $\dfrac{3\sqrt{x}}{4\sqrt{y^3}}$ **36.** $\dfrac{-2\sqrt{x^2y}}{5\sqrt{xy}}$

For Problems 37–46, find the products and express your answers in simplest radical form.

37. $(\sqrt{6})(\sqrt{12})$ **38.** $(2\sqrt{3})(3\sqrt{6})$

39. $(-5\sqrt{8})(2\sqrt{2})$ **40.** $(2\sqrt[3]{7})(5\sqrt[3]{4})$

41. $\sqrt[3]{2}(\sqrt[3]{3} + \sqrt[3]{4})$ **42.** $3\sqrt{5}(\sqrt{8} - 2\sqrt{12})$

43. $(\sqrt{3} + \sqrt{5})(\sqrt{3} + \sqrt{7})$

44. $(2\sqrt{3} + 3\sqrt{2})(\sqrt{3} - 5\sqrt{2})$

45. $(\sqrt{6} + 2\sqrt{7})(3\sqrt{6} - \sqrt{7})$

46. $(3 + 2\sqrt{5})(4 - 3\sqrt{5})$

For Problems 47–50, rationalize the denominators and simplify.

47. $\dfrac{5}{\sqrt{7} - \sqrt{5}}$ **48.** $\dfrac{\sqrt{6}}{\sqrt{3} - \sqrt{2}}$

49. $\dfrac{2}{3\sqrt{2} - \sqrt{6}}$ **50.** $\dfrac{\sqrt{6}}{3\sqrt{7} + 2\sqrt{10}}$

For Problems 51–56, simplify each radical expression.

51. $2\sqrt{50} + 3\sqrt{72} - 5\sqrt{8}$

52. $\sqrt{8x} - 3\sqrt{18x}$ **53.** $9\sqrt[3]{2} - 5\sqrt[3]{16}$

54. $3\sqrt{10} + \sqrt{\dfrac{2}{5}}$ **55.** $4\sqrt{20} - \dfrac{3}{\sqrt{5}} + \sqrt{45}$

56. $\sqrt{\dfrac{2}{3}} - 2\sqrt{54}$

For Problems 57–62, solve each of the equations.

57. $\sqrt{5x + 6} = 6$ **58.** $\sqrt{6x + 1} = \sqrt{3x + 13}$

59. $3\sqrt{n} = n$ **60.** $\sqrt{y + 5} = y + 5$

61. $\sqrt{-3a + 10} = a - 2$

62. $3 - \sqrt{2x - 1} = 2$

For Problems 63–65, use a calculator to evaluate each expression.

63. $\sqrt{2116}$ **64.** $\sqrt{4356}$ **65.** $\sqrt{5184}$

For Problems 66–68, use a calculator to find a whole number approximation for each expression.

66. $\sqrt{690}$ **67.** $\sqrt{2185}$ **68.** $\sqrt{5500}$

Chapter 9 Test

1. Evaluate $-\sqrt{\frac{64}{49}}$.

2. Evaluate $\sqrt{0.0025}$.

For Problems 3–5, use the fact that $\sqrt{2} = 1.41$, to the nearest hundredth, and evaluate each of the following to the nearest tenth.

3. $\sqrt{8}$

4. $-\sqrt{32}$

5. $\frac{3}{\sqrt{2}}$

For Problems 6–14, change each radical expression to simplest radical form. All variables represent positive real numbers.

6. $\sqrt{45}$

7. $-4\sqrt[3]{54}$

8. $\frac{2\sqrt{3}}{3\sqrt{6}}$

9. $\sqrt{\frac{25}{2}}$

10. $\frac{\sqrt{24}}{\sqrt{36}}$

11. $\sqrt{\frac{5}{8}}$

12. $\sqrt[3]{-250x^4y^3}$

13. $\frac{\sqrt{3x}}{\sqrt{5y}}$

14. $\frac{3}{4}\sqrt{48x^3y^2}$

For Problems 15–18, find the indicated products and express the answers in simplest radical form.

15. $(\sqrt{8})(\sqrt{12})$

16. $(6\sqrt[3]{5})(4\sqrt[3]{2})$

17. $\sqrt{6}(2\sqrt{12} - 3\sqrt{8})$

18. $(2\sqrt{5} + \sqrt{3})(\sqrt{5} - 3\sqrt{3})$

19. Rationalize the denominator and simplify: $\frac{\sqrt{6}}{\sqrt{12} + \sqrt{2}}$.

20. Simplify $2\sqrt{24} - 4\sqrt{54} + 3\sqrt{96}$.

21. Find a whole number approximation for $\sqrt{500}$.

For Problems 22–25, solve each equation.

22. $\sqrt{3x + 1} = 4$

23. $\sqrt{2x - 5} = -4$

24. $\sqrt{n - 3} = 3 - n$

25. $\sqrt{3x + 6} = x + 2$

Chapters

For Problems 1–6, evaluate each of the numerical expressions.

1. -2^6

2. $\left(\frac{1}{4}\right)^{-3}$

3. $\left(\frac{1}{3} - \frac{1}{4}\right)^{-2}$

4. $-\sqrt{64}$

5. $\sqrt{\frac{4}{9}}$

6. $3^0 + 3^{-1} + 3^{-2}$

For Problems 7–10, evaluate each algebraic expression for the given values of the variables.

7. $3(2x - 1) - 4(2x + 3) - (x + 6)$ for $x = -4$

8. $(3x^2 - 4x - 6) - (3x^2 + 3x + 1)$ for $x = 6$

9. $2(a - b) - 3(2a + b) + 2(a - 3b)$ for $a = -2$ and $b = 3$

10. $x^2 - 2xy + y^2$ for $x = 5$ and $y = -2$

For Problems 11–25, perform the indicated operations and express your answers in simplest form using positive exponents only.

11. $\frac{3}{4x} + \frac{5}{2x} - \frac{7}{x}$

12. $\frac{3}{x - 2} - \frac{4}{x + 3}$

13. $\frac{3x}{7y} \div \frac{6x}{35y^2}$

14. $\frac{x - 2}{x^2 + x - 6} \cdot \frac{x^2 + 6x + 9}{x^2 - x - 12}$

15. $\frac{7}{x^2 + 3x - 18} - \frac{8}{x - 3}$

16. $(-3xy)(-4y^2)(5x^3y)$

17. $(-4x^{-5})(2x^3)$

18. $\frac{-12a^{-2}b^3}{4a^{-5}b^4}$

19. $(3n^4)^{-1}$

20. $(9x - 2)(3x + 4)$

21. $(-x - 1)(5x + 7)$

22. $(3x + 1)(2x^2 - x - 4)$

23. $\frac{15x^6y^8 - 20x^3y^5}{5x^3y^2}$

24. $(10x^3 - 8x^2 - 17x - 3) \div (5x + 1)$

25. $\frac{\frac{1}{x} - \frac{1}{y}}{\frac{1}{xy}}$

26. If 2 gallons of paint will cover 1500 square feet of walls, how many gallons are needed for 3500 square feet?

27. 18 is what percent of 72?

28. Solve $V = \frac{1}{3}Bh$ for B if $V = 432$ and $h = 12$.

29. How many feet of fencing are needed to enclose a rectangular garden that measures 25 feet by 40 feet?

30. Find the total surface area of a sphere that has a radius 5 inches long. Use 3.14 as an approximation for π.

31. Write each number in scientific notation.

(a) 85,000 **(b)** 0.0009

(c) 0.00000104 **(d)** 53,000,000

For Problems 32–37, factor each expression completely.

32. $12x^3 + 14x^2 - 40x$

33. $12x^2 - 27$

34. $xy + 3x - 2y - 6$

35. $30 + 19x - 5x^2$

36. $4x^4 - 4$

37. $21x^2 + 22x - 8$

For Problems 38–43, change each radical expression to simplest radical form.

38. $4\sqrt{28}$

39. $-\sqrt{45}$

40. $\sqrt{\frac{36}{5}}$

41. $\frac{5\sqrt{8}}{6\sqrt{12}}$

42. $\sqrt{72xy^5}$

43. $\frac{-2\sqrt{ab^2}}{5\sqrt{b}}$

For Problems 44–46, find each product and express your answer in simplest radical form.

44. $(3\sqrt{8})(4\sqrt{2})$

45. $6\sqrt{2}(9\sqrt{8} - 3\sqrt{12})$

46. $(3\sqrt{2} - \sqrt{7})(3\sqrt{2} + \sqrt{7})$

For Problems 47 and 48, rationalize the denominator and simplify.

47. $\dfrac{4}{\sqrt{3}+\sqrt{2}}$

48. $\dfrac{-6}{3\sqrt{5}-\sqrt{6}}$

For Problems 49 and 50, simplify each of the radical expressions.

49. $3\sqrt{50}-7\sqrt{72}+4\sqrt{98}$

50. $\dfrac{2}{3}\sqrt{20}-\dfrac{3}{4}\sqrt{45}+\sqrt{80}$

For Problems 51–55, graph each of the equations.

51. $3x - 6y = -6$

52. $y = \dfrac{1}{3}x + 4$

53. $y = -\dfrac{2}{5}x + 3$

54. $y - 2x = 0$

55. $y = -x$

For Problems 56–58, graph each linear inequality.

56. $y \geq 2x - 6$

57. $3x - 2y < -6$

58. $-2x - 4y > 8$

For Problems 59–64, solve each of the problems.

59. Find the slope of the line determined by the points $(-3, 6)$ and $(2, -4)$.

60. Find the slope of the line determined by the equation $4x - 7y = 12$.

61. Write the equation of the line that has a slope of $\dfrac{2}{3}$ and contains the point $(7, 2)$.

62. Write the equation of the line that contains the points $(-4, 1)$ and $(-1, -3)$.

63. Write the equation of the line that has a slope of $-\dfrac{1}{4}$ and a y intercept of -3.

64. Find the slope of a line whose equation is $3x - 2y = 12$.

For Problems 65–68, solve each of the systems by using either the substitution method or the addition method.

65. $\begin{pmatrix} y = 3x - 5 \\ 3x + 4y = -5 \end{pmatrix}$

66. $\begin{pmatrix} 4x - 3y = -20 \\ 3x + 5y = 14 \end{pmatrix}$

67. $\begin{pmatrix} \dfrac{1}{2}x - \dfrac{2}{3}y = -11 \\ \dfrac{1}{3}x + \dfrac{5}{6}y = 8 \end{pmatrix}$

68. $\begin{pmatrix} 2x + 7y = 22 \\ 4x - 5y = -13 \end{pmatrix}$

For Problems 69–80, solve each equation.

69. $-2(n - 1) + 4(2n - 3) = 4(n + 6)$

70. $\dfrac{4}{x-1} = \dfrac{-1}{x+6}$

71. $\dfrac{t-1}{3} - \dfrac{t+2}{4} = -\dfrac{5}{12}$

72. $-7 - 2n - 6n = 7n - 5n + 12$

73. $\dfrac{n-5}{2} = 3 - \dfrac{n+4}{5}$

74. $0.11x + 0.14(x + 400) = 181$

75. $\dfrac{x}{60-x} = 7 + \dfrac{4}{60-x}$

76. $1 + \dfrac{x+1}{2x} = \dfrac{3}{4}$

77. $x^2 + 4x - 12 = 0$

78. $2x^2 - 8 = 0$

79. $\sqrt{3x-6} = 9$

80. $\sqrt{3n} - 2 = 7$

For Problems 81–86, solve each of the inequalities.

81. $-3n - 4 \leq 11$

82. $-5 > 3n - 4 - 7n$

83 $2(x - 2) + 3(x + 4) > 6$

84. $\dfrac{1}{2}n - \dfrac{2}{3}n < -1$

85. $\dfrac{x+1}{2} + \dfrac{x-2}{6} < \dfrac{3}{8}$

86. $\dfrac{x-3}{7} - \dfrac{x-2}{4} \leq \dfrac{9}{14}$

For Problems 87–97, set up an equation, an inequality, or a system of equations to help solve each problem.

87. If two angles are supplementary and the larger angle is 15° less than twice the smaller angle, find the measure of each angle.

88. The sum of two numbers is 50. If the larger number is 2 less than three times the smaller number, find the numbers.

89. The sum of the squares of the two consecutive odd whole numbers is 130. Find the numbers.

90. Suppose that Nick has 47 coins consisting of nickels, dimes, and quarters. The number of dimes is 1 more than twice the number of nickels, and the number of quarters is 4 more than three times the number of nickels. Find the number of coins of each denomination.

91. If a home valued at $140,000 is assessed $2940 in real estate taxes; at the same rate, what would the taxes be on a home assessed at $180,000?

92. A retailer has some skirts that cost her $30 each. She wants to sell them at a profit of 60% of the cost. What price should she charge for the skirts?

93. Rosa leaves a town traveling in her car at a rate of 45 miles per hour. One hour later Polly leaves the same town traveling the same route at a rate of 55 miles per hour. How long will it take Polly to overtake Rosa?

94. How many milliliters of pure acid must be added to 100 milliliters of a 10% acid solution to obtain a 20% solution?

95. Suppose that Andy has scores of 85, 90, and 86 on his first three algebra tests. What score must he get on the fourth algebra test to have an average of 88 or higher for the four tests?

96. The Cubs have won 70 games and lost 72 games. They have 20 more games to play. To win more than 50% of all their games, how many of the 20 games remaining must they win?

97. Seth can do a job in 20 minutes. Butch can do the same job in 30 minutes. If they work together, how long will it take them to complete the job?

For additional word problems see Appendix B. All Appendix problems with references to chapters 3–8 would be appropriate.

10

Quadratic Equations

The quadratic equation $t(2t - 8) = 330$ can be used to determine the number of trees per row and the number of rows, given that the number of trees per row is eight less than twice the number of rows in an orchard of 330 trees.

The area of a tennis court for singles play is 2106 square feet. The length of the court is $\frac{26}{9}$ times the width. Find the length and width of the court. We can use the quadratic equation $x\left(\frac{26}{9}x\right) = 2106$ to determine that the court is 27 feet wide and 78 feet long. Solving equations has been a central theme of this text. We now pause for a moment and reflect on the different types of equations that we have solved.

Type of equation	Examples
First-degree equations of one variable	$4x + 3 = 7x + 1$; $3(x - 6) = 9$
Second-degree equations of one variable that are factorable	$x^2 + 3x = 0$; $x^2 + 5x + 6 = 0$; $x^2 - 4 = 0$; $x^2 + 10x + 25 = 0$
Fractional equations	$\frac{3}{x} + \frac{2}{x} = 4$; $\frac{5}{a-2} = \frac{6}{a+3}$; $\frac{2}{x^2-4} + \frac{5}{x+2} = \frac{6}{x-2}$

(*continued*)

Type of equation	Examples
Radical equations	$\sqrt{x} = 4$; $\sqrt{y+2} = 3$; $\sqrt{a+1} = \sqrt{2a-7}$
Systems of equations	$\begin{pmatrix} 2x + 3y = 4 \\ 5x - \ \ y = 7 \end{pmatrix}$; $\begin{pmatrix} 3a + 5b = 9 \\ 7a - 9b = 12 \end{pmatrix}$

As indicated in the chart, we have learned how to solve some second-degree equations, but only those for which the quadratic polynomial is factorable. In this chapter, we will expand our work to include more general types of second-degree equations in one variable and thus broaden our problem-solving capabilities.

10.1 Quadratic Equations

A second-degree equation in one variable contains the variable with an exponent of 2, but no higher power. Such equations are also called **quadratic equations**. Here are some examples of quadratic equations.

$$x^2 = 25 \qquad y^2 + 6y = 0 \qquad x^2 + 7x - 4 = 0$$

$$4y^2 + 2y - 1 = 0 \qquad 5x^2 + 2x - 1 = 2x^2 + 6x - 5$$

We can also define a quadratic equation in the variable x as any equation that can be written in the form $ax^2 + bx + c = 0$, where a, b, and c are real numbers and $a \neq 0$. We refer to the form $ax^2 + bx + c = 0$ as the **standard form** of a quadratic equation.

In Chapter 6, you solved quadratic equations (we didn't use the term "quadratic" at that time) by factoring and applying the property, $ab = 0$ if and only if $a = 0$ or $b = 0$. Let's review a few examples of that type.

EXAMPLE 1

Solve $x^2 - 13x = 0$.

Solution

$$x^2 - 13x = 0$$

$$x(x - 13) = 0 \qquad \text{Factor left side of equation}$$

$$x = 0 \quad \text{or} \quad x - 13 = 0 \qquad \text{Apply } ab = 0 \text{ if and only if } a = 0 \text{ or } b = 0$$

$$x = 0 \quad \text{or} \quad x = 13$$

The solution set is $\{0, 13\}$. Don't forget to check these solutions! ■

EXAMPLE 2 Solve $n^2 + 2n - 24 = 0$.

Solution

$$n^2 + 2n - 24 = 0$$

$$(n + 6)(n - 4) = 0 \qquad \text{Factor left side}$$

$$n + 6 = 0 \quad \text{or} \quad n - 4 = 0 \qquad \text{Apply } ab = 0 \text{ if and only if } a = 0 \text{ or } b = 0$$

$$n = -6 \quad \text{or} \quad n = 4$$

The solution set is $\{-6, 4\}$. ■

EXAMPLE 3 Solve $x^2 + 6x + 9 = 0$.

Solution

$$x^2 + 6x + 9 = 0$$

$$(x + 3)(x + 3) = 0 \qquad \text{Factor left side}$$

$$x + 3 = 0 \quad \text{or} \quad x + 3 = 0 \qquad \text{Apply } ab = 0 \text{ if and only if } a = 0 \text{ or } b = 0$$

$$x = -3 \quad \text{or} \quad x = -3$$

The solution set is $\{-3\}$. ■

EXAMPLE 4 Solve $y^2 = 49$.

Solution

$$y^2 = 49$$

$$y^2 - 49 = 0$$

$$(y + 7)(y - 7) = 0 \qquad \text{Factor left side}$$

$$y + 7 = 0 \quad \text{or} \quad y - 7 = 0 \qquad \text{Apply } ab = 0 \text{ if and only if } a = 0 \text{ or } b = 0$$

$$y = -7 \quad \text{or} \quad y = 7$$

The solution set is $\{-7, 7\}$. ■

Note the type of equation that we solved in Example 4. We can generalize from that example and consider the equation $x^2 = a$, where a is any nonnegative real number. We can solve this equation as follows:

$$x^2 = a$$

$$x^2 = (\sqrt{a})^2 \qquad (\sqrt{a})^2 = a$$

$$x^2 - (\sqrt{a})^2 = 0$$

$$(x - \sqrt{a})(x + \sqrt{a}) = 0 \qquad \text{Factor left side}$$

$$x - \sqrt{a} = 0 \quad \text{or} \quad x + \sqrt{a} = 0 \qquad \text{Apply } ab = 0 \text{ if and only if } a = 0 \text{ or } b = 0$$

$$x = \sqrt{a} \quad \text{or} \quad x = -\sqrt{a}$$

The solutions are $\sqrt{a}$ and $-\sqrt{a}$. We shall consider this result as a general property and use it to solve certain types of quadratic equations.

Property 10.1

For any nonnegative real number a,

$$x^2 = a \quad \text{if and only if } x = \sqrt{a} \text{ or } x = -\sqrt{a}$$

(The statement "$x = \sqrt{a}$ or $x = -\sqrt{a}$" can be written as $x = \pm\sqrt{a}$.)

Property 10.1 is sometimes referred to as the **square-root property**. This property, along with our knowledge of square roots, makes it very easy to solve quadratic equations of the form $x^2 = a$.

EXAMPLE 5

Solve $x^2 = 81$.

Solution

$$x^2 = 81$$

$$x = \pm\sqrt{81} \qquad \text{Apply Property 10.1}$$

$$x = \pm 9$$

The solution set is $\{-9, 9\}$. ■

EXAMPLE 6

Solve $x^2 = 8$.

Solution

$$x^2 = 8$$

$$x = \pm\sqrt{8}$$

$$x = \pm 2\sqrt{2} \qquad \sqrt{8} = \sqrt{4}\sqrt{2} = 2\sqrt{2}$$

The solution set is $\{-2\sqrt{2}, 2\sqrt{2}\}$. ■

EXAMPLE 7 Solve $5n^2 = 12$.

Solution

$$5n^2 = 12$$

$$n^2 = \frac{12}{5} \qquad \text{Divided both sides by 5}$$

$$n = \pm\sqrt{\frac{12}{5}}$$

$$n = \pm\frac{2\sqrt{15}}{5} \qquad \sqrt{\frac{12}{5}} = \frac{\sqrt{12}}{\sqrt{5}} = \frac{\sqrt{12}}{\sqrt{5}} \cdot \frac{\sqrt{5}}{\sqrt{5}} = \frac{\sqrt{60}}{5} = \frac{2\sqrt{15}}{5}$$

The solution set is $\left\{-\frac{2\sqrt{15}}{5}, \frac{2\sqrt{15}}{5}\right\}$. ■

EXAMPLE 8 Solve $(x - 2)^2 = 16$.

Solution

$$(x - 2)^2 = 16$$

$$x - 2 = \pm 4$$

$$x - 2 = 4 \quad \text{or} \quad x - 2 = -4$$

$$x = 6 \quad \text{or} \quad x = -2$$

The solution set is $\{-2, 6\}$. ■

EXAMPLE 9 Solve $(x + 5)^2 = 27$.

Solution

$$(x + 5)^2 = 27$$

$$x + 5 = \pm\sqrt{27}$$

$$x + 5 = \pm 3\sqrt{3} \qquad \sqrt{27} = \sqrt{9}\sqrt{3} = 3\sqrt{3}$$

$$x + 5 = 3\sqrt{3} \quad \text{or} \quad x + 5 = -3\sqrt{3}$$

$$x = -5 + 3\sqrt{3} \quad \text{or} \quad x = -5 - 3\sqrt{3}$$

The solution set is $\{-5 - 3\sqrt{3}, -5 + 3\sqrt{3}\}$. ■

It may be necessary to change the form before we can apply Property 10.1. The next example illustrates this procedure.

EXAMPLE 10 Solve $3(2x - 3)^2 + 8 = 44$.

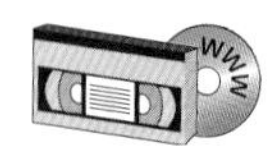

Solution

$$3(2x - 3)^2 + 8 = 44$$

$$3(2x - 3)^2 = 36$$

$$(2x - 3)^2 = 12$$

$$2x - 3 = \pm\sqrt{12} \quad \text{Apply Property 10.1}$$

$$2x - 3 = \sqrt{12} \quad \text{or} \quad 2x - 3 = -\sqrt{12}$$

$$2x = 3 + \sqrt{12} \quad \text{or} \quad 2x = 3 - \sqrt{12}$$

$$x = \frac{3 + \sqrt{12}}{2} \quad \text{or} \quad x = \frac{3 - \sqrt{12}}{2}$$

$$x = \frac{3 + 2\sqrt{3}}{2} \quad \text{or} \quad x = \frac{3 - 2\sqrt{3}}{2} \quad \sqrt{12} = \sqrt{4}\sqrt{3} = 2\sqrt{3}$$

The solution set is $\left\{\frac{3 - 2\sqrt{3}}{2}, \frac{3 + 2\sqrt{3}}{2}\right\}$. ■

Note that quadratic equations of the form $x^2 = a$, where a is a *negative* number, have no real number solutions. For example, $x^2 = -4$ has no real number solutions, because any real number squared is nonnegative. In a like manner, an equation such as $(x + 3)^2 = -14$ has no real number solutions.

■ Using the Pythagorean Theorem

Our work with radicals, Property 10.1, and the Pythagorean theorem merge to form a basis for solving a variety of problems that pertain to right triangles. First, let's restate the Pythagorean theorem.

Pythagorean Theorem

If for a right triangle, a and b are the measures of the legs, and c is the measure of the hypotenuse, then

$$a^2 + b^2 = c^2$$

(The *hypotenuse* is the side opposite the right angle, and the *legs* are the other two sides as shown in Figure 10.1.)

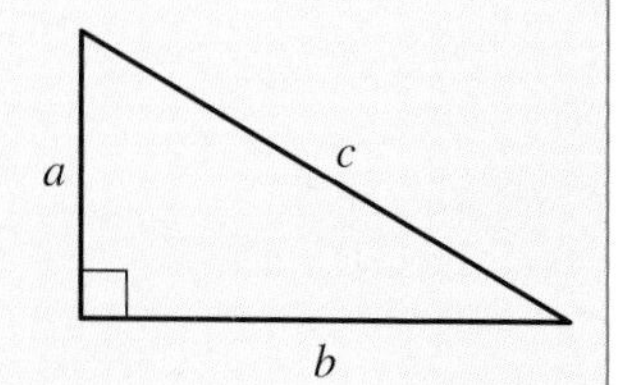

Figure 10.1

PROBLEM 1

Find c in Figure 10.2.

3 centimeters

c

4 centimeters

Figure 10.2

Solution

Applying the Pythagorean theorem, we have

$$c^2 = a^2 + b^2$$
$$c^2 = 3^2 + 4^2$$
$$c^2 = 9 + 16$$
$$c^2 = 25$$
$$c = 5$$

The length of c is 5 centimeters. ■

Remark: Don't forget that the equation $c^2 = 25$ does have two solutions, 5 and -5. However, because we are finding the lengths of line segments, we can disregard the negative solutions.

PROBLEM 2

A 50-foot rope hangs from the top of a flagpole. When pulled taut to its full length, the rope reaches a point on the ground 18 feet from the base of the pole. Find the height of the pole to the nearest tenth of a foot.

p represents the height of the flagpole.

Figure 10.3

Solution

We can sketch Figure 10.3 and record the given information. Using the Pythagorean theorem, we solve for p as follows:

$$p^2 + 18^2 = 50^2$$
$$p^2 + 324 = 2500$$
$$p^2 = 2176$$
$$p = \sqrt{2176} = 46.6 \quad \text{to the nearest tenth}$$

The height of the flagpole is approximately 46.6 feet. ■

An isosceles triangle has two sides of the same length. Thus an **isosceles right triangle** is a right triangle that has both legs of the same length. The next example considers a problem involving an isosceles right triangle.

PROBLEM 3

Find the length of each leg of an isosceles right triangle if the hypotenuse is 8 meters long.

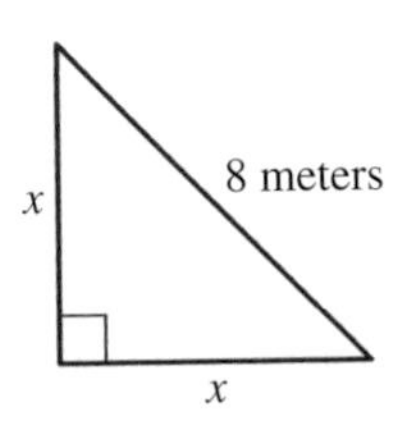

Figure 10.4

Solution

We sketch an isosceles right triangle in Figure 10.4 and let x represent the length of each leg. Then we can determine x by applying the Pythagorean theorem.

$$x^2 + x^2 = 8^2$$
$$2x^2 = 64$$

$$x^2 = 32$$

$$x = \sqrt{32} = \sqrt{16}\sqrt{2} = 4\sqrt{2}$$

Each leg is $4\sqrt{2}$ meters long. ■

Remark: In Problem 2, we made no attempt to express $\sqrt{2176}$ in simplest radical form because the answer was to be given as a rational approximation to the nearest tenth. However, in Problem 3 we left the final answer in radical form and therefore expressed it in simplest radical form.

Another special kind of right triangle is one that contains acute angles of 30° and 60°. In such a right triangle, often referred to as a 30°–60° right triangle, *the side opposite the 30° angle is equal in length to one-half the length of the hypotenuse.* This relationship, along with the Pythagorean theorem, provides us with another problem-solving technique.

PROBLEM 4

Suppose that a 20-foot ladder is leaning against a building and makes an angle of 60° with the ground. How far up on the building does the top of the ladder reach? Express your answer to the nearest tenth of a foot.

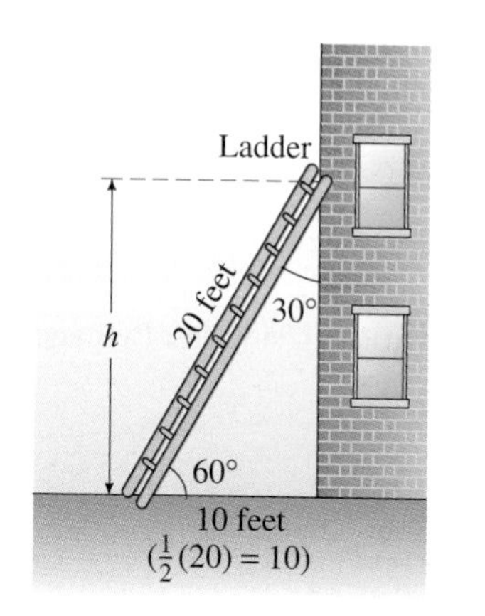

Figure 10.5

Solution

Figure 10.5 illustrates this problem. The side opposite the 30° angle equals one-half of the hypotenuse, so the length of that side is $\frac{1}{2}(20) = 10$ feet. Now we can apply the Pythagorean theorem.

$$h^2 + 10^2 = 20^2$$

$$h^2 + 100 = 400$$

$$h^2 = 300$$

$$h = \sqrt{300} = 17.3 \quad \text{to the nearest tenth}$$

The top of the ladder touches the building at approximately 17.3 feet from the ground. ■

Problem Set 10.1

For Problems 1–18, solve each quadratic equation by factoring and applying the property $ab = 0$ if and only if $a = 0$ or $b = 0$.

1. $x^2 + 15x = 0$
2. $x^2 - 11x = 0$
3. $n^2 = 12n$
4. $n^2 = -21n$
5. $3y^2 = 15y$
6. $8y^2 = -56y$
7. $x^2 - 9x + 8 = 0$
8. $x^2 + 16x + 48 = 0$
9. $x^2 - 5x - 14 = 0$
10. $x^2 - 5x - 36 = 0$
11. $n^2 + 5n - 6 = 0$
12. $n^2 + 3n - 28 = 0$
13. $6y^2 + 7y - 5 = 0$
14. $4y^2 - 21y - 18 = 0$
15. $30x^2 - 37x + 10 = 0$
16. $42x^2 + 67x + 21 = 0$
17. $4x^2 - 4x + 1 = 0$
18. $9x^2 + 12x + 4 = 0$

For Problems 19–56, use Property 10.1 to help solve each quadratic equation. Express irrational solutions in simplest radical form.

19. $x^2 = 64$

20. $x^2 = 169$

21. $x^2 = \frac{25}{9}$

22. $x^2 = \frac{4}{81}$

23. $4x^2 = 64$

24. $5x^2 = 500$

25. $n^2 = 14$

26. $n^2 = 22$

27. $n^2 + 16 = 0$

28. $n^2 = 24$

29. $y^2 = 32$

30. $y^2 + 25 = 0$

31. $3x^2 - 54 = 0$

32. $4x^2 - 108 = 0$

33. $2x^2 = 9$

34. $3x^2 = 16$

35. $8n^2 = 25$

36. $12n^2 = 49$

37. $(x - 1)^2 = 4$

38. $(x - 2)^2 = 9$

39. $(x + 3)^2 = 25$

40. $(x + 5)^2 = 36$

41. $(3x - 2)^2 = 49$

42. $(4x + 3)^2 = 1$

43. $(x + 6)^2 = 5$

44. $(x - 7)^2 = 6$

45. $(n - 1)^2 = 8$

46. $(n + 1)^2 = 12$

47. $(2n + 3)^2 = 20$

48. $(3n - 2)^2 = 28$

49. $(4x - 1)^2 = -2$

50. $(5x + 3)^2 - 32 = 0$

51. $(3x - 5)^2 - 40 = 0$

52. $(2x + 9)^2 + 6 = 0$

53. $2(7x - 1)^2 + 5 = 37$

54. $3(4x - 5)^2 - 50 = 25$

55. $2(x + 8)^2 - 9 = 91$

56. $2(x - 7)^2 - 7 = 101$

For Problems 57–62, a and b represent the lengths of the legs of a right triangle, and c represents the length of the hypotenuse. Express your answers in simplest radical form.

57. Find c if $a = 1$ inch and $b = 7$ inches.

58. Find c if $a = 2$ inches and $b = 6$ inches.

59. Find a if $c = 8$ meters and $b = 6$ meters.

60. Find a if $c = 11$ centimeters and $b = 7$ centimeters.

61. Find b if $c = 12$ feet and $a = 10$ feet.

62. Find b if $c = 10$ yards and $a = 4$ yards.

For Problems 63–68, use Figure 10.6. Express your answers in simplest radical form.

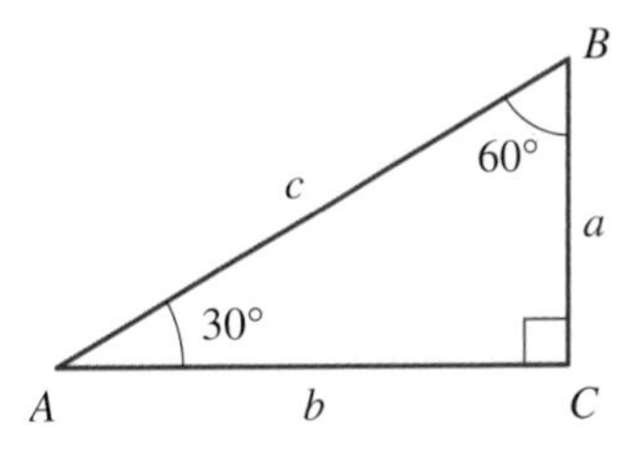

Figure 10.6

63. If $c = 8$ inches, find a and b.

64. If $c = 6$ inches, find a and b.

65. If $a = 6$ feet, find b and c.

66. If $a = 5$ feet, find b and c.

67. If $b = 12$ meters, find a and c.

68. If $b = 5$ centimeters, find a and c.

For Problems 69–72, use the isosceles right triangle in Figure 10.7. Express the answers in simplest radical form.

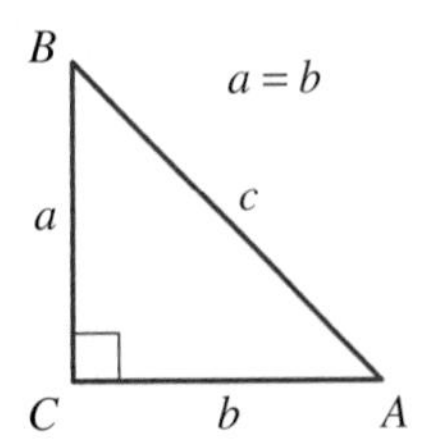

Figure 10.7

69. If $b = 10$ inches, find a and c.

70. If $a = 7$ inches, find b and c.

71. If $c = 9$ meters, find a and b.

72. If $c = 5$ meters, find a and b.

73. An 18-foot ladder resting against a house reaches a windowsill 16 feet above the ground. How far is the base of the ladder from the foundation of the house? Express your answer to the nearest tenth of a foot.

74. A 42-foot guy-wire makes an angle of 60° with the ground and is attached to a telephone pole, as in Figure 10.8. Find the distance from the base of the pole to the point on the pole where the wire is attached. Express your answer to the nearest tenth of a foot.

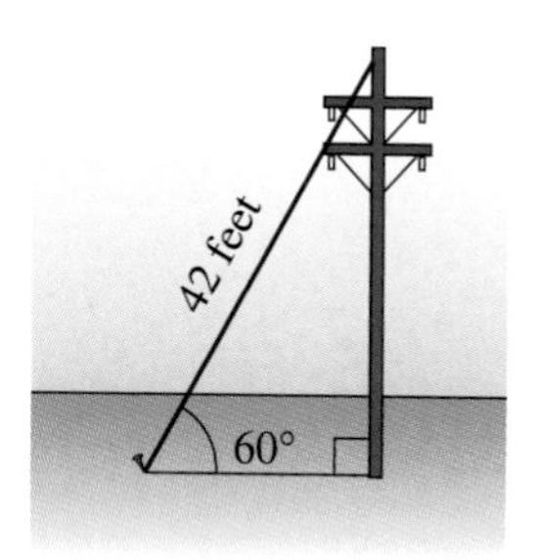

Figure 10.8

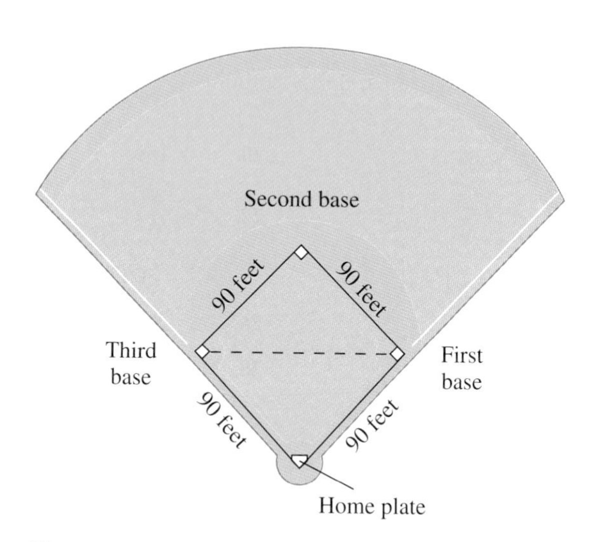

Figure 10.9

75. A rectangular plot measures 18 meters by 24 meters. Find the distance, to the nearest meter, from one corner of the plot to the diagonally opposite corner.

76. Consecutive bases of a square-shaped baseball diamond are 90 feet apart (see Figure 10.9). Find the distance, to the nearest tenth of a foot, from first base diagonally across the diamond to third base.

77. A diagonal of a square parking lot is 50 meters. Find, to the nearest meter, the length of a side of the lot.

THOUGHTS INTO WORDS

78. Explain why the equation $(x - 4)^2 + 14 = 2$ has no real number solutions.

79. Suppose that your friend solved the equation $(x + 3)^2 = 25$ as follows:

$$(x + 3)^2 = 25$$

$$x^2 + 6x + 9 = 25$$

$$x^2 + 6x - 16 = 0$$

$$(x + 8)(x - 2) = 0$$

$$x + 8 = 0 \quad \text{or} \quad x - 2 = 0$$

$$x = -8 \quad \text{or} \quad x = 2$$

Is this a correct approach to the problem? Can you suggest an easier approach to the problem?

FURTHER INVESTIGATIONS

80. Sometimes we simply need to determine whether a particular radical expression is positive or negative. For example, is $-6 + \sqrt{39}$ a positive or negative number? We can determine this by approximating a value for $\sqrt{39}$. Since $6^2 = 36$, we know that $\sqrt{39}$ is a little larger than 6. Therefore, $-6 + \sqrt{39}$ has to be positive.

Determine whether each of the following is positive or negative.

(a) $-8 + \sqrt{56}$ **(b)** $-7 + \sqrt{47}$

(c) $9 - \sqrt{77}$ **(d)** $12 - \sqrt{130}$

(e) $-6 + 5\sqrt{2}$ **(f)** $-10 + 6\sqrt{3}$

(g) $-13 + \sqrt{150}$ **(h)** $-14 + \sqrt{200}$

81. Find the length of an altitude of an equilateral triangle if each side of the triangle is 6 centimeters long. Express your answer to the nearest tenth of a centimeter.

82. Suppose that we are given a cube with each edge being 12 centimeters. Find the length of a diagonal from a lower corner to the diagonally opposite upper corner. Express your answer to the nearest tenth of a centimeter.

83. Suppose that we are given a rectangular box with a length of 8 centimeters, a width of 6 centimeters, and a height of 4 centimeters. Find the length of a diagonal from a lower corner to the diagonally opposite upper corner. Express your answer to the nearest tenth of a centimeter.

84. The converse of the Pythagorean theorem is also true. It states that "if the measures a, b, and c of the sides of a triangle are such that $a^2 + b^2 = c^2$, then the triangle is a right triangle with a and b the measures of the legs and c the measure of the hypotenuse." Use the converse of the Pythagorean theorem to determine which of the triangles having sides with the following measures are right triangles.

(a) 9, 40, 41 **(b)** 20, 48, 52

(c) 19, 21, 26 **(d)** 32, 37, 49

(e) 65, 156, 169 **(f)** 21, 72, 75

10.2 Completing the Square

Thus far we have solved quadratic equations by factoring or by applying Property 10.1 (if $x^2 = a$, then $x = \sqrt{a}$ or $x = -\sqrt{a}$). In this section, we will consider another method called **completing the square**, which will give us the power to solve *any* quadratic equation.

We studied a factoring technique in Chapter 6 that was based on recognizing **perfect square trinomials**. In each of the following equations, the trinomial on the right side, which is the result of squaring a binomial on the left side, is a perfect square trinomial.

$$(x + 5)^2 = x^2 + 10x + 25$$

$$(x + 7)^2 = x^2 + 14x + 49$$

$$(x - 3)^2 = x^2 - 6x + 9$$

$$(x - 6)^2 = x^2 - 12x + 36$$

We need to pay attention to the following special relationship. In each of these perfect square trinomials, **the constant term is equal to the square of one-half of the coefficient of the x term**. For example,

$$x^2 + 10x + 25 \qquad \frac{1}{2}(10) = 5 \quad \text{and} \quad 5^2 = 25$$

$$x^2 - 12x + 36 \qquad \frac{1}{2}(12) = 6 \quad \text{and} \quad 6^2 = 36$$

This relationship allows us to form a perfect square trinomial by adding the proper constant term. For example, suppose that we want to form a perfect square trinomial from $x^2 + 8x$. Because $\frac{1}{2}(8) = 4$ and $4^2 = 16$, we can form the perfect square trinomial $x^2 + 8x + 16$.

Now we can use the preceding ideas to help solve some quadratic equations.

EXAMPLE 1

Solve $x^2 + 8x - 1 = 0$ by the method of completing the square.

Solution

$$x^2 + 8x - 1 = 0$$

$$x^2 + 8x = 1 \qquad \text{Isolated the } x^2 \text{ and } x \text{ terms}$$

$$\frac{1}{2}(8) = 4 \quad \text{and} \quad 4^2 = 16 \qquad \text{Took } \frac{1}{2} \text{ of the coefficient of the } x \text{ term and then squared the result}$$

$$x^2 + 8x + 16 = 1 + 16 \qquad \text{Added 16 to both sides of the equation}$$

$$(x + 4)^2 = 17 \qquad \text{Factored the perfect-square trinomial}$$

Now we can proceed as we did with similar equations in the last section.

$$x + 4 = \pm\sqrt{17}$$

$$x + 4 = \sqrt{17} \quad \text{or} \quad x + 4 = -\sqrt{17}$$

$$x = -4 + \sqrt{17} \quad \text{or} \quad x = -4 - \sqrt{17}$$

The solution set is $\{-4 - \sqrt{17}, -4 + \sqrt{17}\}$. ■

Observe that the method of completing the square to solve a quadratic equation is just what the name implies. We form a perfect square trinomial; then we change the equation to the necessary form for using the property, *if* $x^2 = a$, *then* $x = \sqrt{a}$ *or* $x = -\sqrt{a}$. Let's consider another example.

EXAMPLE 2

Solve $x^2 - 2x - 11 = 0$ by the method of completing the square.

Solution

$$x^2 - 2x - 11 = 0$$

$$x^2 - 2x = 11 \qquad \text{Isolated the } x^2 \text{ and } x \text{ terms}$$

$$\frac{1}{2}(2) = 1 \quad \text{and} \quad 1^2 = 1 \qquad \text{Took } \frac{1}{2} \text{ of the coefficient of the } x \text{ term and then squared the result}$$

$$x^2 - 2x + 1 = 11 + 1 \qquad \text{Added 1 to both sides of the equation}$$

$$(x - 1)^2 = 12 \qquad \text{Factored the perfect-square trinomial}$$

$$x - 1 = \pm\sqrt{12}$$

$$x - 1 = \pm 2\sqrt{3}$$

$$x - 1 = 2\sqrt{3} \quad \text{or} \quad x - 1 = -2\sqrt{3}$$

$$x = 1 + 2\sqrt{3} \quad \text{or} \quad x = 1 - 2\sqrt{3}$$

The solution set is $\{1 - 2\sqrt{3}, 1 + 2\sqrt{3}\}$. ■

In the next example, the coefficient of the x term is odd, which means that taking one-half of it puts us in the realm of fractions. The use of common fractions rather than decimals makes our previous work with radicals applicable.

EXAMPLE 3 Solve $x^2 - 3x + 1 = 0$ by the method of completing the square.

Solution

$$x^2 - 3x + 1 = 0$$

$$x^2 - 3x = -1$$

$$\frac{1}{2}(3) = \frac{3}{2} \quad \text{and} \quad \left(\frac{3}{2}\right)^2 = \frac{9}{4}$$ Took $\frac{1}{2}$ of the coefficient of the x term and then squared the result

$$x^2 - 3x + \frac{9}{4} = -1 + \frac{9}{4}$$ Added $\frac{9}{4}$ to both sides of the equation

$$\left(x - \frac{3}{2}\right)^2 = \frac{5}{4}$$ Factored the perfect-square trinomial

$$x - \frac{3}{2} = \pm\sqrt{\frac{5}{4}}$$

$$x - \frac{3}{2} = \pm\frac{\sqrt{5}}{2}$$

$$x - \frac{3}{2} = \frac{\sqrt{5}}{2} \quad \text{or} \quad x - \frac{3}{2} = -\frac{\sqrt{5}}{2}$$

$$x = \frac{3}{2} + \frac{\sqrt{5}}{2} \quad \text{or} \quad x = \frac{3}{2} - \frac{\sqrt{5}}{2}$$

$$x = \frac{3 + \sqrt{5}}{2} \quad \text{or} \quad x = \frac{3 - \sqrt{5}}{2}$$

The solution set is $\left\{\frac{3 - \sqrt{5}}{2}, \frac{3 + \sqrt{5}}{2}\right\}$. ■

The relationship for a perfect square trinomial that states **the constant term is equal to the square of one-half of the coefficient of the x term** holds only if the coefficient of x^2 is 1. Thus we need to make a slight adjustment when solving quadratic equations that have a coefficient of x^2 other than 1. The next example shows how to make this adjustment.

EXAMPLE 4 Solve $2x^2 + 12x - 3 = 0$ by the method of completing the square.

Solution

$$2x^2 + 12x - 3 = 0$$

$$2x^2 + 12x = 3$$

$$x^2 + 6x = \frac{3}{2} \qquad \text{Multiply both sides by } \frac{1}{2}$$

$$x^2 + 6x + 9 = \frac{3}{2} + 9 \qquad \left[\frac{1}{2}(6)\right]^2 = 3^2 = 9\text{; add 9 to both sides of the equation}$$

$$(x + 3)^2 = \frac{21}{2}$$

$$x + 3 = \pm\sqrt{\frac{21}{2}}$$

$$x + 3 = \pm\frac{\sqrt{42}}{2} \qquad \sqrt{\frac{21}{2}} = \frac{\sqrt{21}}{\sqrt{2}} = \frac{\sqrt{21}}{\sqrt{2}} \cdot \frac{\sqrt{2}}{\sqrt{2}} = \frac{\sqrt{42}}{2}$$

$$x + 3 = \frac{\sqrt{42}}{2} \quad \text{or} \quad x + 3 = -\frac{\sqrt{42}}{2}$$

$$x = -3 + \frac{\sqrt{42}}{2} \quad \text{or} \quad x = -3 - \frac{\sqrt{42}}{2}$$

$$x = \frac{-6 + \sqrt{42}}{2} \quad \text{or} \quad x = \frac{-6 - \sqrt{42}}{2}$$

The solution set is $\left\{\frac{-6 - \sqrt{42}}{2}, \frac{-6 + \sqrt{42}}{2}\right\}$. ■

As we mentioned earlier, we can use the method of completing the square to solve *any* quadratic equation. To illustrate this point, we will use this method to solve an equation that we could also solve by factoring.

EXAMPLE 5

Solve $x^2 + 2x - 8 = 0$ by the method of completing the square and by factoring.

Solution A

By completing the square:

$$x^2 + 2x - 8 = 0$$

$$x^2 + 2x = 8$$

$$x^2 + 2x + 1 = 8 + 1 \qquad \left[\frac{1}{2}(2)\right]^2 = 1^2 = 1\text{; add 1 to both sides of the equation}$$

$$(x + 1)^2 = 9$$

$$x + 1 = \pm 3$$

$$x + 1 = 3 \quad \text{or} \quad x + 1 = -3$$

$$x = 2 \quad \text{or} \quad x = -4$$

The solution set is $\{-4, 2\}$.

Solution B

By factoring:

$$x^2 + 2x - 8 = 0$$

$$(x + 4)(x - 2) = 0$$

$$x + 4 = 0 \quad \text{or} \quad x - 2 = 0$$

$$x = -4 \quad \text{or} \quad x = 2$$

The solution set is $\{-4, 2\}$. ■

We don't claim that using the method of completing the square with an equation such as the one in Example 5 is easier than the factoring technique. However, it is important for you to recognize that the method of completing the square will work with any quadratic equation.

Our final example of this section demonstrates that the method of completing the square will identify those quadratic equations that have no real number solutions.

EXAMPLE 6

Solve $x^2 + 10x + 30 = 0$ by the method of completing the square.

Solution

$$x^2 + 10x + 30 = 0$$

$$x^2 + 10x = -30$$

$$x^2 + 10x + 25 = -30 + 25$$

$$(x + 5)^2 = -5$$

We can stop here and reason as follows: Any value of x will yield a nonnegative value for $(x + 5)^2$; thus, it cannot equal -5. The original equation, $x^2 + 10x + 30 = 0$, has no solutions in the set of real numbers. ■

Problem Set 10.2

For Problems 1–32, use the method of completing the square to help solve each quadratic equation.

1. $x^2 + 8x - 1 = 0$

2. $x^2 - 4x - 1 = 0$

3. $x^2 + 10x + 2 = 0$

4. $x^2 + 8x + 3 = 0$

5. $x^2 - 4x - 4 = 0$

6. $x^2 + 6x - 11 = 0$

7. $x^2 + 6x + 12 = 0$

8. $n^2 - 10n = 7$

9. $n^2 + 2n = 17$

10. $n^2 + 12n + 40 = 0$

11. $x^2 + x - 3 = 0$

12. $x^2 + 3x - 5 = 0$

13. $a^2 - 5a = 2$

14. $a^2 - 7a = 4$

15. $2x^2 + 8x - 3 = 0$

16. $2x^2 - 12x + 1 = 0$

17. $3x^2 + 12x - 2 = 0$

18. $3x^2 - 6x + 2 = 0$

19. $2t^2 - 4t + 1 = 0$

20. $4t^2 + 8t + 5 = 0$

21. $5n^2 + 10n + 6 = 0$

22. $2n^2 + 5n - 1 = 0$

23. $-n^2 + 9n = 4$

24. $-n^2 - 7n = 2$

25. $2x^2 + 3x - 1 = 0$

26. $3x^2 - x - 3 = 0$

27. $3x^2 + 2x - 2 = 0$

28. $9x = 3x^2 - 1$

29. $n(n + 2) = 168$

30. $n(n + 4) = 140$

31. $n(n - 4) = 165$

32. $n(n - 2) = 288$

For Problems 33–42, solve each quadratic equation by using (a) the factoring method and (b) the method of completing the square.

33. $x^2 + 4x - 12 = 0$

34. $x^2 - 6x - 40 = 0$

35. $x^2 + 12x + 27 = 0$

36. $x^2 + 18x + 77 = 0$

37. $n^2 - 3n - 40 = 0$

38. $n^2 + 9n - 36 = 0$

39. $2n^2 - 9n + 4 = 0$

40. $6n^2 - 11n - 10 = 0$

41. $4n^2 + 4n - 15 = 0$

42. $4n^2 + 12n - 7 = 0$

THOUGHTS INTO WORDS

43. Give a step-by-step description of how to solve the equation $3x^2 + 10x - 8 = 0$ by completing the square.

44. An error has been made in the following solution. Find it and explain how to correct it.

$$4x^2 - 4x + 1 = 0$$

$$4x^2 - 4x = -1$$

$$4x^2 - 4x + 4 = -1 + 4$$

$$(2x - 2)^2 = 3$$

$$2x - 2 = \pm\sqrt{3}$$

$$2x - 2 = \sqrt{3} \quad \text{or} \quad 2x - 2 = -\sqrt{3}$$

$$2x = 2 + \sqrt{3} \quad \text{or} \quad 2x = 2 - \sqrt{3}$$

$$x = \frac{2 + \sqrt{3}}{2} \quad \text{or} \quad x = \frac{2 - \sqrt{3}}{2}$$

The solution set is $\left\{\frac{2 + \sqrt{3}}{2}, \frac{2 - \sqrt{3}}{2}\right\}$.

FURTHER INVESTIGATIONS

45. Use the method of completing the square to solve $ax^2 + bx + c = 0$ for x, where a, b, and c are real numbers and $a \neq 0$.

46. Suppose that in Example 4 we wanted to express the solutions to the nearest tenth. Then we would probably proceed from the step $x + 3 = \pm\sqrt{\frac{21}{2}}$ as follows:

$$x + 3 = \pm\sqrt{\frac{21}{2}}$$

$$x = -3 \pm \sqrt{\frac{21}{2}}$$

$$x = -3 + \sqrt{\frac{21}{2}} \quad \text{or} \quad x = -3 - \sqrt{\frac{21}{2}}$$

Now use your calculator to evaluate each of these expressions to the nearest tenth. The solution set is $\{-6.2, 0.2\}$.

Solve each of the following equations and express the solutions to the nearest tenth.

(a) $x^2 - 6x - 4 = 0$

(b) $x^2 - 8x + 4 = 0$

(c) $x^2 + 4x - 4 = 0$

(d) $x^2 + 2x - 5 = 0$

(e) $x^2 - 14x - 2 = 0$

(f) $x^2 + 12x - 1 = 0$

10.3 Quadratic Formula

We can use the method of completing the square to solve any quadratic equation. The equation $ax^2 + bx + c = 0$, where a, b, and c are real numbers with $a \neq 0$, can represent *any* quadratic equation. These two ideas merge to produce the *quadratic formula*, a formula that we can use to solve any quadratic equation. The merger is accomplished by using the method of completing the square to solve the equation $ax^2 + bx + c = 0$ as follows:

$$ax^2 + bx + c = 0$$

$$ax^2 + bx = -c$$

$$x^2 + \frac{b}{a}x = -\frac{c}{a}$$ Multiply both sides by $\frac{1}{a}$

$$x^2 + \frac{b}{a}x + \frac{b^2}{4a^2} = -\frac{c}{a} + \frac{b^2}{4a^2}$$ Complete the square by adding $\frac{b^2}{4a^2}$ to both sides

$$\left(x + \frac{b}{2a}\right)^2 = \frac{b^2 - 4ac}{4a^2}$$ The right side is combined into a single term with LCD $4a^2$

$$x + \frac{b}{2a} = \pm\sqrt{\frac{b^2 - 4ac}{4a^2}}$$

$$x + \frac{b}{2a} = \pm\frac{\sqrt{b^2 - 4ac}}{\sqrt{4a^2}}$$

$$x + \frac{b}{2a} = \pm\frac{\sqrt{b^2 - 4ac}}{2a}$$ $\sqrt{4a^2} = |2a|$ but $2a$ can be used because of $\pm$

$$x = -\frac{b}{2a} \pm \frac{\sqrt{b^2 - 4ac}}{2a}$$

$$x = \frac{-b \pm \sqrt{b^2 - 4ac}}{2a}$$

The solutions are $\frac{-b + \sqrt{b^2 - 4ac}}{2a}$ and $\frac{-b - \sqrt{b^2 - 4ac}}{2a}$.

We usually state the **quadratic formula** as follows:

$$x = \frac{-b \pm \sqrt{b^2 - 4ac}}{2a}$$

We can use it to solve any quadratic equation by expressing the equation in standard form and by substituting the values for a, b, and c into the formula. Consider the following examples.

EXAMPLE 1

Solve $x^2 + 7x + 10 = 0$ by using the quadratic formula.

Solution

The given equation is in standard form with $a = 1$, $b = 7$, and $c = 10$. Let's substitute these values into the quadratic formula and simplify.

$$x = \frac{-b \pm \sqrt{b^2 - 4ac}}{2a}$$

$$x = \frac{-7 \pm \sqrt{7^2 - 4(1)(10)}}{2(1)}$$

$$x = \frac{-7 \pm \sqrt{9}}{2}$$

$$x = \frac{-7 \pm 3}{2}$$

$$x = \frac{-7 + 3}{2} \quad \text{or} \quad x = \frac{-7 - 3}{2}$$

$$x = -2 \quad \text{or} \quad x = -5$$

The solution set is $\{-5, -2\}$. ■

EXAMPLE 2

Solve $x^2 - 3x = 1$ by using the quadratic formula.

Solution

First we need to change the equation to the standard form of $ax^2 + bx + c = 0$.

$$x^2 - 3x = 1$$

$$x^2 - 3x - 1 = 0$$

We need to think of $x^2 - 3x - 1 = 0$ as $x^2 + (-3)x + (-1) = 0$ to determine the values $a = 1$, $b = -3$, and $c = -1$. Let's substitute these values into the quadratic formula and simplify.

$$x = \frac{-(-3) \pm \sqrt{(-3)^2 - 4(1)(-1)}}{2(1)}$$

$$x = \frac{3 \pm \sqrt{9 + 4}}{2}$$

$$x = \frac{3 \pm \sqrt{13}}{2}$$

The solution set is $\left\{\frac{3 - \sqrt{13}}{2}, \frac{3 + \sqrt{13}}{2}\right\}$. ■

EXAMPLE 3

Solve $15n^2 - n - 2 = 0$ by using the quadratic formula.

Solution

Remember that although we commonly use the variable x in the statement of the quadratic formula, any variable could be used. Writing the equation as $15n^2 + (-1)n + (-2) = 0$ gives us the standard form of $an^2 + bn + c = 0$ with $a = 15$, $b = -1$, and $c = -2$. Now we can solve the equation by substituting into the quadratic formula and simplifying.

$$n = \frac{-(-1) \pm \sqrt{(-1)^2 - 4(15)(-2)}}{2(15)}$$

$$n = \frac{1 \pm \sqrt{1 + 120}}{30}$$

$$n = \frac{1 \pm \sqrt{121}}{30}$$

$$n = \frac{1 \pm 11}{30}$$

$$n = \frac{1 + 11}{30} \quad \text{or} \quad n = \frac{1 - 11}{30}$$

$$n = \frac{12}{30} \quad \text{or} \quad n = \frac{-10}{30}$$

$$n = \frac{2}{5} \quad \text{or} \quad n = -\frac{1}{3}$$

The solution set is $\left\{-\frac{1}{3}, \frac{2}{5}\right\}$. ■

EXAMPLE 4

Solve $t^2 - 5t - 84 = 0$ by using the quadratic formula.

Solution

Writing the equation as $t^2 + (-5)t + (-84) = 0$ gives us the standard form of $at^2 + bt + c = 0$ with $a = 1$, $b = -5$, and $c = -84$. Now we can solve the equation by substituting into the quadratic formula and simplifying.

$$t = \frac{-(-5) \pm \sqrt{(-5)^2 - 4(1)(-84)}}{2(1)}$$

$$t = \frac{5 \pm \sqrt{25 + 336}}{2}$$

$$t = \frac{5 \pm \sqrt{361}}{2}$$

$$t = \frac{5 \pm 19}{2}$$

$$t = \frac{5 + 19}{2} \quad \text{or} \quad t = \frac{5 - 19}{2}$$

$$t = \frac{24}{2} \quad \text{or} \quad t = \frac{-14}{2}$$

$$t = 12 \quad \text{or} \quad t = -7$$

The solution set is $\{-7, 12\}$. ■

We can easily identify quadratic equations that have no real number solutions when we use the quadratic formula. The final example of this section illustrates this point.

EXAMPLE 5 Solve $x^2 - 2x + 8 = 0$ by using the quadratic formula.

Solution

$$x = \frac{-(-2) \pm \sqrt{(-2)^2 - 4(1)(8)}}{2(1)}$$

$$x = \frac{2 \pm \sqrt{4 - 32}}{2}$$

$$x = \frac{2 \pm \sqrt{-28}}{2}$$

Since $\sqrt{-28}$ is not a real number, we conclude that the given equation has no real number solutions. (We do more work with this type of equation in Section 11.5.) ■

Problem Set 10.3

Use the quadratic formula to solve each of the following quadratic equations.

1. $x^2 - 5x - 6 = 0$
2. $x^2 + 3x - 4 = 0$
3. $x^2 + 5x = 36$
4. $x^2 - 8x = -12$
5. $n^2 - 2n - 5 = 0$
6. $n^2 - 4n - 1 = 0$
7. $a^2 - 5a - 2 = 0$
8. $a^2 + 3a + 1 = 0$
9. $x^2 - 2x + 6 = 0$
10. $x^2 - 8x + 16 = 0$
11. $y^2 + 4y + 2 = 0$
12. $n^2 + 6n + 11 = 0$
13. $x^2 - 6x = 0$
14. $x^2 + 8x = 0$
15. $2x^2 = 7x$
16. $3x^2 = -10x$
17. $n^2 - 34n + 288 = 0$
18. $n^2 + 27n + 182 = 0$
19. $x^2 + 2x - 80 = 0$
20. $x^2 - 15x + 54 = 0$
21. $t^2 + 4t + 4 = 0$
22. $t^2 + 6t - 5 = 0$

23. $6x^2 + x - 2 = 0$

24. $4x^2 - x - 3 = 0$

25. $5x^2 + 3x - 2 = 0$

26. $6x^2 - x - 2 = 0$

27. $12x^2 + 19x = -5$

28. $2x^2 + 7x - 6 = 0$

29. $2x^2 + 5x - 6 = 0$

30. $2x^2 + 3x - 3 = 0$

31. $3x^2 + 4x - 1 = 0$

32. $3x^2 + 2x - 4 = 0$

33. $16x^2 + 24x + 9 = 0$

34. $9x^2 - 30x + 25 = 0$

35. $4n^2 + 8n - 1 = 0$

36. $4n^2 + 6n - 1 = 0$

37. $6n^2 + 9n + 1 = 0$

38. $5n^2 + 8n + 1 = 0$

39. $2y^2 - y - 4 = 0$

40. $3t^2 + 6t + 5 = 0$

41. $4t^2 + 5t + 3 = 0$

42. $5x^2 + x - 1 = 0$

43. $7x^2 + 5x - 4 = 0$

44. $6x^2 + 2x - 3 = 0$

45. $7 = 3x^2 - x$

46. $-2x^2 + 3x = -4$

47. $n^2 + 23n = -126$

48. $n^2 + 2n = 195$

THOUGHTS INTO WORDS

49. Explain how to use the quadratic formula to solve the equation $x^2 = 2x + 6$.

50. Your friend states that the equation $-x^2 - 6x + 16 = 0$ must be changed to $x^2 + 6x - 16 = 0$ (by multiplying both sides by -1) before the quadratic formula can be applied. Is he right about this, and if not how would you convince him that he is wrong?

51. Another of your friends claims that the quadratic formula can be used to solve the equation $x^2 - 4 = 0$. How would you react to this claim?

FURTHER INVESTIGATIONS

Use the quadratic formula to solve each of the following equations. Express the solutions to the nearest hundredth.

52. $x^2 - 7x - 13 = 0$

53. $x^2 - 5x - 19 = 0$

54. $x^2 + 9x - 15 = 0$

55. $x^2 + 6x - 17 = 0$

56. $2x^2 + 3x - 7 = 0$

57. $3x^2 + 7x - 13 = 0$

58. $5x^2 - 11x - 14 = 0$

59. $4x^2 - 9x - 19 = 0$

60. $-3x^2 + 2x + 11 = 0$

61. $-5x^2 + x + 21 = 0$

62. Let x_1 and x_2 be the two solutions of $ax^2 + bx + c = 0$ obtained by the quadratic formula. Thus, we have

$$x_1 = \frac{-b + \sqrt{b^2 - 4ac}}{2a}$$

$$x_2 = \frac{-b - \sqrt{b^2 - 4ac}}{2a}$$

Find the sum $x_1 + x_2$ and the product $(x_1)(x_2)$. Your answers should be

$$x_1 + x_2 = -\frac{b}{a} \quad \text{and} \quad (x_1)(x_2) = \frac{c}{a}$$

These relationships provide another way of checking potential solutions when solving quadratic equations. For example, back in Example 3, we solved the equation $15n^2 - n - 2 = 0$ and obtained solutions of $-\frac{1}{3}$ and $\frac{2}{5}$. Let's check these solutions using the sum and product relationships.

Sum of solutions $\quad -\frac{1}{3} + \frac{2}{5} = -\frac{5}{15} + \frac{6}{15} = \frac{1}{15}$

$$-\frac{b}{a} = -\frac{-1}{15} = \frac{1}{15}$$

Product of solutions $\quad \left(-\frac{1}{3}\right)\left(\frac{2}{5}\right) = -\frac{2}{15}$

$$\frac{c}{a} = \frac{-2}{15} = -\frac{2}{15}$$

Use the sum and product relationships to check at least ten of the problems that you worked in this problem set.

10.4 Solving Quadratic Equations—Which Method?

We now summarize the three basic methods of solving quadratic equations presented in this chapter by solving a specific quadratic equation using each technique. Consider the equation $x^2 + 4x - 12 = 0$.

Factoring Method

$$x^2 + 4x - 12 = 0$$

$$(x + 6)(x - 2) = 0$$

$$x + 6 = 0 \quad \text{or} \quad x - 2 = 0$$

$$x = -6 \quad \text{or} \quad x = 2$$

The solution set is $\{-6, 2\}$.

Completing the Square Method

$$x^2 + 4x - 12 = 0$$

$$x^2 + 4x = 12$$

$$x^2 + 4x + 4 = 12 + 4$$

$$(x + 2)^2 = 16$$

$$x + 2 = \pm\sqrt{16}$$

$$x + 2 = 4 \quad \text{or} \quad x + 2 = -4$$

$$x = 2 \quad \text{or} \quad x = -6$$

The solution set is $\{-6, 2\}$.

Quadratic Formula Method

$$x^2 + 4x - 12 = 0$$

$$x = \frac{-4 \pm \sqrt{4^2 - 4(1)(-12)}}{2(1)}$$

$$x = \frac{-4 \pm \sqrt{64}}{2}$$

$$x = \frac{-4 \pm 8}{2}$$

$$x = \frac{-4 + 8}{2} \quad \text{or} \quad x = \frac{-4 - 8}{2}$$

$$x = 2 \quad \text{or} \quad x = -6$$

The solution set is $\{-6, 2\}$.

We have also discussed the use of the property $x^2 = a$ *if and only if* $x = \pm\sqrt{a}$ for certain types of quadratic equations. For example, we can solve $x^2 = 4$ easily by applying the property and obtaining $x = \sqrt{4}$ or $x = -\sqrt{4}$; thus, the solutions are 2 and -2.

Which method should you use to solve a particular quadratic equation? Let's consider some examples in which the different techniques are used. Keep in mind that this is a decision you must make as the need arises. So become as familiar as you can with the strengths and weaknesses of each method.

EXAMPLE 1

Solve $2x^2 + 12x - 54 = 0$.

Solution

First, it is very helpful to recognize a factor of 2 in each of the terms on the left side.

$$2x^2 + 12x - 54 = 0$$

$$x^2 + 6x - 27 = 0 \qquad \text{Multiply both sides by } \frac{1}{2}$$

Now you should recognize that the left side can be factored. Thus we can proceed as follows.

$$(x + 9)(x - 3) = 0$$

$$x + 9 = 0 \quad \text{or} \quad x - 3 = 0$$

$$x = -9 \quad \text{or} \quad x = 3$$

The solution set is $\{-9, 3\}$. ■

EXAMPLE 2

Solve $(4x + 3)^2 = 16$.

Solution

The form of this equation lends itself to the use of the property $x^2 = a$ *if and only if* $x = \pm\sqrt{a}$.

$$(4x + 3)^2 = 16$$

$$4x + 3 = \pm\sqrt{16}$$

$$4x + 3 = 4 \quad \text{or} \quad 4x + 3 = -4$$

$$4x = 1 \quad \text{or} \quad 4x = -7$$

$$x = \frac{1}{4} \quad \text{or} \quad x = -\frac{7}{4}$$

The solution set is $\left\{-\frac{7}{4}, \frac{1}{4}\right\}$. ■

EXAMPLE 3

Solve $n + \frac{1}{n} = 5$.

Solution

First, we need to *clear the equation of fractions* by multiplying both sides by n.

$$n + \frac{1}{n} = 5, \qquad n \neq 0$$

$$n\left(n + \frac{1}{n}\right) = 5(n)$$

$$n^2 + 1 = 5n$$

Now we can change the equation to standard form.

$$n^2 - 5n + 1 = 0$$

Because the left side cannot be factored using integers, we must solve the equation by using either the method of completing the square or the quadratic formula. Using the formula, we obtain

$$n = \frac{-(-5) \pm \sqrt{(-5)^2 - 4(1)(1)}}{2(1)}$$

$$n = \frac{5 \pm \sqrt{21}}{2}$$

The solution set is $\left\{\frac{5 - \sqrt{21}}{2}, \frac{5 + \sqrt{21}}{2}\right\}$. ■

EXAMPLE 4

Solve $t^2 = \sqrt{2}t$.

Solution

A quadratic equation without a constant term can be solved easily by the factoring method.

$$t^2 = \sqrt{2}t$$

$$t^2 - \sqrt{2}t = 0$$

$$t(t - \sqrt{2}) = 0$$

$$t = 0 \quad \text{or} \quad t - \sqrt{2} = 0$$

$$t = 0 \quad \text{or} \quad t = \sqrt{2}$$

The solution set is $\{0, \sqrt{2}\}$. (Check each of these solutions in the given equation.)

■

EXAMPLE 5

Solve $x^2 - 28x + 192 = 0$.

Solution

Determining whether or not the left side is factorable presents a bit of a problem because of the size of the constant term. Therefore, let's not concern ourselves with trying to factor; instead we will use the quadratic formula.

$$x^2 - 28x + 192 = 0$$

$$x = \frac{-(-28) \pm \sqrt{(-28)^2 - 4(1)(192)}}{2(1)}$$

$$x = \frac{28 \pm \sqrt{784 - 768}}{2}$$

$$x = \frac{28 \pm \sqrt{16}}{2}$$

$$x = \frac{28 + 4}{2} \quad \text{or} \quad x = \frac{28 - 4}{2}$$

$$x = 16 \quad \text{or} \quad x = 12$$

The solution set is $\{12, 16\}$. ■

EXAMPLE 6

Solve $x^2 + 12x = 17$.

Solution

The form of this equation and the fact that the coefficient of x is even make the method of completing the square a reasonable approach.

$$x^2 + 12x = 17$$

$$x^2 + 12x + 36 = 17 + 36$$

$$(x + 6)^2 = 53$$

$$x + 6 = \pm\sqrt{53}$$

$$x = -6 \pm \sqrt{53}$$

The solution set is $\{-6 - \sqrt{53}, -6 + \sqrt{53}\}$. ■

Problem Set 10.4

Solve each of the following quadratic equations using the method that seems most appropriate to you.

1. $x^2 + 4x = 45$
2. $x^2 + 4x = 60$
3. $(5n + 6)^2 = 49$
4. $(3n - 1)^2 = 25$
5. $t^2 - t - 2 = 0$
6. $t^2 + 2t - 3 = 0$
7. $8x = 3x^2$
8. $5x^2 = 7x$
9. $9x^2 - 6x + 1 = 0$
10. $4x^2 + 36x + 81 = 0$
11. $5n^2 = \sqrt{8}n$
12. $\sqrt{3}n = 2n^2$
13. $n^2 - 14n = 19$
14. $n^2 - 10n = 14$
15. $5x^2 - 2x - 7 = 0$
16. $3x^2 - 4x - 2 = 0$
17. $15x^2 + 28x + 5 = 0$
18. $20y^2 - 7y - 6 = 0$
19. $x^2 - \sqrt{8}x - 7 = 0$
20. $x^2 + \sqrt{5}x - 5 = 0$
21. $y^2 + 5y = 84$
22. $y^2 + 7y = 60$
23. $2n = 3 + \frac{3}{n}$
24. $n + \frac{1}{n} = 7$
25. $3x^2 - 9x - 12 = 0$
26. $2x^2 + 10x - 28 = 0$
27. $2x^2 - 3x + 7 = 0$
28. $3x^2 - 2x + 5 = 0$
29. $n(n - 46) = -480$
30. $n(n + 42) = -432$
31. $n - \frac{3}{n} = -1$
32. $n - \frac{2}{n} = \frac{3}{4}$
33. $x + \frac{1}{x} = \frac{25}{12}$
34. $x + \frac{1}{x} = \frac{65}{8}$
35. $t^2 + 12t + 36 = 49$
36. $t^2 - 10t + 25 = 16$
37. $x^2 - 28x + 187 = 0$
38. $x^2 - 33x + 266 = 0$
39. $\frac{x^2}{3} - x = -\frac{1}{2}$
40. $\frac{2x - 1}{3} = \frac{5}{x + 2}$
41. $\frac{2}{x + 2} - \frac{1}{x} = 3$
42. $\frac{3}{x - 1} + \frac{2}{x} = \frac{3}{2}$
43. $\frac{2}{3n - 1} = \frac{n + 2}{6}$
44. $\frac{x^2}{2} = x + \frac{1}{4}$
45. $(n - 2)(n + 4) = 7$
46. $(n + 3)(n - 8) = -30$

THOUGHTS INTO WORDS

47. Which method would you use to solve the equation $x^2 + 30x = -216$? Explain your reasons for making this choice.
48. Explain how you would solve the equation $0 = -x^2 - x + 6$.
49. How can you tell by inspection that the equation $x^2 + x + 4 = 0$ has no real number solutions?

10.5 Solving Problems Using Quadratic Equations

The following diagram indicates our approach in this text.

Develop skills ⟶ Use skills to solve equations ⟶ Use equations to solve word problems

Now you should be ready to use your skills relative to solving systems of equations (Chapter 8) and quadratic equations to help with additional types of word problems. Before you consider such problems, let's review and update the problem-solving suggestions we offered in Chapter 3.

Suggestions for Solving Word Problems

1. Read the problem carefully and make certain that you understand the meanings of all the words. Be especially alert for any technical terms used in the statement of the problem.
2. Read the problem a second time (perhaps even a third time) to get an overview of the situation being described and to determine the known facts as well as what is to be found.
3. Sketch any figure, diagram, or chart that might be helpful in analyzing the problem.

*4. Choose *meaningful* variables to represent the unknown quantities. Use one or two variables, whichever seems easiest. The term "meaningful" refers to the choice of letters to use as variables. Choose letters that have some significance for the problem under consideration. For example, if the problem deals with the length and width of a rectangle, then l and w are natural choices for the variables.

*5. Look for *guidelines* that you can use to help set up equations. A guideline might be a formula such as *area of a rectangular region equals length times width*, or a statement of a relationship such as *the product of the two numbers is 98*.

*6. **(a)** Form an equation, which contains the variable, that translates the conditions of the guideline from English into algebra; or
(b) form two equations, which contain the two variables, that translate the guidelines from English into algebra.

*7. Solve the equation (system of equations) and use the solution (solutions) to determine all facts requested in the problem.

8. **Check all answers back in the original statement of the problem.**

The asterisks indicate those suggestions that have been revised to include using systems of equations to solve problems. Keep these suggestions in mind as you study the examples and work the problems in this section.

PROBLEM 1

The length of a rectangular region is 2 centimeters more than its width. The area of the region is 35 square centimeters. Find the length and width of the rectangle.

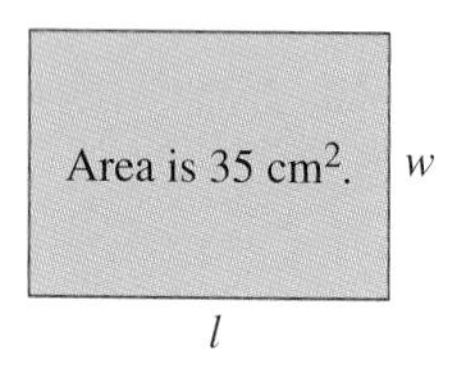

Figure 10.10

Solution

We let l represent the length, and we let w represent the width (see Figure 10.10). We can use the area formula for a rectangle, $A = lw$, and the statement "the length of a rectangular region is 2 centimeters greater than its width" as guidelines to form a system of equations.

$$\begin{pmatrix} lw = 35 \\ l = w + 2 \end{pmatrix}$$

The second equation indicates that we can substitute $w + 2$ for l. Making this substitution in the first equation yields

$$(w + 2)(w) = 35$$

Solving this quadratic equation by factoring, we get

$$w^2 + 2w = 35$$

$$w^2 + 2w - 35 = 0$$

$$(w + 7)(w - 5) = 0$$

$$w + 7 = 0 \quad \text{or} \quad w - 5 = 0$$

$$w = -7 \quad \text{or} \quad w = 5$$

The width of a rectangle cannot be a negative number, so we discard the solution -7. Thus the width of the rectangle is 5 centimeters and the length $(w + 2)$ is 7 centimeters. ■

PROBLEM 2 Find two consecutive whole numbers whose product is 506.

Solution

We let n represent the smaller whole number. Then $n + 1$ represents the next larger whole number. The phrase "whose product is 506" translates into the equation

$$n(n + 1) = 506$$

Changing this quadratic equation into standard form produces

$$n^2 + n = 506$$

$$n^2 + n - 506 = 0$$

Because of the size of the constant term, let's not try to factor; instead, we use the quadratic formula.

$$n = \frac{-1 \pm \sqrt{1^2 - 4(1)(-506)}}{2(1)}$$

$$n = \frac{-1 \pm \sqrt{2025}}{2}$$

$$n = \frac{-1 \pm 45}{2} \qquad \sqrt{2025} = 45$$

$$n = \frac{-1 + 45}{2} \quad \text{or} \quad n = \frac{-1 - 45}{2}$$

$$n = 22 \quad \text{or} \quad n = -23$$

Since we are looking for whole numbers, we discard the solution -23. Therefore, the whole numbers are 22 and 23. ■

PROBLEM 3

The perimeter of a rectangular lot is 100 meters, and its area is 616 square meters. Find the length and width of the lot.

Solution

We let l represent the length, and we let w represent the width (see Figure 10.11).

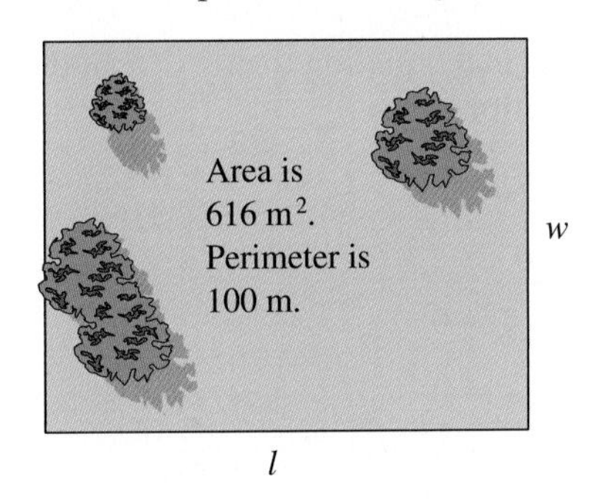

Figure 10.11

Then

$$\begin{pmatrix} lw = 616 \\ 2l + 2w = 100 \end{pmatrix} \quad \begin{matrix} \leftarrow \text{Area is 616 m}^2 \\ \leftarrow \text{Perimeter is 100 m} \end{matrix}$$

Multiplying the second equation by $\frac{1}{2}$ produces $l + w = 50$, which can be changed to $l = 50 - w$. Substituting $50 - w$ for l in the first equation produces the quadratic equation

$$(50 - w)(w) = 616$$

$$50w - w^2 = 616$$

$$w^2 - 50w = -616$$

Using the method of completing the square, we have

$$w^2 - 50w + 625 = -616 + 625$$

$$(w - 25)^2 = 9$$

$$w - 25 = \pm 3$$

$$w - 25 = 3 \quad \text{or} \quad w - 25 = -3$$
$$w = 28 \quad \text{or} \quad w = 22$$

If $w = 28$, then $l = 50 - w = 22$. If $w = 22$, then $l = 50 - w = 28$. The rectangle is 28 meters by 22 meters or 22 meters by 28 meters. ■

PROBLEM 4

Find two numbers such that their sum is 2 and their product is -1.

Solution

We let n represent one of the numbers, and we let m represent the other number.

$$\begin{pmatrix} n + m = 2 \\ nm = -1 \end{pmatrix} \quad \begin{matrix} \leftarrow \text{Their sum is 2} \\ \leftarrow \text{Their product is } -1 \end{matrix}$$

We can change the first equation to $m = 2 - n$; then we can substitute $2 - n$ for m in the second equation.

$$n(2 - n) = -1$$
$$2n - n^2 = -1$$
$$-n^2 + 2n + 1 = 0$$
$$n^2 - 2n - 1 = 0 \quad \text{Multiply both sides by } -1$$
$$n = \frac{-(-2) \pm \sqrt{(-2)^2 - 4(1)(-1)}}{2(1)}$$
$$n = \frac{2 \pm \sqrt{8}}{2} = \frac{2 \pm 2\sqrt{2}}{2} = 1 \pm \sqrt{2}$$

If $n = 1 + \sqrt{2}$, then
$$m = 2 - (1 + \sqrt{2}) = 2 - 1 - \sqrt{2} = 1 - \sqrt{2}$$

If $n = 1 - \sqrt{2}$, then
$$m = 2 - (1 - \sqrt{2}) = 2 - 1 + \sqrt{2} = 1 + \sqrt{2}$$

The numbers are $1 + \sqrt{2}$ and $1 - \sqrt{2}$.

Perhaps you should check these numbers in the original statement of the problem! ■

Finally, let's consider a uniform motion problem similar to those we solved in Chapter 7. Now we have the flexibility of using two equations in two variables.

PROBLEM 5

Larry drove 156 miles in 1 hour more than it took Mike to drive 108 miles. Mike drove at an average rate of 2 miles per hour faster than Larry. How fast did each one travel?

Solution

We can represent the unknown rates and times like this:

Let r represent Larry's rate.
Let t represent Larry's time.
Then $r + 2$ represents Mike's rate,
and $t - 1$ represents Mike's time.

Because *distance equals rate times time*, we can set up the following system:

$$\begin{pmatrix} rt = 156 \\ (r + 2)(t - 1) = 108 \end{pmatrix}$$

Solving the first equation for r produces $r = \dfrac{156}{t}$. Substituting $\dfrac{156}{t}$ for r in the second equation and simplifying, we obtain

$$\left(\frac{156}{t} + 2\right)(t - 1) = 108$$

$$156 - \frac{156}{t} + 2t - 2 = 108$$

$$2t - \frac{156}{t} + 154 = 108$$

$$2t - \frac{156}{t} + 46 = 0$$

$$2t^2 - 156 + 46t = 0 \qquad \text{Multiply both sides by } t,\ t \neq 0$$

$$2t^2 + 46t - 156 = 0$$

$$t^2 + 23t - 78 = 0$$

We can solve this quadratic equation by factoring.

$$(t + 26)(t - 3) = 0$$

$$t + 26 = 0 \quad \text{or} \quad t - 3 = 0$$

$$t = -26 \quad \text{or} \quad t = 3$$

We must disregard the negative solution. So Mike's time is $3 - 1 = 2$ hours. Larry's rate is $\dfrac{156}{3} = 52$ miles per hour, and Mike's rate is $52 + 2 = 54$ miles per hour.

■

Problem Set 10.5

Solve each of the following problems.

1. Find two consecutive whole numbers whose product is 306.

2. Find two consecutive whole numbers whose product is 702.

3. Suppose that the sum of two positive integers is 44 and their product is 475. Find the integers.

4. Two positive integers differ by 6. Their product is 616. Find the integers.

5. Find two numbers such that their sum is 6 and their product is 4.

6. Find two numbers such that their sum is 4 and their product is 1.

7. The sum of a number and its reciprocal is $\frac{3\sqrt{2}}{2}$. Find the number.

8. The sum of a number and its reciprocal is $\frac{73}{24}$. Find the number.

9. Each of three consecutive even whole numbers is squared. The three results are added and the sum is 596. Find the numbers.

10. Each of three consecutive whole numbers is squared. The three results are added, and the sum is 245. Find the three whole numbers.

11. The sum of the square of a number and the square of one-half of the number is 80. Find the number.

12. The difference between the square of a positive number, and the square of one-half the number is 243. Find the number.

13. Find the length and width of a rectangle if its length is 4 meters less than twice the width, and the area of the rectangle is 96 square meters.

14. Suppose that the length of a rectangular region is 4 centimeters greater than its width. The area of the region is 45 square centimeters. Find the length and width of the rectangle.

15. The perimeter of a rectangle is 80 centimeters, and its area is 375 square centimeters. Find the length and width of the rectangle.

16. The perimeter of a rectangle is 132 yards and its area is 1080 square yards. Find the length and width of the rectangle.

17. The area of a tennis court is 2106 square feet (see Figure 10.12). The length of the court is $\frac{26}{9}$ times the width. Find the length and width of a tennis court.

Figure 10.12

18. The area of a badminton court is 880 square feet. The length of the court is 2.2 times the width. Find the length and width of the court.

19. An auditorium in a local high school contains 300 seats. There are 5 fewer rows than the number of seats per row. Find the number of rows and the number of seats per row.

20. Three hundred seventy-five trees were planted in rows in an orchard. The number of trees per row was 10 more than the number of rows. How many rows of trees are in the orchard?

21. The area of a rectangular region is 63 square feet. If the length and width are each increased by 3 feet, the area is increased by 57 square feet. Find the length and width of the original rectangle.

22. The area of a circle is numerically equal to twice the circumference of the circle. Find the length of a radius of the circle.

23. The sum of the lengths of the two legs of a right triangle is 14 inches. If the length of the hypotenuse is 10 inches, find the length of each leg.

24. A page for a magazine contains 70 square inches of type. The height of a page is twice the width. If the margin around the type is to be 2 inches uniformly, what are the dimensions of the page?

25. A 5-by-7-inch picture is surrounded by a frame of uniform width (see Figure 10.13). The area of the picture and frame together is 80 square inches. Find the width of the frame.

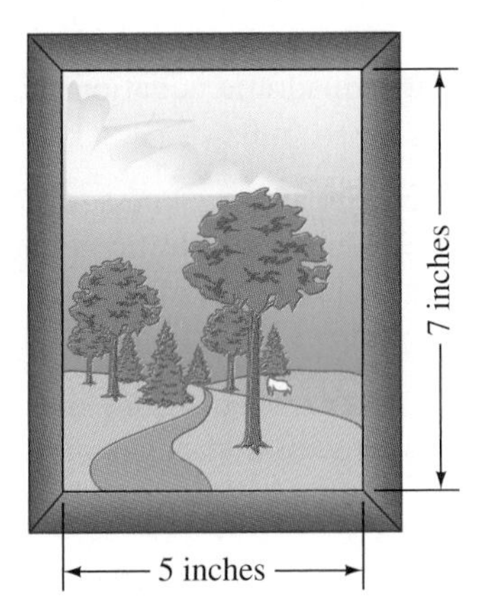

Figure 10.13

26. A rectangular piece of cardboard is 3 inches longer than it is wide. From each corner, a square piece 2 inches on a side is cut out. The flaps are then turned up to form an open box that has a volume of 140 cubic inches. Find the length and width of the original piece of cardboard.

27. A class trip was to cost \$3000. If there had been ten more students, it would have cost each student \$25 less. How many students took the trip?

28. Simon mowed some lawns and earned \$40. It took him 3 hours longer than he anticipated, and thus he earned \$3 per hour less than he anticipated. How long did he expect it to take him?

29. A piece of wire 56 inches long is cut into two pieces and each piece is bent into the shape of a square. If the sum of the areas of the two squares is 100 square inches, find the length of each piece of wire.

30. Suppose that by increasing the speed of a car by 10 miles per hour, it is possible to make a trip of 200 miles in 1 hour less time. What was the original speed for the trip?

31. On a 50-mile bicycle ride, Irene averaged 4 miles per hour faster for the first 36 miles than she did for the last 14 miles. The entire trip of 50 miles took 3 hours. Find her rate for the first 36 miles.

32. One side of a triangle is 1 foot more than twice the length of the altitude to that side. If the area of the triangle is 18 square feet, find the length of a side and the length of the altitude to that side.

A selection of additional word problems is in Appendix B. All Appendix problems that are referenced as (10.5) are appropriate for this section.

THOUGHTS INTO WORDS

33. Return to Problem 1 of this section and explain how the problem could be solved using one variable and one equation.

34. Write a page or two on the topic "using algebra to solve problems."

Chapter 10 Summary

(10.1) A **quadratic equation** in the variable x is any equation that can be written in the form $ax^2 + bx + c = 0$, where a, b, and c are real numbers and $a \neq 0$. We can solve quadratic equations that are factorable using integers by factoring and applying the property $ab = 0$ *if and only if* $a = 0$ *or* $b = 0$.

The property $x^2 = a$ *if and only if* $x = \pm\sqrt{a}$ can be used to solve certain types of quadratic equations.

This property can be used when working with the Pythagorean theorem when the resulting equation is of the form, $x^2 = a$.

Don't forget:

1. In an isosceles right triangle, the lengths of the two legs are equal; and
2. In a 30°–60° right triangle, the length of the leg opposite the 30° angle is one-half the length of the hypotenuse.

(10.2) You should be able to solve quadratic equations by the method of **completing the square**. To review this method, look back over the examples in Section 10.2.

(10.3) We usually state the **quadratic formula** as

$$x = \frac{-b \pm \sqrt{b^2 - 4ac}}{2a}$$

We can use it to solve any quadratic equation that is written in the form $ax^2 + bx + c = 0$.

(10.4) To review the strengths and weaknesses of the three basic methods for solving a quadratic equation (factoring, completing the square, the quadratic formula), go back over the examples in Section 10.4.

(10.5) Our knowledge of systems of equations and quadratic equations provides us with a stronger basis for solving word problems.

Chapter 10 Review Problem Set

For Problems 1–22, solve each quadratic equation.

1. $(2x + 7)^2 = 25$

2. $x^2 + 8x = -3$

3. $21x^2 - 13x + 2 = 0$

4. $x^2 = 17x$

5. $n - \frac{4}{n} = -3$

6. $n^2 - 26n + 165 = 0$

7. $3a^2 + 7a - 1 = 0$

8. $4x^2 - 4x + 1 = 0$

9. $5x^2 + 6x + 7 = 0$

10. $3x^2 + 18x + 15 = 0$

11. $3(x - 2)^2 - 2 = 4$

12. $x^2 + 4x - 14 = 0$

13. $y^2 = 45$

14. $x(x - 6) = 27$

15. $x^2 = x$

16. $n^2 - 4n - 3 = 6$

17. $n^2 - 44n + 480 = 0$

18. $\frac{x^2}{4} = x + 1$

19. $\frac{5x - 2}{3} = \frac{2}{x + 1}$

20. $\frac{-1}{3x - 1} = \frac{2x + 1}{-2}$

21. $\frac{5}{x - 3} + \frac{4}{x} = 6$

22. $\frac{1}{x + 2} - \frac{2}{x} = 3$

For Problems 23–32, set up an equation or a system of equations to help solve each problem.

23. The perimeter of a rectangle is 42 inches, and its area is 108 square inches. Find the length and width of the rectangle.

24. Find two consecutive whole numbers whose product is 342.

25. Each of three consecutive odd whole numbers is squared. The three results are added and the sum is 251. Find the numbers.

26. The combined area of two squares is 50 square meters. Each side of the larger square is three times as long as a side of the smaller square. Find the lengths of the sides of each square.

27. The difference in the lengths of the two legs of a right triangle is 2 yards. If the length of the hypotenuse is $2\sqrt{13}$ yards, find the length of each leg.

28. Tony bought a number of shares of stock for a total of $720. A month later the value of the stock increased by $8 per share, and he sold all but 20 shares and regained his original investment plus a profit of $80. How many shares did Tony sell and at what price per share?

29. A company has a rectangular parking lot 40 meters wide and 60 meters long. They plan to increase the area of the lot by 1100 square meters by adding a strip of equal width to one side and one end. Find the width of the strip to be added.

30. Jay traveled 225 miles in 2 hours less time than it took Jean to travel 336 miles. If Jay's rate was 3 miles per hour slower than Jean's rate, find each rate.

31. The length of the hypotenuse of an isosceles right triangle is 12 inches. Find the length of each leg.

32. In a 30°–60° right triangle, the side opposite the 60° angle is 8 centimeters long. Find the length of the hypotenuse.

For more practice on word problems, consult Appendix B. All Appendix problems that have a chapter 10 reference would be appropriate.

Chapter 10 Test

1. The two legs of a right triangle are 4 inches and 6 inches long. Find the length of the hypotenuse. Express your answer in simplest radical form.

2. A diagonal of a rectangular plot of ground measures 14 meters. If the width of the rectangle is 5 meters, find the length to the nearest meter.

3. A diagonal of a square piece of paper measures 10 inches. Find, to the nearest inch, the length of a side of the square.

4. In a 30°–60° right triangle, the side opposite the 30° angle is 4 centimeters long. Find the length of the side opposite the 60° angle. Express your answer in simplest radical form.

For Problems 5–20, solve each equation.

5. $(3x + 2)^2 = 49$

6. $4x^2 = 64$

7. $8x^2 - 10x + 3 = 0$

8. $x^2 - 3x - 5 = 0$

9. $n^2 + 2n = 9$

10. $(2x - 1)^2 = -16$

11. $y^2 + 10y = 24$

12. $2x^2 - 3x - 4 = 0$

13. $\dfrac{x - 2}{3} = \dfrac{4}{x + 1}$

14. $\dfrac{2}{x - 1} + \dfrac{1}{x} = \dfrac{5}{2}$

15. $n(n - 28) = -195$

16. $n + \dfrac{3}{n} = \dfrac{19}{4}$

17. $(2x + 1)(3x - 2) = -2$

18. $(7x + 2)^2 - 4 = 21$

19. $(4x - 1)^2 = 27$

20. $n^2 - 5n + 7 = 0$

For Problems 21–25, set up an equation or a system of equations to help solve each problem.

21. A room contains 120 seats. The number of seats per row is 1 less than twice the number of rows. Find the number of seats per row.

22. Abu rode his bicycle 56 miles in 2 hours less time than it took Stan to ride his bicycle 72 miles. If Abu's rate was 2 miles per hour faster than Stan's rate, find Abu's rate.

23. Find two consecutive odd whole numbers whose product is 255.

24. The combined area of two squares is 97 square feet. Each side of the larger square is 1 foot more than twice the length of a side of the smaller square. Find the length of a side of the larger square.

25. Dee bought a number of shares of stock for a total of \$160. Two weeks later, the value of the stock had increased \$2 per share, and she sold all but 4 shares and regained her initial investment of \$160. How many shares did Dee originally buy?

11

Additional Topics

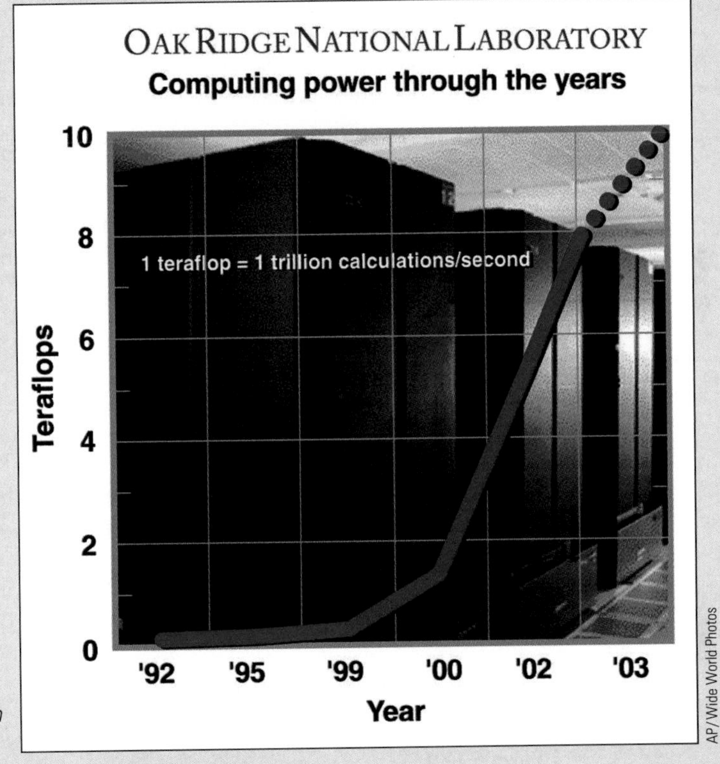

AP/Wide World Photos

Be sure that you understand the meaning of the markings on both the horizontal and vertical axes.

We include this chapter to give you the opportunity to expand your knowledge of topics presented in earlier chapters. From the list of section titles, the topics may appear disconnected; however, each section is a continuation of a topic presented in a previous chapter.

Section 11.1 continues the development of techniques for solving equations and inequalities, which was the focus of Chapter 3. Section 11.2 uses the method of elimination-by-addition from Section 8.6 to solve systems containing three linear equations in three variables. Section 11.3 is an extension of the work we did with exponents in Section 5.6 and with radicals in Chapter 9. Sections 11.4 and 11.5 enhance the study of quadratic equations from Chapter 10. Sections 11.6–11.8 extend our work with coordinate geometry from Chapter 8.

11.1 Equations and Inequalities Involving Absolute Value

In Chapter 1, we used the concept of absolute value to explain the addition and multiplication of integers. We defined the absolute value of a number to be the distance between the number and zero on a number line. For example, the absolute value of 3 is 3, and the absolute value of -3 is 3. The absolute value of 0 is 0. Symbolically, absolute value is denoted with vertical bars.

$$|3| = 3 \qquad |-3| = 3 \qquad |0| = 0$$

In general, we can say that the absolute value of any number except 0 is positive.

If we interpret absolute value as distance on a number line, we can solve a variety of equations and inequalities that involve absolute value. We will first consider some equations.

EXAMPLE 1

Solve and graph the solutions for $|x| = 2$.

Solution

When we think in terms of the distance between the number and zero, we can see that x must be 2 or -2. Thus $|x| = 2$ implies

$$x = 2 \qquad \text{or} \qquad x = -2$$

The solution set is $\{-2, 2\}$, and its graph is shown in Figure 11.1.

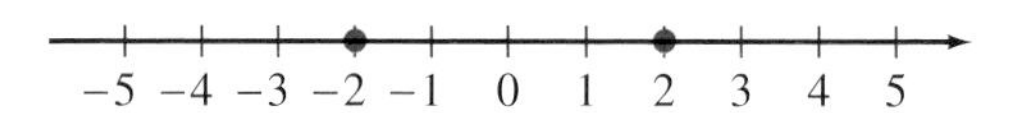

Figure 11.1

EXAMPLE 2

Solve and graph the solutions for $|x + 4| = 1$.

Solution

The number $x + 4$ must be 1 or -1. Thus $|x + 4| = 1$ implies

$$x + 4 = 1 \qquad \text{or} \qquad x + 4 = -1$$

$$x = -3 \qquad \text{or} \qquad x = -5$$

The solution set is $\{-5, -3\}$, and its graph is shown in Figure 11.2.

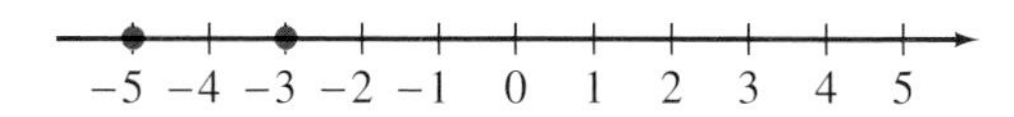

Figure 11.2

EXAMPLE 3

Solve and graph the solutions for $|3x - 2| = 4$.

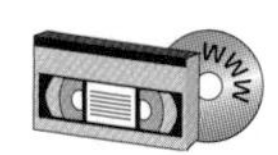

Solution

The number $3x - 2$ must be 4 or -4. Thus $|3x - 2| = 4$ implies

$$3x - 2 = 4 \quad \text{or} \quad 3x - 2 = -4$$

$$3x = 6 \quad \text{or} \quad 3x = -2$$

$$x = 2 \quad \text{or} \quad x = -\frac{2}{3}$$

The solution set is $\left\{-\frac{2}{3}, 2\right\}$, and its graph is shown in Figure 11.3.

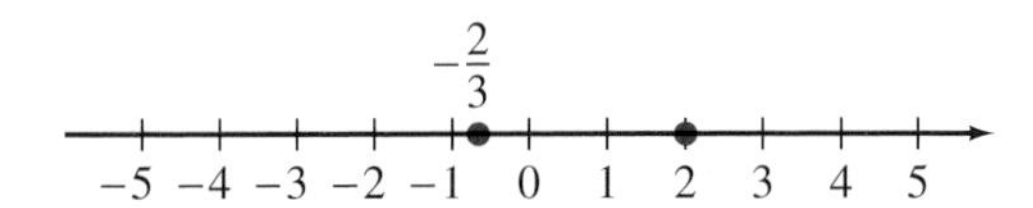

Figure 11.3

The "distance interpretation" for absolute value also provides a good basis for solving inequalities involving absolute value. Consider the following examples.

EXAMPLE 4

Solve and graph the solutions for $|x| < 2$.

Solution

The number x must be *less than two units away from zero*. Thus, $|x| < 2$ implies

$$x > -2 \quad \text{and} \quad x < 2$$

The solution set is $\{x|x > -2 \text{ and } x < 2\}$, and its graph is shown in Figure 11.4. The solution set is $(-2, 2)$ written in interval notation.

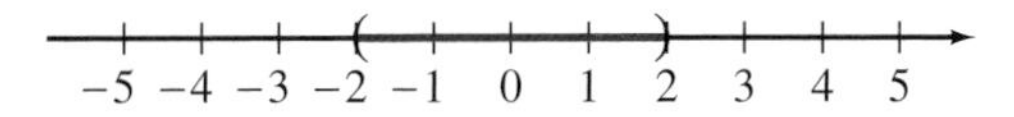

Figure 11.4

EXAMPLE 5

Solve and graph the solutions for $|x - 1| < 2$.

Solution

The number $x - 1$ must be *less than two units away from zero*. Thus $|x - 1| < 2$ implies

$$x - 1 > -2 \quad \text{and} \quad x - 1 < 2$$

$$x > -1 \quad \text{and} \quad x < 3$$

The solution set is $\{x|x > -1 \text{ and } x < 3\}$, and its graph is shown in Figure 11.5. The solution set is $(-1, 3)$ written in interval notation.

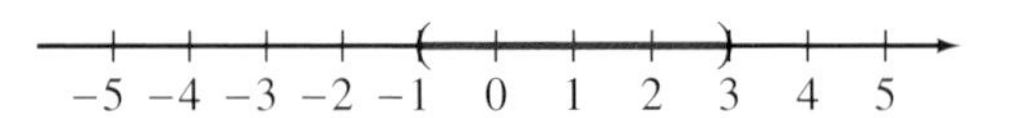

Figure 11.5

EXAMPLE 6 Solve and graph the solutions for $|2x + 5| \leq 1$.

Solution

The number $2x + 5$ must be *equal to or less than one unit away from zero*. Therefore $|2x + 5| \leq 1$ implies

$$2x + 5 \geq -1 \quad \text{and} \quad 2x + 5 \leq 1$$
$$2x \geq -6 \quad \text{and} \quad 2x \leq -4$$
$$x \geq -3 \quad \text{and} \quad x \leq -2$$

The solution set is $\{x|x \geq -3 \text{ and } x \leq -2\}$, and its graph is shown in Figure 11.6. The solution set is $[-3, -2]$ written in interval notation.

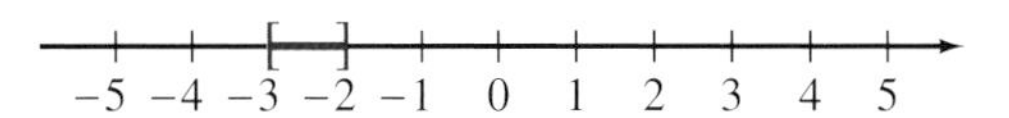

Figure 11.6

EXAMPLE 7 Solve and graph the solutions for $|x| > 2$.

Solution

The number x must be *more than two units away from zero*. Thus $|x| > 2$ implies

$$x < -2 \quad \text{or} \quad x > 2$$

The solution set is $\{x|x < -2 \text{ or } x > 2\}$, and its graph is shown in Figure 11.7. The solution set is $(-\infty, -2) \cup (2, \infty)$ written in interval notation.

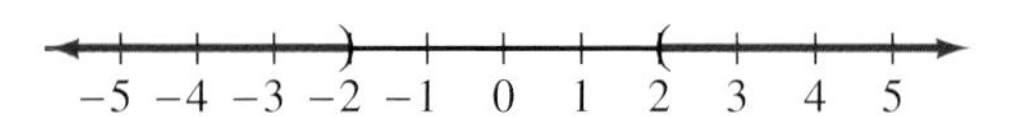

Figure 11.7

EXAMPLE 8 Solve and graph the solutions for $|3x - 1| > 4$.

Solution

The number $3x - 1$ must be *more than four units away from zero*. Thus $|3x - 1| > 4$ implies

$$3x - 1 < -4 \quad \text{or} \quad 3x - 1 > 4$$

$$3x < -3 \quad \text{or} \quad 3x > 5$$

$$x < -1 \quad \text{or} \quad x > \frac{5}{3}$$

The solution set is $\left\{x|x < -1 \text{ or } x > \frac{5}{3}\right\}$, and its graph is shown in Figure 11.8.

The solution set is $(-\infty, -1) \cup \left(\frac{5}{3}, \infty\right)$ written in interval notation.

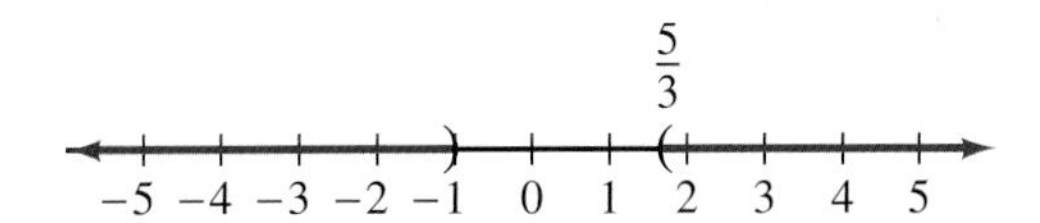

Figure 11.8

The solutions for equations and inequalities such as $|3x - 7| = -4$, $|x + 5| < -3$, and $|2x - 3| > -7$ can be found by *inspection*. Notice that in each of these examples the right side is a negative number. Therefore, using the fact that the **absolute value of any number is nonnegative**, we can reason as follows:

$|3x - 7| = -4$ has *no solutions* because the absolute value of a number cannot be negative.

$|x + 5| < -3$ has *no solutions* because we cannot obtain an absolute value less than -3.

$|2x - 3| > -7$ is *satisfied by all real numbers* because the absolute value of $2x - 3$, regardless of what number is substituted for x, will always be greater than -7.

Problem Set 11.1

For Problems 1–26, solve the equation or inequality. Graph the solutions.

1. $|x| = 4$

2. $|x| = 3$

3. $|x| < 1$

4. $|x| < 4$

5. $|x| \geq 2$

6. $|x| \geq 1$

7. $|x + 2| = 1$

8. $|x + 3| = 2$

9. $|x - 1| = 2$

10. $|x - 2| = 1$

11. $|x - 2| \leq 2$

12. $|x + 1| \leq 3$

13. $|x + 1| > 3$

14. $|x - 3| > 1$

15. $|2x + 1| = 3$

16. $|3x - 1| = 5$

17. $|5x - 2| = 4$

18. $|4x + 3| = 8$

19. $|2x - 3| \geq 1$

20. $|2x + 1| \geq 3$

21. $|4x + 3| < 2$

22. $|5x - 2| < 8$

23. $|3x + 6| = 0$

24. $|4x - 3| = 0$

25. $|3x - 2| > 0$

26. $|2x + 7| < 0$

For Problems 27–42, solve each of the following.

27. $|3x - 1| = 17$

28. $|4x + 3| = 27$

29. $|2x + 1| > 9$

30. $|3x - 4| > 20$

31. $|3x - 5| < 19$

32. $|5x + 3| < 14$

33. $|-3x - 1| = 17$

34. $|-4x + 7| = 26$

35. $|4x - 7| \leq 31$

36. $|5x - 2| \leq 21$

37. $|5x + 3| \geq 18$

38. $|2x - 11| \geq 4$

39. $|-x - 2| < 4$

40. $|-x - 5| < 7$

41. $|-2x + 1| > 6$

42. $|-3x + 2| > 8$

For Problems 43–50, solve each equation or inequality *by inspection*.

43. $|7x| = 0$

44. $|3x - 1| = -4$

45. $|x - 6| > -4$

46. $|3x + 1| > -3$

47. $|x + 4| < -7$

48. $|5x - 2| < -2$

49. $|x + 6| \leq 0$

50. $|x + 7| > 0$

THOUGHTS INTO WORDS

51. Explain why the equation $|3x + 2| = -6$ has no real number solutions.

52. Explain why the inequality $|x + 6| < -4$ has no real number solutions.

FURTHER INVESTIGATIONS

A conjunction such as $x > -2$ and $x < 4$ can be written in a more compact form $-2 < x < 4$, which is read as "-2 is less than x, and x is less than 4." In other words, x is clamped between -2 and 4. The compact form is very convenient for solving conjunctions as follows:

$$-3 < 2x - 1 < 5$$

$$-2 < 2x < 6 \quad \text{Add 1 to the left side, middle, and right side.}$$

$$-1 < x < 3 \quad \text{Divide through by 2.}$$

Thus, the solution set can be expressed as $\{x|-1 < x < 3\}$.

For Problems 53–62, solve the compound inequalities using the compact form.

53. $-2 < x - 6 < 8$

54. $-1 < x + 3 < 9$

55. $1 \leq 2x + 3 \leq 11$

56. $-2 \leq 3x - 1 \leq 14$

57. $-4 < \dfrac{x - 1}{3} < 2$

58. $2 < \dfrac{x + 1}{4} < 5$

59. $|x + 4| < 3$
[*Hint*: $|x + 4| < 3$ implies $-3 < x + 4 < 3$]

60. $|x - 6| < 5$

61. $|2x - 5| < 7$

62. $|3x + 2| < 14$

11.2 3 × 3 Systems of Equations

When we find the solution set of an equation in two variables, such as $2x + y = 9$, we are finding the ordered pairs that make the equation a true statement. Plotted in two dimensions, the graph of the solution set is a line.

Now consider an equation with three variables, such as $2x - y + 4z = 8$. A solution set of this equation is an ordered triple, (x, y, z), that makes the equation a true statement. For example, the ordered triple $(3, 2, 1)$ is a solution of $2x - y + 4z = 8$

because $2(3) - 2 + 4(1) = 8$. The graph of the solution set of an equation in three variables is a plane, not a line. In fact, graphing equations in three variables requires the use of a three-dimensional coordinate system.

A 3×3 (read "3 by 3") system of equations is a system of three linear equations in three variables. To solve a 3×3 system such as

$$\begin{pmatrix} 2x - y + 4z = 5 \\ 3x + 2y + 5z = 4 \\ 4x - 3y - z = 11 \end{pmatrix}$$

means to find all the ordered triples that satisfy all three equations. In other words, the solution set of the system is the intersection of the solution sets of all three equations in the system. Using a graphing approach to solve systems of three linear equations in three variables is not at all practical. However, a graphic analysis will provide insight into the types of possible solutions.

In general, each linear equation in three variables produces a plane. A system of three such equations produces three planes. There are various ways that the planes can intersect. For our purposes at this time, however, you need to realize that a system of three linear equations in three variables produces one of the following possible solution sets.

1. There is *one ordered triple* that satisfies all three equations. The three planes have a common point of intersection as indicated in Figure 11.9.

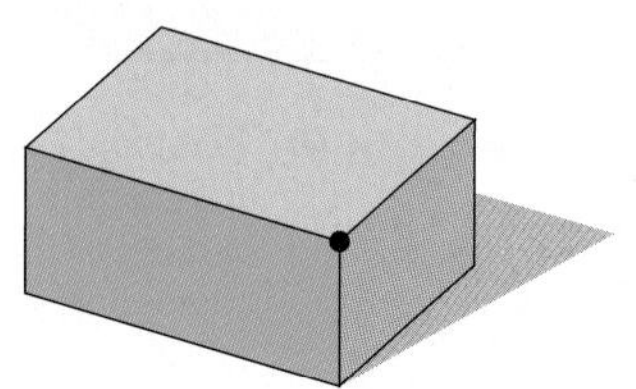

Figure 11.9

2. There are *infinitely many* ordered triples in the solution set, all of which are coordinates of points on a line common to the planes. This can happen when three planes have a common line of intersection, as in Figure 11.10(a), or when two of the planes coincide and the third plane intersects them, as in Figure 11.10(b).

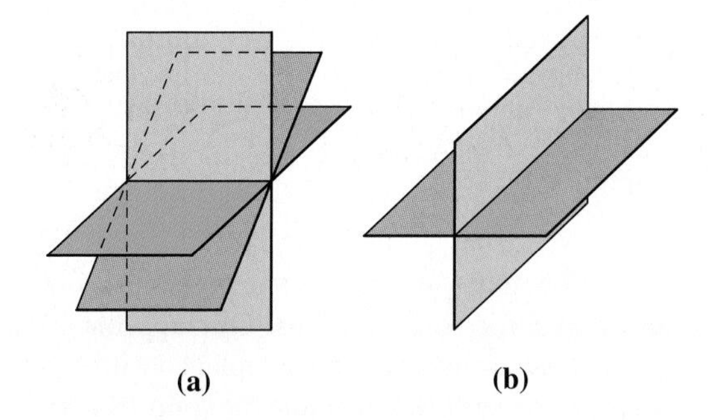

Figure 11.10

3. There are *infinitely many* ordered triples in the solution set, all of which are coordinates of points on a plane. This happens when the three planes coincide, as illustrated in Figure 11.11.

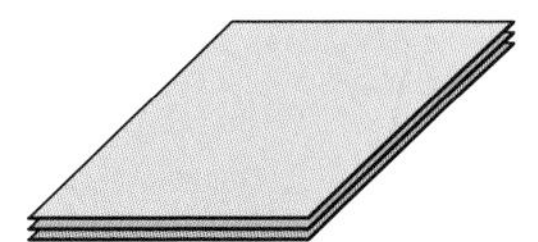

Figure 11.11

4. The solution set is *empty*; it is $\emptyset$. This can happen in various ways, as you can see in Figure 11.12. Notice that in each situation there are no points common to all three planes.

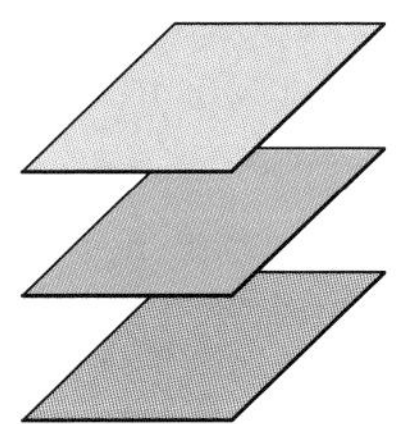

(a) Three parallel planes

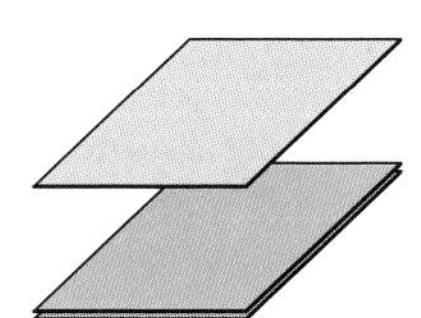

(b) Two planes coincide and the third one is parallel to the coinciding planes.

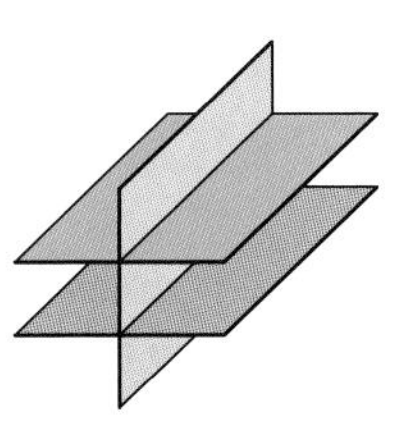

(c) Two planes are parallel and the third intersects them in parallel lines.

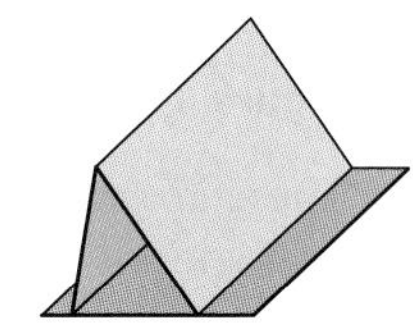

(d) No two planes are parallel, but two of them intersect in a line that is parallel to the third plane.

Figure 11.12

Now that you know what possibilities exist, we can consider finding the solution sets for some systems. Our approach will be the elimination-by-addition method, whereby systems are replaced with equivalent systems until we get a system where we can easily determine the solution set. We will start with an example that allows us to determine the solution set without changing to another equivalent system.

EXAMPLE 1

Solve the system

$$\begin{pmatrix} 4x + 2y - z = -5 \\ 3y + z = -1 \\ 2z = 10 \end{pmatrix} \quad \begin{matrix} (1) \\ (2) \\ (3) \end{matrix}$$

Solution

From equation (3), we can find the value of z.

$$2z = 10$$

$$\boxed{z = 5}$$

Now we can substitute 5 for z in equation (2).

$$3y + z = -1$$

$$3y + 5 = -1$$

$$3y = -6$$

$$\boxed{y = -2}$$

Finally, we can substitute -2 for y and 5 for z in equation (1).

$$4x + 2y - z = -5$$

$$4x + 2(-2) - 5 = -5$$

$$4x - 9 = -5$$

$$4x = 4$$

$$\boxed{x = 1}$$

The solution set is $\{(1, -2, 5)\}$. ■

Notice the format of the equations in the system in Example 1. The first equation contains all three variables, the second equation has only two variables, and the third equation has only one variable. This allowed us to solve the third equation and then use "back substitution" to find the values of the other variables. Let's consider another example where we have to replace one equation to make an equivalent system.

EXAMPLE 2

Solve the system

$$\begin{pmatrix} 2x + 4y - 5z = -8 \\ y + 4z = 7 \\ 5y + 3z = 1 \end{pmatrix} \quad \begin{matrix} (1) \\ (2) \\ (3) \end{matrix}$$

Solution

In order to achieve the same format as in Example 1, we will need to eliminate the term with the y variable in equation (3). Using the concept of elimination from

Section 8.4, we can replace equation (3) with an equivalent equation we form by multiplying equation (2) by -5 and then adding that result to equation (3). The equivalent system is

$$\begin{pmatrix} 2x + 4y - 5z = -8 \\ y + 4z = 7 \\ -17z = -34 \end{pmatrix} \quad \begin{matrix} (4) \\ (5) \\ (6) \end{matrix}$$

From equation (6) we can find the value of z.

$$-17z = -34$$

$$z = 2$$

Now we can substitute 2 for z in equation (5).

$$y + 4z = 7$$

$$y + 4(2) = 7$$

$$y = -1$$

Finally, we can substitute -1 for y and 2 for z in equation (4).

$$2x + 4y - 5z = -8$$

$$2x + 4(-1) - 5(2) = -8$$

$$2x - 4 - 10 = -8$$

$$2x - 14 = -8$$

$$2x = 6$$

$$x = 3$$

The solution set is $\{(3, -1, 2)\}$. ■

Now let's consider some examples where we have to replace more than one equation to make an equivalent system.

EXAMPLE 3

Solve the system

$$\begin{pmatrix} x + 2y - 3z = -1 \\ 3x - y + 2z = -13 \\ 2x + 3y - 5z = -4 \end{pmatrix} \quad \begin{matrix} (1) \\ (2) \\ (3) \end{matrix}$$

Solution

We start by picking a pair of equations to form a new equation by eliminating a variable. We will use equations (1) and (2) to form a new equation while eliminating the x variable. We can replace equation (2) with an equation formed by multiplying equation (1) by −3 and adding the result to equation (2). The equivalent system is

$$\begin{pmatrix} x + 2y - 3z = -1 \\ -7y + 11z = -10 \\ 2x + 3y - 5z = -4 \end{pmatrix} \quad \begin{matrix} (4) \\ (5) \\ (6) \end{matrix}$$

Now we take equation (4) and equation (6) and eliminate the same variable, x. We can replace equation (6) with a new equation formed by multiplying equation (4) by −2 and adding the result to equation (6). The equivalent system is

$$\begin{pmatrix} x + 2y - 3z = -1 \\ -7y + 11z = -10 \\ -y + z = -2 \end{pmatrix} \quad \begin{matrix} (7) \\ (8) \\ (9) \end{matrix}$$

Now we take equations (8) and (9) and form a new equation by eliminating a variable. Either y or z can be eliminated. For this example we will eliminate y. We can replace equation (8) with a new equation formed by multiplying equation (9) by −7 and adding the result to equation (8). The equivalent system is

$$\begin{pmatrix} x + 2y - 3z = -1 \\ 4z = 4 \\ -y + z = -2 \end{pmatrix} \quad \begin{matrix} (10) \\ (11) \\ (12) \end{matrix}$$

From equation (11), we can find the value of z.

$$4z = 4$$

$$z = 1$$

Now we substitute 1 for z in equation (12) and determine the value of y.

$$-y + z = -2$$

$$-y + 1 = -2$$

$$-y = -3$$

$$y = 3$$

Finally, we can substitute 3 for y and 1 for z in equation (10).

$$x + 2y - 3z = -1$$

$$x + 2(3) - 3(1) = -1$$

$$x + 6 - 3 = -1$$

$$x + 3 = -1$$

$$x = -4$$

The solution set is $\{(-4, 3, 1)\}$. ■

EXAMPLE 4

Solve the system

$$\begin{pmatrix} 2x + 3y - z = 8 \\ 5x + 2y - 3z = 21 \\ 3x - 4y + 2z = 5 \end{pmatrix} \quad \begin{matrix} (1) \\ (2) \\ (3) \end{matrix}$$

Solution

Studying the coefficients in the system indicates that eliminating the z terms from equations (2) and (3) would be easy to do. We can replace equation (2) with an equation formed by multiplying equation (1) by -3 and adding the result to equation (2). The equivalent system is

$$\begin{pmatrix} 2x + 3y - z = 8 \\ -x - 7y = -3 \\ 3x - 4y + 2z = 5 \end{pmatrix} \quad \begin{matrix} (4) \\ (5) \\ (6) \end{matrix}$$

Now we replace equation (6) with an equation formed by multiplying equation (4) by 2 and adding the result to equation (6). The equivalent system is

$$\begin{pmatrix} 2x + 3y - z = 8 \\ -x - 7y = -3 \\ 7x + 2y = 21 \end{pmatrix} \quad \begin{matrix} (7) \\ (8) \\ (9) \end{matrix}$$

Now we can eliminate the x term from equation (9). We replace equation (9) with an equation formed by multiplying equation (8) by 7 and adding the result to equation (9). The equivalent system is

$$\begin{pmatrix} 2x + 3y - z = 8 \\ -x - 7y = -3 \\ -47y = 0 \end{pmatrix} \quad \begin{matrix} (10) \\ (11) \\ (12) \end{matrix}$$

From equation (12), we can determine the value of y.

$$-47y = 0$$

$$y = 0$$

Now we can substitute 0 for y in equation (11) and find the value of x.

$$-x - 7y = -3$$

$$-x - 7(0) = -3$$

$$-x = -3$$

$$x = 3$$

Finally, we can substitute 3 for x and 0 for y in equation (10).

$$2x + 3y - z = 8$$

$$2(3) + 3(0) - z = 8$$

$$6 - z = 8$$
$$-z = 2$$
$$z = -2$$

The solution set is $\{(3, 0, -2)\}$. ■

EXAMPLE 5

Solve the system

$$\begin{pmatrix} x + 3y - 2z = 3 \\ 3x - 4y - z = 4 \\ 2x + 6y - 4z = 9 \end{pmatrix} \begin{matrix} (1) \\ (2) \\ (3) \end{matrix}$$

Solution

Studying the coefficients indicates that it would be easy to eliminate the x terms from equations (2) and (3). We can replace equation (2) with an equation formed by multiplying equation (1) by -3 and adding the result to equation (2). Likewise, we can replace equation (3) with an equation formed by multiplying equation (1) by -2 and adding the result to equation (3). The equivalent system is

$$\begin{pmatrix} x + 3y - 2z = 3 \\ -13y + 5z = -5 \\ 0 + 0 + 0 = 3 \end{pmatrix} \begin{matrix} (4) \\ (5) \\ (6) \end{matrix}$$

The false statement $0 = 3$ in equation (6) indicates that the system is inconsistent, and therefore the solution set is $\varnothing$. [If you were to graph this system, equations (1) and (3) would produce parallel planes, which is the situation depicted in Figure 11.12(c).] ■

EXAMPLE 6

Solve the system

$$\begin{pmatrix} x + y + z = 6 \\ 3x + y - z = 2 \\ 5x + y - 3z = -2 \end{pmatrix} \begin{matrix} (1) \\ (2) \\ (3) \end{matrix}$$

Solution

Studying the coefficients indicates that it would be easy to eliminate the y terms from equations (2) and (3). We can replace equation (2) with an equation formed by multiplying equation (1) by -1 and adding the result to equation (2). Likewise, we can replace equation (3) with an equation formed by multiplying equation (1) by -1 and adding the result to equation (3). The equivalent system is

$$\begin{pmatrix} x + y + z = 6 \\ 2x - 2z = -4 \\ 4x - 4z = -8 \end{pmatrix} \begin{matrix} (4) \\ (5) \\ (6) \end{matrix}$$

Now we replace equation (6) with an equation formed by multiplying equation (5) by -2 and adding the result to equation (6). The equivalent system is

$$\begin{pmatrix} x + y + z = 6 \\ 2x - 2z = -4 \\ 0 + 0 = 0 \end{pmatrix} \quad \begin{matrix} (7) \\ (8) \\ (9) \end{matrix}$$

The true numerical statement $0 + 0 = 0$ in equation (9) indicates that the system has *infinitely many solutions*. ■

Now we will use the techniques we have presented to solve a geometric problem.

PROBLEM 1

In a certain triangle, the measure of $\angle A$ is 5° more than twice the measure of $\angle B$. The sum of the measures of $\angle B$ and $\angle C$ is 10° more than the measure of $\angle A$. Find the measures of all three angles.

Solution

We can solve this problem by setting up a system of three linear equations in three variables. We let

$x =$ measure of $\angle A$

$y =$ measure of $\angle B$

$z =$ measure of $\angle C$

Knowing that the sum of the measures of the angles in a triangle is 180° gives us the equation $x + y + z = 180$. The information "the measure of $\angle A$ is 5° more than twice the measure of $\angle B$" gives us the equation $x = 2y + 5$ or an equivalent form $x - 2y = 5$. The information "the sum of the measures of $\angle B$ and $\angle C$ is 10° more than the measure of $\angle A$" gives us the equation $y + z = x + 10$ or an equivalent form $x - y - z = -10$. Putting the three equations together, we get the system of equations

$$\begin{pmatrix} x + y + z = 180 \\ x - 2y = 5 \\ x - y - z = -10 \end{pmatrix} \quad \begin{matrix} (1) \\ (2) \\ (3) \end{matrix}$$

To solve the system, we first replace equation (3) with an equation formed by adding equation (1) and equation (3). The equivalent system is

$$\begin{pmatrix} x + y + z = 180 \\ x - 2y = 5 \\ 2x = 170 \end{pmatrix} \quad \begin{matrix} (4) \\ (5) \\ (6) \end{matrix}$$

From equation (6), we can determine that $x = 85$.

Now we can substitute 85 for x in equation (5) and find the value of y.

$$x - 2y = 5$$

$$85 - 2y = 5$$

$$-2y = -80$$
$$y = 40$$

Finally, we can substitute 40 for y and 85 for x in equation (4) and find the value of z.

$$x + y + z = 180$$
$$85 + 40 + z = 180$$
$$125 + z = 180$$
$$z = 55$$

The measures of the angles are $\angle A = 85°$, $\angle B = 40°$, and $\angle C = 55°$. ■

Problem Set 11.2

For Problems 1–16, solve each system of equations.

1. $\begin{pmatrix} 3x + y + 2z = 6 \\ 6y + 5z = -4 \\ -4z = 8 \end{pmatrix}$

2. $\begin{pmatrix} 2x - 3y + z = -20 \\ 2y - 5z = -8 \\ 3z = 12 \end{pmatrix}$

3. $\begin{pmatrix} x + 2y - z = 1 \\ y + 2z = 11 \\ 2y - z = 2 \end{pmatrix}$

4. $\begin{pmatrix} 3x + y - 2z = 0 \\ x + 3z = -5 \\ 2x - 5z = 1 \end{pmatrix}$

5. $\begin{pmatrix} 4x + 3y - 2z = 9 \\ 2x + y = 7 \\ 3x - 2y = 21 \end{pmatrix}$

6. $\begin{pmatrix} 2x - 5y + 3z = 43 \\ 3x - y = 25 \\ x + 3y = -5 \end{pmatrix}$

7. $\begin{pmatrix} x + 2y - 3z = -11 \\ 2x - y + 2z = 3 \\ 4x + 3y + z = 6 \end{pmatrix}$

8. $\begin{pmatrix} x - 2y + z = 0 \\ 3x + y + 2z = 15 \\ 2x + 3y - 3z = 19 \end{pmatrix}$

9. $\begin{pmatrix} 4x - 3y + z = 14 \\ 2x + y - 3z = 16 \\ 3x - 4y + 2z = 9 \end{pmatrix}$

10. $\begin{pmatrix} 5x + 2y + z = -13 \\ 2x - 3y + 2z = -15 \\ x - y - 3z = -10 \end{pmatrix}$

11. $\begin{pmatrix} 2x + y + 4z = 5 \\ 5x - 2y + z = -10 \\ 3x + 3y - 2z = 4 \end{pmatrix}$

12. $\begin{pmatrix} 4x - y + 2z = 1 \\ 3x + 2y - z = 11 \\ 2x + 3y + 4z = -5 \end{pmatrix}$

13. $\begin{pmatrix} x + 3y - 4z = 11 \\ 3x - y + 2z = 5 \\ 2x + 5y - z = 8 \end{pmatrix}$

14. $\begin{pmatrix} x - 2y + 3z = 13 \\ 4x + y - 2z = -8 \\ 2x + 3y + z = -3 \end{pmatrix}$

15. $\begin{pmatrix} 3x + y - 2z = 3 \\ 2x - 3y + 4z = -2 \\ 4x + z = 6 \end{pmatrix}$

16. $\begin{pmatrix} x + y + 3z = 11 \\ 2x - 2y + 4z = 0 \\ 3x + 2z = 11 \end{pmatrix}$

For Problems 17–26, solve each problem by setting up and solving a system of three linear equations in three variables.

17. Brooks has 20 coins consisting of quarters, dimes, and nickels worth $3.40. The sum of the number of dimes and nickels is equal to the number of quarters. How many coins of each kind are there?

18. One binder, 2 reams of paper, and 5 spiral notebooks cost \$14.82. Three binders, 1 ream of paper, and 4 spiral notebooks cost \$14.32. Two binders, 3 reams of paper, and 3 spiral notebooks cost \$19.82. Find the cost for each item.

19. In a certain triangle, the measure of $\angle A$ is five times the measure of $\angle B$. The sum of the measures of $\angle B$ and $\angle C$ is $60°$ less than the measure of $\angle A$. Find the measure of each angle.

20. Shannon purchased a skirt, blouse, and sweater for \$72. The cost of the skirt and sweater was \$2 more than six times the cost of the blouse. The skirt cost twice the sum of the costs of the blouse and sweater. Find the cost of each item.

21. The wages for a crew consisting of a plumber, an apprentice, and a laborer are \$80 an hour. The plumber earns \$20 an hour more than the sum of the wages of the apprentice and the laborer. The plumber earns five times as much as the laborer. Find the hourly wage of each.

22. Martha has 24 bills consisting of \$1, \$5, and \$20 bills worth \$150. The number of \$20 bills is twice the number of \$5 bills. How many bills of each kind are there?

23. Two pounds of peaches, 1 pound of cherries, and 3 pounds of pears cost \$5.64. One pound of peaches, 2 pounds of cherries, and 2 pounds of pears cost \$4.65. Two pounds of peaches, 4 pounds of cherries, and 1 pound of pears cost \$7.23. Find the price per pound for each item.

24. In a certain triangle, the measure of $\angle C$ is $40°$ more than the sum of the measures of $\angle A$ and $\angle B$. The measure of $\angle A$ is $20°$ less than the measure of $\angle B$. Find the measure of each angle.

25. Mike bought a motorcycle helmet, jacket, and gloves for \$650. The jacket costs \$100 more than the helmet. The cost of the helmet and gloves together was \$50 less than the cost of the jacket. How much did each item cost?

26. A catering group that has a chef, a salad maker, and a server costs the customer \$70 per hour. The salad maker costs \$5 per hour more than the server. The chef costs the same as the salad maker and the server cost together. Find the cost per hour of each.

THOUGHTS INTO WORDS

27. Give a step-by-step description of how to solve this system of equations.

$$\begin{pmatrix} 2x + 3y - z = 3 \\ 4y + 3z = -2 \\ 6z = -12 \end{pmatrix}$$

28. Describe how you would solve this system of equations.

$$\begin{pmatrix} 2x + 3y - z = 7 \\ x + 2y = 4 \\ 3x - 4y = 2 \end{pmatrix}$$

FURTHER INVESTIGATIONS

For Problems 29–32, solve each system of equations, indicating whether the solution is $\varnothing$ or contains infinitely many solutions.

29. $\begin{pmatrix} x + 2y - 3z = 1 \\ 2x - y + 2z = 3 \\ 3x + y - z = 4 \end{pmatrix}$

30. $\begin{pmatrix} 2x + 3y + z = 7 \\ x - 4y + 2z = -3 \\ 3x - y + 3z = 5 \end{pmatrix}$

31. $\begin{pmatrix} 3x - y + 2z = -1 \\ 2x + 3y + z = 8 \\ 8x + y + 5z = 4 \end{pmatrix}$

32. $\begin{pmatrix} x + 2y - z = 6 \\ 3x - y + 4z = -10 \\ 5x + 3y + 2z = 2 \end{pmatrix}$

11.3 Fractional Exponents

At the beginning of Chapter 9, we defined and discussed the concept of square root. For your convenience and for the sake of continuity, we will repeat some of that material at this time. To **square a number** means to raise it to the second power—that is, to use the number as a factor twice.

$$3^2 = 3 \cdot 3 = 9$$

$$7^2 = 7 \cdot 7 = 49$$

$$(-3)^2 = (-3)(-3) = 9$$

$$(-7)^2 = (-7)(-7) = 49$$

A **square root of a number** is one of its two equal factors. Thus, 3 is a square root of 9 because $3 \cdot 3 = 9$. Likewise, -3 is also a square root of 9 because $(-3)(-3) = 9$. The concept of square root is defined as follows.

Definition 11.1

a is a **square root** of b if $a^2 = b$.

The following generalizations are a direct consequence of Definition 11.1:

1. Every positive real number has two square roots; one is positive and the other is negative. They are opposites of each other.
2. Negative real numbers have no real number square roots. (This follows from Definition 11.1 because any nonzero real number is positive when squared.)
3. The square root of 0 is 0.

We use the symbol $\sqrt{}$, called a **radical sign**, to indicate the nonnegative square root. Consider the next examples.

$\sqrt{49} = 7$ — $\sqrt{49}$ indicates the nonnegative or principal square root of 49

$-\sqrt{49} = -7$ — $-\sqrt{49}$ indicates the negative square root of 49

$\sqrt{0} = 0$ — Zero has only one square root

$\sqrt{-4}$ — $\sqrt{-4}$ is not a real number

$-\sqrt{-4}$ — $-\sqrt{-4}$ is not a real number

To **cube a number** means to raise it to the third power—that is, to use the number as a factor three times.

$$2^3 = 2 \cdot 2 \cdot 2 = 8$$

$$4^3 = 4 \cdot 4 \cdot 4 = 64$$

$$\left(\frac{2}{3}\right)^3 = \frac{2}{3} \cdot \frac{2}{3} \cdot \frac{2}{3} = \frac{8}{27}$$

$$(-2)^3 = (-2)(-2)(-2) = -8$$

A **cube root of a number** is one of its three equal factors. Thus, -2 is a cube root of -8 because $(-2)(-2)(-2) = -8$. In general, the concept of cube root can be defined as follows:

Definition 11.2

a is a **cube root** of b if $a^3 = b$.

The following generalizations are a direct consequence of Definition 11.2:

1. Every positive real number has one positive real number cube root.
2. Every negative real number has one negative real number cube root.
3. The cube root of 0 is 0.

(Technically, every nonzero real number has three cube roots, but only one of them is a real number. The other two cube roots are complex numbers.)

The symbol $\sqrt[3]{\ }$ is used to designate the real number cube root. Thus, we can write

$$\sqrt[3]{8} = 2 \qquad \sqrt[3]{\frac{1}{27}} = \frac{1}{3} \qquad \sqrt[3]{-8} = -2 \qquad \sqrt[3]{-\frac{1}{27}} = -\frac{1}{3}$$

We can extend the concept of *root* to fourth roots, fifth roots, sixth roots, and in general, nth roots. We can make these generalizations:
If n is an even positive integer, then the following statements are true.

1. Every positive real number has exactly two real nth roots, one positive and one negative. For example, the real fourth roots of 16 are 2 and -2. We use the symbol $\sqrt[n]{\ }$ to designate the positive root. Thus we write $\sqrt[4]{16} = 2$.
2. Negative real numbers do not have real nth roots. For example, there are no real fourth roots of -16.

If n is an odd positive integer greater than 1, then the following statements are true.

1. Every real number has exactly one real nth root, and we designate this root by the symbol $\sqrt[n]{\ }$.
2. The real nth root of a positive number is positive. For example, the fifth root of 32 is 2, and we write $\sqrt[5]{32} = 2$.
3. The nth root of a negative number is negative. For example, the fifth root of -32 is -2, and we write $\sqrt[5]{-32} = -2$.

To complete our terminology, we call the n in the radical $\sqrt[n]{b}$ the **index** of the radical. If $n = 2$, we commonly write $\sqrt{b}$ instead of $\sqrt[2]{b}$.

Merging of Exponents and Roots

In Section 5.6 we used the basic properties of positive integer exponents to motivate a definition for the use of zero and negative integers as exponents. Now we can use the properties of integer exponents to motivate definitions for the use of all rational numbers as exponents. This material is commonly referred to as "fractional exponents."

Consider the following comparisons.

From the meaning of root, we know that	If $(b^n)^m = b^{mn}$ is to hold when n equals a rational number of the form $\frac{1}{p}$, where p is a positive integer greater than 1, then
$(\sqrt{5})^2 = 5$	$\left(5^{\frac{1}{2}}\right)^2 = 5^{2\left(\frac{1}{2}\right)} = 5^1 = 5$
$(\sqrt[3]{8})^3 = 8$	$\left(8^{\frac{1}{3}}\right)^3 = 8^{3\left(\frac{1}{3}\right)} = 8^1 = 8$
$(\sqrt[4]{21})^4 = 21$	$\left(21^{\frac{1}{4}}\right)^4 = 21^{4\left(\frac{1}{4}\right)} = 21^1 = 21$

The following definition is motivated by such examples.

Definition 11.3

If b is a real number, n is a positive integer greater than 1, and $\sqrt[n]{b}$ exists, then

$$b^{\frac{1}{n}} = \sqrt[n]{b}$$

Definition 11.3 states that $b^{\frac{1}{n}}$ means the nth root of b. The following examples illustrate this definition.

$$25^{\frac{1}{2}} = \sqrt{25} = 5 \qquad 16^{\frac{1}{4}} = \sqrt[4]{16} = 2$$

$$8^{\frac{1}{3}} = \sqrt[3]{8} = 2 \qquad \left(\frac{36}{49}\right)^{\frac{1}{2}} = \sqrt{\frac{36}{49}} = \frac{6}{7}$$

$$(-27)^{\frac{1}{3}} = \sqrt[3]{-27} = -3 \qquad (-32)^{\frac{1}{5}} = \sqrt[5]{-32} = -2$$

Now the next definition provides the basis for the use of *all* rational numbers as exponents.

Definition 11.4

If $\frac{m}{n}$ is a rational number, where n is a positive integer greater than 1 and b is a real number such that $\sqrt[n]{b}$ exists, then

$$b^{\frac{m}{n}} = \sqrt[n]{b^m} = (\sqrt[n]{b})^m \qquad \frac{m}{n} \text{ is in reduced form}$$

Whether we use the form $\sqrt[n]{b^m}$ or $(\sqrt[n]{b})^m$ for computational purposes depends somewhat on the magnitude of the problem. We use both forms on two problems to illustrate this point.

$$8^{\frac{2}{3}} = \sqrt[3]{8^2} = \sqrt[3]{64} = 4 \qquad \text{or} \qquad 8^{\frac{2}{3}} = (\sqrt[3]{8})^2 = (2)^2 = 4$$

$$27^{\frac{2}{3}} = \sqrt[3]{27^2} = \sqrt[3]{729} = 9 \qquad \text{or} \qquad 27^{\frac{2}{3}} = (\sqrt[3]{27})^2 = 3^2 = 9$$

To compute $8^{\frac{2}{3}}$, either form seems to work about as well as the other one. However, to compute $27^{\frac{2}{3}}$, it should be obvious that $(\sqrt[3]{27})^2$ is much easier to handle than $\sqrt[3]{27^2}$.

Remember that in Section 5.6 we used the definition $b^{-n} = \frac{1}{b^n}$ as a basis for our work with negative integer exponents. Definition 11.4 and this definition, extended to all rational numbers, are used in the following examples.

$$25^{\frac{3}{2}} = (\sqrt{25})^3 = 5^3 = 125$$

$$16^{\frac{3}{4}} = (\sqrt[4]{16})^3 = 2^3 = 8$$

$$36^{-\frac{1}{2}} = \frac{1}{36^{\frac{1}{2}}} = \frac{1}{\sqrt{36}} = \frac{1}{6}$$

$$(-8)^{-\frac{1}{3}} = \frac{1}{(-8)^{\frac{1}{3}}} = \frac{1}{\sqrt[3]{-8}} = \frac{1}{-2} = -\frac{1}{2}$$

$$-8^{\frac{2}{3}} = -(\sqrt[3]{8})^2 = -(2)^2 = -4$$

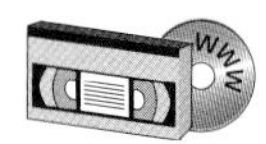

The basic properties of exponents we discussed in Chapter 5 are true for all rational numbers. They provide the basis for simplifying algebraic expressions that contain rational exponents, as the next examples illustrate. Our objective is to simplify and express the final result using only positive exponents.

1. $x^{\frac{1}{2}} \cdot x^{\frac{1}{3}} = x^{\frac{1}{2}+\frac{1}{3}} = x^{\frac{3}{6}+\frac{2}{6}} = x^{\frac{5}{6}}$

2. $\left(2a^{\frac{3}{4}}\right)\left(3a^{\frac{1}{3}}\right) = 2 \cdot 3 \cdot a^{\frac{3}{4}} \cdot a^{\frac{1}{3}} = 6a^{\frac{3}{4}+\frac{1}{3}} = 6a^{\frac{9}{12}+\frac{4}{12}} = 6a^{\frac{13}{12}}$

3. $\left(3n^{\frac{1}{2}}\right)\left(5n^{-\frac{1}{6}}\right) = 3 \cdot 5 \cdot n^{\frac{1}{2}} \cdot n^{-\frac{1}{6}}$

$$= 15n^{\frac{1}{2}+\left(-\frac{1}{6}\right)}$$

$$= 15n^{\frac{3}{6}-\frac{1}{6}}$$

$$= 15n^{\frac{2}{6}}$$

$$= 15n^{\frac{1}{3}}$$

4. $\left(2x^{\frac{1}{2}}y^{\frac{1}{3}}\right)^2 = (2)^2\left(x^{\frac{1}{2}}\right)^2\left(y^{\frac{1}{3}}\right)^2 = 4xy^{\frac{2}{3}}$ $\quad (b^n)^m = b^{mn}$

5. $\dfrac{x^{\frac{1}{3}}}{x^{\frac{1}{2}}} = x^{\frac{1}{3}-\frac{1}{2}} = x^{\frac{2}{6}-\frac{3}{6}} = x^{-\frac{1}{6}} = \dfrac{1}{x^{\frac{1}{6}}}$ $\quad \dfrac{b^n}{b^m} = b^{n-m}$

6. $\dfrac{12x^{\frac{3}{4}}}{2x^{-\frac{1}{4}}} = \dfrac{12}{2}x^{\frac{3}{4}-\left(-\frac{1}{4}\right)} = 6x^{\frac{4}{4}} = 6x$

7. $\left(\dfrac{2x^{\frac{1}{2}}}{3y^{\frac{2}{3}}}\right)^2 = \dfrac{(2)^2\left(x^{\frac{1}{2}}\right)^2}{(3)^2\left(y^{\frac{2}{3}}\right)^2} = \dfrac{4x}{9y^{\frac{4}{3}}}$

8. $3^{\frac{1}{2}} \cdot 3^{\frac{1}{3}} = 3^{\frac{1}{2}+\frac{1}{3}} = 3^{\frac{3}{6}+\frac{2}{6}} = 3^{\frac{5}{6}}$

9. $\dfrac{2^{\frac{3}{4}}}{2^{\frac{1}{4}}} = 2^{\frac{3}{4}-\frac{1}{4}} = 2^{\frac{2}{4}} = 2^{\frac{1}{2}}$

Problem Set 11.3

For Problems 1–50, evaluate each of the numerical expressions.

1. $\sqrt{81}$ **2.** $\sqrt{\dfrac{49}{4}}$ **3.** $-\sqrt{100}$

4. $-\sqrt{121}$ **5.** $\sqrt[3]{125}$ **6.** $\sqrt[3]{\dfrac{27}{8}}$

7. $\sqrt[3]{-64}$ **8.** $\sqrt[3]{-\dfrac{8}{27}}$ **9.** $\dfrac{\sqrt[3]{64}}{\sqrt{49}}$

10. $\dfrac{\sqrt{144}}{\sqrt[3]{8}}$ **11.** $\sqrt[4]{81}$ **12.** $\sqrt[4]{\dfrac{16}{81}}$

13. $\sqrt[5]{-243}$ **14.** $\sqrt[5]{1}$ **15.** $64^{\frac{1}{2}}$

16. $64^{\frac{1}{3}}$ **17.** $64^{\frac{2}{3}}$ **18.** $(-27)^{\frac{1}{3}}$

19. $(-64)^{\frac{2}{3}}$ **20.** $16^{\frac{3}{2}}$ **21.** $4^{\frac{5}{2}}$

22. $16^{\frac{1}{4}}$ **23.** $32^{-\frac{1}{5}}$ **24.** $-16^{\frac{1}{2}}$

25. $-27^{\frac{1}{3}}$ **26.** $8^{-\frac{2}{3}}$ **27.** $16^{-\frac{3}{4}}$

28. $\left(\dfrac{1}{2}\right)^{-2}$ **29.** $\left(\dfrac{2}{3}\right)^{-3}$ **30.** $\left(-\dfrac{1}{8}\right)^{-\frac{1}{3}}$

31. $\left(\dfrac{16}{64}\right)^{-\frac{1}{2}}$ **32.** $81^{\frac{3}{4}}$ **33.** $125^{\frac{4}{3}}$

34. $-16^{\frac{5}{2}}$ **35.** $-16^{\frac{5}{4}}$ **36.** $\left(\dfrac{1}{27}\right)^{-\frac{2}{3}}$

37. $\left(\dfrac{1}{32}\right)^{\frac{3}{5}}$ **38.** $(-8)^{\frac{4}{3}}$ **39.** $2^{\frac{1}{3}} \cdot 2^{\frac{2}{3}}$

40. $2^{\frac{3}{4}} \cdot 2^{\frac{5}{4}}$ **41.** $3^{\frac{4}{3}} \cdot 3^{\frac{5}{3}}$ **42.** $2^{\frac{3}{2}} \cdot 2^{-\frac{1}{2}}$

43. $\dfrac{2^{\frac{1}{2}}}{2^{\frac{1}{2}}}$ **44.** $\dfrac{3^{\frac{1}{3}}}{3^{-\frac{2}{3}}}$ **45.** $\dfrac{3^{-\frac{2}{3}}}{3^{\frac{1}{3}}}$

46. $\dfrac{2^{\frac{3}{4}}}{2^{-\frac{1}{4}}}$ **47.** $\dfrac{2^{\frac{9}{4}}}{2^{\frac{1}{4}}}$ **48.** $\dfrac{3^{-\frac{1}{2}}}{3^{-\frac{1}{2}}}$

49. $\dfrac{7^{\frac{4}{3}}}{7^{-\frac{2}{3}}}$ **50.** $\dfrac{5^{\frac{3}{5}}}{5^{-\frac{7}{5}}}$

For Problems 51–80, simplify each of the following and express the final results using positive exponents only; for example,

$$\left(2x^{\frac{1}{3}}\right)\left(3x^{\frac{1}{2}}\right) = 6x^{\frac{5}{6}}$$

51. $x^{\frac{1}{2}} \cdot x^{\frac{1}{4}}$ **52.** $x^{\frac{1}{3}} \cdot x^{\frac{1}{4}}$ **53.** $a^{\frac{2}{3}} \cdot a^{\frac{3}{4}}$

54. $a^{\frac{1}{3}} \cdot a^{-\frac{1}{4}}$ **55.** $\left(3x^{\frac{1}{4}}\right)\left(5x^{\frac{1}{3}}\right)$ **56.** $\left(2x^{\frac{1}{2}}\right)\left(5x^{\frac{1}{3}}\right)$

57. $\left(4x^{\frac{2}{3}}\right)\left(6x^{\frac{1}{4}}\right)$ **58.** $\left(3x^{\frac{2}{5}}\right)\left(6x^{\frac{1}{4}}\right)$ **59.** $\left(2y^{\frac{2}{3}}\right)\left(y^{-\frac{1}{4}}\right)$

60. $\left(y^{-\frac{1}{3}}\right)\left(4y^{\frac{2}{5}}\right)$ **61.** $\left(5n^{\frac{3}{4}}\right)\left(2n^{-\frac{1}{2}}\right)$ **62.** $\left(7n^{-\frac{1}{3}}\right)\left(8n^{\frac{5}{6}}\right)$

63. $\left(2x^{\frac{1}{3}}\right)\left(x^{-\frac{1}{2}}\right)$ **64.** $\left(x^{\frac{2}{5}}\right)\left(3x^{-\frac{1}{2}}\right)$ **65.** $\left(5x^{\frac{1}{2}}y\right)^2$

66. $\left(2x^{\frac{1}{3}}y^2\right)^3$ **67.** $\left(4x^{\frac{1}{4}}y^{\frac{1}{2}}\right)^3$ **68.** $\left(9x^2y^4\right)^{\frac{1}{2}}$

69. $(8x^6y^3)^{\frac{1}{3}}$

70. $\dfrac{18x^{\frac{1}{2}}}{9x^{\frac{1}{3}}}$

71. $\dfrac{24x^{\frac{3}{5}}}{6x^{\frac{1}{3}}}$

72. $\dfrac{56a^{\frac{1}{6}}}{7a^{\frac{1}{4}}}$

73. $\dfrac{48b^{\frac{1}{3}}}{12b^{\frac{3}{4}}}$

74. $\dfrac{16n^{\frac{1}{3}}}{8n^{-\frac{2}{3}}}$

75. $\dfrac{27n^{-\frac{1}{3}}}{9n^{-\frac{1}{3}}}$

76. $\dfrac{5x^{\frac{2}{5}}}{3x^{-\frac{1}{3}}}$

77. $\left(\dfrac{3x^{\frac{1}{3}}}{2x^{\frac{1}{2}}}\right)^2$

78. $\left(\dfrac{2a^{\frac{1}{2}}}{3a^{\frac{1}{4}}}\right)^3$

79. $\left(\dfrac{5x^{\frac{1}{2}}}{6y^{\frac{1}{3}}}\right)^3$

80. $\left(\dfrac{4y^{\frac{2}{5}}}{3x^{\frac{1}{3}}}\right)^2$

THOUGHTS INTO WORDS

81. Why is $\sqrt[4]{-16}$ not a real number?

82. Explain how you would evaluate $-4^{\frac{7}{2}}$.

FURTHER INVESTIGATIONS

83. Use a calculator to evaluate each expression.

(a) $\sqrt[3]{21{,}952}$ **(b)** $\sqrt[3]{42{,}875}$

(c) $\sqrt[4]{83{,}521}$ **(d)** $\sqrt[4]{3{,}111{,}696}$

84. Use a calculator to evaluate each number.

(a) $16^{\frac{5}{2}}$ **(b)** $36^{\frac{3}{2}}$

(c) $16^{\frac{7}{4}}$ **(d)** $27^{\frac{5}{3}}$

(e) $343^{\frac{4}{3}}$ **(f)** $81^{\frac{3}{4}}$

85. Use a calculator to estimate each expression to the nearest hundredth.

(a) $5^{\frac{3}{2}}$ **(b)** $8^{\frac{4}{5}}$

(c) $17^{\frac{2}{5}}$ **(d)** $19^{\frac{5}{2}}$

(e) $12^{\frac{3}{4}}$ **(f)** $14^{\frac{2}{3}}$

11.4 Complex Numbers

In Chapter 10 we presented some quadratic equations that have no real number solutions. For example, the equation $x^2 = -4$ has no real number solutions because $2^2 = 4$ and $(-2)^2 = 4$. In this section we will consider a set of numbers that contains some numbers whose squares are negative real numbers. Then in the next section, we will show that this set of numbers, called the set of **complex numbers**, provides solutions not only for equations such as $x^2 = -4$ but also for any quadratic equation with real number coefficients in one variable.

Our work with complex numbers is based on the following definition.

Definition 11.5

The number i is such that

$$i = \sqrt{-1} \qquad \text{and} \qquad i^2 = -1$$

The number i is not a real number and is often called the **imaginary unit**, but the number i^2 is the real number -1.

In Chapter 9 we used the property $\sqrt{a}\sqrt{b} = \sqrt{ab}$ to multiply radicals and to express radicals in simplest radical form. This property also holds if *only one of a or b* is negative. Thus we can simplify square root radicals that contain negative numbers as radicands as follows.

$$\sqrt{-4} = \sqrt{-1}\sqrt{4} = i(2) \quad \text{Usually written as } 2i$$

$$\sqrt{-13} = \sqrt{-1}\sqrt{13} = i\sqrt{13}$$

$$\sqrt{-12} = \sqrt{-1}\sqrt{12} = i\sqrt{12} = i\sqrt{4}\sqrt{3} = 2i\sqrt{3}$$

$$\sqrt{-18} = \sqrt{-1}\sqrt{18} = i\sqrt{18} = i\sqrt{9}\sqrt{2} = 3i\sqrt{2}$$

The imaginary unit i is used to define a complex number as follows.

Definition 11.6

A **complex number** is any number that can be expressed in the form

$$a + bi$$

where a and b are real numbers.

The form $a + bi$ is called the **standard form** of a complex number. We call the real number a the **real part** of the complex number, and we call b the **imaginary part**. The next examples demonstrate this terminology.

1. The number $3 + 4i$ is a complex number in standard form that has a real part of 3 and an imaginary part of 4.
2. We can write the number $-5 - 2i$ in the standard form $-5 + (-2i)$; thus it is a complex number that has a real part of -5 and an imaginary part of -2. [We often use the form $-5 - 2i$, knowing that it means $-5 + (-2i)$.]
3. We can write the number $-7i$ in the standard form $0 + (-7i)$; thus it is a complex number that has a real part of 0 and an imaginary part of -7.
4. We can write the number 9 in the standard form $9 + 0i$; thus, it is a complex number that has a real part of 9 and an imaginary part of 0.

Number 4 shows us that all real numbers can be considered complex numbers.

The commutative, associative, and distributive properties hold for all complex numbers and enable us to manipulate with complexes. The following two statements describe addition and subtraction of complex numbers.

$$(a + bi) + (c + di) = (a + c) + (b + d)i \quad \text{Addition}$$

$$(a + bi) - (c + di) = (a - c) + (b - d)i \quad \text{Subtraction}$$

To add complex numbers, we add their real parts and add their imaginary parts. To subtract complex numbers, we subtract their real parts and subtract their imaginary parts. Consider these examples.

1. $(3 + 5i) + (4 + 7i) = (3 + 4) + (5 + 7)i = 7 + 12i$

2. $(-5 + 2i) + (7 - 8i) = (-5 + 7) + [2 + (-8)]i$

$= 2 + (-6i)$ Usually written as $2 - 6i$

3. $\left(\frac{1}{2} + \frac{3}{4}i\right) + \left(\frac{2}{3} + \frac{1}{5}i\right) = \left(\frac{1}{2} + \frac{2}{3}\right) + \left(\frac{3}{4} + \frac{1}{5}\right)i$

$= \left(\frac{3}{6} + \frac{4}{6}\right) + \left(\frac{15}{20} + \frac{4}{20}\right)i$

$= \frac{7}{6} + \frac{19}{20}i$

4. $(8 + 9i) - (5 + 4i) = (8 - 5) + (9 - 4)i = 3 + 5i$

5. $(2 - 3i) - (5 + 9i) = (2 - 5) + (-3 - 9)i = -3 + (-12i)$

$= -3 - 12i$

6. $(-1 + 6i) - (-2 - i) = [-1 - (-2)] + [6 - (-1)]i = 1 + 7i$

Multiplying Complex Numbers

Since complex numbers have a *binomial form*, we find the product of two complex numbers in the same way that we find the product of two binomials. We present some examples.

EXAMPLE 1

Multiply $(3 + 2i)(4 + 5i)$.

Solution

When we multiply each term of the first number times each term of the second number and simplify, we get

$$(3 + 2i)(4 + 5i) = 3(4) + 3(5i) + 2i(4) + 2i(5i)$$
$$= 12 + 15i + 8i + 10i^2$$
$$= 12 + (15 + 8)i + 10(-1) \qquad i^2 = -1$$
$$= 2 + 23i$$

■

EXAMPLE 2

Multiply $(6 - 3i)(-5 + 2i)$.

Solution

$$(6 - 3i)(-5 + 2i) = 6(-5) + 6(2i) - (3i)(-5) - (3i)(2i)$$
$$= -30 + 12i + 15i - 6i^2$$

$$= -30 + 27i - 6(-1)$$
$$= -24 + 27i$$ ■

EXAMPLE 3 Find the indicated product $(5 - i)^2$.

Solution

Remember that $(a - b)^2$ means $(a - b)(a - b)$.

$$(5 - i)^2 = (5 - i)(5 - i) = 5(5) - 5(i) - i(5) - (i)(-i)$$
$$= 25 - 5i - 5i + i^2$$
$$= 25 - 10i + (-1)$$
$$= 24 - 10i$$ ■

To find products such as $(4i)(3i)$ or $2i(4 + 6i)$, we could change each number to standard *binomial form* and then proceed as in the preceding examples, but it is easier to handle them as follows:

$$(4i)(3i) = 4 \cdot 3 \cdot i \cdot i = 12i^2 = 12(-1) = -12 \quad \text{or} \quad -12 + 0i$$
$$2i(4 + 6i) = 2i(4) + 2i(6i)$$
$$= 8i + 12i^2$$
$$= 8i + 12(-1)$$
$$= -12 + 8i$$

Problem Set 11.4

For Problems 1–12, write each radical in terms of i and simplify; for example, $\sqrt{-20} = i\sqrt{20} = i\sqrt{4} \cdot \sqrt{5} = 2i\sqrt{5}$.

1. $\sqrt{-64}$ **2.** $\sqrt{-81}$ **3.** $\sqrt{-\frac{25}{9}}$

4. $\sqrt{-\frac{49}{16}}$ **5.** $\sqrt{-11}$ **6.** $\sqrt{-17}$

7. $\sqrt{-50}$ **8.** $\sqrt{-32}$ **9.** $\sqrt{-48}$

10. $\sqrt{-45}$ **11.** $\sqrt{-54}$ **12.** $\sqrt{-28}$

For Problems 13–34, add or subtract the complex numbers as indicated.

13. $(3 + 8i) + (5 + 9i)$

14. $(7 + 10i) + (2 + 3i)$

15. $(7 - 6i) + (3 - 4i)$

16. $(8 - 2i) + (-7 + 3i)$

17. $(10 + 4i) - (6 + 2i)$

18. $(12 + 7i) - (5 + i)$

19. $(5 + 2i) - (7 + 8i)$

20. $(3 + i) - (7 + 4i)$

21. $(-2 - i) - (3 - 4i)$

22. $(-7 - 3i) - (8 - 9i)$

23. $(-4 - 7i) + (-8 - 9i)$

24. $(-1 - 2i) + (-6 - 6i)$

25. $(0 - 6i) + (-10 + 2i)$

26. $(4 - i) + (4 + i)$

27. $(-9 + 7i) - (-8 - 5i)$

28. $(-12 + 6i) - (-7 + 2i)$

29. $(-10 - 4i) - (10 + 4i)$

30. $(-8 - 2i) - (8 + 2i)$

31. $\left(\frac{1}{2} + \frac{2}{3}i\right) + \left(\frac{1}{3} - \frac{1}{4}i\right)$

32. $\left(\frac{3}{4} - \frac{1}{2}i\right) + \left(\frac{1}{2} + i\right)$

33. $\left(\frac{3}{5} - \frac{1}{4}i\right) - \left(\frac{2}{3} - \frac{5}{6}i\right)$

34. $\left(\frac{1}{6} + \frac{5}{4}i\right) - \left(-\frac{1}{3} - \frac{3}{5}i\right)$

For Problems 35–54, find each product and express it in the standard form of a complex number $(a + bi)$.

35. $(7i)(8i)$ **36.** $(-6i)(7i)$

37. $2i(6 + 3i)$ **38.** $3i(-4 + 9i)$

39. $-4i(-5 - 6i)$ **40.** $-5i(7 - 8i)$

41. $(2 + 3i)(5 + 4i)$ **42.** $(4 + 2i)(6 + 5i)$

43. $(7 - 3i)(8 + i)$ **44.** $(9 - 3i)(2 - 5i)$

45. $(-2 - 3i)(6 - 3i)$ **46.** $(-3 - 8i)(1 - i)$

47. $(-1 - 4i)(-2 - 7i)$

48. $(-3 - 7i)(-6 - 10i)$

49. $(4 + 5i)^2$ **50.** $(7 - 2i)^2$

51. $(5 - 6i)(5 + 6i)$ **52.** $(7 - 3i)(7 + 3i)$

53. $(-2 + i)(-2 - i)$ **54.** $(-5 - 8i)(-5 + 8i)$

THOUGHTS INTO WORDS

55. Why is the set of real numbers a subset of the set of complex numbers?

56. Is it possible for the product of two nonreal complex numbers to be a real number? Defend your answer.

11.5 Quadratic Equations: Complex Solutions

As we stated in the preceding section, the set of complex numbers provides solutions for all quadratic equations that have real number coefficients. In other words, every quadratic equation of the form $ax^2 + bx + c = 0$, where a, b, and c are real numbers and $a \neq 0$, has a solution (or solutions) from the set of complex numbers.

To find solutions for quadratic equations, we continue to use the techniques of factoring, completing the square, the quadratic formula, and the property *if* $x^2 = a$, *then* $x = \pm\sqrt{a}$. Let's consider some examples.

EXAMPLE 1 Solve $x^2 = -4$.

Solution

Use the property *if* $x^2 = a$, *then* $x = \pm\sqrt{a}$, and proceed as follows:

$$x^2 = -4$$

$$x = \pm\sqrt{-4}$$

$$x = \pm 2i \qquad \sqrt{-4} = \sqrt{-1}\sqrt{4} = 2i$$

Check

$$x^2 = -4 \qquad x^2 = -4$$
$$(2i)^2 \stackrel{?}{=} -4 \qquad (-2i)^2 \stackrel{?}{=} -4$$
$$4i^2 \stackrel{?}{=} -4 \qquad 4i^2 \stackrel{?}{=} -4$$
$$4(-1) \stackrel{?}{=} -4 \qquad 4(-1) \stackrel{?}{=} -4$$
$$-4 = -4 \qquad -4 = -4$$

The solution set is $\{-2i, 2i\}$. ■

EXAMPLE 2

Solve $(x - 2)^2 = -7$.

Solution

$$(x - 2)^2 = -7$$
$$x - 2 = \pm\sqrt{-7}$$
$$x - 2 = \pm i\sqrt{7} \qquad \sqrt{-7} = \sqrt{-1}\sqrt{7} = i\sqrt{7}$$
$$x = 2 \pm i\sqrt{7}$$

Check

$$(x - 2)^2 = -7 \qquad (x - 2)^2 = -7$$
$$(2 + i\sqrt{7} - 2)^2 \stackrel{?}{=} -7 \qquad (2 - i\sqrt{7} - 2)^2 \stackrel{?}{=} -7$$
$$(i\sqrt{7})^2 \stackrel{?}{=} -7 \qquad (-i\sqrt{7})^2 \stackrel{?}{=} -7$$
$$7i^2 \stackrel{?}{=} -7 \qquad 7i^2 \stackrel{?}{=} -7$$
$$7(-1) \stackrel{?}{=} -7 \qquad 7(-1) \stackrel{?}{=} -7$$
$$-7 = -7 \qquad -7 = -7$$

The solution set is $\{2 - i\sqrt{7}, 2 + i\sqrt{7}\}$. ■

EXAMPLE 3

Solve $x^2 + 2x = -10$.

Solution

The form of the equation lends itself to *completing the square*, so we proceed as follows:

$$x^2 + 2x = -10$$
$$x^2 + 2x + 1 = -10 + 1$$
$$(x + 1)^2 = -9$$
$$x + 1 = \pm 3i$$
$$x = -1 \pm 3i$$

Check

$$\begin{aligned} x^2 + 2x &= -10 \\ (-1 + 3i)^2 + 2(-1 + 3i) &\stackrel{?}{=} -10 \\ 1 - 6i + 9i^2 - 2 + 6i &\stackrel{?}{=} -10 \\ -1 + 9i^2 &\stackrel{?}{=} -10 \\ -1 + 9(-1) &\stackrel{?}{=} -10 \\ -10 &= -10 \end{aligned}$$

$$\begin{aligned} x^2 + 2x &= -10 \\ (-1 - 3i)^2 + 2(-1 - 3i) &\stackrel{?}{=} -10 \\ 1 + 6i + 9i^2 - 2 - 6i &\stackrel{?}{=} -10 \\ -1 + 9i^2 &\stackrel{?}{=} -10 \\ -1 + 9(-1) &\stackrel{?}{=} -10 \\ -10 &= -10 \end{aligned}$$

The solution set is $\{-1 - 3i, -1 + 3i\}$. ■

EXAMPLE 4 Solve $x^2 - 2x + 2 = 0$.

Solution

We use the *quadratic formula* to obtain the solutions.

$$x^2 - 2x + 2 = 0$$

$$x = \frac{2 \pm \sqrt{(-2)^2 - 4(1)(2)}}{2} \qquad x = \frac{-b \pm \sqrt{b^2 - 4ac}}{2a}$$

$$x = \frac{2 \pm \sqrt{4 - 8}}{2}$$

$$x = \frac{2 \pm \sqrt{-4}}{2}$$

$$x = \frac{2 \pm 2i}{2}$$

$$x = \frac{2(1 \pm i)}{2}$$

$$x = 1 \pm i$$

Check

$$\begin{aligned} x^2 - 2x + 2 &= 0 \\ (1 + i)^2 - 2(1 + i) + 2 &\stackrel{?}{=} 0 \\ 1 + 2i + i^2 - 2 - 2i + 2 &= 0 \\ 1 + i^2 &\stackrel{?}{=} 0 \\ 1 - 1 &\stackrel{?}{=} 0 \\ 0 &= 0 \end{aligned}$$

$$\begin{aligned} x^2 - 2x + 2 &= 0 \\ (1 - i)^2 - 2(1 - i) + 2 &\stackrel{?}{=} 0 \\ 1 - 2i + i^2 - 2 + 2i + 2 &\stackrel{?}{=} 0 \\ 1 + i^2 &\stackrel{?}{=} 0 \\ 1 - 1 &\stackrel{?}{=} 0 \\ 0 &= 0 \end{aligned}$$

The solution set is $\{1 - i, 1 + i\}$. ■

EXAMPLE 5 Solve $x^2 + 3x - 10 = 0$.

Solution

We can *factor* $x^2 + 3x - 10$ and proceed as follows:

$$x^2 + 3x - 10 = 0$$

$$(x + 5)(x - 2) = 0$$

$$x + 5 = 0 \quad \text{or} \quad x - 2 = 0$$

$$x = -5 \quad \text{or} \quad x = 2$$

The solution set is $\{-5, 2\}$. (Don't forget that all real numbers are complex numbers; that is, -5 and 2 can be written as $-5 + 0i$ and $2 + 0i$.) ■

To summarize our work with quadratic equations in Chapter 10 and this section, we suggest the following approach to solve a quadratic equation.

1. If the equation is in a form where the property *if* $x^2 = a$, *then* $x = \pm\sqrt{a}$ applies, use it. (See Examples 1 and 2.)
2. If the quadratic expression can be factored using integers, factor it and apply the property *if* $ab = 0$, *then* $a = 0$ *or* $b = 0$. (See Example 5.)
3. If numbers 1 and 2 don't apply, use either the quadratic formula or the process of completing the square. (See Examples 3 and 4.)

Problem Set 11.5

Solve each of the following quadratic equations, and check your solutions.

1. $x^2 = -64$

2. $x^2 = -49$

3. $(x - 2)^2 = -1$

4. $(x + 3)^2 = -16$

5. $(x + 5)^2 = -13$

6. $(x - 7)^2 = -21$

7. $(x - 3)^2 = -18$

8. $(x + 4)^2 = -28$

9. $(5x - 1)^2 = 9$

10. $(7x + 3)^2 = 1$

11. $a^2 - 3a - 4 = 0$

12. $a^2 + 2a - 35 = 0$

13. $t^2 + 6t = -12$

14. $t^2 - 4t = -9$

15. $n^2 - 6n + 13 = 0$

16. $n^2 - 4n + 5 = 0$

17. $x^2 - 4x + 20 = 0$

18. $x^2 + 2x + 5 = 0$

19. $3x^2 - 2x + 1 = 0$

20. $2x^2 + x + 1 = 0$

21. $2x^2 - 3x - 5 = 0$

22. $3x^2 - 5x - 2 = 0$

23. $y^2 - 2y = -19$

24. $y^2 + 8y = -24$

25. $x^2 - 4x + 7 = 0$

26. $x^2 - 2x + 3 = 0$

27. $4x^2 - x + 2 = 0$

28. $5x^2 + 2x + 1 = 0$

29. $6x^2 + 2x + 1 = 0$

30. $7x^2 + 3x + 3 = 0$

THOUGHTS INTO WORDS

31. Which method would you use to solve the equation $x^2 + 4x = -5$? Explain your reasons for making that choice.

32. Explain why the expression $b^2 - 4ac$ from the quadratic formula will determine whether or not the solutions of a particular quadratic equation are imaginary.

11.6 Pie, Bar, and Line Graphs

The saying "a picture is worth a thousand words" is also true for mathematics. In mathematics, a graph or chart is often used to show information. In this section, you will see various types of graphs and how they are used.

Pie Chart

A **pie chart** or **circle graph** is used to illustrate the parts of a whole. The pie or circle represents the whole, and the sectors of the pie or circle represent the parts of the whole. The parts are usually expressed in percents.

A student activities group is planning a ski trip and knows that the expenses, per person, are as follows:

Expenses for a Ski Trip, per person

Category	Expense	Percent
Food	$155	15.5%
Transportation	360	36%
Equipment rental	200	20%
Lift tickets	85	8.5%
Lodging	200	20%
Total	$1000	

The pie chart in Figure 11.13 is a useful graph to show the ski club how the expenses of the trip are broken down.

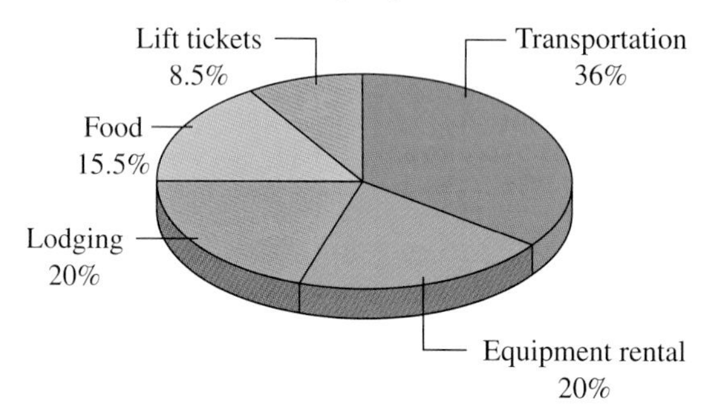

Figure 11.13

A glance at a pie chart can often tell the story and eliminate the need to compare numbers. Consider the pie chart in Figure 11.14. Without any numerical values presented, the graph gives us the information that a majority of voters are in favor of the proposition because more than half the circle is shaded for yes.

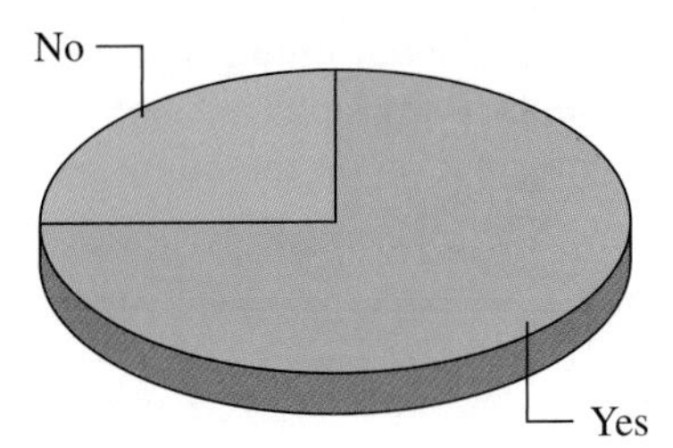

Figure 11.14

EXAMPLE 1

Use the pie chart shown in Figure 11.15 to answer the questions.

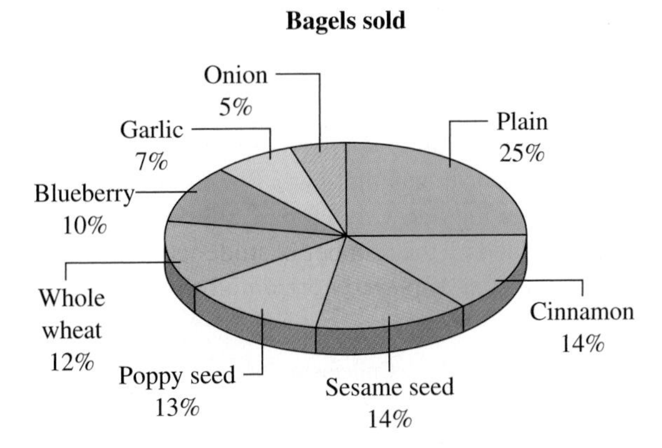

Figure 11.15

(a) What percent of the bagels sold were plain, sesame seed, or poppy seed?
(b) What percent of the bagels sold were not onion or garlic?
(c) Were more than half the bagels sold either plain or whole wheat?

Solution

(a) Of the bagels sold, 25% were plain, 14% were sesame seed, and 13% were poppy seed. Together these add up to 52% of the bagels sold.
(b) Five percent were onion bagels, and 7% were garlic bagels. Therefore, onion bagels together with garlic bagels accounted for 12% of the sales. Since the pie chart represents the whole, or 100%, of the bagels sold, the percent of bagels sold that were not onion or garlic is 100% − 12%, or 88%.

(c) By inspection the sectors for plain and whole wheat do not make up more than half of the circle. Mathematically, plain bagels were 25% of the sales, and whole wheat bagels were 12% of the sales. Together, plain and whole wheat bagels made up 37% of the sales, which is less than half. ■

Bar Graph

Another type of graph is the **bar graph**. Bars are drawn either vertically or horizontally to show amounts. Bar graphs are very useful for comparisons.

Consider this information on the number of students in certain college majors at a university:

College majors	Number of students
Business	2400
Computer science	850
Natural science	700
English	1800
Fine arts	400
Education	1000

The information is displayed in the bar graph in Figure 11.16. The graph has a title, a vertical axis that gives the numbers of students, and a horizontal axis that shows the majors. Vertical bars are drawn for each major; the height of the bar is determined by the number of students (per the scale on the vertical axis). The bars also could be displayed horizontally.

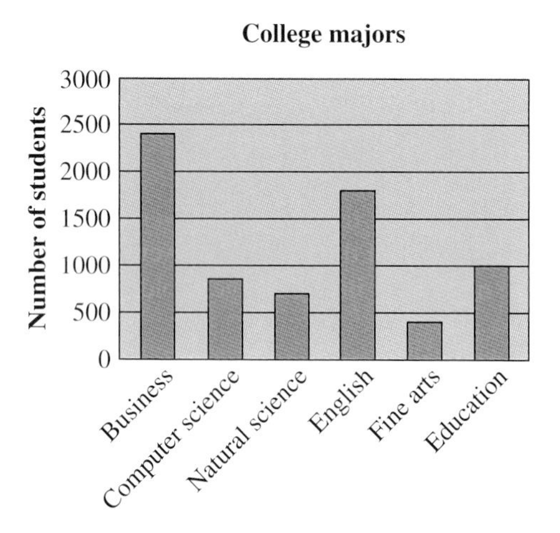

Figure 11.16

Bar graphs can display and be used to compare information about two or more groups by using multiple bars. Suppose you are trying to decide whether there is a difference in the number of music CDs purchased by men and those purchased by women in the various age groups. The following table contains the information, and the bar graph in Figure 11.17 displays this information.

Ages	Men	Women
18–21	35	31
22–35	28	12
36–50	8	14

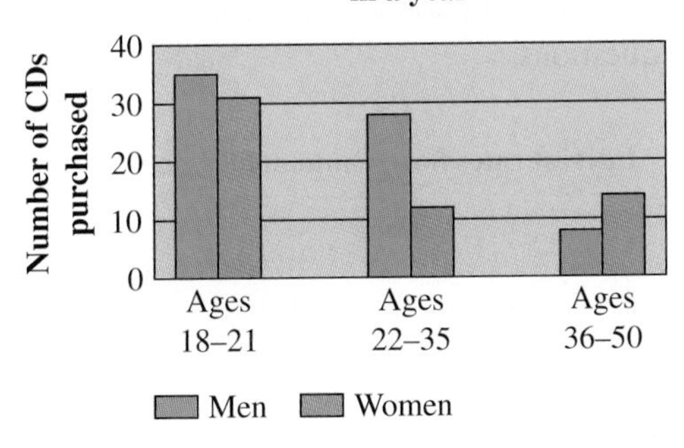

Figure 11.17

■ Line Graph

Line graphs are used to show the relationship between two variables. Each graph has two perpendicular axes with a variable assigned to each axis. Line graphs are useful for indicating trends. Consider the following information regarding a corporation's profit over the years shown. The profit is shown in millions of dollars.

Year	2000	2001	2002	2003	2004	2005
Profit	258	110	165	205	224	185

A line graph for this information is shown in Figure 11.18. From the graph, you can notice the trends in profits. There was a large decrease in profits between 2000 and 2001. After that, profits rose from 2001 to 2004. Then there was a decrease in profits from 2004 to 2005.

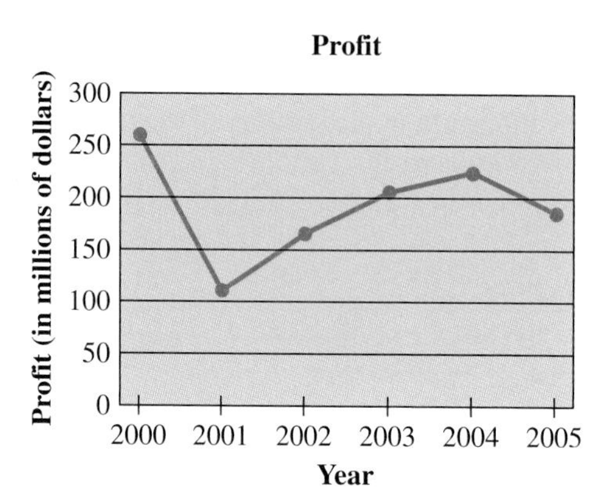

Figure 11.18

EXAMPLE 2

The graph in Figure 11.19 displays information about the interest rates charged by a bank for an automobile loan over the last 7 months. Use the graph to answer these questions.

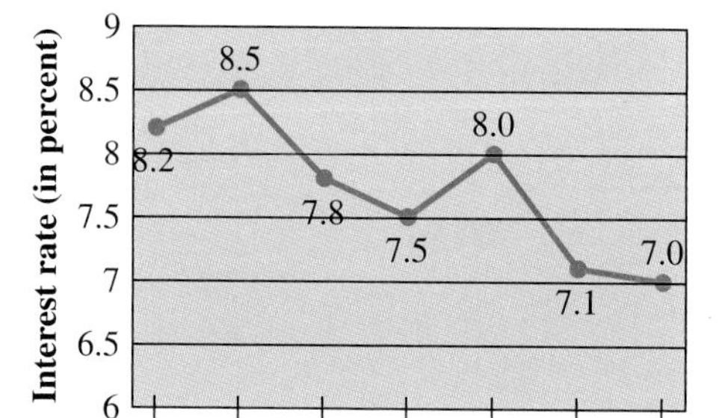

Figure 11.19

(a) The largest decrease in interest rates was between which two months?
(b) What was the change in interest rates between July and August?
(c) If you wanted the lowest possible interest rate, in what month should you have gotten a loan?

Solution

(a) The largest decrease was between October and November.
(b) The change between July and August was $7.8 - 8.5$, or -0.7. So the interest rate dropped by 0.7 point.
(c) The interest rate was the lowest in December. ■

One of the challenging aspects of creating graphs is deciding which type of graph to use. The information in the following table about a budget is displayed in both a pie chart (Figure 11.20) and a bar graph (Figure 11.21). Each graph displays the information in a different format; which graph is considered more appropriate is sometimes a personal preference.

Budget Expenses

Item	Percent of income
Automobile	26
Rent	33
Groceries	12
Utilities	10
Phones	5
Entertainment	6
Clothes	8

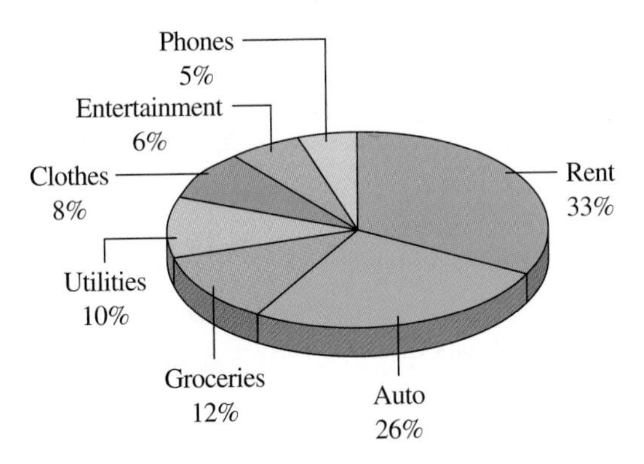

Figure 11.20

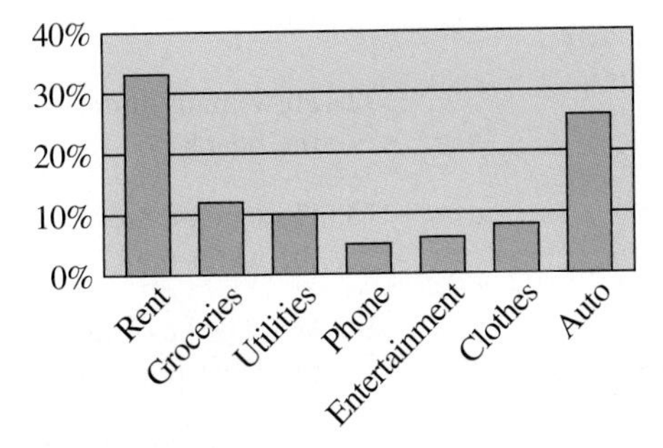

Figure 11.21

Problem Set 11.6

For Problems 1–5, use the pie chart in Figure 11.22.

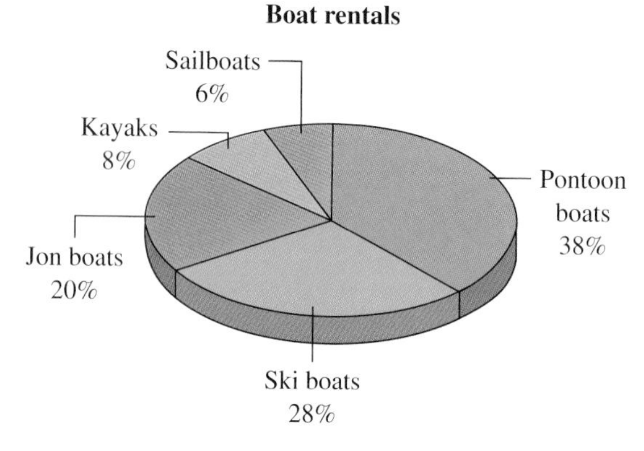

Figure 11.22

1. What percent of the boat rentals were kayaks or sailboats?
2. Were more than half the rentals pontoon or ski boats?
3. If there were 2400 rentals, how many times were Jon boats rented?
4. Is the ratio of sailboat rentals to kayak rentals the same as the ratio of ski boat rentals to pontoon boat rentals? (Justify your answer.)
5. What percent of the rentals were not sailboats or ski boats?

For Problems 6–11, use the pie chart in Figure 11.23.

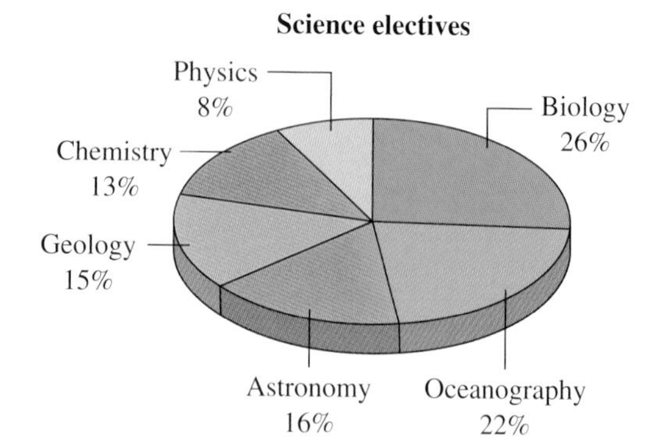

Figure 11.23

6. Which science elective is the most popular?
7. Which science elective is the least popular?
8. What percent of the students chose biology or geology?
9. What percent of the students chose chemistry or physics?
10. What percent of the students did not choose biology or oceanography?
11. What percent of the students did not choose oceanography or astronomy?

For Problems 12–15, use the bar graph in Figure 11.24.

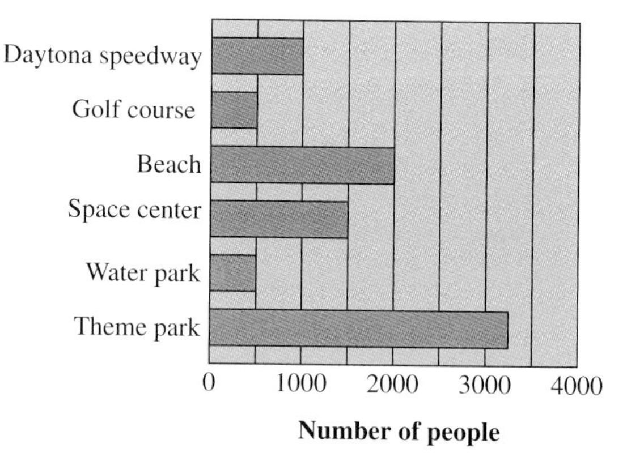

Figure 11.24

12. How many more people preferred the theme park to the beach?
13. How many more people preferred the space center to the water park?
14. What is the order of activities from the most popular to the least popular?
15. What is the difference in the number of people who chose the beach over the golf course?

For Problems 16–22, use Figure 11.25.

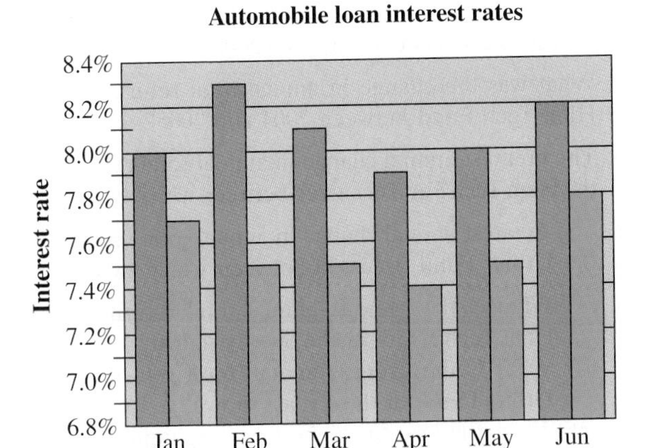

Figure 11.25

16. The greatest difference in the interest rates between banks and credit unions occurs in which month?

17. Between which two months did the interest rate for banks go up while the interest rate for credit unions went down?

18. Between which two months did the banks raise the interest rate by 0.2%?

19. Between which two months did the credit union keep the interest rate constant?

20. For which month and which type of institution was the interest rate the lowest?

21. In June what was the difference in the interest rates between the bank and credit union?

22. What was the change in interest rates for the bank between May and June?

For Problems 23–28, use the graph in Figure 11.26.

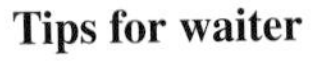

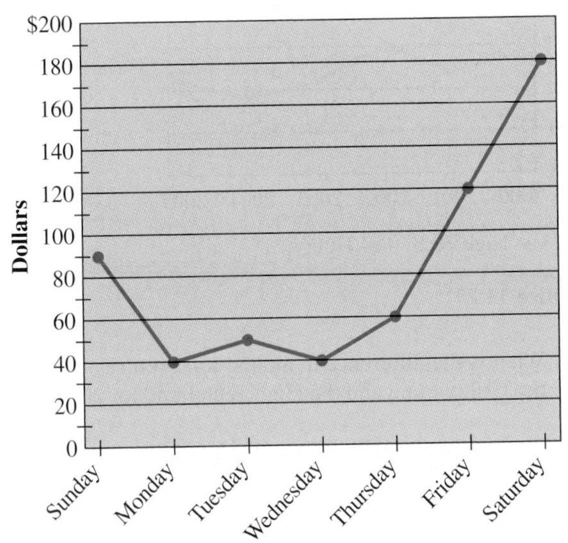

Figure 11.26

23. What is the total amount the waiter earned in tips for Friday, Saturday, and Sunday?

24. What is the total amount the waiter earned in tips for Monday through Thursday?

25. To avoid loss of income, what would be the best day for the waiter to take off work (according to the information in the graph)?

26. What was the difference in tips between Saturday and Wednesday?

27. How much did the waiter earn in tips for the week?

28. What was the average daily amount of tips?

For Problems 29–35, use the graph in Figure 11.27.

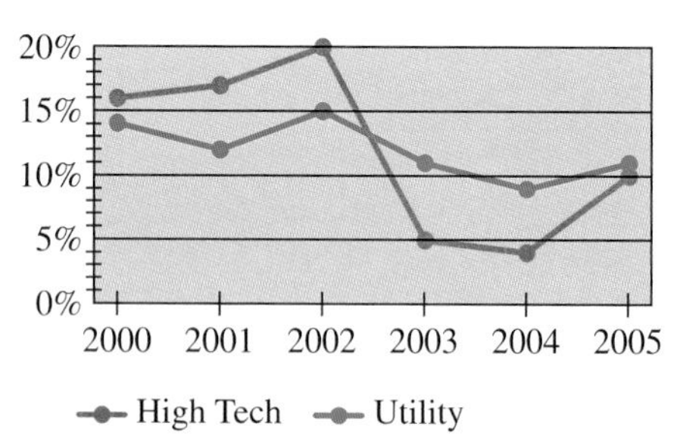

Figure 11.27

29. What is the difference in annual total returns between the Utility Fund and the High Tech Fund for 2000?

30. What is the difference in annual total returns between the Utility Fund and the High Tech Fund for 2002?

31. What was the change in annual total returns for the Utility Fund between 2001 and 2002?

32. What was the change in annual total returns for the High Tech Fund between 2004 and 2005?

33. The greatest annual change in annual total returns for the High Tech Fund occurred between which two years?

34. The greatest annual change in annual total returns for the Utility Fund occurred between which two years?

35. **(a)** What is the average annual total return for the High Tech Fund for the 6 years shown?

(b) What is the average annual total return for the Utility Fund for the 6 years shown?

(c) Which fund has the highest average annual total return?

THOUGHTS INTO WORDS

36. Mrs. Guenther says that each of her classes performed quite differently on the final exam according to the graph in Figure 11.28. Do you agree or disagree with her statement?

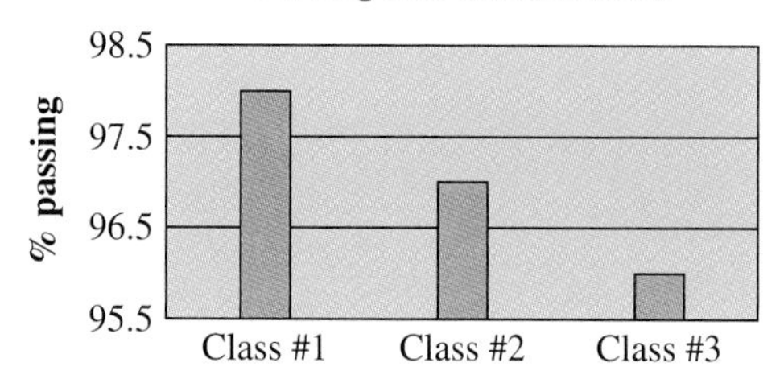

Figure 11.28

37. Lauren's boss says her sales over the past 5 years have been dropping significantly and steadily and shows her the graph in Figure 11.29. Lauren shows her boss the graph in Figure 11.30 and claims that there has been only a slight decrease. If you were the boss, would you accept the portrayal of Lauren's sales as shown in her graph?

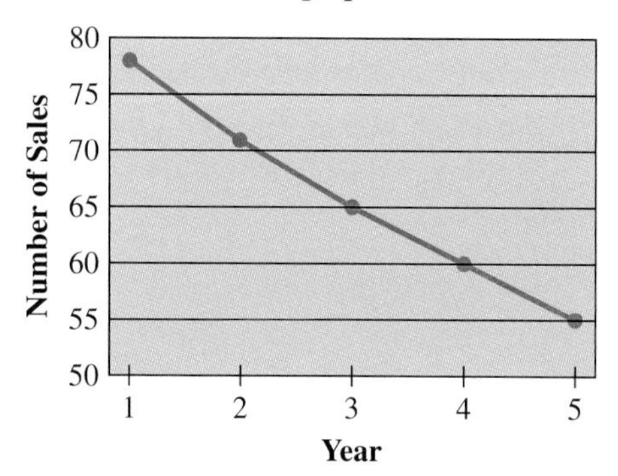

Figure 11.29

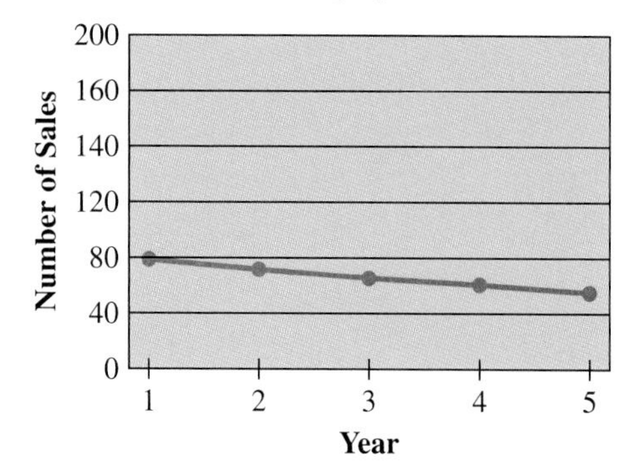

Figure 11.30

FURTHER INVESTIGATIONS

The first step in constructing a pie chart using a compass and protractor is to find the degrees for each sector. Remember, we are dividing up the 360° of the entire circle. Consider the information in the table below.

Pizza Sales for Saturday

Type	Number sold
Pepperoni	60
Sausage	45
Supreme	40
Veggie	15
Cheese	20
Total	180

$$\text{Degrees for any type} = \frac{\text{Number sold of that type}}{\text{Total number sold}} \times 360^\circ$$

$$\text{Degrees for pepperoni} = \frac{60}{180} \times 360^\circ = 120^\circ$$

$$\text{Degrees for sausage} = \frac{45}{180} \times 360^\circ = 90^\circ$$

$$\text{Degrees for supreme} = \frac{40}{180} \times 360^\circ = 80^\circ$$

$$\text{Degrees for veggie} = \frac{15}{180} \times 360^\circ = 30^\circ$$

$$\text{Degrees for cheese} = \frac{20}{180} \times 360^\circ = 40^\circ$$

Using a compass, draw a circle with a diameter of your choosing. From the center of the circle draw a radius. With the protractor located at the center of the circle and the radius as a side of the angle, draw the angle for the desired number of degrees in the first sector. Continue until all the sectors are shown. Be sure to label the pie chart and the sectors as in Figure 11.31.

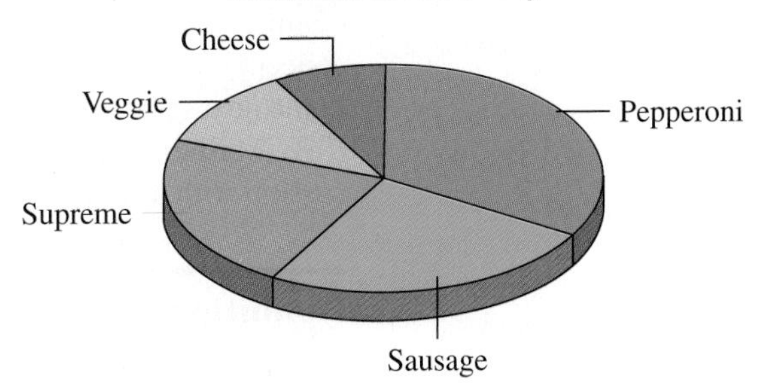

Figure 11.31

38. Construct a pie chart for the following data:

Marching Rams Membership

Category	Number of students
Band	105
Colorguard	42
Majorettes	33

39. Construct a pie chart for these data:

Investment Portfolio for Mr. Jordan

Category	Dollars invested
Stocks	$ 8,000
Mutual funds	14,000
Bonds	6,000
Annuities	5,000
Gold	3,000

40. If you have access to a computer with a spreadsheet application, try producing the pie charts in Problems 38 and 39 using the software.

11.7 Relations and Functions

In the next two sections of this chapter, we will work with a concept that has an important role throughout mathematics—namely, the concept of a *function*. Functions are used to unify mathematics and also to serve as a meaningful way of applying mathematics to many real-world problems. They provide us with a means of studying quantities that vary with one another—that is, when a change in one quantity causes a corresponding change in another. We will consider the general concept of a function in this section and then deal with the application of functions in the next section.

Mathematically, a function is a special kind of **relation**, so we begin our discussion with a simple definition of a relation.

Definition 11.7

A **relation** is a set of ordered pairs.

Thus a set of ordered pairs such as $\{(1, 2), (3, 7), (8, 14)\}$ is a relation. The **domain** of a relation is the set of all first components of the ordered pair. The **range** of a relation is the set of all second components of the ordered pair. The relation $\{(1, 2), (3, 7), (8, 14)\}$ has a domain of $\{1, 3, 8\}$ and a range of $\{2, 7, 14\}$.

EXAMPLE 1

Here are four examples of relations. State the domain and range for each relation.

(a) The presidential election year and who won the election:
{(1928, Hoover), (1960, Kennedy), (1964, Johnson), (1976, Carter), (1992, Clinton), (1996, Clinton)}

(b) A radioactive element and its half-life in hours:
{(Iodine-133, 20.9), (Barium-135, 28.7), (Technetium-99m, 6)}

(c) A natural number less than 5 and its opposite:
$\{(1, -1), (2, -2), (3, -3), (4, -4)\}$

(d) A number and its square:
$\{(-3, 9), (-2, 4), (-1, 1), (0, 0), (1, 1), (2, 4), (3, 9)\}$

Solution

(a) Domain = {1928, 1960, 1964, 1976, 1992, 1996}; Range = {Hoover, Kennedy, Johnson, Carter, Clinton}

(b) Domain = {Iodine-133, Barium-135, Technetium-99m}; Range = {6, 20.9, 28.7}

(c) Domain = $\{1, 2, 3, 4\}$; Range = $\{-1, -2, -3, -4\}$

(d) Domain = $\{-3, -2, -1, 0, 1, 2, 3\}$; Range = $\{0, 1, 4, 9\}$ ■

The ordered pairs we refer to in Definition 11.7 may be generated in various ways, such as from a graph, a chart, or a description of the relation. However, one of the most common ways of generating ordered pairs is from equations. Since the solution set of an equation in two variables is a set of ordered pairs, such an equation describes a relation. Each of the following equations describes a relation between the variables x and y. We list *some* of the infinitely many ordered pairs (x, y) of each relation.

1. $x^2 + y^2 = 4$: $(0, 2), (0, -2), (2, 0), (-2, 0)$

2. $x = y^2$: $(16, 4), (16, -4), (25, 5), (25, -5)$

3. $2x - y = -3$: $(-2, -1), (-1, 1), (0, 3), (1, 5), (2, 7)$

4. $y = \dfrac{1}{x - 2}$: $\left(-2, -\dfrac{1}{4}\right), \left(-1, -\dfrac{1}{3}\right), \left(0, -\dfrac{1}{2}\right), (1, -1), (3, 1)$

5. $y = x^2$: $(-2, 4), (-1, 1), (0, 0), (1, 1), (2, 4)$

Now direct your attention to the ordered pairs of the last three relations. These relations are a special type called **functions**.

Definition 11.8

A **function** is a relation in which each member of the domain is assigned one and only one member of the range. A function is a relation in which no two different ordered pairs have the same first component.

Notice that the relation described by equation 1 is not a function because two different ordered pairs, $(0, 2)$ and $(0, -2)$, have the same first component. Likewise, the relation described by equation 2 is not a function because $(16, 4)$ and $(16, -4)$ have the same first component.

EXAMPLE 2

Specify the domain and range for each relation, and state whether or not the relation is a function.

(a) $\{(-3, 3), (-2, 2), (-1, 1), (0, 0), (1, 1), (2, 2), (3, 3)\}$
(b) $\{(1, 73), (2, 73), (3, 73)\}$
(c) $\{(5, 10), (6, 20), (5, -10), (6, -20)\}$
(d) {(O'Neil, Magic), (O'Neil, Heat), (O'Neil, Lakers)}

Solution

(a) Domain $= \{-3, -2, -1, 0, 1, 2, 3\}$; Range $= \{0, 1, 2, 3\}$. It is a function.
(b) Domain $= \{1, 2, 3\}$; Range $= \{73\}$. It is a function.
(c) Domain $= \{5, 6\}$; Range $= \{-20, -10, 10, 20\}$. It is not a function because $(5, 10)$ and $(5, -10)$ have the same first component.
(d) Domain = {O'Neil}; Range = {Magic, Heat, Lakers}. It is not a function because (O'Neil, Magic) and (O'Neil, Heat) have the same first component. ■

The domain of a function is frequently of more concern than the range. You should be aware of any necessary restrictions on x. Consider the next examples.

EXAMPLE 3

Specify the domain for each relation:

(a) $y = 2x + 3$ **(b)** $y = \dfrac{1}{x - 3}$ **(c)** $y = \dfrac{5x}{3x - 4}$

Solution

(a) The domain of the relation described by $y = 2x + 3$ is the set of all real numbers because we can substitute any real number for x.

(b) We can replace x with any real number except 3 because 3 makes the denominator zero. Thus the domain is all real numbers except 3.

(c) We need to find the value of x that makes the denominator equal to zero. To do that we set the denominator equal to zero and solve for x.

$$3x - 4 = 0$$

$$3x = 4$$

$$x = \frac{4}{3}$$

Since $\frac{4}{3}$ makes the denominator zero, the domain is all real numbers except $\frac{4}{3}$. ■

■ Functional Notation

Thus far we have been using the regular notation for writing equations to describe functions; that is, we have used equations such as $y = x^2$ and $y = \frac{1}{x - 2}$, where y is expressed in terms of x, to specify certain functions. There is a special **functional notation** that is very convenient to use when working with the function concept.

The notation $f(x)$ is read "f of x" and is defined to be the value of the function f at x. [Do not interpret $f(x)$ to mean f times x!] Instead of writing $y = x^2$, we can write $f(x) = x^2$. Therefore, $f(2)$ means the value of the function f at 2, which is $2^2 = 4$. So we write $f(2) = 4$. This is a convenient way of expressing various values of the function. We illustrate that idea with another example.

EXAMPLE 4

If $f(x) = x^2 - 6$, find $f(0)$, $f(1)$, $f(2)$, $f(3)$, $f(-1)$, and $f(h)$.

Solution

$$\mathbf{f(x) = x^2 - 6}$$

$$f(0) = 0^2 - 6 = -6$$

$$f(1) = 1^2 - 6 = -5$$

$$f(2) = 2^2 - 6 = -2$$

$$f(3) = 3^2 - 6 = 3$$

$$f(-1) = (-1)^2 - 6 = -5$$

$$f(h) = h^2 - 6$$

■

When we are working with more than one function in the same problem, we use different letters to designate the different functions, as the next example demonstrates.

EXAMPLE 5 If $f(x) = 2x + 5$ and $g(x) = x^2 - 2x + 1$, find $f(2)$, $f(-3)$, $g(-1)$, and $g(4)$.

Solution

$$f(x) = 2x + 5 \qquad g(x) = x^2 - 2x + 1$$

$$f(2) = 2(2) + 5 = 9 \qquad g(-1) = (-1)^2 - 2(-1) + 1 = 4$$

$$f(-3) = 2(-3) + 5 = -1 \qquad g(4) = 4^2 - 2(4) + 1 = 9$$

■

Problem Set 11.7

For Problems 1–14, specify the domain and the range for each relation. Also state whether or not the relation is a function.

1. $\{(4, 7), (6, 11), (8, 20), (10, 28)\}$
2. $\{(3, 6), (5, 7), (7, 8)\}$
3. $\{(-2, 4), (-1, 3), (0, 2), (1, 1)\}$
4. $\{(10, -1), (8, -2), (6, -3), (4, -2)\}$
5. $\{(9, -3), (9, 3), (4, 2), (4, -2)\}$
6. $\{(0, 2), (1, 5), (0, -2), (3, 7)\}$
7. $\{(3, 15), (4, 15), (5, 15), (6, 15)\}$
8. {(1995, Clinton), (1996, Clinton), (1997, Clinton)}
9. {(Carol, 22400), (Carol, 23700), (Carol, 25200)}
10. $\{(4, 6), (4, 8), (4, 10), (4, 12)\}$
11. $\{(-6, 1), (-6, 2), (-6, 3), (-6, 4)\}$
12. $\{(-7, -1), (-5, 0), (-2, 2), (0, 3)\}$
13. $\{(-2, 4), (-1, 1), (0, 0), (1, 1), (2, 4)\}$
14. $\{(-1, 7), (-2, 4), (0, 0), (1, 7), (2, 4)\}$

For Problems 15–30, specify the domain for each function.

15. $y = 3x - 2$
16. $y = -4x + 1$
17. $y = \dfrac{1}{x + 8}$
18. $y = \dfrac{1}{x - 6}$
19. $y = x^2 + 4x - 7$
20. $y = 3x^2 - 2x - 5$
21. $y = \dfrac{3}{2x - 10}$
22. $y = \dfrac{4}{7 - x}$
23. $y = \dfrac{3x}{5x - 8}$
24. $y = \dfrac{6x}{7x + 3}$
25. $y = \dfrac{2x - 6}{5}$
26. $y = \dfrac{6x - 12}{7}$
27. $y = \dfrac{2x + 3}{x}$
28. $y = \dfrac{x - 4}{2x}$
29. $y = x^3$
30. $y = -2x^4$
31. If $f(x) = 3x + 4$, find $f(0)$, $f(1)$, $f(-1)$, and $f(6)$.
32. If $f(x) = -2x + 5$, find $f(2)$, $f(-2)$, $f(-3)$, and $f(5)$.
33. If $f(x) = -5x - 1$, find $f(3)$, $f(-4)$, $f(-5)$, and $f(t)$.
34. If $f(x) = 7x - 3$, find $f(-1)$, $f(0)$, $f(4)$, and $f(a)$.
35. If $g(x) = \dfrac{2}{3}x + \dfrac{3}{4}$, find $g(3)$, $g\left(\dfrac{1}{2}\right)$, $g\left(-\dfrac{1}{3}\right)$, and $g(-2)$.
36. If $g(x) = -\dfrac{1}{2}x + \dfrac{5}{6}$, find $g(1)$, $g(-1)$, $g\left(\dfrac{2}{3}\right)$, and $g\left(-\dfrac{1}{3}\right)$.
37. If $f(x) = x^2 - 4$, find $f(2)$, $f(-2)$, $f(7)$, and $f(0)$.
38. If $f(x) = 2x^2 + x - 1$, find $f(2)$, $f(-3)$, $f(4)$, and $f(-1)$.
39. If $f(x) = -x^2 + 1$, find $f(-1)$, $f(2)$, $f(-2)$, and $f(-3)$.
40. If $f(x) = -2x^2 - 3x - 1$, find $f(1)$, $f(0)$, $f(-1)$, and $f(-2)$.
41. If $f(x) = 4x + 3$ and $g(x) = x^2 - 2x$, find $f(5)$, $f(-6)$, $g(-1)$, and $g(4)$.
42. If $f(x) = -2x - 7$ and $g(x) = 2x^2 + 1$, find $f(-2)$, $f(4)$, $g(-2)$, and $g(4)$.
43. If $f(x) = 3x^2 - x + 4$ and $g(x) = -3x + 5$, find $f(-1)$, $f(4)$, $g(-1)$, and $g(4)$.
44. If $f(x) = -2x^2 + 3x + 1$ and $g(x) = -6x$, find $f(3)$, $f(-5)$, $g(2)$, and $g(-4)$.

THOUGHTS INTO WORDS

45. Are all functions also relations? Are all relations also functions?

46. Give two examples of a function that describes a real-world situation where the domain is restricted.

47. What meanings do we give to the word "function" in our everyday activities? Are any of these meanings closely related to the use of *function* in mathematics?

11.8 Applications of Functions

In the preceding section, we presented functions and their domain and range. In this section, we will consider some applications that use the concept of function to connect mathematics to the real world.

Functions are an ideal way of studying quantities that vary with one another. You are already aware of many relationships that are functions but never thought of them in mathematical terms. For example, the area of a circle varies with the radius of the circle and can be written as a function, $f(r) = \pi r^2$. The distance that is traveled in a car driving at a constant rate of speed—say, 50 miles per hour—can be expressed as a function of the time t as $f(t) = 50t$.

EXAMPLE 1

A car rental agency charges a fixed amount per day plus an amount per mile for renting a car. The daily charges can be expressed as the function $f(x) = 0.20x + 25$, where x is the number of miles driven. Find the charges if the car is driven 158 miles.

Solution

Find the value of the function when $x = 158$.

$$\begin{aligned} f(x) &= 0.20x + 25 \\ f(158) &= 0.20(158) + 25 \\ &= 56.60 \end{aligned}$$

The charges are \$56.60 when the car is driven 158 miles. ■

EXAMPLE 2

A retailer has some items that she wants to sell at a profit of 40% of the cost of each item. The selling price can be represented as a function of the cost by $f(c) = 1.4c$, where c is the cost of an item. Make a table showing the cost and the selling price of items that cost \$8, \$12, \$15, \$22, and \$30.

Solution

Find the values of $f(c) = 1.4c$ when c is \$8, \$12, \$15, \$22, and \$30.

$$\begin{aligned} f(8) &= 1.4(8) = 11.20 \\ f(12) &= 1.4(12) = 16.80 \\ f(15) &= 1.4(15) = 21.00 \end{aligned}$$

$$f(22) = 1.4(22) = 30.80$$

$$f(30) = 1.4(30) = 42.00$$

Cost	\$8	\$12	\$15	\$22	\$30
Selling price	\$11.20	\$16.80	\$21.00	\$30.80	\$42.00

The function from Example 2, $f(c) = 1.4c$, is a linear relationship. To graph this linear relationship, we can label the horizontal axis c and the vertical axis $f(c)$. The domain is restricted to nonnegative values for cost. If the cost is 0, then the selling price is 0, so the graph starts at the origin. Then we use one other point in the table in Example 2 to draw the straight line in Figure 11.32.

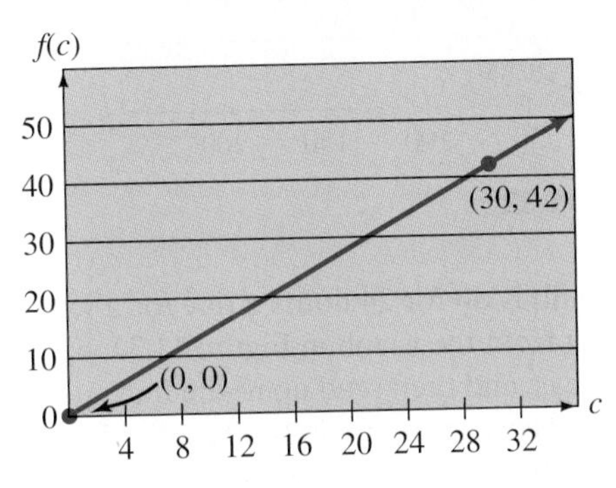

Figure 11.32

■

From the graph, we can approximate the selling price based on a given cost. For example, if $c = 25$, then by reading up from 25 on the c axis to the line, and then across to the $f(c)$ axis, we see that the selling price is approximately \$35.

EXAMPLE 3

The cost for burning a 60-watt lightbulb is given by the function $c(h) = 0.0036h$, where h represents the number of hours that the bulb is burning.

(a) How much does it cost to burn a 60-watt bulb for 30 days when the bulb is burning 3 hours per day?

(b) Graph the linear function $c(h) = 0.0036h$.

(c) Suppose that a 60-watt lightbulb is left burning in a closet for a week before it is discovered and turned off. Use the graph from part (b) to approximate the cost of allowing the bulb to burn for a week. Then use the function to find the exact cost.

Solution

(a)

$$c(h) = 0.0036h$$

$$c(90) = 0.0036(90)$$

$$= 0.324$$

The cost, to the nearest cent, is 32 cents.

(b) We can draw the graph with horizontal axis h and vertical axis $c(h)$. The domain is the nonnegative numbers. Since $c(0) = 0$ and $c(100) = 0.36$, we can use the points (0, 0) and (100, 0.36) to graph the linear function $c(h) = 0.0036h$. The graph is shown in Figure 11.33.

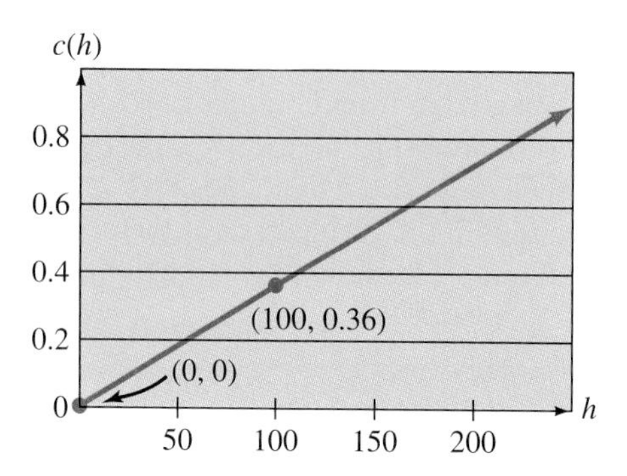

Figure 11.33

(c) If the bulb is on for 24 hours a day for a week, it burns for $24(7) = 168$ hours. Reading from the graph in Figure 11.33, we can approximate 168 on the horizontal axis, and then read up to the line and across to the vertical axis. It looks as if it will cost approximately 60 cents. Using $c(h) = 0.0036h$, we obtain exactly $c(168) = 0.0036(168) = 0.6048$. ■

Problem Set 11.8

1. The function $A(s) = s^2$ expresses the area of a square as a function of the length of a side s of the square. Compute $A(3)$, $A(17)$, $A(8.5)$, $A(20.75)$, and $A(11.25)$.

2. A boat rental agency charges \$50 plus \$35 per hour to rent a fishing boat. Therefore, the charge for renting a boat is a function of the number of hours h the boat is rented and can be expressed as $f(h) = 50 + 35h$. Compute $f(5)$, $f(10)$, $f(1)$, and $f(3)$.

3. The cost, in dollars, of manufacturing clocks is a function of the number of clocks n produced and can be expressed as $C(n) = 12n + 44{,}500$. Find the cost to produce 35,000 clocks.

4. The function $A(r) = \pi r^2$ expresses the area of a circle as a function of the radius of the circle. Compute $A(5)$, $A(7.5)$, $A(10)$, and $A(12)$. (Use 3.14 to approximate π.)

5. A retailer has some items that he wants to sell and make a profit of 50% of the cost of each item. The function $s(c) = 1.5c$, where c represents the cost of an item, can be used to determine the selling price. Find the selling price of items that cost \$4.50, \$6.75, \$9.00, and \$16.40.

6. "SALE—All Items 20% Off Marked Price" is a sign at the clothing store. The function $f(p) = 0.80p$, where p is the marked price, can be used to determine the sale price. Find the sale price of a \$22 t-shirt, a \$38 pair of jeans, a \$45 sweater, and a \$5.50 pair of socks.

7. Mike has a job delivering pizzas for Pizza City. He gets paid \$15 an evening plus \$0.75 for every pizza he delivers. His pay, in dollars, can be expressed as the function $f(n) = 0.75n + 15$, where n is the number of pizzas he delivers. Find his pay for delivering 20 pizzas, 0 pizzas, and 16 pizzas.

8. The ABC Car Rental uses the function $f(x) = 26$ for the charges, in dollars, for the daily use of a car driven up to and including 200 miles. When a car is driven more than 200 miles per day, they use the function $g(x) = 26 + 0.15(x - 200)$ to determine the charges. How much would they charge for daily driving of 150 miles, 230 miles, 360 miles, and 430 miles?

9. To charge, in dollars, for children's meals, a restaurant uses the function $f(x) = 0$ when the child is less than 2 years old, and the function $g(x) = 0.50x$ when the child's age is 2 or more, where x is the age of the child. How much does a child's meal cost for an 8-year-old child, a 3-year-old child, a 1-year-old child, and a 12-year-old child?

10. The volume, in cubic inches, of a right circular cylinder with a fixed base can be expressed as the function $V(h) = 201h$, where h is the height of the cylinder in inches. Determine the volume of a cylinder that is 12 inches high.

11. Mario's pay, in dollars, can be expressed as a function of the number of hours h worked. The function $f(h) = 10.50h$ represents his pay if h, the hours worked, is less than or equal to 40. The function $g(h) = 15.75h - 210$ represents his pay if h is greater than 40. How much is Mario's pay if he works 35 hours, 40 hours, 50 hours, and 20 hours?

12. A wholesale shopping club's discounts are dependent on the total amount purchased p. The chart below shows the functions used to calculate the cost based on the amount purchased. Find the cost when the amount purchased is \$75, when the amount purchased is \$530, when the amount purchased is \$270, and when the amount purchased is \$156.

13. A hotel charges \$98.50 for a room with one guest. If there is more than one guest, then the charge for the room is expressed as the function $f(n) = 98.50 + 20(n - 1)$, where n is the number of guests. What are the charges when there are 2 guests, 3 guests, 1 guest, and 4 guests?

14. Wesley owns 1000 shares of stock. He is considering selling some of it, and he knows his profit can be represented by the function $f(x) = 28x - 150$, where x is the number of shares sold. Make a table showing the profit for selling 100, 200, 400, 500, or 600 shares.

15. The probability that an event will occur can be determined from the function $f(x) = \frac{x}{10}$, where $x = 0$, 1, 2, 3, 4. Make a table showing all the values of x and $f(x)$.

16. The annual cost, in dollars, of \$100,000 of life insurance is a function of a person's age x and can be expressed as $c(x) = 354 + 10(x - 30)$. Make a table showing the annual cost of insurance for a 20-year-old, a 30-year-old, a 35-year-old, a 40-year-old, a 50-year-old, and a 55-year-old.

17. A pool service company adds neutralizer to pools on a monthly basis. The number of pounds of neutralizer used can be expressed as $f(n) = 0.0003n$, where n is the number of gallons of water in the pool. Make a table showing the number of pounds of neutralizer used for pools with 10,000 gallons of water, 15,000 gallons of water, 20,000 gallons of water, 25,000 gallons of water, and 30,000 gallons of water.

18. The equation $I(r) = 750r$ expresses the amount of simple interest earned by an investment of \$750 in 1 year as a function of the rate of interest r. Compute $I(0.075)$, $I(0.0825)$, $I(0.0875)$, and $I(0.095)$.

19. The function $P(s) = 4s$ expresses the perimeter of a square as a function of the length of a side s of the square.

(a) Find the perimeter of a square whose sides are 3 feet long.

(b) Find the perimeter of a square whose sides are 5 feet long.

(c) Graph the linear function $P(s) = 4s$.

(d) Use the graph from part (c) to approximate the perimeter of a square whose sides are 4.25 feet long. Then use the function to find the exact perimeter.

Amount Purchased	$p \leq 100$	$100 < p \leq 200$	$200 < p \leq 300$	$300 < p \leq 400$	$p > 400$
Function for cost	$f(p) = 0.98p$	$f(p) = 0.96p$	$f(p) = 0.94p$	$f(p) = 0.92p$	$f(p) = 0.90p$

20. An antique dealer assumes that an item appreciates the same amount each year. Suppose an antique costs \$2500 and it appreciates \$200 each year for t years. Then we can express the value of the antique after t years by the function $V(t) = 2500 + 200t$.

(a) Find the value of the antique after 5 years.
(b) Find the value of the antique after 8 years.
(c) Graph the linear function $V(t) = 2500 + 200t$.
(d) Use the graph from part (c) to approximate the value of the antique after 10 years. Then use the function to find the exact value.
(e) Use the graph to approximate how many years it will take for the value of the antique to become \$3750.
(f) Use the function to determine exactly how long it will take for the value of the antique to become \$3750.

21. The linear depreciation method assumes that an item depreciates the same amount each year. Suppose a new piece of machinery costs \$32,500 and it depreciates \$1950 each year for t years. Then we can express the value of the machinery after t years by the function $V(t) = 32{,}500 - 1950t$.

(a) Find the value of the machinery after 6 years.
(b) Find the value of the machinery after 9 years.
(c) Graph the linear function $V(t) = 32{,}500 - 1950t$.
(d) Use the graph from part (c) to approximate the value of the machinery after 10 years. Then use the function to find the exact value.
(e) Use the graph to approximate how many years it will take for the value of the machinery to become zero.
(f) Use the function to determine exactly how long it will take for the value of the machinery to become zero.

22. The function $f(C) = \frac{9}{5}C + 32$ expresses the temperature in degrees Fahrenheit as a function of the temperature in degrees Celsius.

(a) Use the function to complete the table.

C	0	10	15	−5	−10	−15	−25
f(C)							

(b) Graph the linear function $f(C) = \frac{9}{5}C + 32$.
(c) Use the graph from part (b) to approximate the temperature in degrees Fahrenheit when the temperature is 20°C. Then use the function to find the exact value.

23. The function $f(t) = \frac{5}{9}(t - 32)$ expresses the temperature in degrees Celsius as a function of the temperature in degrees Fahrenheit.

(a) Use the function to complete the table.

t	50	41	−4	212	95	77	59
f(t)							

(b) Graph the linear function $f(t) = \frac{5}{9}(t - 32)$.
(c) Use the graph from part (b) to approximate the temperature in degrees Celsius when the temperature is 20°F. Then use the function to find the exact value.

THOUGHTS INTO WORDS

24. Describe a situation that occurs in your life where a function could be used. Write the equation for the function and evaluate it for three values in the domain.

25. Suppose you are driving at a constant speed of 60 miles per hour. Explain how the distance you drive is a linear function of the time you drive.

FURTHER INVESTIGATIONS

26. Most spreadsheet applications for computers allow you to enter a function, and when you highlight a list of values, the program evaluates the function for those values. If you have access to spreadsheet software, try evaluating the function $f(x) = 10x + 20$ for $x = 10$, 25, 30, 40, and 50.

Chapter 11 Summary

(11.1) Interpreting **absolute value** to mean *the distance between a number and zero on a number line* allows us to solve a variety of equations and inequalities involving absolute value.

(11.2) Solving **a system of three linear equations in three variables** produces one of the following results.

1. There is *one ordered triple* that satisfies all three equations.
2. There are *infinitely many ordered triples* in the solution set, all of which are coordinates of points on a line common to the planes.
3. There are *infinitely many ordered triples* in the solution set, all of which are coordinates of points on a plane.
4. The solution set is empty; it is $\varnothing$.

(11.3) The following definitions merge the concepts of **root** and **fractional exponents**.

1. a is a square root of b if $a^2 = b$.
2. a is a cube root of b if $a^3 = b$.
3. a is an nth root of b if $a^n = b$.
4. $b^{\frac{1}{n}}$ means $\sqrt[n]{b}$.
5. $b^{\frac{m}{n}}$ means $\sqrt[n]{b^m}$, which equals $(\sqrt[n]{b})^m$.

(11.4) A **complex number** is any number that we can express in the form $a + bi$, where a and b are real numbers and $i = \sqrt{-1}$.

The property $\sqrt{ab} = \sqrt{a}\sqrt{b}$ holds if only one of a or b is negative, and therefore we use it as a basis for simplifying radicals containing negative radicands.

We describe **addition** and **subtraction** of complex numbers as follows:

$$(a + bi) + (c + di) = (a + c) + (b + d)i$$

$$(a + bi) - (c + di) = (a - c) + (b - d)i$$

We find the **product** of two complex numbers in the same way that we find the product of two binomials.

(11.5) Every quadratic equation of the form $ax^2 + bx + c = 0$, where a, b, and c are real numbers and $a \neq 0$, has a solution (solutions) in the set of complex numbers.

(11.6) Graphs are used to convey mathematical information at a glance. A **pie chart** clearly illustrates the parts of a whole. A **bar graph** is used to compare various categories. A **line graph** is helpful for indicating trends.

(11.7) A **relation** is a set of ordered pairs. The **domain** is the set of all first components of the ordered pairs. The **range** is the set of all second components of the ordered pairs.

A **function** is a relation in which no two different ordered pairs have the same first component.

In **functional notation** $f(x)$ is read "f of x" and we define it to be the value of the function f at x.
Therefore, if

$$f(x) = 2x + 3,$$

then

$$f(1) = 2(1) + 3 = 5$$

(11.8) Functions can be used to describe real-world problems. We can evaluate functions for various values in the domain to produce a table of ordered pairs. From the table we can draw a graph of the function. Graphs of linear functions can be used to approximate the value of the function for a specific value in the domain.

Chapter 11 Review Problem Set

For Problems 1–3, use the graph in Figure 11.34.

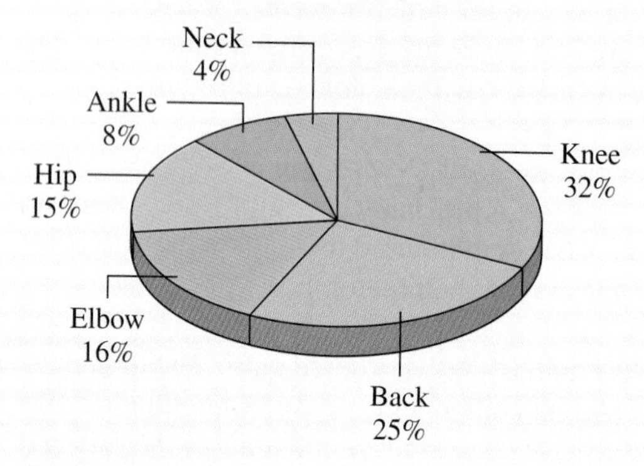

Figure 11.34

1. What percent of the patients received therapy for the ankle, hip, or knee?
2. What percent of the patients received therapy for the hip, neck, or back?
3. What percent of the patients were not seen for the knee or elbow?

For Problems 4–6, use the graph in Figure 11.35.

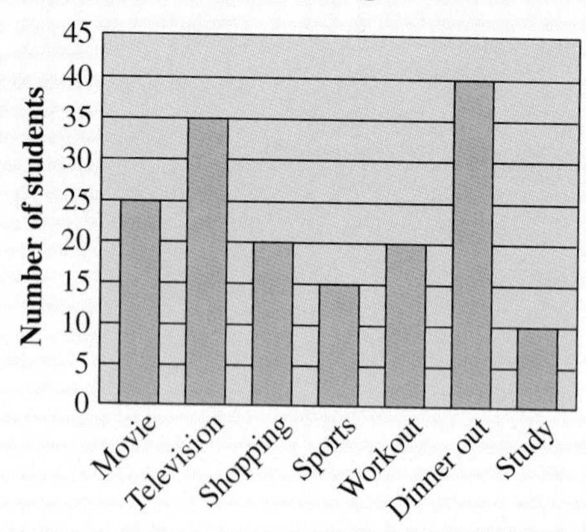

Figure 11.35

4. How many more students spent Friday evening watching television than studying?
5. Which two activities were equally popular?
6. Arrange the activities in order from most popular to least popular.

For Problems 7–10, use the graph in Figure 11.36.

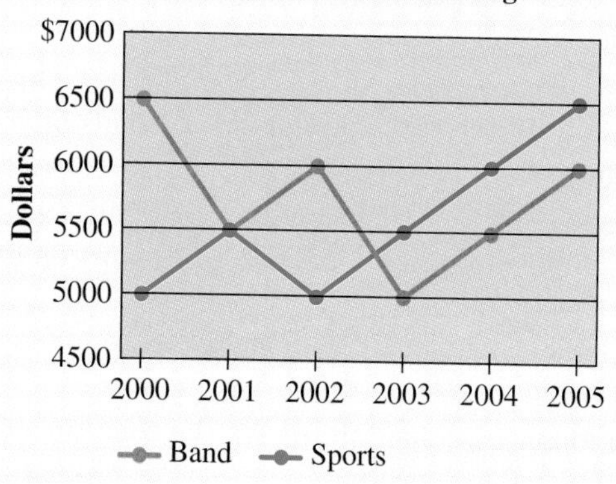

Figure 11.36

7. In what year was the fundraising for the band equal to the fundraising for sports?
8. What was the largest difference in fundraising between band and sports in a single year?
9. What is the average amount of fundraising for the band over the 6 years?
10. What is the average amount of fundraising for sports over the 6 years?

For Problems 11–16, solve and graph the solutions for each problem.

11. $|3x - 5| = 7$
12. $|x - 4| < 1$
13. $|2x - 1| \geq 3$
14. $|3x - 2| \leq 4$
15. $|2x - 1| = 9$
16. $|5x - 2| \geq 6$

For Problems 17–31, evaluate each of the numerical expressions.

17. $\sqrt{\dfrac{64}{36}}$
18. $-\sqrt{1}$
19. $\sqrt[3]{\dfrac{27}{64}}$

20. $\sqrt[3]{-125}$

21. $\sqrt[4]{\frac{81}{16}}$

22. $25^{\frac{3}{2}}$

23. $8^{\frac{5}{3}}$

24. $(-8)^{\frac{5}{3}}$

25. 4^{-2}

26. $4^{-\frac{1}{2}}$

27. $(32)^{-\frac{2}{5}}$

28. $\left(\frac{2}{3}\right)^{-1}$

29. $2^{\frac{7}{4}} \cdot 2^{\frac{5}{4}}$

30. $3^{\frac{1}{3}} \cdot 3^{\frac{5}{3}}$

31. $\frac{3^{\frac{1}{3}}}{3^{\frac{4}{3}}}$

For Problems 32–39, simplify and express the final results with positive exponents only.

32. $x^{\frac{5}{6}} \cdot x^{\frac{5}{6}}$

33. $\left(3x^{\frac{1}{4}}\right)\left(2x^{\frac{3}{5}}\right)$

34. $\left(9a^{\frac{1}{2}}\right)\left(4a^{-\frac{1}{3}}\right)$

35. $\left(3x^{\frac{1}{3}}y^{\frac{2}{3}}\right)^3$

36. $\left(25x^4y^6\right)^{\frac{1}{2}}$

37. $\frac{39n^{\frac{3}{5}}}{3n^{\frac{1}{4}}}$

38. $\frac{64n^{\frac{5}{8}}}{16n^{\frac{7}{8}}}$

39. $\left(\frac{6x^{\frac{2}{7}}}{3x^{-\frac{5}{7}}}\right)^3$

40. Solve the system of equations

$$\begin{pmatrix} x + 3y - z = 1 \\ 2x - y + z = 3 \\ 3x + y + 2z = 12 \end{pmatrix}.$$

41. Solve the system of equations

$$\begin{pmatrix} 2x + 3y - z = 4 \\ x + 2y + z = 7 \\ 3x + y + 2z = 13 \end{pmatrix}.$$

For Problems 42–55, perform the indicated operations on complex numbers.

42. $(5 - 7i) + (-4 + 9i)$

43. $(-3 + 2i) + (-4 - 7i)$

44. $(6 - 9i) - (4 - 5i)$

45. $(-5 + 3i) - (-8 + 7i)$

46. $(7 - 2i) - (6 - 4i) + (-2 + i)$

47. $(-4 + i) - (-4 - i) - (6 - 8i)$

48. $(2 + 5i)(3 + 8i)$

49. $(4 - 3i)(1 - 2i)$

50. $(-1 + i)(-2 + 6i)$

51. $(-3 - 3i)(7 + 8i)$

52. $(2 + 9i)(2 - 9i)$

53. $(-3 + 7i)(-3 - 7i)$

54. $(-3 - 8i)(3 + 8i)$

55. $(6 + 9i)(-1 - i)$

For Problems 56–65, solve each quadratic equation.

56. $(x - 6)^2 = -25$

57. $n^2 + 2n = -7$

58. $x^2 - 2x + 17 = 0$

59. $x^2 - x + 7 = 0$

60. $2x^2 - x + 3 = 0$

61. $6x^2 - 11x + 3 = 0$

62. $-x^2 + 5x - 7 = 0$

63. $-2x^2 - 3x - 6 = 0$

64. $3x^2 + x + 5 = 0$

65. $x(4x + 1) = -3$

For Problems 66–71, state the domain and range of the relation, and specify whether it is a function.

66. $\left\{\left(\text{red}, \frac{1}{4}\right), \left(\text{blue}, \frac{1}{8}\right), \left(\text{green}, \frac{5}{8}\right)\right\}$

67. $\{(3, 5), (4, 7), (5, 9)\}$

68. $\{(1, 8), (1, -8), (2, 16), (2, -16)\}$

69. $\{(2, 10), (3, 10), (4, 10), (5, 10)\}$

70. $\{(-2, 4), (-1, 1), (0, 0), (1, 1), (2, 4)\}$

71. $\{(1, 4), (2, 8), (3, 15), (2, 10)\}$

For Problems 72–81, determine the domain.

72. $f(x) = \frac{1}{x - 6}$

73. $f(x) = x^2 + 4$

74. $f(x) = 5x - 7$

75. $f(x) = \frac{3}{x + 4}$

76. $f(x) = \frac{5}{2x - 1}$

77. $f(x) = \frac{2}{3x + 1}$

78. If $f(x) = 3x - 2$, find $f(-4)$, $f(0)$, $f(5)$, and $f(a)$.

79. If $f(x) = \frac{6}{x - 4}$, find $f(-4)$, $f(0)$, $f(1)$, and $f(2)$.

80. If $f(x) = \frac{x}{2x + 1}$, find $f(-3)$, $f(0)$, $f(2)$, and $f(3)$.

81. If $f(x) = x^2 + 4x - 3$, find $f(-1)$, $f(0)$, $f(1)$, and $f(2)$.

82. A mail center charges, in dollars, according to the function $f(x) = 0.20 + 0.30x$, where x is the weight of the letter in ounces.

(a) Find the charge for mailing a 2-ounce letter.

(b) Find the charge for mailing a 5-ounce letter.

(c) Graph the linear function $f(x) = 0.20 + 0.30x$.

(d) Using the graph from part (c), approximate the charge for 4 ounces, and then use the function to find the exact charge.

83. To determine how many gallons of cooking oil are needed, Fast Fry Company uses the function $f(x) = 12 + 0.2x$, where x is the pounds of food to be fried.

(a) Find the number of gallons of cooking oil needed to fry 50 pounds of food.

(b) Find the number of gallons of cooking oil needed to fry 100 pounds of food.

(c) Graph the linear function $f(x) = 12 + 0.2x$.

(d) Use the graph from part (c) to approximate the number of gallons of cooking oil needed to fry 80 pounds of food, and then use the function to find the exact amount.

Chapter 11 Test

For Problems 1–4, use the graph in Figure 11.37.

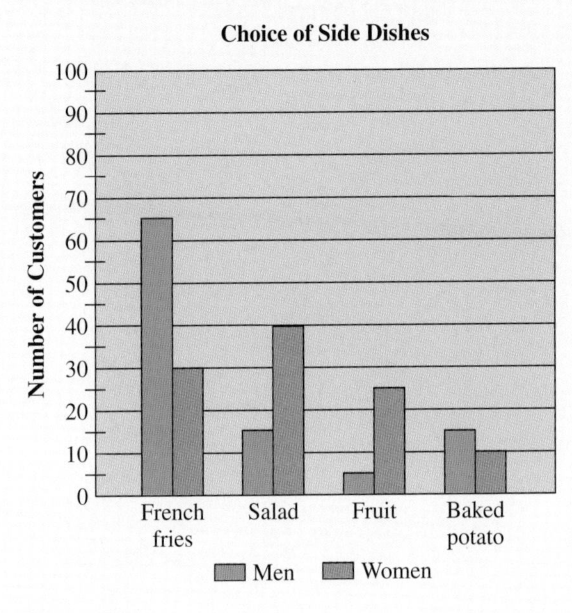

Figure 11.37

1. Which side item was picked most often by men? women?

2. For the women, how many more chose salad instead of baked potato?

3. For the men, how many more chose French fries instead of fruit?

4. How many more men than women chose French fries as their side item?

5. Multiply and simplify $(3 + i)(2 - 5i)$.

6. Express $\sqrt{-75}$ in terms of i and simplify.

7. Evaluate $36^{\frac{3}{2}}$.

8. Evaluate $\left(\frac{2}{3}\right)^{-3}$.

9. Simplify $\left(2x^{\frac{1}{4}}\right)\left(5x^{\frac{2}{3}}\right)$.

10. Simplify $\frac{30n^{\frac{1}{2}}}{6n^{\frac{2}{5}}}$.

For Problems 11–19, solve each equation.

11. $(x - 2)^2 = -16$

12. $x^2 - 2x + 3 = 0$

13. $x^2 + 6x = -21$

14. $x^2 - 3x + 5 = 0$

15. $|x - 2| = 6$

16. $|4x + 5| = 2$

17. $|3x - 1| = -4$

18. $2x^2 - x + 1 = 0$

19. $3x^2 + 5x - 28 = 0$

20. Solve the inequality $|x + 3| \geq 2$.

21. Solve the inequality $|2x - 1| < 7$.

22. Determine the domain of $f(x) = \frac{5}{2x - 7}$.

23. If $f(x) = 5x - 6$, find $f(2)$, $f(0)$, $f(-3)$, and $f(4)$.

24. The equation $P(x) = 20 + 0.10x$ is used to determine the pay for concession stand workers, where x is the amount of sales. Find the pay when sales are \$4270.

25. A real estate broker uses the function $f(x) = 0.07x$ for finding commissions when x, the selling price, is \$200,000 or less. If the selling price is greater than \$200,000, the broker uses the function $g(x) = 0.05x + 4000$. How much is the commission for a selling price of \$165,000? A selling price of \$245,000?

Appendixes

A Table of Squares and Approximate Square Roots

Squares and Approximate Square Roots

n	n^2	$\sqrt{n}$	n	n^2	$\sqrt{n}$
1	1	1.000	26	676	5.099
2	4	1.414	27	729	5.196
3	9	1.732	28	784	5.292
4	16	2.000	29	841	5.385
5	25	2.236	30	900	5.477
6	36	2.449	31	961	5.568
7	49	2.646	32	1024	5.657
8	64	2.838	33	1089	5.745
9	81	3.000	34	1156	5.831
10	100	3.162	35	1225	5.916
11	121	3.317	36	1296	6.000
12	144	3.464	37	1369	6.083
13	169	3.606	38	1444	6.164
14	196	3.742	39	1521	6.245
15	225	3.873	40	1600	6.325
16	256	4.000	41	1681	6.403
17	289	4.123	42	1764	6.481
18	324	4.243	43	1849	6.557
19	361	4.359	44	1936	6.633
20	400	4.472	45	2025	6.708
21	441	4.583	46	2116	6.782
22	484	4.690	47	2209	6.856
23	529	4.796	48	2304	6.928
24	576	4.899	49	2401	7.000
25	625	5.000	50	2500	7.071

(continued)

Squares and Approximate Square Roots (*continued*)

n	n^2	$\sqrt{n}$	n	n^2	$\sqrt{n}$
51	2601	7.141	76	5776	8.718
52	2704	7.211	77	5929	8.775
53	2809	7.280	78	6084	8.832
54	2916	7.348	79	6241	8.888
55	3025	7.416	80	6400	8.944
56	3136	7.483	81	6561	9.000
57	3249	7.550	82	6724	9.055
58	3364	7.616	83	6889	9.110
59	3481	7.681	84	7056	9.165
60	3600	7.746	85	7225	9.220
61	3721	7.810	86	7396	9.274
62	3844	7.874	87	7569	9.327
63	3969	7.937	88	7744	9.381
64	4096	8.000	89	7921	9.434
65	4225	8.062	90	8100	9.487
66	4356	8.124	91	8281	9.539
67	4489	8.185	92	8464	9.592
68	4624	8.246	93	8649	9.644
69	4761	8.307	94	8836	9.695
70	4900	8.367	95	9025	9.747
71	5041	8.426	96	9216	9.798
72	5184	8.485	97	9409	9.849
73	5329	8.544	98	9604	9.899
74	5476	8.602	99	9801	9.950
75	5625	8.660	100	10000	10.000

From the table we can find rational approximations as follows.

$\sqrt{31} = 5.568$ Rounded to three decimal places

Locate 31 in the column labeled n.

Locate this value for $\sqrt{31}$ in the column labeled $\sqrt{n}$.

Be sure that you agree with the following values taken from the table. Each of these is rounded to three decimal places.

$\sqrt{14} = 3.742, \quad \sqrt{46} = 6.782, \quad \sqrt{65} = 8.062$

The column labeled n^2 contains the squares of the whole numbers from 1 through 100, inclusive. For example, from the table we obtain $25^2 = 625$, $62^2 = 3844$, and $89^2 = 7921$. Since $25^2 = 625$, we can state that $\sqrt{625} = 25$. Thus the column labeled n^2 also provides us with additional square root facts.

$$\sqrt{1936} = 44$$

Locate 1936 in the column labeled n^2

Locate this value for in the column labeled n

Now suppose that you want to find a value for $\sqrt{500}$. Scanning the column labeled n^2, we see that 500 does not appear; however, it is between two values given in the table. We can reason as follows.

$$22^2 = 484$$

← 500

$$23^2 = 529$$

Since 500 is closer to 484 than to 529, we approximate $\sqrt{500}$ to be closer to 22 than to 23. Thus, we write $\sqrt{500} = 22$, to the nearest whole number. This is merely a whole number approximation, but for some purposes it may be sufficiently precise.

B Extra Word Problems

These problems are categorized and cross referenced regarding when they can be used. For example, the first problem is a *number problem* and can be used anytime after Section 3.2 has been studied.

Number Problems

1. (3.2) Seventeen added to a certain number is 33. Find the number.

2. (3.2) Six subtracted from a certain number produces 19. Find the number.

3. (3.2) If 3 is added to four times a certain number, the result is 23. Find the number.

4. (3.2) If 2 is subtracted from five times a certain number, the result is 4. What is the number?

5. (3.2) If six times a certain number is subtracted from 15, the result is 3. Find the number.

6. (3.3) Find two consecutive whole numbers whose sum is 67.

7. (3.3) Find two consecutive integers whose sum is -83.

8. (3.3) Find three consecutive even numbers whose sum is 72.

9. (3.3) Find three consecutive odd numbers whose sum is 147.

10. (3.3) One more than two times a certain number is five less than four times the number. Find the number.

11. (3.3) Two less than three times a certain number is one more than four times the number. Find the number.

12. (3.3) Find two consecutive whole numbers such that three times the smaller number plus four times the larger number totals 137.

13. (3.4) Find three consecutive whole numbers such that if the largest number is subtracted from the sum of the other two numbers, the result is 48.

14. (3.4) The sum of a number and two-thirds of the number is 15. Find the number.

15. (3.4) The sum of one-half of a number and three-fourths of the number is 40. Find the number.

16. (3.4) If three-eighths of a number is subtracted from five-sixths of the number, the result is 22. Find the number.

17. (3.4) If the sum of one-eighth of a number and one-sixth of the number is subtracted from three-fourths of the number, the result is 33. Find the number.

18. (6.5) Find two numbers whose product is 28 such that one of the numbers is one less than twice the other number.

19. (6.5) Find two numbers whose product is 24 such that one of the numbers is two less than five times the other number.

20. (6.5) Suppose that the sum of the squares of two consecutive integers is 61. Find the integers.

21. (6.5) Suppose that the sum of the squares of two consecutive odd whole numbers is 34. Find the whole numbers.

22. (7.5) The sum of two numbers is 57. If the larger number is divided by the smaller number, the quotient is 2 and the remainder is 3. Find the numbers.

23. (7.5) The difference of two numbers is 47. If the larger number is divided by the smaller number, the quotient is 6 and the remainder is 2. Find the numbers.

24. (7.5) The denominator of a fraction is one larger than the numerator. If you double the numerator and add 4 to the denominator, the resulting fraction is equivalent to 1. Find the original fraction.

25. (7.5) What number must be subtracted from both the numerator and the denominator of $\frac{28}{37}$ to produce a fraction equivalent to $\frac{8}{11}$?

26. (7.6) If the reciprocal of a number subtracted from the number yields $4\frac{4}{5}$, find the number.

27. (7.6) Suppose that the reciprocal of a number subtracted from the number yields $-\frac{7}{12}$. Find the number.

28. (7.6) The sum of a number and three times its reciprocal is $6\frac{1}{2}$. Find the number.

29. (7.6) The reciprocal of a number is $\frac{5}{6}$ larger than the number. Find the number.

30. (8.7) The sum of two numbers is 78 and the smaller number subtracted from the larger number produces 6. Find the numbers.

31. (8.7) The difference of two numbers is 27. If three times the smaller number is subtracted from the larger number the result is -1. Find the numbers.

32. (8.7) The sum of two numbers is 102. If the larger number is subtracted from 6 times the smaller number, the result is equal to the smaller number. Find the numbers.

33. (8.7) The sum of two numbers is 99, and the difference of the numbers is 35. Find the numbers.

34. (8.7) Find two numbers such that 4 times the larger number minus the smaller number is equal to 64, and twice the larger number plus the smaller is 176.

35. (10.5) Find two numbers such that their sum is 10 and their product is 22.

36. (10.5) Find two consecutive whole numbers such that the sum of their squares is 145.

37. (10.5) Suppose that the sum of two whole numbers is 9 and the sum of their reciprocals is $\frac{1}{2}$. Find the numbers.

Age Problems

38. (3.2) Six years ago, Sean was 14 years old. What is his present age?

39. (3.2) Fourteen years from now, Nancy will be 37 years old. What is her present age?

40. (4.5) At the present time, Nikki is one-third as old as Kaitlin. In 10 years, Kaitlin's age will be 6 years less than twice Nikki's age at that time. Find the present ages of both girls.

41. (4.5) Annilee's present age is two-thirds of Jessie's present age. In 12 years the sum of their ages will be 54 years. Find their present ages.

42. (4.5) The sum of the present ages of Angie and her mother is 64 years. In 8 years Angie will be three-fifths as old as her mother at that time. Find the present ages of Angie and her mother.

Inequality Problems

43. (3.6) Four less than three times a number is greater than 9. Find all numbers that satisfy this relationship.

44. (3.6) Five more than two times a number is less than or equal to 21. Find the numbers that satisfy this relationship.

45. (3.6) One less than six times a number is less than twice the number plus 11. Find the numbers that satisfy this relationship.

46. (3.6) Jose has scores of 87, 89, and 92 on his first three algebra exams. What score must he get on the fourth exam to have an average of 90 or better for the four exams?

47. (3.6) Olga has scores of 89, 92, 96, and 98 on her first four biology exams. What score must she get on the fifth exam to have an average of 95 or better for the five exams?

48. (3.6) Justin shot rounds of 81 and 82 on the first two days of a golf tournament. What must he shoot on the third day to average 80 or less for the three days?

49. (3.6) Gabrielle bowled 148 and 166 in her first two games. What must she bowl in the third game to have an average of at least 160 for the three games?

Geometric Problems

50. (3.2) The length of a rectangular box is 34 centimeters. This length is 4 centimeters more than three times the width. Find the width of the box.

51. (3.3) The width of a rectangular sheet of paper is 8 inches. This width is 1 inch more than one-half of the length. Find the length of the sheet of paper.

52. (3.3) One angle of a triangle has a measure of 75°. Find the measures of the other two angles if one of them is twice as large as the other one.

53. (3.3) If two angles are supplementary, and the difference between their measures is 56°, find the measure of each angle.

54. (3.3) One of two complementary angles is 6° less than three times the other angle. Find the measure of each angle.

55. (3.4) The sum of the complement of an angle and one-half of the supplement of that angle is 90°. Find the measure of the angle.

56. (3.4) In triangle ABC, the measure of angle C is 6° less than three times the measure of angle A, and the measure of angle B is 26° more than the measure of angle A. Find the measure of each angle of the triangle.

57. (3.4) The supplement of an angle is 15° less than four times its complement. Find the measure of the angle.

58. (4.4) The length of a rectangle is 1 inch more than twice the width. If the perimeter of the rectangle is 44 inches, find the length and width.

59. (4.4) The width of a rectangle is 7 centimeters less than the length. If the perimeter of the rectangle is 70 centimeters, find the length and width.

60. (4.4) A triangular plot of ground is enclosed with 34 yards of fencing. The longest side of the triangle is 1 yard more than twice the shortest side. The other side is 5 yards longer than the shortest side. Find the length of each side of the triangular plot.

61. (4.4) Suppose that a square and an equilateral triangle have the same perimeter. If each side of the equilateral triangle is 4 inches longer than each side of the square, find the length of each side of the square.

62. (4.4) Suppose that the perimeter of a square is 17 centimeters more than the perimeter of an equilateral triangle. If each side of the equilateral triangle is 3 centimeters shorter than each side of the square, find the length of each side of the equilateral triangle.

63. (6.5) The area of a rectangular garden plot is 40 square meters. The length of the plot is 3 meters more than its width. Find the length and width of the plot.

64. (6.5) The area of a triangular sheet of paper is 18 square inches. The length of one side of the triangle is 1 inch more than twice the altitude to that side. Find the length of that side and the length of the altitude to that side.

65. (10.5) The sum of the lengths of the two legs of a right triangle is 21 inches. If the length of the hypotenuse is 15 inches, find the length of each leg.

66. (10.5) The length of a rectangular floor is 1 meter less than twice its width. If a diagonal of the rectangle is 17 meters, find the length and width of the floor.

67. (10.5) A rectangular plot of ground measuring 12 meters by 20 meters is surrounded by a sidewalk of a uniform width. The area of the sidewalk is 68 square meters. Find the width of the walk.

68. (10.5) The perimeter of a rectangle is 44 inches, and its area is 112 square inches. Find the length and width of the rectangle.

69. (10.5) A rectangular piece of cardboard is 2 units longer than it is wide. From each of its corners a square piece 2 units on a side is cut out. The flaps are then turned up to form an open box that has a volume of 70 cubic units. Find the length and width of the original piece of cardboard.

Investment Problems

70. (4.5) Eva invested a certain amount of money at 5% interest, and $1500 more than that amount at 6%. Her total yearly interest income was $420. How much did she invest at each rate?

71. (4.5) A total of $4000 was invested, part of it at 8% and the remainder at 9%. If the total interest income amounted to $350, how much was invested at each rate?

72. (4.5) If $500 is invested at 6% interest, how much additional money must be invested at 9% so that the total return for both investments averages 8%?

73. (4.5) A sum of $2000 is split between two investments, one paying 7% interest and the other 8%. If the return on the 8% investment exceeds that on the 7% investment by $40 per year, how much is invested at each rate?

74. (10.5) Tony bought a number of shares of stock for $720. A month later the value of the stock increased by $8 per share, and he sold all but 20 shares and regained his original investment plus a profit of $80. How many shares did he sell and at what price per share?

75. (10.5) Barry bought a number of shares of stock for \$600. A week later the value of the stock had increased \$3 per share, and he sold all but 10 shares and regained his original investment of \$600. How many shares did he sell and at what price per share?

76. (10.5) A businesswoman bought a parcel of land on speculation for \$120,000. She subdivided the land into lots, and when she had sold all but 18 lots at a profit of \$6000 per lot, she regained the entire cost of the land. How many lots were sold and at what price per lot?

Ratio and Proportion Problems

77. (4.1) A blueprint has a scale where 1 inch represents 4 feet. Find the dimensions of a rectangular room that measures 3.5 inches by 4.25 inches on the blueprint.

78. (4.1) On a certain map, 1 inch represents 20 miles. If two cities are 6.5 inches apart on the map, find the number of miles between the cities.

79. (4.1) If a car travels 200 miles using 10 gallons of gasoline, how far will it travel on 15 gallons of gasoline?

80. (4.1) A home valued at \$150,000 is assessed \$3000 in real estate taxes. At the same rate, how much are the taxes on a home with a value of \$200,000?

81. (4.1) If 20 pounds of fertilizer will cover 1400 square feet of grass, how many pounds are needed for 1750 square feet?

82. (4.1) A board 24 feet long is cut into two pieces whose lengths are in the ratio of 2 to 3. Find the lengths of the two pieces.

83. (4.1) On a certain map $1\frac{1}{2}$ inches represents 25 miles. If two cities are $5\frac{1}{4}$ inches apart on the map, find the number of miles between the cities.

84. (4.1) A sum of \$750 is to be divided between two people in the ratio of three to two. How much does each person receive?

85. (4.1) The ratio of female students to male students at a certain junior college is five to three. If there is a total of 4400 students, find the number of female students and the number of male students.

Percent Problems

86. (4.1) Twenty-one is what percent of 70?

87. (4.1) Forty-six is what percent of 40?

88. (4.1) Eighty percent of what number is 60?

89. (4.1) One hundred and ten percent of what number is 60.5?

90. (4.1) Two hundred and twenty percent of what number is 198?

91. (4.2) Louise bought a dress for \$36.40, which represents a 30% discount of the original price. Find the original price of the dress.

92. (4.2) Find the cost of a \$75 shirt that is on sale for 20% off.

93. (4.2) Ely bought a putter for \$72 that was originally listed for \$120. What rate of discount did he receive?

94. (4.2) Dominic bought a suit for \$162.50 that was originally listed for \$250. What rate of discount did he receive?

95. (4.2) A retailer has some golf gloves that cost him \$5 each. He wants to sell them at a profit of 60% of the cost. What should be the selling price of the golf gloves?

96. (4.2) A retailer has some skirts that cost her \$25 each. She wants to sell them at a profit of 45% of the cost. What should be the selling price of the skirts?

97. (4.2) A bookstore manager buys some textbooks for \$22 each. He wants to sell them at a profit of 20% based on the selling price. What should be the selling price of the books?

98. (4.2) A supermarket manager buys some apples at \$.80 per pound. He wants to sell them at a profit of 50% based on the selling price. What should be the selling price of the apples?

Uniform Motion Problems

99. (4.4) Two cars start from the same place traveling in opposite directions. One car travels 7 miles per hour faster than the other car. If at the end of 3 hours they are 369 miles apart, find the speed of each car.

100. (4.4) Two cities, A and B, are 406 miles apart. Billie starts at city A in her car traveling toward city B at 52 miles per hour. At the same time, using the same route, Zorka leaves in her car from city B traveling toward city A at 64 miles per hour. How long will it be before the cars meet?

101. (4.4) Macrina starts jogging at 4 miles per hour. One-half hour later, Gordon starts jogging on the same route at 6 miles per hour. How long will it take Gordon to catch Macrina?

102. (4.4) In $\frac{1}{4}$ of an hour more time, Jack, riding his bicycle at 12 miles per hour, rode 6 miles farther than Nikki, who was riding her bicycle at 10 miles per hour. How long did Jack ride?

103. (7.5) Kent rides his bicycle 36 miles in the same time that it takes Kaitlin to ride her bicycle 27 miles. If Kent rides 3 miles per hour faster than Kaitlin, find the rate of each.

104. (7.5) To walk 7 miles it takes Dave $1\frac{1}{2}$ hours longer than it takes Simon to walk 6 miles. If Simon walks at a rate 1 mile per hour faster than Dave, find the times and rates of both boys.

105. (7.6) Debbie rode her bicycle out into the country for a distance of 24 miles. On the way back, she took a much shorter route of 12 miles and made the return trip in one-half hour less time. If her rate out into the country was 4 miles per hour faster than her rate on the return trip, find both rates.

106. (7.6) Felipe jogs for 10 miles and then walks another 10 miles. He jogs $2\frac{1}{2}$ miles per hour faster than he walks, and the entire distance of 20 miles takes 6 hours. Find the rate at which he walks and the rate at which he jogs.

107. (10.5) On a 570-mile trip, Andy averaged 5 miles per hour faster for the last 240 miles than he did for the first 330 miles. The entire trip took 10 hours. How fast did he travel for the first 330 miles?

108. (10.5) On a 135-mile bicycle excursion, Maria averaged 5 miles per hour faster for the first 60 miles than she did for the last 75 miles. The entire trip took 8 hours. Find her rate for the first 60 miles.

109. (10.5) Charlotte traveled 250 miles in 1 hour more time than it took Lorraine to travel 180 miles. Charlotte drove 5 miles per hour faster than Lorraine. How fast did each one travel?

Mixture Problems

110. (4.5) How many liters of pure alcohol must be added to 10 liters of a 30% solution to obtain a 50% solution?

111. (4.5) How many cups of grapefruit juice must be added to 40 cups of punch that contains 5% grapefruit juice to obtain a punch that is 10% grapefruit juice?

112. (4.5) How many milliliters of pure acid must be added to 150 milliliters of a 30% solution of acid to obtain a 40% solution?

113. (4.5) How much water needs to be removed from 20 quarts of a 20% salt solution to change it to a 30% salt solution?

114. (4.5) How many gallons of a 12% salt solution must be mixed with 6 gallons of a 20% salt solution to obtain a 15% salt solution?

115. (4.5) Suppose that 10 gallons of a 30% salt solution is mixed with 20 gallons of a 50% salt solution. What is the percent of salt in the final solution?

116. (4.5) Suppose that you have a supply of a 30% alcohol solution and a 70% alcohol solution. How many quarts of each should be mixed to produce 20 quarts that are 40% alcohol?

117. (4.5) A 16-quart radiator contains a 50% solution of antifreeze. How much needs to be drained out and replaced with pure antifreeze to obtain a 60% antifreeze solution?

Work Problems

118. (7.6) Susan can do a job in 20 minutes and Ellen can do the same job in 30 minutes. If they work together, how long should it take them to do the job?

119. (7.6) Working together Ramon and Sean can mow a lawn in 15 minutes. Ramon could mow the lawn by himself in 25 minutes. How long would it take Sean to mow the lawn by himself?

120. (7.6) It takes two pipes $1\frac{1}{5}$ hours to fill a water tank. Pipe B can fill the tank alone in 1 hour less time than it would take pipe A to fill the tank alone. How long would it take each pipe to fill the tank by itself?

121. (7.6) An inlet pipe can fill a tank in 10 minutes. A drain can empty the tank in 12 minutes. If the tank is empty and both the inlet pipe and drain are open, how long will it take before the tank overflows?

122. (10.5) It takes Terry 2 hours longer to do a certain job than it takes Tom. They worked together for 3 hours; then Tom left and Terry finished the job in 1 hour. How long would it take each of them to do the job alone?

123. (10.5) Suppose that Arlene can mow the entire lawn in 40 minutes less time with the power mower than she can with the push mower. One day the power mower broke down after she had been mowing for 30 minutes. She finished the lawn with the push mower in 20 minutes. How long does it take Arlene to mow the entire lawn with the power mower?

Coin Problems

124. (3.4) Simon has 40 coins consisting of pennies, nickels, and dimes. He has three times as many nickels as pennies and four times as many dimes as pennies. How many coins of each kind does he have?

125. (3.4) Carson has 30 coins consisting of nickels and dimes amounting to $2.30. How many coins of each kind does Carson have?

126. (3.4) Karl has 104 coins consisting of pennies, nickels, and dimes. The number of dimes is two-thirds of the number of pennies, and the number of nickels is one-half the number of pennies. How many coins of each kind does he have?

127. (3.4) Amanda has \$12.50 consisting of nickels, dimes, and quarters. She has one-half as many dimes as quarters, and the number of nickels is one-fourth of the number of quarters. Find the number of coins of each kind.

128. (3.4) Pierre has 95 coins consisting of pennies, nickels, and dimes. The number of nickels is 10 less than the number of pennies, and the number of dimes is 15 less than twice the number of pennies. How many coins of each kind does Pierre have?

129. (3.4) Mona has 23 coins consisting of dimes and quarters amounting to \$4.55. The number of quarters is one less than twice the number of dimes. How many coins of each kind does she have?

130. (3.4) Chen has some nickels, dimes, and quarters amounting to \$11.85. The number of dimes is one more than twice the number of nickels, and there are 10 more quarters than dimes. How many coins of each kind does he have?

Miscellaneous Problems

131. (3.2) Cheryl worked 7 hours on Saturday, and she earned \$66.50. How much per hour did she earn?

132. (3.2) A mathematics textbook is priced at \$45. This is \$5 more than four times the price of an accompanying workbook. Find the price of the workbook.

133. (3.2) A lawnmower repair bill without tax was \$75. This included \$15 for parts and 2 hours of labor. Find the price per hour of labor.

134. (3.2) On sale at a supermarket one could buy four 12-packs of soda for \$10. This represented a savings of \$4.76 from the regular price of four 12-packs. Find the regular price of a 12-pack.

135. (3.3) A calculator and a mathematics textbook are priced at \$85 in the college bookstore. The price of the textbook is \$5 more than three times the price of the calculator. Find the price of the textbook.

136. (3.3) In a recent county election 80 more Democrats voted than Republicans. Together there were 1380 Republicans and Democrats that voted. How many Democrats voted?

137. (3.3) In the local elementary school there are 40 more girls than boys. The total number of students is 680. How many girls are there in the school?

138. (3.3) Zorka sold some stock at \$37 per share. This was \$9 a share less than twice what she paid for it. What price did she pay for the stock?

139. (3.3) Michael is paid two times his normal hourly rate for each hour that he works over 40 hours in a week. Last week he earned \$552 for 43 hours of work. What is his normal hourly rate?

140. (3.4) Nicole is paid time and a half for each hour of work over 40 hours in a week. Last week she worked 45 hours and was paid $484.50. Find her normal hourly rate.

141. (6.5) Nate plants 65 apple trees in a rectangular array such that the number of trees per row is two less than three times the number of rows. Find the number of rows and the number of trees per row.

142. (6.5) In a classroom of 48 desks, the number of desks per row is two more than the number of rows. Find the number of rows and the number of desks per row.

143. (8.7) Two pounds of Gala apples and 3 pounds of Fuji apples cost $6.55. One pound of Gala apples and 4 pounds of Fuji apples cost $5.75. Find the price per pound for each type of apple.

144. (8.7) Six rolls of tape and one package of manila envelopes cost $12.03. Four rolls of tape and 3 packages of envelopes cost $12.43. Find the price of a roll of tape.

145. (8.7) Two boxes of Corn flakes and 3 boxes of Wheat flakes cost $13.25. Three boxes of Corn flakes and two boxes of Wheat flakes cost $12.65. Find the price per box of Corn flakes and the price per box of Wheat flakes.

146. (10.5) A man did a job for $360. It took him 6 hours longer than he expected, and therefore he earned $2 per hour less than he anticipated. How long did he expect that it would take to do the job?

147. (10.5) A group of students agreed that each would chip in the same amount to pay for a party that would cost $100. Then they found 5 more students interested in the party and in sharing the expenses. This decreased the amount each had to pay by $1. How many students were involved in the party and how much did each student have to pay?

148. (10.5) A group of customers agreed that each would contribute the same amount to buy their favorite waitress a $100 birthday gift. At the last minute two of the people decided not to chip in. This increased the amount that the remaining people had to pay by $2.50 per person. How many people actually contributed to the gift?

149. (10.5) A retailer bought a number of special mugs for $48. Two of the mugs were broken in the store, but by selling each of the other mugs $3 above the original cost per mug, she made a total profit of $22. How many mugs did she buy and at what price per mug did she sell them?

150. (10.5) At a point 16 yards from the base of a tower, the distance to the top of the tower is 4 yards more than the height of the tower. Find the height of the tower.

Answers to Odd-Numbered Problems and All Chapter Review, Chapter Test, Cumulative Review Problems, and Appendix B Problems

CHAPTER 1

Problem Set 1.1 (page 7)

1. 16 **3.** 35 **5.** 51 **7.** 72 **9.** 82 **11.** 55 **13.** 60 **15.** 66 **17.** 26 **19.** 2 **21.** 47 **23.** 21 **25.** 11 **27.** 15 **29.** 14 **31.** 79 **33.** 6 **35.** 74 **37.** 12 **39.** 187 **41.** 884 **43.** 9 **45.** 18 **47.** 55 **49.** 99 **51.** 72 **53.** 11 **55.** 48 **57.** 21 **59.** 40 **61.** 170 **63.** 164 **65.** 153 **71.** $36 + 12 \div (3 + 3) + 6 \cdot 2$ **73.** $36 + (12 \div 3 + 3) + 6 \cdot 2$

Problem Set 1.2 (page 13)

1. True **3.** False **5.** True **7.** True **9.** True **11.** False **13.** True **15.** False **17.** True **19.** False **21.** 3 and 8 **23.** 2 and 12 **25.** 4 and 9 **27.** 5 and 10 **29.** 1 and 9 **31.** Prime **33.** Prime **35.** Composite **37.** Prime **39.** Composite **41.** $2 \cdot 59$ **43.** $3 \cdot 67$ **45.** $5 \cdot 17$ **47.** $3 \cdot 3 \cdot 13$ **49.** $3 \cdot 43$ **51.** $2 \cdot 13$ **53.** $2 \cdot 2 \cdot 3 \cdot 3$ **55.** $7 \cdot 7$ **57.** $2 \cdot 2 \cdot 2 \cdot 7$ **59.** $2 \cdot 2 \cdot 2 \cdot 3 \cdot 5$ **61.** $3 \cdot 3 \cdot 3 \cdot 5$ **63.** 4 **65.** 8 **67.** 9 **69.** 12 **71.** 18 **73.** 12 **75.** 24 **77.** 48 **79.** 140 **81.** 392 **83.** 168 **85.** 90 **89.** All other even numbers are divisible by 2. **91.** 61 **93.** x **95.** xy

Problem Set 1.3 (page 21)

1. 2 **3.** −4 **5.** −7 **7.** 6 **9.** −6 **11.** 8 **13.** −11 **15.** −15 **17.** −7 **19.** −31 **21.** −19 **23.** 9 **25.** −61 **27.** −18 **29.** −92 **31.** −5 **33.** −13 **35.** 12 **37.** 6 **39.** −1 **41.** −45 **43.** −29 **45.** 27 **47.** −65 **49.** −29 **51.** −11 **53.** −1 **55.** −8 **57.** −13 **59.** −35 **61.** −15 **63.** −32 **65.** 2 **67.** −4 **69.** −31 **71.** −9 **73.** 18 **75.** 8 **77.** −29 **79.** −7 **81.** 15 **83.** 1 **85.** 36 **87.** −39 **89.** −24 **91.** 7 **93.** −1 **95.** 10 **97.** 9 **99.** −17 **101.** −3 **103.** −10 **105.** −3 **107.** 11 **109.** 5 **111.** −65 **113.** −100 **115.** −25 **117.** 130 **119.** 80 **121.** $-17 + 14 = -3$ **123.** $3 + (-2) + (-3) + (-5) = -7$ **125.** $-2 + 1 + 3 + 1 + (-2) = 1$

Problem Set 1.4 (page 27)

1. −30 **3.** −9 **5.** 7 **7.** −56 **9.** 60 **11.** −12 **13.** −126 **15.** 154 **17.** −9 **19.** 11 **21.** 225 **23.** −14 **25.** 0 **27.** 23 **29.** −19 **31.** 90 **33.** 14 **35.** Undefined **37.** −4 **39.** −972 **41.** −47 **43.** 18 **45.** 69 **47.** 4 **49.** 4 **51.** −6 **53.** 31 **55.** 4 **57.** 28 **59.** −7 **61.** 10 **63.** −59 **65.** 66 **67.** 7 **69.** 69 **71.** −7 **73.** 126 **75.** −70 **77.** 15 **79.** −10 **81.** −25 **83.** 77 **85.** 104 **87.** 14 **89.** $800(19) + 800(2) + 800(4)(-1) = 13{,}600$ **91.** $5 + 4(-3) = -7$

Problem Set 1.5 (page 36)

1. Distributive property
3. Associative property of addition
5. Commutative property of multiplication
7. Additive inverse property
9. Identity property of addition
11. Associative property of multiplication **13.** 56 **15.** 7 **17.** 1800 **19.** −14,400 **21.** −3700 **23.** 5900 **25.** −338 **27.** −38 **29.** 7 **31.** $-5x$ **33.** $-3m$ **35.** $-11y$ **37.** $-3x - 2y$ **39.** $-16a - 4b$ **41.** $-7xy + 3x$ **43.** $10x + 5$ **45.** $6xy - 4$ **47.** $-6a - 5b$ **49.** $5ab - 11a$ **51.** $8x + 36$ **53.** $11x + 28$ **55.** $8x + 44$ **57.** $5a + 29$ **59.** $3m + 29$ **61.** $-8y + 6$ **63.** −5 **65.** −40 **67.** 72 **69.** −18 **71.** 37 **73.** −74 **75.** 180 **77.** 34 **79.** −65 **85.** Equivalent **87.** Not equivalent **89.** Not equivalent

Chapter 1 Review Problem Set (page 39)

1. −3 **2.** −25 **3.** −5 **4.** −15 **5.** −1 **6.** 2 **7.** −156 **8.** 252 **9.** 6 **10.** −13 **11.** Prime **12.** Composite

13. Composite **14.** Composite **15.** Composite
16. $2 \cdot 2 \cdot 2 \cdot 3$ **17.** $3 \cdot 3 \cdot 7$ **18.** $3 \cdot 19$
19. $2 \cdot 2 \cdot 2 \cdot 2 \cdot 2 \cdot 2$ **20.** $2 \cdot 2 \cdot 3 \cdot 7$ **21.** 18
22. 12 **23.** 180 **24.** 945 **25.** 66 **26.** -7 **27.** -2
28. 4 **29.** -18 **30.** 12 **31.** -34 **32.** -27 **33.** -38
34. -93 **35.** 2 **36.** 3 **37.** 35 **38.** 27 **39.** 175°F
40. 20,602 feet **41.** $2(6) - 4 + 3(8) - 1 = 31$ **42.** $3444
43. $8x$ **44.** $-5y - 9$ **45.** $-5x + 4y$ **46.** $13a - 6b$
47 $-ab - 2a$ **48.** $-3xy - y$ **49.** $10x + 74$ **50.** $2x + 7$
51. $-7x - 18$ **52.** $-3x + 12$ **53.** $-2a + 4$
54. $-2a - 4$ **55.** -59 **56.** -57 **57.** 2 **58.** 1 **59.** 12
60. 13 **61.** 22 **62.** 32 **63.** -9 **64.** 37 **65.** -39
66. -32 **67.** 9 **68.** -44

Chapter 1 Test (page 42)

1. 7 **2.** 45 **3.** 38 **4.** -11 **5.** -58 **6.** -58 **7.** 4
8. -1 **9.** -20 **10.** -7 **11.** -6°F **12.** 26 **13.** -36
14. 9 **15.** -57 **16.** -47 **17.** -4 **18.** Prime
19. $2 \cdot 2 \cdot 2 \cdot 3 \cdot 3 \cdot 5$ **20.** 12 **21.** 72
22. Associative property of addition
23. Distributive property **24.** $-13x + 6y$
25. $-13x - 21$

CHAPTER 2

Problem Set 2.1 (page 50)

1. $\frac{2}{3}$ **3.** $\frac{2}{3}$ **5.** $\frac{5}{3}$ **7.** $-\frac{1}{6}$ **9.** $-\frac{3}{4}$ **11.** $\frac{27}{28}$ **13.** $\frac{6}{11}$
15. $\frac{3x}{7y}$ **17.** $\frac{2x}{5}$ **19.** $-\frac{5a}{13c}$ **21.** $\frac{8z}{7x}$ **23.** $\frac{5b}{7}$ **25.** $\frac{15}{28}$
27. $\frac{10}{21}$ **29.** $\frac{3}{10}$ **31.** $-\frac{4}{3}$ **33.** $\frac{7}{5}$ **35.** $-\frac{3}{10}$ **37.** $\frac{1}{4}$
39. -27 **41.** $\frac{35}{27}$ **43.** $\frac{8}{21}$ **45.** $-\frac{5}{6y}$ **47.** $2a$ **49.** $\frac{2}{5}$
51. $\frac{y}{2x}$ **53.** $\frac{20}{13}$ **55.** $-\frac{7}{9}$ **57.** $\frac{2}{9}$ **59.** $\frac{2}{5}$ **61.** $\frac{13}{28}$
63. $\frac{8}{5}$ **65.** -4 **67.** $\frac{36}{49}$ **69.** 1 **71.** $\frac{2}{3}$ **73.** $\frac{20}{9}$
75. $\frac{1}{4}$ **77.** $2\frac{1}{4}$ cups **79.** $1\frac{3}{4}$ cups **81.** 65 yards
87. **(a)** larger **(c)** smaller **(e)** larger

Problem Set 2.2 (page 59)

1. $\frac{5}{7}$ **3.** $\frac{5}{9}$ **5.** 3 **7.** $\frac{2}{3}$ **9.** $-\frac{1}{2}$ **11.** $\frac{2}{3}$ **13.** $\frac{15}{x}$
15. $\frac{2}{y}$ **17.** $\frac{8}{15}$ **19.** $\frac{9}{16}$ **21.** $\frac{37}{30}$ **23.** $\frac{59}{96}$ **25.** $-\frac{19}{72}$
27. $-\frac{1}{24}$ **29.** $-\frac{1}{3}$ **31.** $-\frac{1}{6}$ **33.** $-\frac{31}{7}$ **35.** $-\frac{21}{4}$
37. $\frac{3y + 4x}{xy}$ **39.** $\frac{7b - 2a}{ab}$ **41.** $\frac{11}{2x}$ **43.** $\frac{4}{3x}$ **45.** $-\frac{2}{5x}$
47. $\frac{19}{6y}$ **49.** $\frac{1}{24y}$ **51.** $-\frac{17}{24n}$ **53.** $\frac{5y + 7x}{3xy}$
55. $\frac{32y + 15x}{20xy}$ **57.** $\frac{63y - 20x}{36xy}$ **59.** $\frac{-6y - 5x}{4xy}$
61. $\frac{3x + 2}{x}$ **63.** $\frac{4x - 3}{2x}$ **65.** $\frac{1}{4}$ **67.** $\frac{37}{30}$ **69.** $\frac{1}{3}$
71. $-\frac{12}{5}$ **73.** $-\frac{1}{30}$ **75.** 14 **77.** 68 **79.** $\frac{7}{26}$ **81.** $\frac{11}{15}x$
83. $\frac{5}{24}a$ **85.** $\frac{4}{3}x$ **87.** $\frac{13}{20}n$ **89.** $\frac{20}{9}n$ **91.** $-\frac{79}{36}n$
93. $\frac{13}{14}x + \frac{9}{8}y$ **95.** $-\frac{11}{45}x - \frac{9}{20}y$ **97.** $2\frac{1}{8}$ yards
99. $10\frac{3}{4}$ feet **101.** $1\frac{3}{4}$ miles **103.** $36\frac{2}{3}$ yards

Problem Set 2.3 (page 71)

1. Real, rational, integer, and negative
3. Real, irrational, and positive
5. Real, rational, and positive
7. Real, rational, and negative
9. 0.62 **11.** 1.45 **13.** 3.8 **15.** -3.3 **17.** 7.5 **19.** 7.8
21. -0.9 **23.** -7.8 **25.** 1.16 **27.** -0.272 **29.** -24.3
31. 44.8 **33.** 0.0156 **35.** 1.2 **37.** -7.4 **39.** 0.38
41. 7.2 **43.** -0.42 **45.** 0.76 **47.** 4.7 **49.** 4.3
51. -14.8 **53.** 1.3 **55.** $-1.2x$ **57.** $3n$ **59.** $0.5t$
61. $-5.8x + 2.8y$ **63.** $0.1x + 1.2$ **65.** $-3x - 2.3$
67. $4.6x - 8$ **69.** $\frac{11}{12}$ **71.** $\frac{4}{3}$ **73.** 17.3 **75.** -97.8
77. 2.2 **79.** 13.75 **81.** 0.6 **83.** $10,002
85. 19.1 centimeters **87.** 4.7 centimeters **89.** $6.55
91. 322. 58 miles **97.** **(a)** $0.\overline{142857}$ **(c)** $0.\overline{4}$ **(e)** $0.\overline{27}$

Problem Set 2.4 (page 78)

1. 64 **3.** 81 **5.** -8 **7.** -9 **9.** 16 **11.** $\frac{16}{81}$ **13.** $-\frac{1}{8}$
15. $\frac{9}{4}$ **17.** 0.027 **19.** -1.44 **21.** -47 **23.** -33
25. 11 **27.** -75 **29.** -60 **31.** 31 **33.** -13 **35.** $9x^2$
37. $12xy^2$ **39.** $-18x^4y$ **41.** $15xy$ **43.** $12x^4$ **45.** $8a^5$
47. $-8x^2$ **49.** $4y^3$ **51.** $-2x^2 + 6y^2$ **53.** $-\frac{11}{60}n^2$
55. $-2x^2 - 6x$ **57.** $7x^2 - 3x + 8$ **59.** $\frac{3y}{5}$ **61.** $\frac{11}{3y}$
63. $\frac{7b^2}{17a}$ **65.** $-\frac{3ac}{4}$ **67.** $\frac{x^2y^2}{4}$ **69.** $\frac{4x}{9}$ **71.** $\frac{5}{12ab}$

73. $\frac{6y^2 + 5x}{xy^2}$ **75.** $\frac{5 - 7x^2}{x^4}$ **77.** $\frac{3 + 12x^2}{2x^3}$ **79.** $\frac{13}{12x^2}$
81. $\frac{11b^2 - 14a^2}{a^2b^2}$ **83.** $\frac{3 - 8x}{6x^3}$ **85.** $\frac{3y - 4x - 5}{xy}$ **87.** 79
89. $\frac{23}{36}$ **91.** $\frac{25}{4}$ **93.** -64 **95.** -25 **97.** -33 **99.** 0.45

Problem Set 2.5 (page 84)

Answers may vary somewhat for Problems 1–11.
1. The difference of a and b
3. One-third of the product of B and h
5. Two times the quantity, ℓ plus w
7. The quotient of A divided by w
9. The quantity, a plus b, divided by 2
11. Two more than three times y **13.** $\ell + w$ **15.** ab
17. $\frac{d}{t}$ **19.** ℓwh **21.** $y - x$ **23.** $xy + 2$ **25.** $7 - y^2$
27. $\frac{x - y}{4}$ **29.** $10 - x$ **31.** $10(n + 2)$ **33.** $xy - 7$
35. $xy - 12$ **37.** $35 - n$ **39.** $n + 45$ **41.** $y + 10$
43. $2x - 3$ **45.** $10d + 25q$ **47.** $\frac{d}{t}$ **49.** $\frac{d}{p}$ **51.** $\frac{d}{12}$
53. $n + 1$ **55.** $n + 2$ **57.** $3y - 2$ **59.** $36y + 12f$ **61.** $\frac{f}{3}$
63. $8w$ **65.** $3\ell - 4$ **67.** $48f + 72$ **69.** $2w^2$ **71.** $9s^2$

Chapter 2 Review Problem Set (page 87)

1. 64 **2.** -27 **3.** -16 **4.** 125 **5.** $-\frac{1}{4}$ **6.** $\frac{9}{16}$ **7.** $\frac{49}{36}$
8. 0.216 **9.** 0.0144 **10.** 0.0036 **11.** $-\frac{8}{27}$ **12.** $\frac{1}{16}$
13. $-\frac{1}{64}$ **14.** $\frac{4}{9}$ **15.** $\frac{19}{24}$ **16.** $\frac{39}{70}$ **17.** $\frac{1}{15}$
18. $\frac{14y + 9x}{2xy}$ **19.** $\frac{5x - 8y}{x^2y}$ **20.** $\frac{7y}{20}$ **21.** $\frac{4x^3}{5y^2}$
22. $\frac{2}{7}$ **23.** 1 **24.** $\frac{27n^2}{28}$ **25.** $\frac{1}{24}$ **26.** $-\frac{13}{8}$ **27.** $\frac{7}{9}$
28. $\frac{29}{12}$ **29.** $\frac{1}{2}$ **30.** 0.67 **31.** 0.49 **32.** 2.4 **33.** -0.11
34. 1.76 **35.** 36 **36.** 1.92 **37.** $\frac{5}{56}x^2 + \frac{7}{20}y^2$
38. $-0.58ab + 0.36bc$ **39.** $\frac{11x}{24}$ **40.** $2.2a + 1.7b$
41. $-\frac{1}{10}n$ **42.** $\frac{41}{20}n$ **43.** $\frac{19}{42}$ **44.** $-\frac{1}{72}$ **45.** -0.75
46. -0.35 **47.** $\frac{1}{17}$ **48.** -8 **49.** $72 - n$
50. $p + 10d$ **51.** $\frac{x}{60}$ **52.** $2y - 3$ **53.** $5n + 3$
54. $36y + 12f$ **55.** $100m$ **56.** $5n + 10d + 25q$
57. $n - 5$ **58.** $5 - n$ **59.** $10(x - 2)$ **60.** $10x - 2$
61. $x - 3$ **62.** $\frac{d}{r}$ **63.** $x^2 + 9$ **64.** $(x + 9)^2$
65. $x^3 + y^3$ **66.** $xy - 4$

Chapter 2 Test (page 89)

1. (a) 81 **(b)** -64 **(c)** 0.008 **2.** $\frac{7}{9}$ **3.** $\frac{9xy}{16}$ **4.** -2.6
5. 3.04 **6.** -0.56 **7.** $\frac{1}{256}$ **8.** $\frac{2}{9}$ **9.** $-\frac{5}{24}$
10. $\frac{187}{60}$ or $3\frac{7}{60}$ **11.** $-\frac{13}{48}$ **12.** $\frac{4y}{5}$ **13.** $2x^2$
14. $\frac{4y^2 - 5x}{xy^2}$ **15.** $\frac{8}{3x}$ **16.** $\frac{35y + 27}{21y^2}$ **17.** $\frac{10a^2b}{9}$
18. $-x + 5xy$ **19.** $-3a^2 - 2b^2$ **20.** $\frac{37}{36}$ **21.** -0.48
22. $-\frac{31}{40}$ **23.** 2.85 **24.** $5n + 10d + 25q$ **25.** $4n - 3$

Cumulative Review Problem Set (page 90)

1. 10 **2.** -30 **3.** 1 **4.** -26 **5.** -29 **6.** 17 **7.** $\frac{1}{2}$
8. $-\frac{7}{6}$ **9.** $\frac{1}{36}$ **10.** -64 **11.** 200 **12.** 0.173
13. -142 **14.** 136 **15.** $\frac{19}{9}$ **16.** -0.01 **17.** -2.4
18. $\frac{79}{40}$ **19.** $\frac{7}{50}$ **20.** $\frac{3}{5}$ **21.** $2 \cdot 3 \cdot 3 \cdot 3$
22. $2 \cdot 3 \cdot 13$ **23.** $7 \cdot 13$ **24.** $3 \cdot 3 \cdot 17$ **25.** 14
26. 9 **27.** 4 **28.** 6 **29.** 140 **30.** 200 **31.** 108
32. 80 **33.** $-\frac{1}{12}x - \frac{11}{12}y$ **34.** $-\frac{1}{15}n$
35. $-3a + 1.9b$ **36.** $-2n + 6$ **37.** $-x - 15$
38. $-9a - 13$ **39.** $\frac{11}{48}$ **40.** $-\frac{31}{36}$ **41.** $\frac{5 - 2y + 3x}{xy}$
42. $\frac{-7y + 9x}{x^2y}$ **43.** $\frac{2x}{3}$ **44.** $\frac{8a^2}{21b}$ **45.** $\frac{4x^2}{3y}$ **46.** $-\frac{27}{16}$
47. $p + 5n + 10d$ **48.** $4n - 5$ **49.** $36y + 12f + i$
50. $200x + 200y$ or $200(x + y)$

CHAPTER 3

Problem Set 3.1 (page 98)

1. $\{8\}$ **3.** $\{-6\}$ **5.** $\{-9\}$ **7.** $\{-6\}$ **9.** $\{13\}$ **11.** $\{48\}$
13. $\{23\}$ **15.** $\{-7\}$ **17.** $\left\{\frac{17}{12}\right\}$ **19.** $\left\{-\frac{4}{15}\right\}$
21. $\{0.27\}$ **23.** $\{-3.5\}$ **25.** $\{-17\}$ **27.** $\{-35\}$
29. $\{-8\}$ **31.** $\{-17\}$ **33.** $\left\{\frac{37}{5}\right\}$ **35.** $\{-3\}$ **37.** $\left\{\frac{13}{2}\right\}$
39. $\{144\}$ **41.** $\{24\}$ **43.** $\{-15\}$ **45.** $\{24\}$ **47.** $\{-35\}$
49. $\left\{\frac{3}{10}\right\}$ **51.** $\left\{-\frac{9}{10}\right\}$ **53.** $\left\{\frac{1}{2}\right\}$ **55.** $\left\{-\frac{1}{3}\right\}$
57. $\left\{\frac{27}{32}\right\}$ **59.** $\left\{-\frac{5}{14}\right\}$ **61.** $\left\{-\frac{7}{5}\right\}$ **63.** $\left\{-\frac{1}{12}\right\}$
65. $\left\{-\frac{3}{20}\right\}$ **67.** $\{0.3\}$ **69.** $\{9\}$ **71.** $\{-5\}$

Problem Set 3.2 (page 104)

1. $\{4\}$ **3.** $\{6\}$ **5.** $\{8\}$ **7.** $\{11\}$ **9.** $\left\{\frac{17}{6}\right\}$ **11.** $\left\{\frac{19}{2}\right\}$
13. $\{6\}$ **15.** $\{-1\}$ **17.** $\{-5\}$ **19.** $\{-6\}$ **21.** $\left\{\frac{11}{2}\right\}$
23. $\{-2\}$ **25.** $\left\{\frac{10}{7}\right\}$ **27.** $\{18\}$ **29.** $\left\{-\frac{25}{4}\right\}$ **31.** $\{-7\}$
33. $\left\{-\frac{24}{7}\right\}$ **35.** $\left\{\frac{5}{2}\right\}$ **37.** $\left\{\frac{4}{17}\right\}$ **39.** $\left\{-\frac{12}{5}\right\}$
41. 9 **43.** 22 **45.** \$18 **47.** 35 years old **49.** \$6.50
51. 6 **53.** 5 **55.** \$5.25 **57.** 6.1 inches **59.** 3
61. \$300 **63.** 4 meters **65.** 341 million
67. 1.25 hours

Problem Set 3.3 (page 112)

1. $\{5\}$ **3.** $\{-8\}$ **5.** $\left\{\frac{8}{5}\right\}$ **7.** $\{-11\}$ **9.** $\left\{-\frac{5}{2}\right\}$
11. $\{-9\}$ **13.** $\{2\}$ **15.** $\{-3\}$ **17.** $\left\{\frac{13}{2}\right\}$ **19.** $\varnothing$
21. $\{17\}$ **23.** $\left\{-\frac{13}{2}\right\}$ **25.** $\left\{\frac{16}{3}\right\}$ **27.** {All reals}
29. $\left\{-\frac{1}{3}\right\}$ **31.** $\left\{-\frac{19}{10}\right\}$ **33.** 17 **35.** 35 and 37
37. 36, 38, and 40 **39.** $\frac{3}{2}$ **41.** -6 **43.** 32° and 58°
45. 50° and 130° **47.** 65° and 75° **49.** \$42
51. \$9 per hour **53.** 150 males and 450 females
55. \$91 **57.** \$145

Problem Set 3.4 (page 120)

1. $\{1\}$ **3.** $\{10\}$ **5.** $\{-9\}$ **7.** $\left\{\frac{29}{4}\right\}$ **9.** $\left\{-\frac{17}{3}\right\}$
11. $\{10\}$ **13.** $\{44\}$ **15.** $\{26\}$ **17.** {All reals} **19.** $\varnothing$
21. $\{3\}$ **23.** $\{-1\}$ **25.** $\{-2\}$ **27.** $\{16\}$ **29.** $\left\{\frac{22}{3}\right\}$
31. $\{-2\}$ **33.** $\left\{-\frac{1}{6}\right\}$ **35.** $\{-57\}$ **37.** $\left\{-\frac{7}{5}\right\}$ **39.** $\{2\}$
41. $\{-3\}$ **43.** $\left\{\frac{27}{10}\right\}$ **45.** $\left\{\frac{3}{28}\right\}$ **47.** $\left\{\frac{18}{5}\right\}$
49. $\left\{\frac{24}{7}\right\}$ **51.** $\{5\}$ **53.** $\{0\}$ **55.** $\left\{-\frac{51}{10}\right\}$ **57.** $\{-12\}$
59. $\{15\}$ **61.** 7 and 8 **63.** 14, 15, and 16 **65.** 6 and 11
67. 48 **69.** 14 minutes **71.** 8 feet and 12 feet
73. 15 nickels, 20 quarters **75.** 40 nickels, 80 dimes, and 90 quarters **77.** 8 dimes and 10 quarters
79. 4 crabs, 12 fish, and 6 plants **81.** 30° **83.** 20°, 50°, and 110° **85.** 40° **91.** Any three consecutive integers

Problem Set 3.5 (page 130)

1. True **3.** False **5.** False **7.** True **9.** True
11. $\{x|x > -2\}$ or $(-2, \infty)$

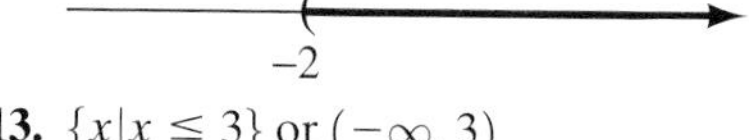

13. $\{x|x \leq 3\}$ or $(-\infty, 3)$

15. $\{x|x > 2\}$ or $(2, \infty)$

17. $\{x|x \leq -2\}$ or $(-\infty, -2]$

19. $\{x|x < -1\}$ or $(-\infty, -1)$

21. $\{x|x < 2\}$ or $(-\infty, 2)$

23. $\{x|x < -20\}$ or $(-\infty, -20)$
25. $\{x|x \geq -9\}$ or $[-9, \infty)$ **27.** $\{x|x > 9\}$ or $(9, \infty)$
29. $\left\{x|x < \frac{10}{3}\right\}$ or $\left(-\infty, \frac{10}{3}\right)$
31. $\{x|x < -8\}$ or $(-\infty, -8)$ **33.** $\{n|n \geq 8\}$ or $[8, \infty)$
35. $\left\{n|n > -\frac{24}{7}\right\}$ or $\left(-\frac{24}{7}, \infty\right)$

37. $\{n|n > 7\}$ or $(7, \infty)$ **39.** $\{x|x > 5\}$ or $(5, \infty)$
41. $\{x|x \leq 6\}$ or $(-\infty, 6]$
43. $\{x|x \leq -21\}$ or $(-\infty, -21]$
45. $\left\{x|x < \frac{8}{3}\right\}$ or $\left(-\infty, \frac{8}{3}\right)$
47. $\left\{x|x < \frac{5}{4}\right\}$ or $\left(-\infty, \frac{5}{4}\right)$
49. $\{x|x < 1\}$ or $(-\infty, 1)$ **51.** $\{t|t \geq 4\}$ or $[4, \infty)$
53. $\{x|x > 14\}$ or $(14, \infty)$ **55.** $\left\{x|x > \frac{3}{2}\right\}$ or $\left(\frac{3}{2}, \infty\right)$
57. $\left\{t|t \geq \frac{1}{4}\right\}$ or $\left[\frac{1}{4}, \infty\right)$
59. $\left\{x|x < -\frac{9}{4}\right\}$ or $\left(-\infty, -\frac{9}{4}\right)$
65. All real numbers **67.** $\varnothing$
69. All real numbers **71.** $\varnothing$

Problem Set 3.6 (page 137)

1. $\{x|x > 2\}$ or $(2, \infty)$ **3.** $\{x|x < -1\}$ or $(-\infty, -1)$
5. $\left\{x|x > -\frac{10}{3}\right\}$ or $\left(-\frac{10}{3}, \infty\right)$
7. $\{n|n \geq -11\}$ or $[-11, \infty)$ **9.** $\{t|t \leq 11\}$ or $(-\infty, 11]$
11. $\left\{x|x > -\frac{11}{5}\right\}$ or $\left(-\frac{11}{5}, \infty\right)$
13. $\left\{x|x < \frac{5}{2}\right\}$ or $\left(-\infty, \frac{5}{2}\right)$ **15.** $\{x|x \leq 8\}$ or $(-\infty, 8]$
17. $\left\{n|n > \frac{3}{2}\right\}$ or $\left(\frac{3}{2}, \infty\right)$ **19.** $\{y|y > -3\}$ or $(-3, \infty)$
21. $\left\{x|x < \frac{5}{2}\right\}$ or $\left(-\infty, \frac{5}{2}\right)$ **23.** $\{x|x < 8\}$ or $(-\infty, 8)$
25. $\{x|x < 21\}$ or $(-\infty, 21)$ **27.** $\{x|x < 6\}$ or $(-\infty, 6)$
29. $\left\{n|n > -\frac{17}{2}\right\}$ or $\left(-\frac{17}{2}, \infty\right)$
31. $\{n|n \leq 42\}$ or $(-\infty, 42]$
33. $\left\{n|n > -\frac{9}{2}\right\}$ or $\left(-\frac{9}{2}, \infty\right)$
35. $\left\{x|x > \frac{4}{3}\right\}$ or $\left(\frac{4}{3}, \infty\right)$ **37.** $\{n|n \geq 4\}$ or $[4, \infty)$
39. $\{t|t > 300\}$ or $(300, \infty)$ **41.** $\{x|x \leq 50\}$ or $(-\infty, 50]$
43. $\{x|x > 0\}$ or $(0, \infty)$ **45.** $\{x|x > 64\}$ or $(64, \infty)$
47. $\left\{n|n > \frac{33}{5}\right\}$ or $\left(\frac{33}{5}, \infty\right)$
49. $\left\{x|x \geq -\frac{16}{3}\right\}$ or $\left[-\frac{16}{3}, \infty\right)$
51. (number line: open interval from −1 to 2) −1 2
53. (number line: shaded left of −2 and right of 1, open endpoints) −2 1
55. (number line: from −2 open to 2 closed) −2 2
57. (number line: shaded right of 2, open) 2
59. (number line: shaded right of −4, open) −4
61. (number line, nothing shaded) $\varnothing$
63. (number line: shaded left of 0 closed and right of 2 closed) 0 2
65. (number line: entire line shaded) All reals
67. All numbers greater than 7 **69.** 15 inches
71. 158 or better **73.** Better than 90
75. More than 250 sales **77.** 77 or less

Chapter 3 Review Problem Set (page 140)

1. $\{-3\}$ **2.** $\{1\}$ **3.** $\left\{-\frac{3}{4}\right\}$ **4.** $\{9\}$ **5.** $\{-4\}$ **6.** $\left\{\frac{40}{3}\right\}$
7. $\left\{\frac{9}{4}\right\}$ **8.** $\left\{-\frac{15}{8}\right\}$ **9.** $\{-7\}$ **10.** $\left\{\frac{2}{41}\right\}$ **11.** $\left\{\frac{19}{7}\right\}$
12. $\left\{\frac{1}{2}\right\}$ **13.** $\{-32\}$ **14.** $\{-12\}$ **15.** $\{21\}$ **16.** $\{-60\}$
17. $\{10\}$ **18.** $\left\{-\frac{11}{4}\right\}$ **19.** $\left\{-\frac{8}{5}\right\}$ **20.** $\left\{\frac{5}{21}\right\}$
21. $\{x|x > 4\}$ or $(4, \infty)$ **22.** $\{x|x > -4\}$ or $(-4, \infty)$
23. $\{x|x \geq 13\}$ or $[13, \infty)$ **24.** $\left\{x|x \geq \frac{11}{2}\right\}$ or $\left[\frac{11}{2}, \infty\right)$
25. $\{x|x > 35\}$ or $(35, \infty)$
26. $\left\{x|x < \frac{26}{5}\right\}$ or $\left(-\infty, \frac{26}{5}\right)$ **27.** $\{n|n < 2\}$ or $(-\infty, 2)$
28. $\left\{n|n > \frac{5}{11}\right\}$ or $\left(\frac{5}{11}, \infty\right)$
29. $\{y|y < 24\}$ or $(-\infty, 24)$ **30.** $\{x|x > 10\}$ or $(10, \infty)$
31. $\left\{n|n < \frac{2}{11}\right\}$ or $\left(-\infty, \frac{2}{11}\right)$
32. $\{n|n > 33\}$ or $(33, \infty)$ **33.** $\{n|n \leq 120\}$ or $(-\infty, 120]$
34. $\left\{n|n \leq -\frac{180}{13}\right\}$ or $\left(-\infty, -\frac{180}{13}\right]$
35. $\left\{x|x > \frac{9}{2}\right\}$ or $\left(\frac{9}{2}, \infty\right)$
36. $\left\{x|x < -\frac{43}{3}\right\}$ or $\left(-\infty, -\frac{43}{3}\right)$

37. 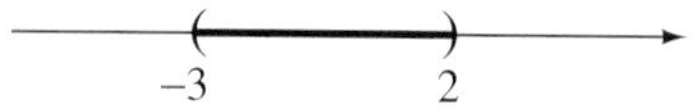 (number line: −3, 2)

38.  (number line: −1, 4)

39. All reals

40. 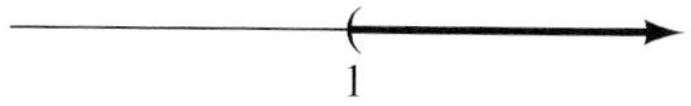 (number line: 1)

41. 24 **42.** 7 **43.** 33 **44.** 8 **45.** 89 or better
46. 16 and 24 **47.** 18 **48.** 88 or better
49. 8 nickels and 22 dimes **50.** 8 nickels, 25 dimes, and 50 quarters **51.** 52° **52.** 700 miles

Chapter 3 Test (page 143)

1. $\{2\}$ **2.** $\{3\}$ **3.** $\{-9\}$ **4.** $\{-5\}$ **5.** $\{-53\}$ **6.** $\{-18\}$
7. $\left\{-\frac{5}{2}\right\}$ **8.** $\left\{\frac{35}{18}\right\}$ **9.** $\{12\}$ **10.** $\left\{\frac{11}{5}\right\}$ **11.** $\{22\}$
12. $\left\{\frac{31}{2}\right\}$ **13.** $\{x|x < 5\}$ or $(-\infty, 5)$
14. $\{x|x \le 1\}$ or $(-\infty, 1]$ **15.** $\{x|x \ge -9\}$ or $[-9, \infty)$
16. $\{x|x < 0\}$ or $(-\infty, 0)$
17. $\left\{x|x > -\frac{23}{2}\right\}$ or $\left(-\frac{23}{2}, \infty\right)$
18. $\{n|n \ge 12\}$ or $[12, \infty)$

19. 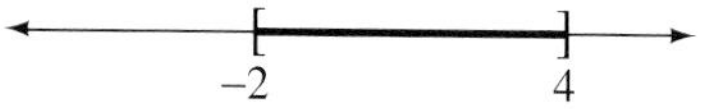 (number line: −2, 4)

20. 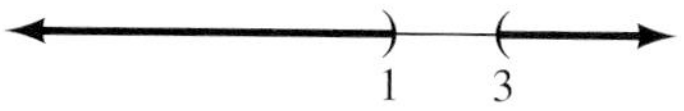 (number line: 1, 3)

21. \$0.35
22. 15 meters, 25 meters, and 30 meters **23.** 96 or better
24. 17 nickels, 33 dimes, and 53 quarters
25. 60°, 30°, 90°

CHAPTER 4

Problem Set 4.1 (page 151)

1. $\{9\}$ **3.** $\{10\}$ **5.** $\left\{\frac{15}{2}\right\}$ **7.** $\{-22\}$ **9.** $\{-4\}$ **11.** $\{6\}$
13. $\{-28\}$ **15.** $\{34\}$ **17.** $\{6\}$ **19.** $\left\{-\frac{8}{5}\right\}$ **21.** $\{7\}$
23. $\left\{\frac{9}{2}\right\}$ **25.** $\left\{-\frac{53}{2}\right\}$ **27.** $\{50\}$ **29.** $\{120\}$ **31.** $\left\{\frac{9}{7}\right\}$
33. 55% **35.** 60% **37.** $16\frac{2}{3}\%$ **39.** $37\frac{1}{2}\%$ **41.** 150%
43. 240% **45.** 2.66 **47.** 42 **49.** 80% **51.** 60
53. 115% **55.** 90 **57.** 15 feet by $19\frac{1}{2}$ feet
59. 330 miles **61.** 60 centimeters **63.** 7.5 pounds
65. $33\frac{1}{3}$ pounds **67.** 90,000 **69.** 137.5 grams **71.** \$300
73. \$150,000 **77.** All real numbers except 2 **79.** $\{0\}$
81. All real numbers

Problem Set 4.2 (page 159)

1. $\{1.11\}$ **3.** $\{6.6\}$ **5.** $\{0.48\}$ **7.** $\{80\}$ **9.** $\{3\}$
11. $\{50\}$ **13.** $\{70\}$ **15.** $\{200\}$ **17.** $\{450\}$ **19.** $\{150\}$
21. $\{2200\}$ **23.** \$50 **25.** \$3600 **27.** \$20.80 **29.** 30%
31. \$8.50 **33.** \$12.40 **35.** \$1000 **37.** 40% **39.** 8%
41. \$4166.67 **43.** \$1900 **45.** \$785.42
49. Yes, if the profit is figured as a percent of the selling price. **51.** Yes **53.** $\{1.625\}$ **55.** $\{350\}$ **57.** $\{0.06\}$
59. $\{15.4\}$

Problem Set 4.3 (page 167)

1. 7 **3.** 500 **5.** 20 **7.** 48 **9.** 9 **11.** 46 centimeters
13. 15 inches **15.** 504 square feet **17.** \$8 **19.** 7 inches
21. 150π square centimeters **23.** $\frac{1}{4}\pi$ square yards
25. $S = 324\pi$ square inches and $V = 972\pi$ cubic inches
27. $V = 1152\pi$ cubic feet and $S = 416\pi$ square feet
29. 12 inches **31.** 8 feet **33.** *E* **35.** *B* **37.** *A* **39.** *C*
41. *F* **43.** $h = \frac{V}{B}$ **45.** $B = \frac{3V}{h}$ **47.** $w = \frac{P - 2l}{2}$
49. $h = \frac{3V}{\pi r^2}$ **51.** $C = \frac{5}{9}(F - 32)$ **53.** $h = \frac{A - 2\pi r^2}{2\pi r}$
55. $x = \frac{9 - 7y}{3}$ **57.** $y = \frac{9x - 13}{6}$ **59.** $x = \frac{11y - 14}{2}$
61. $x = \frac{-y - 4}{3}$ **63.** $y = \frac{3}{2}x$ **65.** $y = \frac{ax - c}{b}$
67. $x = \frac{2y - 22}{5}$ **69.** $y = mx + b$
75. 125.6 square centimeters **77.** 245 square centimeters
79. 65 cubic inches

Problem Set 4.4 (page 174)

1. $\left\{8\frac{1}{3}\right\}$ **3.** $\{16\}$ **5.** $\{25\}$ **7.** $\{7\}$ **9.** $\{24\}$
11. $\{4\}$ **13.** $12\frac{1}{2}$ years **15.** $33\frac{1}{3}$ years
17. The width is 14 inches and the length is 42 inches.
19. The width is 12 centimeters and the length is 34 centimeters. **21.** 80 square inches
23. 24 feet, 31 feet, and 45 feet

25. 6 centimeters, 19 centimeters, 21 centimeters
27. 12 centimeters **29.** 7 centimeters
31. 9 hours **33.** $2\frac{1}{2}$ hours **35.** 55 miles per hour
37. 64 and 72 miles per hour **39.** 60 miles

Problem Set 4.5 (page 181)

1. $\{15\}$ **3.** $\left\{\frac{20}{7}\right\}$ **5.** $\left\{\frac{15}{4}\right\}$ **7.** $\left\{\frac{5}{3}\right\}$ **9.** $\{2\}$
11. $\left\{\frac{33}{10}\right\}$ **13.** 12.5 milliliters **15.** 15 centiliters
17. $7\frac{1}{2}$ quarts of the 30% solution and $2\frac{1}{2}$ quarts of the 50% solution **19.** 5 gallons **21.** 3 quarts
23. 12 gallons **25.** 16.25%
27. The square is 6 inches by 6 inches and the rectangle is 9 inches long and 3 inches wide.
29. 40 minutes **31.** Pam is 9 and Bill is 18.
33. \$5000 at 6%; \$7000 at 8%
35. \$500 at 9%; \$1000 at 10%; \$1500 at 11%
37. \$900 at 3%; \$2150 at 5% **39.** \$6000
41. \$2166.67 at 5%; \$3833.33 at 7%

Chapter 4 Review Problem Set (page 183)

1. $\left\{\frac{17}{12}\right\}$ **2.** $\{5\}$ **3.** $\{800\}$ **4.** $\{16\}$ **5.** $\{73\}$ **6.** 6
7. 25 **8.** $t = \frac{A - P}{Pr}$ **9.** $x = \frac{13 + 3y}{2}$
10. 77 square inches **11.** 6 centimeters **12.** 15 feet
13. 60% **14.** 40 and 56 **15.** 40
16. 6 meters by 17 meters **17.** $1\frac{1}{2}$ hours **18.** 20 liters
19. 15 centimeters by 40 centimeters
20. 29 yards by 10 yards **21.** 20° **22.** 30 gallons
23. \$675 at 3%; \$1425 at 5% **24.** \$40 **25.** 35%
26. 34° and 99° **27.** 5 hours **28.** 18 gallons **29.** 26%
30. \$367.50

Chapter 4 Test (page 185)

1. $\{-22\}$ **2.** $\left\{-\frac{17}{18}\right\}$ **3.** $\{-77\}$ **4.** $\left\{\frac{4}{3}\right\}$ **5.** $\{14\}$
6. $\left\{\frac{12}{5}\right\}$ **7.** $\{100\}$ **8.** $\{70\}$ **9.** $\{250\}$ **10.** $\left\{\frac{11}{2}\right\}$
11. $C = \frac{5F - 160}{9}$ **12.** $x = \frac{y + 8}{2}$ **13.** $y = \frac{9x + 47}{4}$
14. 64π square centimeters **15.** 576 square inches
16. 14 yards **17.** 125% **18.** 70 **19.** \$350 **20.** \$52
21. 40% **22.** 875 women **23.** 10 hours
24. 4 centiliters **25.** 11.1 years

Cumulative Review Problem Set (page 186)

1. $-16x$ **2.** $4a - 6$ **3.** $12x + 27$ **4.** $-5x + 1$
5. $9n - 8$ **6.** $14n - 5$ **7.** $\frac{1}{4}x$ **8.** $-\frac{1}{10}n$ **9.** $-0.1x$
10. $0.7x + 0.2$ **11.** -65 **12.** -51 **13.** 20 **14.** 32
15. $\frac{7}{8}$ **16.** $-\frac{5}{6}$ **17.** -0.28 **18.** $-\frac{1}{4}$ **19.** 5 **20.** $-\frac{1}{4}$
21. 81 **22.** -64 **23.** $\frac{8}{27}$ **24.** $-\frac{1}{32}$ **25.** $\frac{25}{36}$
26. $-\frac{1}{512}$ **27.** $\{-4\}$ **28.** $\{-2\}$ **29.** $\varnothing$ **30.** $\{-8\}$
31. $\left\{\frac{25}{2}\right\}$ **32.** $\left\{-\frac{4}{7}\right\}$ **33.** $\left\{\frac{34}{3}\right\}$ **34.** $\{200\}$
35. {All reals} **36.** $\{11\}$ **37.** $\left\{\frac{3}{2}\right\}$ **38.** $\{0\}$
39. $\{x|x > 7\}$ or $(7, \infty)$ **40.** $\{x|x > -6\}$ or $(-6, \infty)$
41. $\left\{n|n \geq \frac{7}{5}\right\}$ or $\left[\frac{7}{5}, \infty\right)$ **42.** $\{x|x \geq 21\}$ or $[21, \infty)$
43. $\{t|t < 100\}$ or $(-\infty, 100)$
44. $\{x|x < -1\}$ or $(-\infty, -1)$
45. $\{n|n \geq 18\}$ or $[18, \infty)$ **46.** $\left\{x|x < \frac{5}{3}\right\}$ or $\left(-\infty, \frac{5}{3}\right)$
47. \$15,000 **48.** 45° and 135°
49. 8 nickels and 17 dimes **50.** 130 or higher
51. 12 feet and 18 feet **52.** \$40 per pair
53. 45 miles per hour and 50 miles per hour **54.** 5 liters

CHAPTER 5

Problem Set 5.1 (page 193)

1. 3 **3.** 2 **5.** 3 **7.** 2 **9.** $8x + 11$ **11.** $4y + 10$
13. $2x^2 - 2x - 23$ **15.** $17x - 19$ **17.** $6x^2 - 5x - 4$
19. $5n - 6$ **21.** $-7x^2 - 13x - 7$ **23.** $5x + 5$
25. $-2x - 5$ **27.** $-3x + 7$ **29.** $2x^2 + 15x - 6$
31. $5n^2 + 2n + 3$ **33.** $-3x^3 + 2x^2 - 11$ **35.** $9x - 2$
37. $2a + 15$ **39.** $-2x^2 + 5$ **41.** $6x^3 + 12x^2 - 5$
43. $4x^3 - 8x^2 + 13x + 12$ **45.** $x + 11$ **47.** $-3x - 14$
49. $-x^2 - 13x - 12$ **51.** $x^2 - 11x - 8$ **53.** $-10a - 3b$
55. $-n^2 + 2n - 17$ **57.** $8x + 1$ **59.** $-5n - 1$
61. $-2a + 6$ **63.** $11x + 6$ **65.** $6x + 7$ **67.** $-5n + 7$
69. $8x + 6$ **71.** $20x^2$

Problem Set 5.2 (page 200)

1. $45x^2$ **3.** $21x^3$ **5.** $-6x^2y^2$ **7.** $14x^3y$ **9.** $-48a^3b^3$
11. $5x^4y$ **13.** $104a^3b^2c^2$ **15.** $30x^6$ **17.** $-56x^2y^3$

19. $-6a^2b^3$ **21.** $72c^3d^3$ **23.** $\frac{2}{5}x^3y^5$ **25.** $-\frac{2}{9}a^2b^5$
27. $0.28x^8$ **29.** $-6.4a^4b^2$ **31.** $4x^8$ **33.** $9a^4b^6$ **35.** $27x^6$
37. $-64x^{12}$ **39.** $81x^8y^{10}$ **41.** $16x^8y^4$ **43.** $81a^{12}b^8$
45. $x^{12}y^6$ **47.** $15x^2 + 10x$ **49.** $18x^3 - 6x^2$
51. $-28x^3 + 16x$ **53.** $2x^3 - 8x^2 + 12x$
55. $-18a^3 + 30a^2 + 42a$ **57.** $28x^3y - 7x^2y + 35xy$
59. $-9x^3y + 2x^2y + 6xy$ **61.** $13x + 22y$
63. $-2x - 9y$ **65.** $4x^3 - 3x^2 - 14x$ **67.** $-x + 14$
69. $-7x + 12$ **71.** $18x^5$ **73.** $-432x^5$ **75.** $25x^7y^8$
77. $-a^{12}b^5c^9$ **79.** $-16x^{11}y^{17}$ **81.** $7x + 5$ **83.** $3\pi x^2$
89. x^{7n} **91.** x^{6n+1} **93.** x^{6n+3} **95.** $-20x^{10n}$ **97.** $12x^{7n}$

Problem Set 5.3 (page 207)

1. $xy + 3x + 2y + 6$ **3.** $xy + x - 4y - 4$
5. $xy - 6x - 5y + 30$ **7.** $xy + xz + x + 2y + 2z + 2$
9. $6xy + 2x + 9y + 3$ **11.** $x^2 + 10x + 21$
13. $x^2 + 5x - 24$ **15.** $x^2 - 6x - 7$ **17.** $n^2 - 10n + 24$
19. $3n^2 + 19n + 6$ **21.** $15x^2 + 29x - 14$
23. $x^3 + 7x^2 + 21x + 27$ **25.** $x^3 + 3x^2 - 10x - 24$
27. $2x^3 - 7x^2 - 22x + 35$ **29.** $8a^3 - 14a^2 + 23a - 9$
31. $3a^3 + 2a^2 - 8a - 5$
33. $x^4 + 7x^3 + 17x^2 + 23x + 12$
35. $x^4 - 3x^3 - 34x^2 + 33x + 63$ **37.** $x^2 + 11x + 18$
39. $x^2 + 4x - 12$ **41.** $x^2 - 8x - 33$ **43.** $n^2 - 7n + 12$
45. $n^2 + 18n + 72$ **47.** $y^2 - 4y - 21$
49. $y^2 - 19y + 84$ **51.** $x^2 + 2x - 35$
53. $x^2 - 6x - 112$ **55.** $a^2 + a - 90$ **57.** $2a^2 + 13a + 6$
59. $5x^2 + 33x - 14$ **61.** $6x^2 - 11x - 7$
63. $12a^2 - 7a - 12$ **65.** $12n^2 - 28n + 15$
67. $14x^2 + 13x - 12$ **69.** $45 - 19x + 2x^2$
71. $-8x^2 + 22x - 15$ **73.** $-9x^2 + 9x + 4$
75. $72n^2 - 5n - 12$ **77.** $27 - 21x + 2x^2$
79. $20x^2 - 7x - 6$ **81.** **(a)** $x^2 + 8x + 16$
(b) $x^2 - 10x + 25$ **(c)** $x^2 - 12x + 36$
(d) $9x^2 + 12x + 4$ **(e)** $x^2 + 2x + 1$ **(f)** $x^2 - 6x + 9$
83. $x^2 + 14x + 49$ **85.** $25x^2 - 4$ **87.** $x^2 - 2x + 1$
89. $9x^2 + 42x + 49$ **91.** $4x^2 - 12x + 9$ **93.** $4x^2 - 9y^2$
95. $1 - 10n + 25n^2$ **97.** $9x^2 + 24xy + 16y^2$
99. $9 + 24y + 16y^2$ **101.** $1 - 49n^2$
103. $16a^2 - 56ab + 49b^2$ **105.** $x^2 + 16xy + 64y^2$
107. $25x^2 - 121y^2$ **109.** $64x^3 - x$ **111.** $-32x^3 + 2xy^2$
113. $x^3 + 6x^2 + 12x + 8$ **115.** $x^3 - 9x^2 + 27x - 27$
117. $8n^3 + 12n^2 + 6n + 1$ **119.** $27n^3 - 54n^2 + 36n - 8$
123. $4x^3 - 56x^2 + 196x$; $196 - 4x^2$

Problem Set 5.4 (page 212)

1. x^8 **3.** $2x^2$ **5.** $-8n^4$ **7.** -8 **9.** $13xy^2$ **11.** $7ab^2$
13. $18xy^4$ **15.** $-32x^5y^2$ **17.** $-8x^5y^4$ **19.** -1
21. $14ab^2c^4$ **23.** $16yz^4$ **25.** $4x^2 + 6x^3$ **27.** $3x^3 - 8x$
29. $-7n^3 + 9$ **31.** $5x^4 - 8x^3 - 12x$ **33.** $4n^5 - 8n^2 + 13$
35. $5a^6 + 8a^2$ **37.** $-3xy + 5y$ **39.** $-8ab - 10a^2b^3$
41. $-3bc + 13b^2c^4$ **43.** $-9xy^2 + 12x^2y^3$
45. $-3x^4 - 5x^2 + 7$ **47.** $-3a^2 + 7a + 13b$
49. $-1 + 5xy^2 - 7xy^5$

Problem Set 5.5 (page 217)

1. $x + 12$ **3.** $x + 2$ **5.** $x + 8$ with a remainder of 4
7. $x + 4$ with a remainder of -7 **9.** $5n + 4$
11. $8y - 3$ with a remainder of 2 **13.** $4x - 7$
15. $3x + 2$ with a remainder of -6 **17.** $2x^2 + 3x + 4$
19. $5n^2 - 4n - 3$ **21.** $n^2 + 6n - 4$ **23.** $x^2 + 3x + 9$
25. $9x^2 + 12x + 16$ **27.** $3n - 8$ with a remainder of 17
29. $3t + 2$ with a remainder of 6 **31.** $3n^2 - n - 4$
33. $4x^2 - 5x + 5$ with a remainder of -3
35. $x + 4$ with a remainder of $5x - 1$
37. $2x - 12$ with a remainder of $49x - 5$
39. $x^3 - 2x^2 + 4x - 8$

Problem Set 5.6 (page 224)

1. $\frac{1}{9}$ **3.** $\frac{1}{64}$ **5.** $\frac{2}{3}$ **7.** 16 **9.** 1 **11.** $-\frac{27}{8}$ **13.** $\frac{1}{4}$
15. $-\frac{1}{9}$ **17.** $\frac{27}{64}$ **19.** $\frac{1}{8}$ **21.** 27 **23.** 1000
25. $\frac{1}{1000}$ or 0.001 **27.** 18 **29.** 144 **31.** x^5 **33.** $\frac{1}{n^2}$
35. $\frac{1}{a^5}$ **37.** $8x$ **39.** $\frac{27}{x^4}$ **41.** $-\frac{15}{y^3}$ **43.** 96 **45.** x^{10}
47. $\frac{1}{n^4}$ **49.** $2n^2$ **51.** $-\frac{3}{x^4}$ **53.** 4 **55.** x^6 **57.** $\frac{1}{x^4}$
59. $\frac{1}{x^3y^4}$ **61.** $\frac{1}{x^6y^3}$ **63.** $\frac{8}{n^6}$ **65.** $\frac{1}{16n^6}$ **67.** $\frac{81}{a^8}$ **69.** $\frac{x^2}{25}$
71. $\frac{x^2y}{2}$ **73.** $\frac{y}{x^2}$ **75.** a^4b^8 **77.** $\frac{x^2}{y^6}$ **79.** x **81.** $\frac{1}{8x^3}$
83. $\frac{x^4}{4}$ **85.** $4.2(10^{10})$ **87.** $4.3(10^{-3})$ **89.** $8.9(10^4)$
91. $2.5(10^{-2})$ **93.** $1.1(10^7)$ **95.** 8000 **97.** 52,100
99. 11,400,000 **101.** 0.07 **103.** 0.000987
105. 0.00000864 **107.** 0.84 **109.** 450 **111.** 4,000,000
113. 0.0000002 **115.** 0.3 **117.** 0.000007 **119.** \$875
121. 358 feet

Chapter 5 Review Problem Set (page 227)

1. $8x^2 - 13x + 2$ **2.** $3y^2 + 11y - 9$ **3.** $3x^2 + 2x - 9$
4. $-8x^2 + 18$ **5.** $11x + 8$ **6.** $-9x^2 + 8x - 20$
7. $2y^2 - 54y + 18$ **8.** $-13a - 30$ **9.** $-27a - 7$
10. $n - 2$ **11.** $-5n^2 - 2n$ **12.** $17n^2 - 14n - 16$

13. $35x^6$ **14.** $-54x^8$ **15.** $24x^3y^5$ **16.** $-6a^4b^9$
17. $8a^6b^9$ **18.** $9x^2y^4$ **19.** $35x^2 + 15x$ **20.** $-24x^3 + 3x^2$
21. $x^2 + 17x + 72$ **22.** $3x^2 + 10x + 7$
23. $x^2 - 3x - 10$ **24.** $y^2 - 13y + 36$ **25.** $14x^2 - x - 3$
26. $20a^2 - 3a - 56$ **27.** $9a^2 - 30a + 25$
28. $2x^3 + 17x^2 + 26x - 24$ **29.** $30n^2 + 19n - 5$
30. $12n^2 + 13n - 4$ **31.** $4n^2 - 1$ **32.** $16n^2 - 25$
33. $4a^2 + 28a + 49$ **34.** $9a^2 + 30a + 25$
35. $x^3 - 3x^2 + 8x - 12$ **36.** $2x^3 + 7x^2 + 10x - 7$
37. $a^3 + 15a^2 + 75a + 125$ **38.** $a^3 - 18a^2 + 108a - 216$
39. $x^4 + x^3 + 2x^2 - 7x - 5$
40. $n^4 - 5n^3 - 11n^2 - 30n - 4$ **41.** $-12x^3y^3$
42. $7a^3b^4$ **43.** $-3x^2y - 9x^4$ **44.** $10a^4b^9 - 13a^3b^7$
45. $14x^2 - 10x - 8$ **46.** $x + 4, R = -21$ **47.** $7x - 6$
48. $2x^2 + x + 4, R = 4$ **49.** 13 **50.** 25 **51.** $\frac{1}{16}$
52. 1 **53.** -1 **54.** 9 **55.** $\frac{16}{9}$ **56.** $\frac{1}{4}$ **57.** -8 **58.** $\frac{11}{18}$
59. $\frac{5}{4}$ **60.** $\frac{1}{25}$ **61.** $\frac{1}{x^3}$ **62.** $12x^3$ **63.** x^2 **64.** $\frac{1}{x^2}$
65. $8a^6$ **66.** $\frac{4}{n}$ **67.** $\frac{x^2}{y}$ **68.** $\frac{b^6}{a^4}$ **69.** $\frac{1}{2x}$ **70.** $\frac{1}{9n^4}$
71. $\frac{n^3}{8}$ **72.** $-12b$ **73.** 610 **74.** 56,000 **75.** 0.08
76. 0.00092 **77.** $(9)(10^3)$ **78.** $(4.7)(10)$ **79.** $(4.7)(10^{-2})$
80. $(2.1)(10^{-4})$ **81.** 0.48 **82.** 4.2 **83.** 2000
84. 0.00000002

Chapter 5 Test (page 229)

1. $-2x^2 - 2x + 5$ **2.** $-3x^2 - 6x + 20$ **3.** $-13x + 2$
4. $-28x^3y^5$ **5.** $12x^5y^5$ **6.** $x^2 - 7x - 18$
7. $n^2 + 7n - 98$ **8.** $40a^2 + 59a + 21$
9. $9x^2 - 42xy + 49y^2$ **10.** $2x^3 + 2x^2 - 19x - 21$
11. $81x^2 - 25y^2$ **12.** $15x^2 - 68x + 77$ **13.** $8x^2y^4$
14. $-7x + 9y$ **15.** $x^2 + 4x - 5$ **16.** $4x^2 - x + 6$
17. $\frac{27}{8}$ **18.** $1\frac{5}{16}$ **19.** 16 **20.** $-\frac{24}{x^2}$ **21.** $\frac{x^3}{4}$ **22.** $\frac{x^6}{y^{10}}$
23. $(2.7)(10^{-4})$ **24.** 9,200,000 **25.** 0.006

Cumulative Review Problem Set (page 230)

1. 130 **2.** -1 **3.** 27 **4.** -16 **5.** 81 **6.** -32 **7.** $\frac{3}{2}$
8. 16 **9.** 36 **10.** $1\frac{3}{4}$ **11.** 0 **12.** $\frac{13}{40}$ **13.** $-\frac{2}{13}$
14. 5 **15.** -1 **16.** -33 **17.** $-15x^3y^7$ **18.** $12ab^7$
19. $-8x^6y^{15}$ **20.** $-6x^2y + 15xy^2$ **21.** $15x^2 - 11x + 2$
22. $21x^2 + 25x - 4$ **23.** $-2x^2 - 7x - 6$ **24.** $49 - 4y^2$
25. $3x^3 - 7x^2 - 2x + 8$ **26.** $2x^3 - 3x^2 - 13x + 20$
27. $8n^3 + 36n^2 + 54n + 27$ **28.** $1 - 6n + 12n^2 - 8n^3$
29. $2x^4 + x^3 - 4x^2 + 42x - 36$ **30.** $-4x^2y^2$
31. $14ab^2$ **32.** $7y - 8x^2 - 9x^3y^3$ **33.** $2x^2 - 4x - 7$
34. $x^2 + 6x + 4$ **35.** $-\frac{6}{x}$ **36.** $\frac{2}{x}$ **37.** $\frac{xy^2}{3}$ **38.** $\frac{z^2}{x^2y^4}$
39. 0.12 **40.** 0.0000000018 **41.** 200 **42.** $\{11\}$
43. $\{-1\}$ **44.** $\{48\}$ **45.** $\left\{-\frac{3}{7}\right\}$ **46.** $\{9\}$ **47.** $\{13\}$
48. $\left\{\frac{9}{14}\right\}$ **49.** $\{500\}$ **50.** $\{x|x \le -1\}$ or $(-\infty, -1]$
51. $\{x|x > 0\}$ or $(0, \infty)$ **52.** $\left\{x|x < \frac{4}{5}\right\}$ or $\left(-\infty, \frac{4}{5}\right)$
53. $\{x|x < -2\}$ or $(-\infty, -2)$
54. $\left\{x|x \ge \frac{12}{7}\right\}$ or $\left[\frac{12}{7}, \infty\right)$
55. $\{x|x \ge 300\}$ or $[300, \infty)$ **56.** 3 **57.** 40
58. 8 dimes and 10 quarters
59. \$700 at 8% and \$800 at 9% **60.** 3 gallons
61. $3\frac{1}{2}$ hours
62. The length is 15 meters and the width is 7 meters.

CHAPTER 6

Problem Set 6.1 (page 238)

1. $6y$ **3.** $12xy$ **5.** $14ab^2$ **7.** $2x$ **9.** $8a^2b^2$
11. $4(2x + 3y)$ **13.** $7y(2x - 3)$ **15.** $9x(2x + 5)$
17. $6xy(2y - 5x)$ **19.** $12a^2b(3 - 5ab^3)$
21. $xy^2(16y + 25x)$ **23.** $8(8ab - 9cd)$ **25.** $9a^2b(b^3 - 3)$
27. $4x^4y(13y + 15x^2)$ **29.** $8x^2y(5y + 1)$
31. $3x(4 + 5y + 7x)$ **33.** $x(2x^2 - 3x + 4)$
35. $4y^2(11y^3 - 6y - 5)$ **37.** $7ab(2ab^2 + 5b - 7a^2)$
39. $(y + 1)(x + z)$ **41.** $(b - 4)(a - c)$
43. $(x + 3)(x + 6)$ **45.** $(x + 1)(2x - 3)$
47. $(x + y)(5 + b)$ **49.** $(x - y)(b - c)$
51. $(a + b)(c + 1)$ **53.** $(x + 5)(x + 12)$
55. $(x - 2)(x - 8)$ **57.** $(2x + 1)(x - 5)$
59. $(2n - 1)(3n - 4)$ **61.** $\{0, 8\}$ **63.** $\{-1, 0\}$
65. $\{0, 5\}$ **67.** $\left\{0, \frac{3}{2}\right\}$ **69.** $\left\{-\frac{3}{7}, 0\right\}$ **71.** $\{-5, 0\}$
73. $\left\{0, \frac{3}{2}\right\}$ **75.** $\{0, 7\}$ **77.** $\{0, 13\}$ **79.** $\left\{-\frac{5}{2}, 0\right\}$
81. $\{-5, 4\}$ **83.** $\{4, 6\}$ **85.** 0 or 9 **87.** 20 units **89.** $\frac{4}{\pi}$
91. The square is 3 inches by 3 inches and the rectangle is 3 inches by 6 inches. **95.** **(a)** \$116 **(c)** \$750
97. $x = 0$ or $x = \frac{c}{b^2}$ **99.** $y = \frac{c}{1 + a - b}$

Problem Set 6.2 (page 244)

1. $(x-1)(x+1)$ **3.** $(x-10)(x+10)$
5. $(x-2y)(x+2y)$ **7.** $(3x-y)(3x+y)$
9. $(6a-5b)(6a+5b)$ **11.** $(1-2n)(1+2n)$
13. $5(x-2)(x+2)$ **15.** $8(x^2+4)$
17. $2(x-3y)(x+3y)$ **19.** $x(x-5)(x+5)$
21. Not factorable **23.** $9x(5x-4y)$
25. $4(3-x)(3+x)$ **27.** $4a^2(a^2+4)$
29. $(x-3)(x+3)(x^2+9)$ **31.** $x^2(x^2+1)$
33. $3x(x^2+16)$ **35.** $5x(1-2x)(1+2x)$
37. $4(x-4)(x+4)$ **39.** $3xy(5x-2y)(5x+2y)$
41. $\{-3, 3\}$ **43.** $\{-2, 2\}$ **45.** $\left\{-\frac{4}{3}, \frac{4}{3}\right\}$ **47.** $\{-11, 11\}$
49. $\left\{-\frac{2}{5}, \frac{2}{5}\right\}$ **51.** $\{-5, 5\}$ **53.** $\{-4, 0, 4\}$
55. $\{-4, 0, 4\}$ **57.** $\left\{-\frac{1}{3}, \frac{1}{3}\right\}$ **59.** $\{-10, 0, 10\}$
61. $\left\{-\frac{9}{8}, \frac{9}{8}\right\}$ **63.** $\left\{-\frac{1}{2}, 0, \frac{1}{2}\right\}$ **65.** -7 or 7
67. $-4, 0,$ or 4 **69.** 3 inches and 15 inches
71. The length is 20 centimeters and the width is 8 centimeters. **73.** 4 meters and 8 meters
75. 5 centimeters **81.** $(x-2)(x^2+2x+4)$
83. $(n+4)(n^2-4n+16)$
85. $(3a-4b)(9a^2+12ab+16b^2)$
87. $(1+3a)(1-3a+9a^2)$
89. $(2x-y)(4x^2+2xy+y^2)$
91. $(3x-2y)(9x^2+6xy+4y^2)$
93. $(5x+2y)(25x^2-10xy+4y^2)$
95. $(4+x)(16-4x+x^2)$

Problem Set 6.3 (page 253)

1. $(x+4)(x+6)$ **3.** $(x+5)(x+8)$ **5.** $(x-2)(x-9)$
7. $(n-7)(n-4)$ **9.** $(n+9)(n-3)$
11. $(n-10)(n+4)$ **13.** Not factorable
15. $(x-6)(x-12)$ **17.** $(x+11)(x-6)$
19. $(y-9)(y+8)$ **21.** $(x+5)(x+16)$
23. $(x+12)(x-6)$ **25.** Not factorable
27. $(x-2y)(x+5y)$ **29.** $(a-8b)(a+4b)$
31. $\{-7, -3\}$ **33.** $\{3, 6\}$ **35.** $\{-2, 5\}$ **37.** $\{-9, 4\}$
39. $\{-4, 10\}$ **41.** $\{-8, 7\}$ **43.** $\{2, 14\}$ **45.** $\{-12, 1\}$
47. $\{2, 8\}$ **49.** $\{-6, 4\}$ **51.** 7 and 8 or -7 and -8
53. 12 and 14 **55.** $-4, -3, -2,$ and -1 or 7, 8, 9, and 10
57. 4 and 7 or 0 and 3
59. The length is 9 inches and the width is 6 inches.
61. 9 centimeters by 6 centimeters **63.** 7 rows
65. 8 feet, 15 feet, and 17 feet **67.** 6 inches and 8 inches
73. $(x^a+8)(x^a+5)$ **75.** $(x^a+9)(x^a-3)$

Problem Set 6.4 (page 259)

1. $(3x+1)(x+2)$ **3.** $(2x+5)(3x+2)$
5. $(4x-1)(x-6)$ **7.** $(4x-5)(3x-4)$
9. $(5y+2)(y-7)$ **11.** $2(2n-3)(n+8)$
13. Not factorable **15.** $(3x+7)(6x+1)$
17. $(7x-2)(x-4)$ **19.** $(4x+7)(2x-3)$
21. $t(3t+2)(3t-7)$ **23.** $(12y-5)(y+7)$
25. Not factorable **27.** $(7x+3)(2x+7)$
29. $(4x-3)(5x-4)$ **31.** $(4n+3)(4n-5)$
33. $(6x-5)(4x-5)$ **35.** $(2x+9)(x+8)$
37. $(3a+1)(7a-2)$ **39.** $(12a+5)(a-3)$
41. $3(2x+3)(2x+3)$ **43.** $(2x-y)(3x-y)$
45. $(4x+3y)(5x-2y)$ **47.** $(5x-2)(x-6)$
49. $(8x+1)(x-7)$ **51.** $\left\{-6, -\frac{1}{2}\right\}$
53. $\left\{-\frac{2}{3}, -\frac{1}{4}\right\}$ **55.** $\left\{\frac{1}{3}, 8\right\}$ **57.** $\left\{\frac{2}{5}, \frac{7}{3}\right\}$
59. $\left\{-7, \frac{5}{6}\right\}$ **61.** $\left\{-\frac{3}{8}, \frac{3}{2}\right\}$ **63.** $\left\{-\frac{2}{3}, \frac{4}{3}\right\}$
65. $\left\{\frac{2}{5}, \frac{5}{2}\right\}$ **67.** $\left\{-\frac{5}{2}, -\frac{2}{3}\right\}$ **69.** $\left\{-\frac{5}{4}, \frac{1}{4}\right\}$
71. $\left\{-\frac{3}{7}, \frac{7}{5}\right\}$ **73.** $\left\{\frac{5}{4}, 10\right\}$ **75.** $\left\{-7, \frac{3}{7}\right\}$
77. $\left\{-\frac{5}{12}, 4\right\}$ **79.** $\left\{-\frac{7}{2}, \frac{4}{9}\right\}$

Problem Set 6.5 (page 268)

1. $(x+2)^2$ **3.** $(x-5)^2$ **5.** $(3n+2)^2$ **7.** $(4a-1)^2$
9. $(2+9x)^2$ **11.** $(4x-3y)^2$ **13.** $(2x+1)(x+8)$
15. $2x(x-6)(x+6)$ **17.** $(n-12)(n+5)$
19. Not factorable **21.** $8(x^2+9)$ **23.** $(3x+5)^2$
25. $5(x+2)(3x+7)$ **27.** $(4x-3)(6x+5)$
29. $(x+5)(y-8)$ **31.** $(5x-y)(4x+7y)$
33. $3(2x-3)(4x+9)$ **35.** $6(2x^2+x+5)$
37. $5(x-2)(x+2)(x^2+4)$ **39.** $(x+6y)^2$ **41.** $\{0, 5\}$
43. $\{-3, 12\}$ **45.** $\{-2, 0, 2\}$ **47.** $\left\{-\frac{2}{3}, \frac{11}{2}\right\}$
49. $\left\{\frac{1}{3}, \frac{3}{4}\right\}$ **51.** $\{-3, -1\}$ **53.** $\{0, 6\}$ **55.** $\{-4, 0, 6\}$
57. $\left\{\frac{2}{5}, \frac{6}{5}\right\}$ **59.** $\{12, 16\}$ **61.** $\left\{-\frac{10}{3}, 1\right\}$ **63.** $\{0, 6\}$
65. $\left\{\frac{4}{3}\right\}$ **67.** $\{-5, 0\}$ **69.** $\left\{-\frac{4}{3}, \frac{5}{8}\right\}$
71. $\frac{5}{4}$ and 12 or -5 and -3 **73.** -1 and 1 or $-\frac{1}{2}$ and 2
75. 4 and 9 or $-\frac{24}{5}$ and $-\frac{43}{5}$

77. 6 rows and 9 chairs per row
79. One square is 6 feet by 6 feet and the other one is 18 feet by 18 feet.
81. 11 centimeters long and 5 centimeters wide
83. The side is 17 inches long and the altitude to that side is 6 inches long.
85. $1\frac{1}{2}$ inches **87.** 6 inches and 12 inches

Chapter 6 Review Problem Set (page 271)

1. $(x-2)(x-7)$ **2.** $3x(x+7)$
3. $(3x+2)(3x-2)$ **4.** $(2x-1)(2x+5)$ **5.** $(5x-6)^2$
6. $n(n+5)(n+8)$ **7.** $(y+12)(y-1)$ **8.** $3xy(y+2x)$
9. $(x+1)(x-1)(x^2+1)$ **10.** $(6n+5)(3n-1)$
11. Not factorable **12.** $(4x-7)(x+1)$
13. $3(n+6)(n-5)$ **14.** $x(x+y)(x-y)$
15. $(2x-y)(x+2y)$ **16.** $2(n-4)(2n+5)$
17. $(x+y)(5+a)$ **18.** $(7t-4)(3t+1)$
19. $2x(x+1)(x-1)$ **20.** $3x(x+6)(x-6)$
21. $(4x+5)^2$ **22.** $(y-3)(x-2)$
23. $(5x+y)(3x-2y)$ **24.** $n^2(2n-1)(3n-1)$
25. $\{-6, 2\}$ **26.** $\{0, 11\}$ **27.** $\left\{-4, \frac{5}{2}\right\}$ **28.** $\left\{-\frac{8}{3}, \frac{1}{3}\right\}$
29. $\{-2, 2\}$ **30.** $\left\{-\frac{5}{4}\right\}$ **31.** $\{-1, 0, 1\}$ **32.** $\left\{-\frac{2}{7}, -\frac{9}{4}\right\}$
33. $\{-7, 4\}$ **34.** $\{-5, 5\}$ **35.** $\left\{-6, \frac{3}{5}\right\}$ **36.** $\left\{-\frac{7}{2}, 1\right\}$
37. $\{-2, 0, 2\}$ **38.** $\{8, 12\}$ **39.** $\left\{-5, \frac{3}{4}\right\}$ **40.** $\{-2, 3\}$
41. $\left\{-\frac{7}{3}, \frac{5}{2}\right\}$ **42.** $\{-9, 6\}$ **43.** $\left\{-5, \frac{3}{2}\right\}$ **44.** $\left\{\frac{4}{3}, \frac{5}{2}\right\}$
45. $-\frac{8}{3}$ and $-\frac{19}{3}$ or 4 and 7
46. The length is 8 centimeters and the width is 2 centimeters.
47. A 2-by-2-inch square and a 10-by-10-inch square
48. 8 by 15 by 17 **49.** $-\frac{13}{6}$ and -12 or 2 and 13
50. 7, 9, and 11 **51.** 4 shelves
52. A 5-by-5-yard square and a 5-by-40-yard rectangle
53. -18 and -17 or 17 and 18 **54.** 6 units
55. 2 meters and 7 meters **56.** 9 and 11
57. 2 centimeters **58.** 15 feet

Chapter 6 Test (page 273)

1. $(x+5)(x-2)$ **2.** $(x+3)(x-8)$
3. $2x(x+1)(x-1)$ **4.** $(x+9)(x+12)$
5. $3(2n+1)(3n+2)$ **6.** $(x+y)(a+2b)$
7. $(4x-3)(x+5)$ **8.** $6(x^2+4)$
9. $2x(5x-6)(3x-4)$ **10.** $(7-2x)(4+3x)$
11. $\{-3, 3\}$ **12.** $\{-6, 1\}$ **13.** $\{0, 8\}$ **14.** $\left\{-\frac{5}{2}, \frac{2}{3}\right\}$
15. $\{-6, 2\}$ **16.** $\{-12, -4, 0\}$ **17.** $\{5, 9\}$
18. $\left\{-12, \frac{1}{3}\right\}$ **19.** $\left\{-4, \frac{2}{3}\right\}$ **20.** $\{-5, 0, 5\}$ **21.** $\left\{\frac{7}{5}\right\}$
22. 14 inches **23.** 12 centimeters
24. 16 chairs per row **25.** 12 units

CHAPTER 7

Problem Set 7.1 (page 277)

1. $\frac{3x}{7y}$ **3.** $\frac{3y}{8}$ **5.** $-\frac{3xy}{5}$ **7.** $\frac{3y}{4x^2}$ **9.** $-\frac{2b^2}{9}$ **11.** $\frac{4xy}{9z}$
13. $\frac{y}{x-2}$ **15.** $\frac{2x+3y}{3}$ **17.** $\frac{x+2}{x-7}$ **19.** -1 **21.** -3
23. $-4x$ **25.** $\frac{x+1}{3x}$ **27.** $\frac{x+y}{x}$ **29.** $\frac{x(2x-5y)}{2(x+4y)}$
31. $\frac{n}{n+1}$ **33.** $\frac{2n-1}{n-3}$ **35.** $\frac{2x+7}{3x+4}$ **37.** $\frac{3}{4(x-1)}$
39. $\frac{x+9}{x+3}$ **41.** $\frac{2a-1}{3a-1}$ **43.** $\frac{x+3y}{2x+y}$ **45.** $-\frac{x-3}{x}$
47. $\frac{n+7}{8}$ **49.** $\frac{2n-3}{n+1}$ **51.** $\frac{y-12}{y-14}$ **53.** $\frac{1+x}{x}$
55. $\frac{2+x}{4+5x}$ **57.** $-\frac{x+9}{x+6}$ **59.** $-\frac{1}{2}$ **63.** $\frac{y-3}{y+5}$
65. $\frac{x+1}{x+5}$ **67.** $\frac{1}{x^6}$ **69.** $\frac{1}{x^3y^2}$ **71.** $-\frac{4}{a^3}$

Problem Set 7.2 (page 282)

1. $\frac{1}{6}$ **3.** $-\frac{9}{14}$ **5.** $-\frac{17}{19}$ **7.** $\frac{2x^2}{7y}$ **9.** $-\frac{3n^2}{10}$ **11.** $2a$
13. $\frac{10b^3}{27}$ **15.** $\frac{3x^2y}{2}$ **17.** $-\frac{4}{5b^3}$ **19.** $\frac{s}{17}$ **21.** $\frac{x-y}{x}$
23. $\frac{1}{5}$ **25.** $\frac{3a}{14}$ **27.** $\frac{5(x+6)}{x+9}$ **29.** $\frac{3y(2x-y)}{2(x+y)}$
31. $\frac{a}{(5a+2)(3a+1)}$ **33.** $\frac{2x(x+4)}{5y(x+8)}$ **35.** $\frac{5}{x+y}$
37. $\frac{4t+1}{3(4t+3)}$ **39.** 4 **41.** $\frac{y^2}{3}$ **43.** $\frac{x-1}{y^2(1-y)(x-y)}$
45. $\frac{x+6}{x}$

Problem Set 7.3 (page 288)

1. $\frac{17}{x}$ **3.** $\frac{2}{3x}$ **5.** $\frac{4}{n}$ **7.** $-\frac{1}{x^2}$ **9.** $\frac{x+4}{x}$ **11.** $-\frac{3}{x-1}$
13. 1 **15.** $\frac{5t+2}{4}$ **17.** $\frac{3a+8}{3}$ **19.** $\frac{5n+4}{4}$
21. $-n-1$ **23.** $\frac{-3x-5}{7x}$ **25.** $\frac{9}{4}$ **27.** 3 **29.** $-\frac{1}{7x}$
31. $a-2$ **33.** $\frac{3}{x-6}$ **35.** $\frac{13x}{8}$ **37.** $-\frac{3n}{4}$ **39.** $\frac{11y}{12}$
41. $\frac{47x}{21}$ **43.** $\frac{14x}{15}$ **45.** $\frac{13n}{24}$ **47.** $\frac{7x-14}{10}$ **49.** $\frac{4x}{9}$
51. $\frac{18n+11}{12}$ **53.** $\frac{9n-14}{18}$ **55.** $\frac{7x}{24}$ **57.** $\frac{-23x-18}{60}$
59. $\frac{19}{24x}$ **61.** $\frac{1}{18y}$ **63.** $\frac{20x-33}{48x^2}$ **65.** $\frac{25}{12x}$
67. $\frac{10x-35}{x(x-5)}$ **69.** $\frac{-n+3}{n(n-1)}$ **71.** $\frac{-2n+16}{n(n+4)}$
73. $\frac{6}{x(2x+1)}$ **75.** $\frac{10x+12}{(x+4)(x-3)}$ **77.** $\frac{-6x+21}{(x-2)(x+1)}$
79. $\frac{x+7}{(2x-1)(3x+1)}$ **85.** $\frac{4}{x-3}$ **87.** $\frac{-6}{a-1}$
89. $\frac{n+3}{2n-1}$

Problem Set 7.4 (page 296)

1. $\frac{3x-8}{x(x-4)}$ **3.** $\frac{-5x-3}{x(x+2)}$ **5.** $\frac{8n-50}{n(n-6)}$ **7.** $-\frac{4}{n+1}$
9. $\frac{5x-7}{2x(x-1)}$ **11.** $\frac{5x-17}{(x+4)(x-4)}$ **13.** $\frac{4}{x+1}$
15. $\frac{11a-6}{a(a-2)(a+2)}$ **17.** $\frac{12}{x(x-6)(x+6)}$
19. $\frac{n+8}{3(n+4)(n-4)}$ **21.** $\frac{19x}{6(3x+2)}$ **23.** $\frac{-2x+17}{15(x+1)}$
25. $\frac{5x+6}{(x+3)(x+4)(x-3)}$ **27.** $\frac{x^2-10x-20}{(x+2)(x+4)(x-5)}$
29. $\frac{a-b}{ab}$ **31.** $\frac{8x-14}{(x-5)(x+5)}$ **33.** $\frac{15x+4}{x(x-2)(x+2)}$
35. $\frac{4x-4}{(x+5)(x+2)}$ **37.** $\frac{-2x+6}{(3x-5)(x+4)}$ **39.** $\frac{3}{x+4}$
41. $-\frac{1}{2}$ **43.** $\frac{10}{3}$ **45.** $\frac{28}{27}$ **47.** $\frac{y}{3x}$ **49.** $\frac{2y+3x}{5y-x}$
51. $\frac{x^2-4y}{7xy-3x^2}$ **53.** $\frac{6+2x}{3+4x}$ **55.** $\frac{9-24x}{10x+42}$ **57.** $\frac{x^2+2x}{6x+4}$
59. $\frac{-2x+3}{4x-1}$ **61.** $\frac{m}{40}$ **63.** $\frac{k}{r}$ **65.** $\frac{d}{l}$ **67.** $\frac{34}{n}$ **69.** $\frac{47}{l}$
71. $\frac{96}{b}$ **73.** $\frac{-n^2+n-1}{n-1}$ **75.** $\frac{3x^2-4x+2}{4x-2}$

Problem Set 7.5 (page 303)

1. $\{12\}$ **3.** $\left\{-\frac{2}{21}\right\}$ **5.** $\{4\}$ **7.** $\{-1\}$ **9.** $\{3\}$
11. $\{-38\}$ **13.** $\left\{-\frac{13}{8}\right\}$ **15.** $\{2\}$ **17.** $\{6\}$ **19.** $\left\{\frac{5}{18}\right\}$
21. $\left\{-\frac{7}{10}\right\}$ **23.** $\left\{-\frac{1}{3}\right\}$ **25.** $\{8\}$ **27.** $\{37\}$ **29.** $\varnothing$
31. $\{39\}$ **33.** $\left\{-\frac{5}{4}\right\}$ **35.** $\{-6\}$ **37.** $\left\{-\frac{2}{3}\right\}$
39. $\left\{\frac{12}{7}\right\}$ **41.** $\frac{40}{48}$ **43.** 10 **45.** 48° and 72° **47.** 60°
49. 15 miles per hour
51. 50 miles per hour for Dave and 54 miles per hour for Kent **57.** All real numbers except 0
59. All real numbers except -2 and -3

Problem Set 7.6 (page 313)

1. $\{-2\}$ **3.** $\{5\}$ **5.** $\left\{\frac{9}{2}\right\}$ **7.** $\left\{-\frac{49}{10}\right\}$ **9.** $\left\{\frac{2}{3}\right\}$
11. $\left\{\frac{4}{3}\right\}$ **13.** $\left\{\frac{3}{2}\right\}$ **15.** $\left\{\frac{11}{3}\right\}$ **17.** $\left\{\frac{13}{4}\right\}$
19. $\left\{-1, -\frac{5}{8}\right\}$ **21.** $\left\{\frac{1}{4}, 4\right\}$ **23.** $\left\{-\frac{5}{2}, 6\right\}$ **25.** $\{5\}$
27. $\{-21\}$ **29.** $\{2\}$ **31.** $\{-8, 1\}$ **33.** $\frac{1}{2}$ or 4
35. $-\frac{2}{5}$ or $\frac{5}{2}$
37. 17 miles per hour for Tom and 20 miles per hour for Celia
39. 16 miles per hour for the trip out and 12 miles per hour for the return trip
41. 30 minutes **43.** 60 minutes for Mike and 120 minutes for Barry **45.** 9 hours **47.** 4 minutes
49. 10 oysters per minute for Tchaika and 15 oysters per minute for Pachena **53.** $\{0\}$

Chapter 7 Review Problem Set (page 316)

1. $\frac{7x^2}{9y^2}$ **2.** $\frac{x}{x+3}$ **3.** $\frac{3n+5}{n+2}$ **4.** $\frac{4a+3}{5a-2}$ **5.** $\frac{3x}{8}$
6. $x(x-3)$ **7.** $\frac{n-7}{n^2}$ **8.** $\frac{2a+1}{a+6}$ **9.** $\frac{22x-19}{20}$
10. $\frac{43x-3}{12x^2}$ **11.** $\frac{10n-7}{n(n-1)}$ **12.** $\frac{-a+8}{(a-4)(a-2)}$
13. $\frac{5x+9}{4x(x-3)}$ **14.** $\frac{5x-4}{(x+5)(x-5)(x+2)}$
15. $\frac{6x-37}{(x-7)(x+3)}$ **16.** $\frac{3y^2-4x}{4xy+5y^2}$ **17.** $\frac{2y-xy}{3xy+5x}$
18. $\left\{\frac{20}{17}\right\}$ **19.** $\left\{-\frac{61}{60}\right\}$ **20.** $\{9\}$ **21.** $\left\{\frac{28}{3}\right\}$ **22.** $\left\{\frac{3}{4}\right\}$

23. $\{1\}$ **24.** $\{-7\}$ **25.** $\left\{\frac{1}{7}\right\}$ **26.** $\left\{\frac{1}{2}, 2\right\}$

27. $\{-5, 10\}$ **28.** $\left\{-\frac{1}{5}\right\}$ **29.** $\{-1\}$

30. Becky $2\frac{2}{3}$ hours, Nancy 8 hours **31.** 1 or 2 **32.** $\frac{36}{72}$

33. Todd's rate is 15 miles per hour and Lanette's rate is 22 miles per hour. **34.** 8 miles per hour **35.** 60 minutes

Chapter 7 Test (page 318)

1. $\frac{8x^2y}{9}$ **2.** $\frac{x}{x-6}$ **3.** $\frac{2n+1}{3n+4}$ **4.** $\frac{2x-3}{x-5}$ **5.** $2x^2y^2$

6. $\frac{x-2}{x+3}$ **7.** $\frac{(x+4)^2}{x(x+7)}$ **8.** $\frac{6x+5}{24}$ **9.** $\frac{9n-4}{30}$

10. $\frac{41-15x}{18x}$ **11.** $\frac{2n-6}{n(n-1)}$ **12.** $\frac{5x-18}{4x(x+6)}$

13. $\frac{5x-11}{(x-4)(x+8)}$ **14.** $\frac{-13x+43}{(2x-5)(3x+4)(x-6)}$

15. $\{-5\}$ **16.** $\left\{-\frac{19}{16}\right\}$ **17.** $\left\{\frac{4}{3}, 3\right\}$ **18.** $\{-6, 8\}$

19. $\left\{-\frac{1}{5}, 2\right\}$ **20.** $\{-23\}$ **21.** $\{2\}$ **22.** $\left\{-\frac{3}{2}\right\}$

23. $\frac{2}{3}$ or 3 **24.** 14 miles per hour **25.** 12 minutes

Cumulative Review Problem Set (page 319)

1. $\frac{5}{2}$ **2.** 6 **3.** $\frac{17}{12}$ **4.** 0.6 **5.** 20 **6.** 0 **7.** 2 **8.** -3

9. $\frac{1}{27}$ **10.** $\frac{3}{2}$ **11.** 1 **12.** $\frac{12}{7}$ **13.** $-\frac{1}{16}$ **14.** $\frac{9}{4}$

15. $\frac{4}{25}$ **16.** $-\frac{1}{27}$ **17.** $\frac{19}{10x}$ **18.** $\frac{2y}{3x}$ **19.** $\frac{7x-2}{(x-6)(x+4)}$

20. $\frac{-x+12}{x^2(x-4)}$ **21.** $\frac{x-7}{3y}$ **22.** $\frac{-3x-4}{(x-4)(x+3)}$

23. $-35x^5y^5$ **24.** $81a^2b^6$ **25.** $-15n^4 - 18n^3 + 6n^2$

26. $15x^2 + 17x - 4$ **27.** $4x^2 + 20x + 25$

28. $2x^3 + x^2 - 7x - 2$ **29.** $x^4 + x^3 - 6x^2 + x + 3$

30. $-6x^2 + 11x + 7$ **31.** $3xy - 6x^3y^3$ **32.** $7x + 4$

33. $3x(x^2 + 5x + 9)$ **34.** $(x + 10)(x - 10)$

35. $(5x - 2)(x - 4)$ **36.** $(4x + 7)(2x - 9)$

37. $(n + 16)(n + 9)$ **38.** $(x + y)(n - 2)$

39. $3x(x + 1)(x - 1)$ **40.** $2x(x - 9)(x + 6)$

41. $(6x - 5)^2$ **42.** $(3x + y)(x - 2y)$ **43.** $\left\{\frac{16}{3}\right\}$

44. $\{-11, 0\}$ **45.** $\left\{\frac{1}{14}\right\}$ **46.** $\{15\}$ **47.** $\{-1, 1\}$

48. $\{-6, 1\}$ **49.** $\{2\}$ **50.** $\left\{\frac{11}{12}\right\}$ **51.** $\{1, 2\}$

52. $\left\{-\frac{1}{18}\right\}$ **53.** $\left\{-\frac{7}{2}, \frac{1}{3}\right\}$ **54.** $\left\{\frac{1}{2}, 8\right\}$ **55.** $\{-9, 2\}$

56. $\left\{\frac{16}{5}\right\}$ **57.** $\left\{\frac{1}{2}\right\}$ **58.** 6 inches, 8 inches, 10 inches

59. 75 **60.** 100 milliliters **61.** 6 feet by 8 feet

62. 2.5 hours **63.** 7.5 centimeters **64.** 27 gallons

65. 92 or better **66.** $\{1, 2, 3\}$

67. All real numbers greater than $\frac{7}{2}$

CHAPTER 8

Problem Set 8.1 (page 329)

1. Yes **3.** No

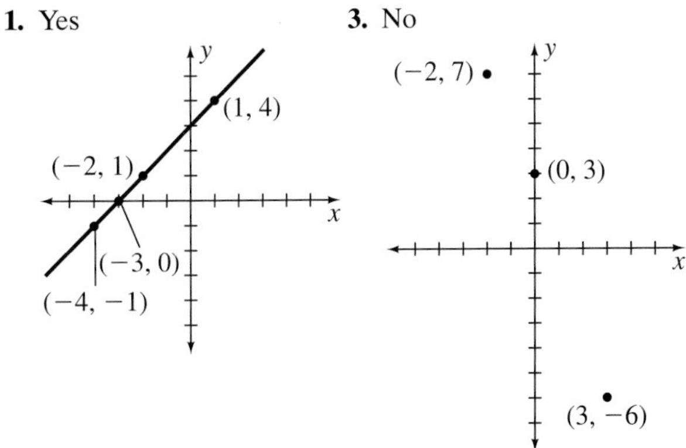

5. Yes **7.** Yes

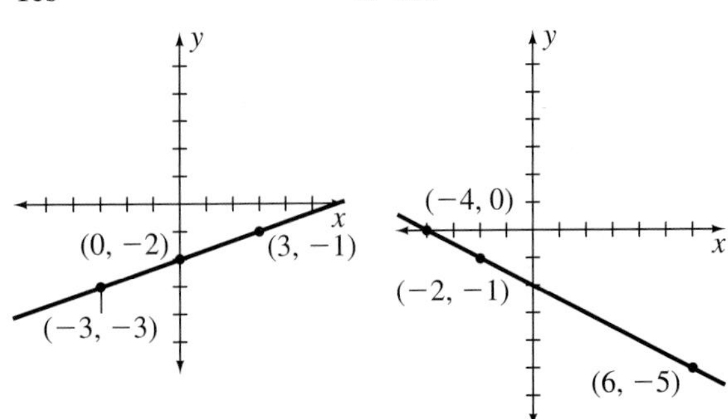

9. No **11.**

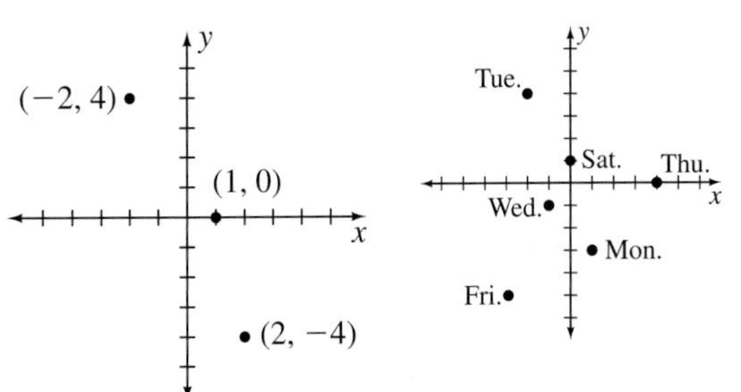

13.

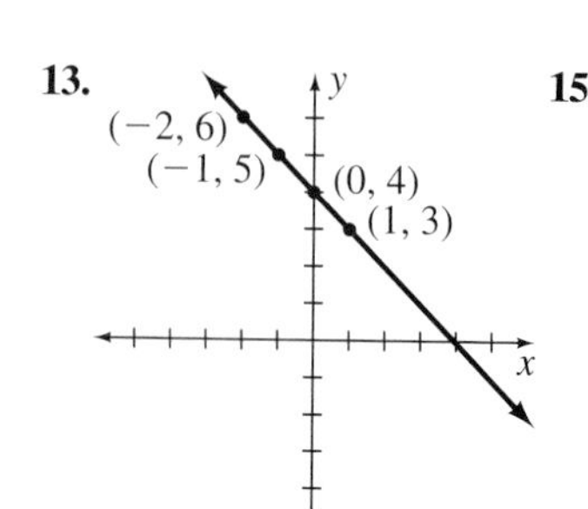

15.

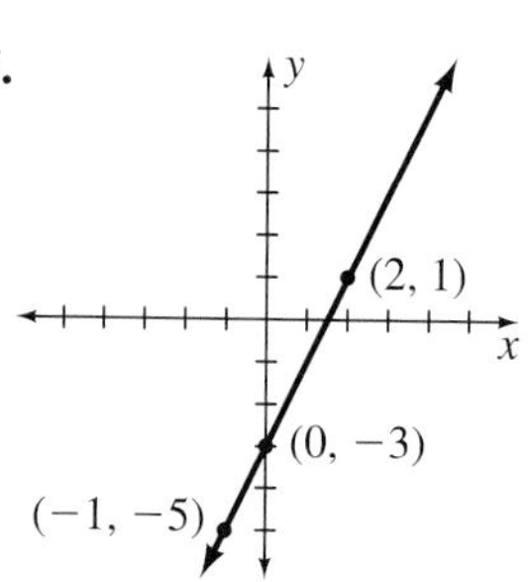

17.

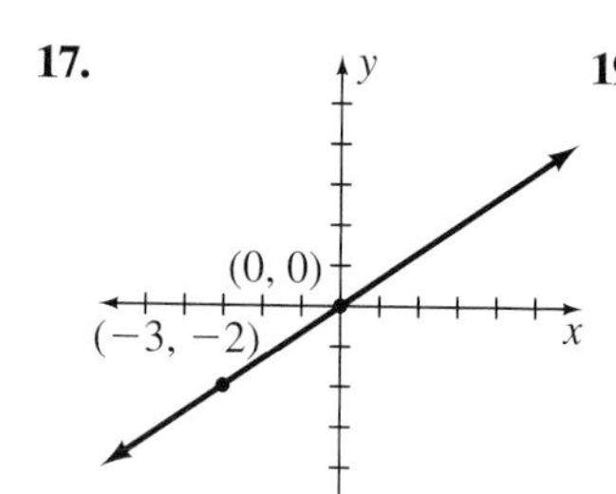

19.

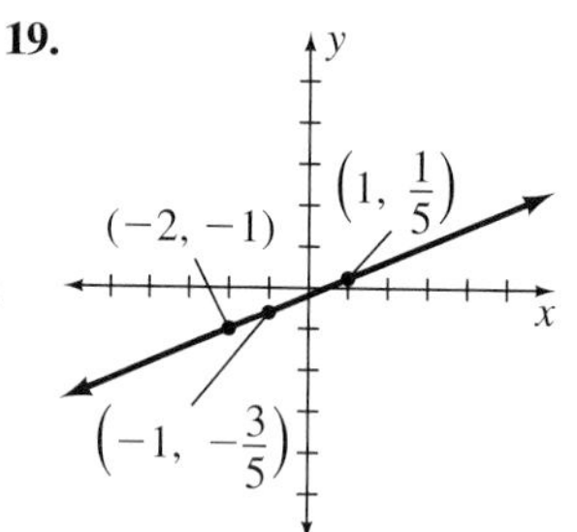

21. $y = \frac{3}{7}x + \frac{13}{7}$ **23.** $x = 3y + 9$ **25.** $y = \frac{1}{5}x + \frac{14}{5}$

27. $x = \frac{1}{3}y - \frac{7}{3}$ **29.** $y = \frac{2}{3}x - \frac{5}{3}$

31. **33.**

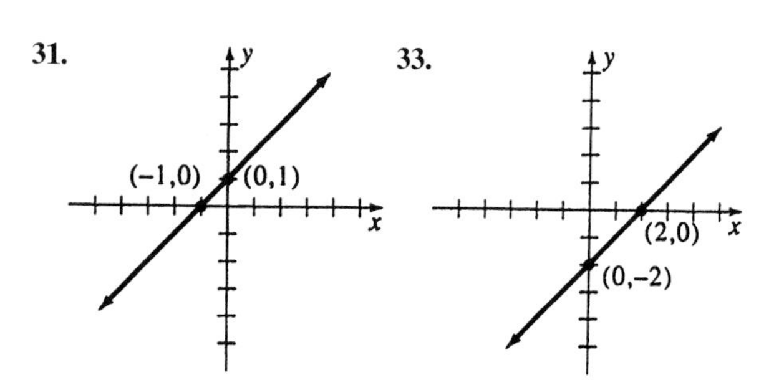

35. **37.**

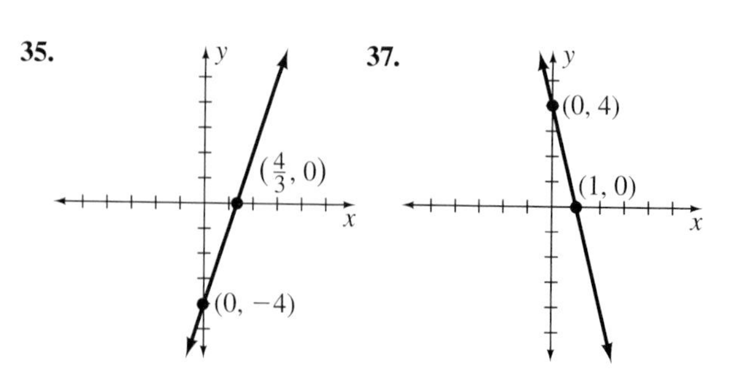

39.

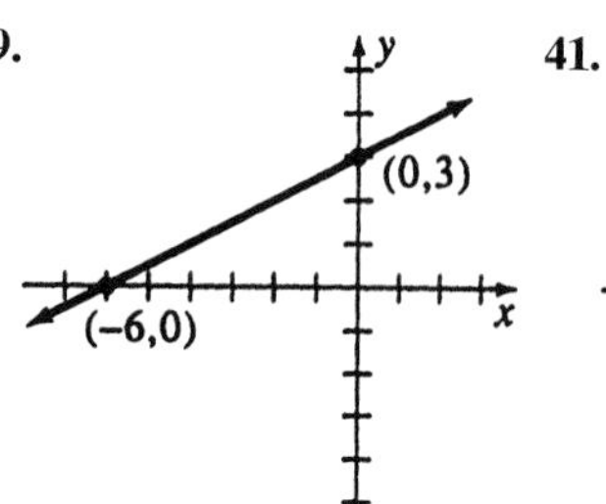

41.

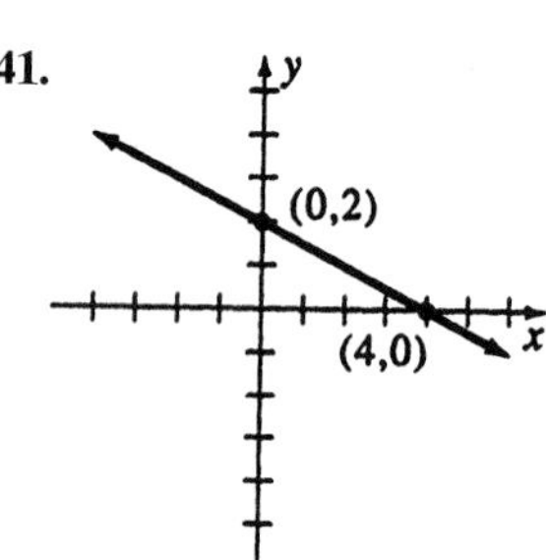

43.

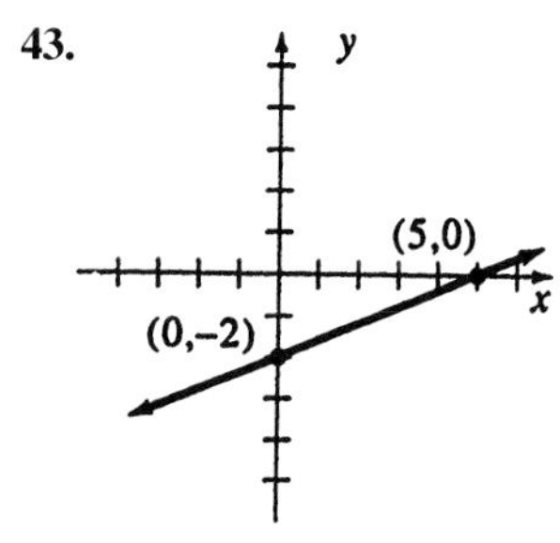

45.

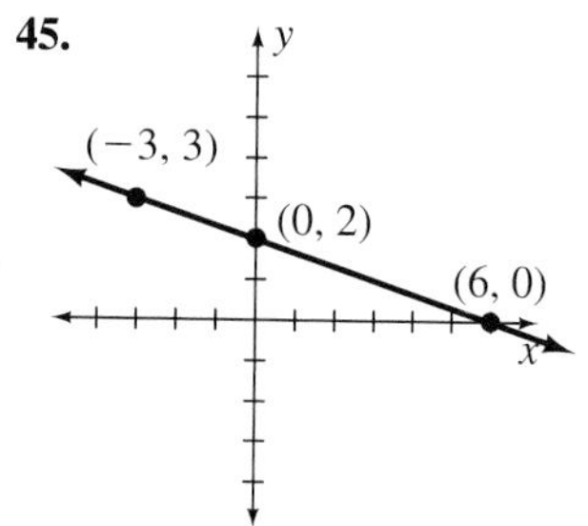

47. **49.**

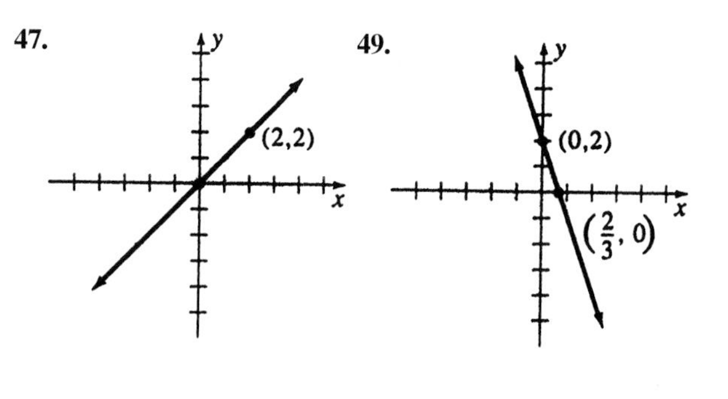

53. **55.**

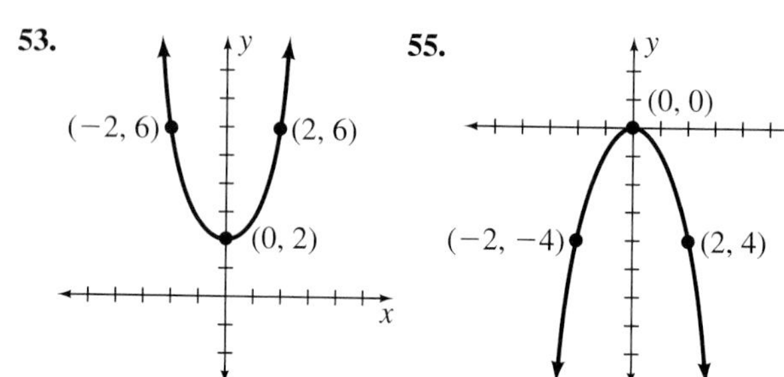

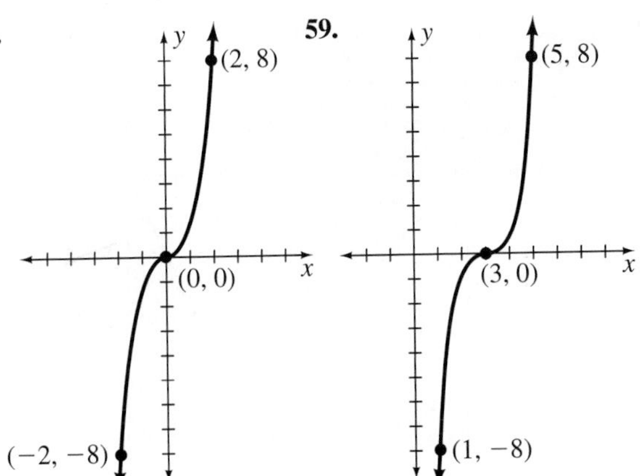

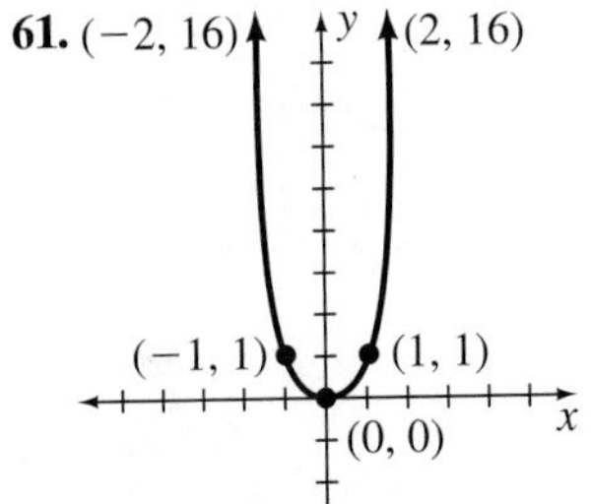

Problem Set 8.2 (page 339)

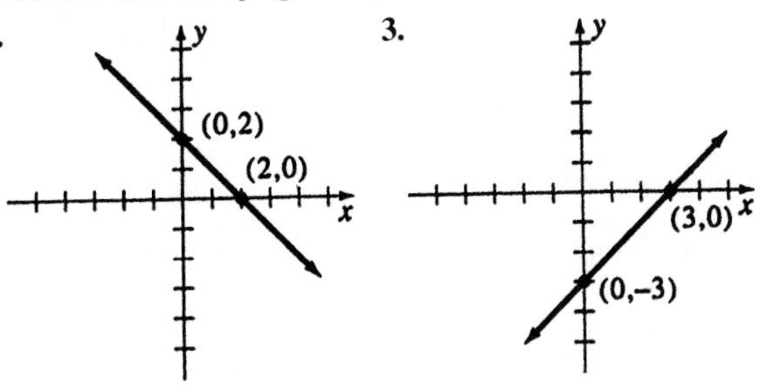

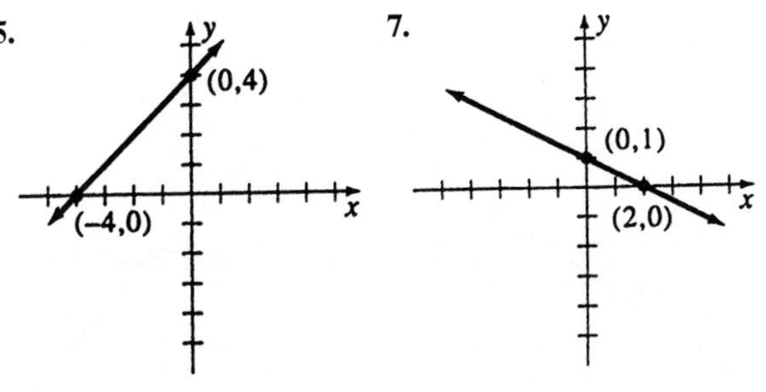

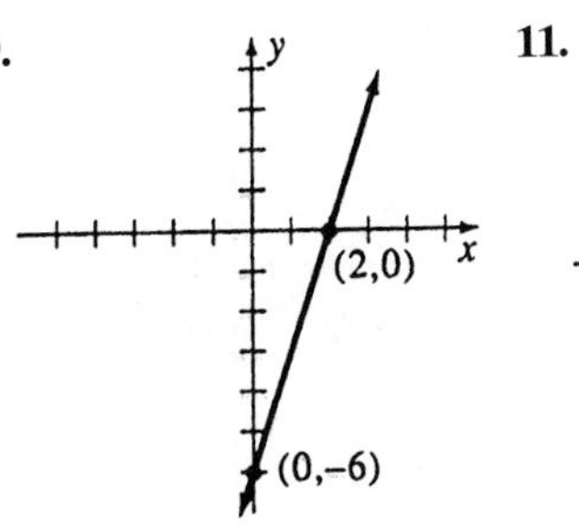

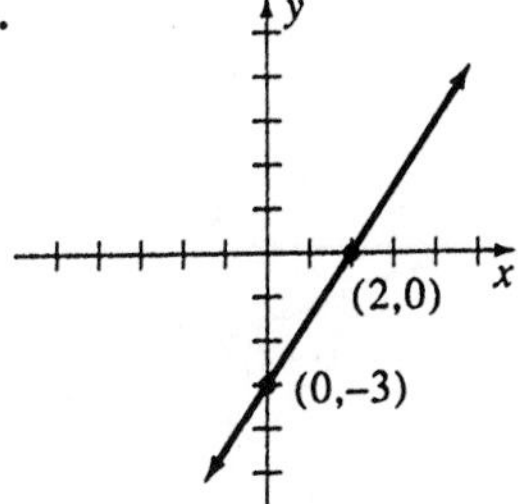

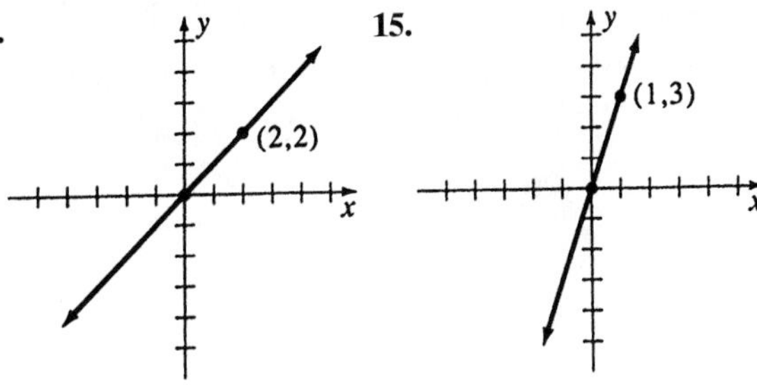

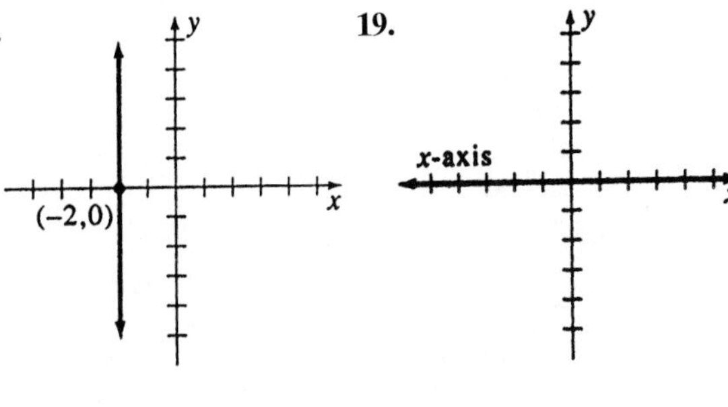

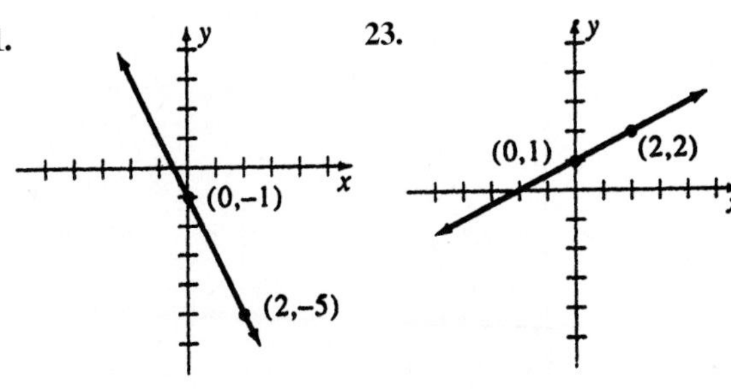

25.

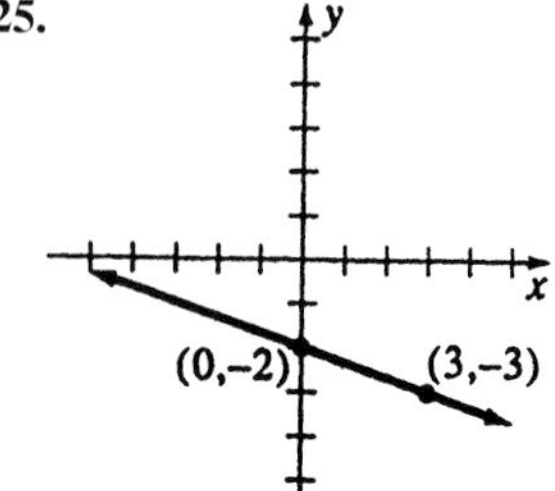

27.

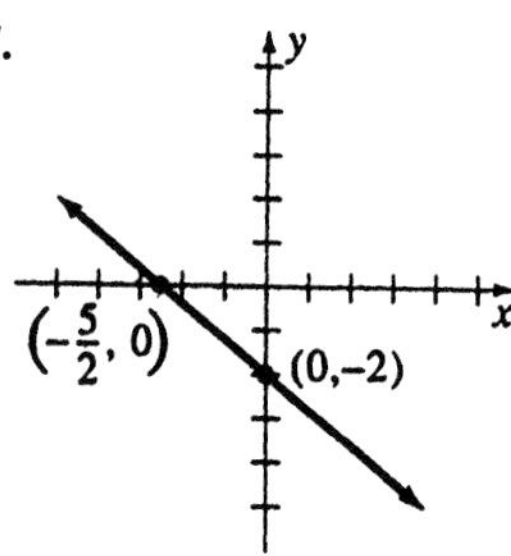

29. **31.**

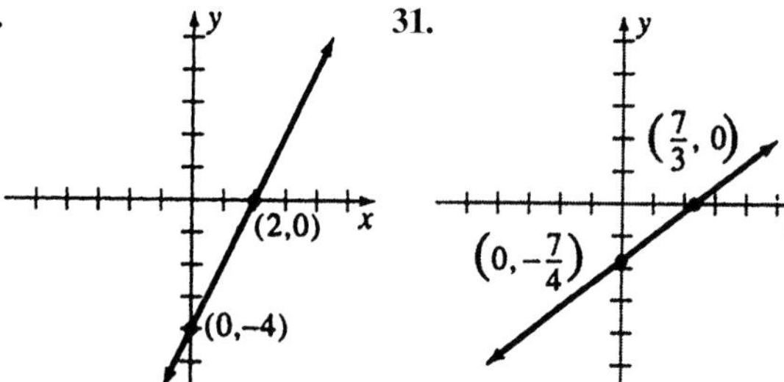

33. **35.**

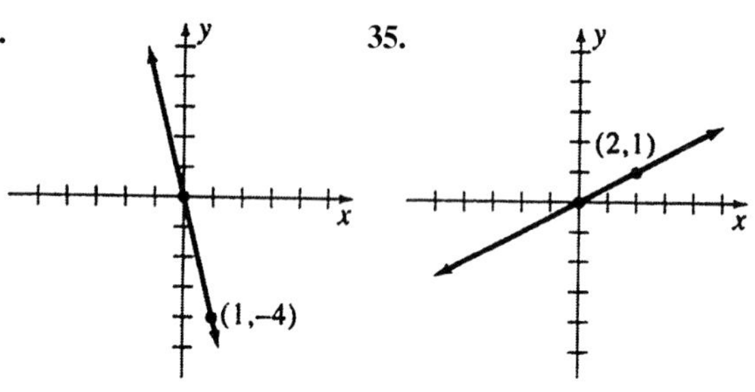

37. **39.**

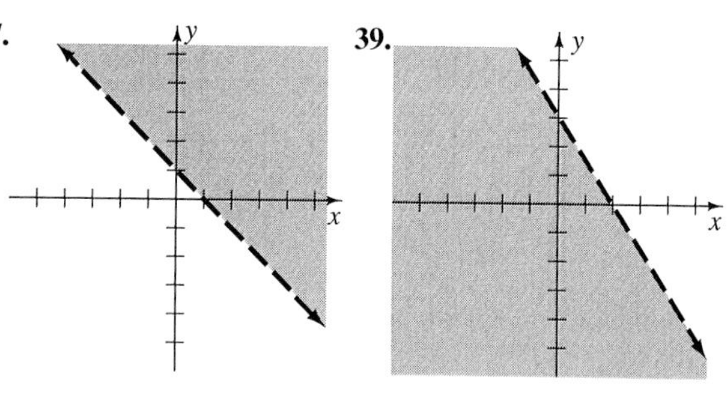

41. **43.**

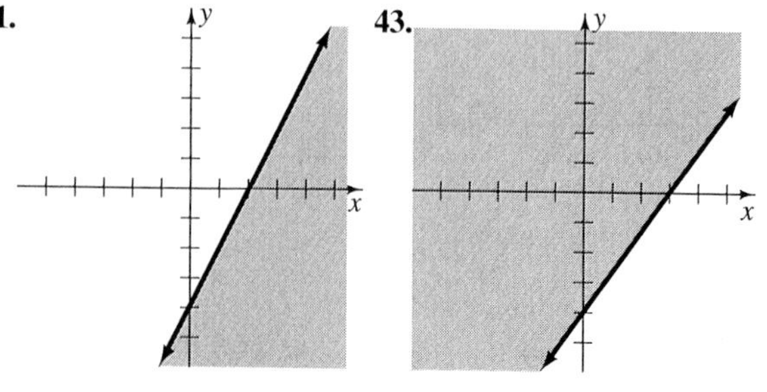

45.

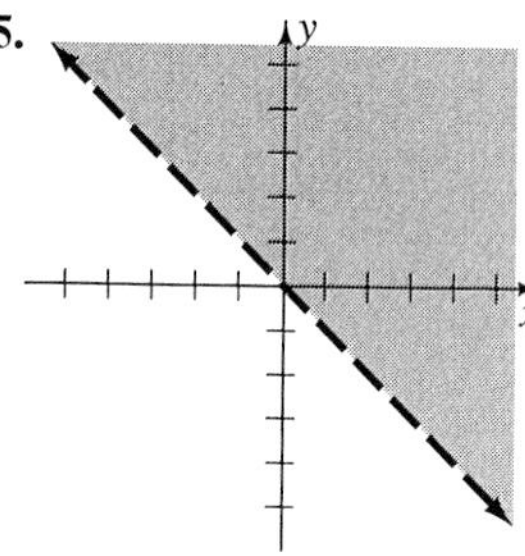

47.

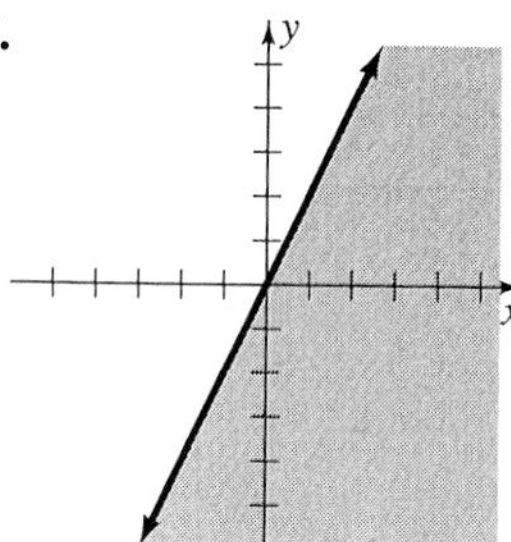

49.

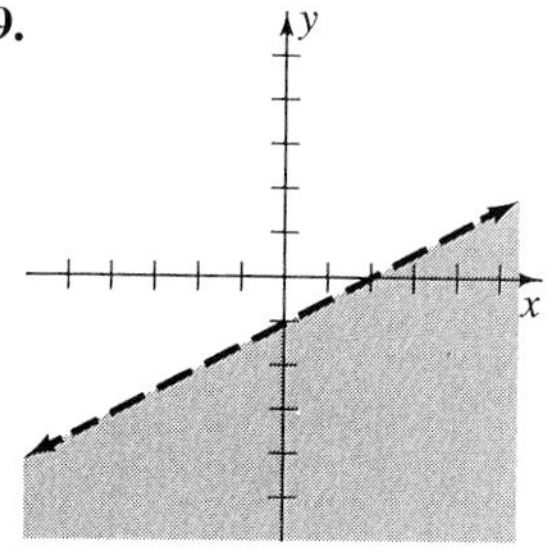

51.

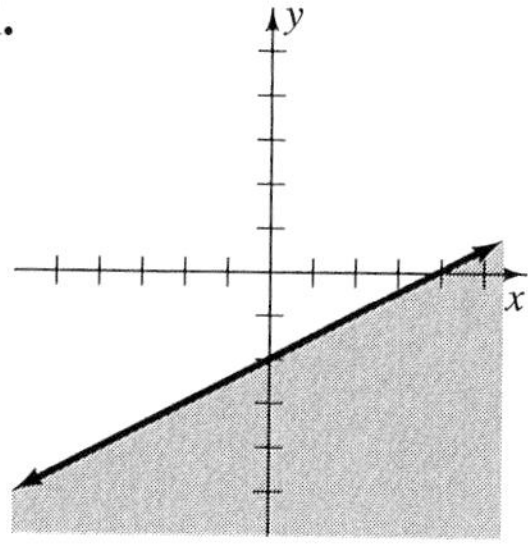

53.

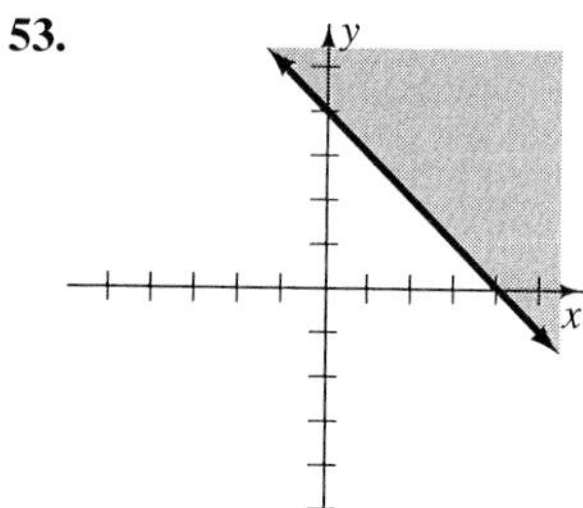

55.

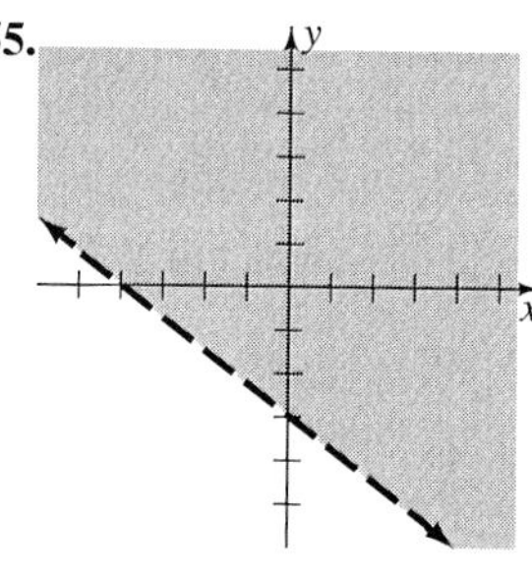

Problem Set 8.3 (page 347)

1. $\frac{3}{4}$ **3.** $\frac{7}{5}$ **5.** $-\frac{6}{5}$ **7.** $-\frac{10}{3}$ **9.** $\frac{3}{4}$ **11.** 0 **13.** $-\frac{3}{2}$

15. Undefined **17.** 1 **19.** $\frac{b-d}{a-c}$ or $\frac{d-b}{c-a}$ **21.** 4

23. -6 **25–31.** Answers will vary. **33.** Negative

35. Positive **37.** Zero **39.** Negative

41.

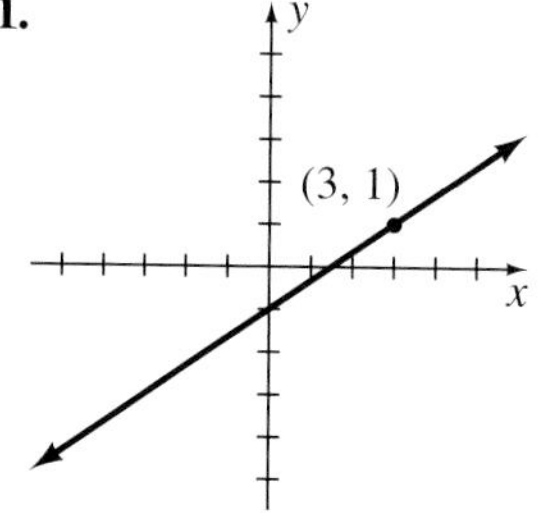

43.

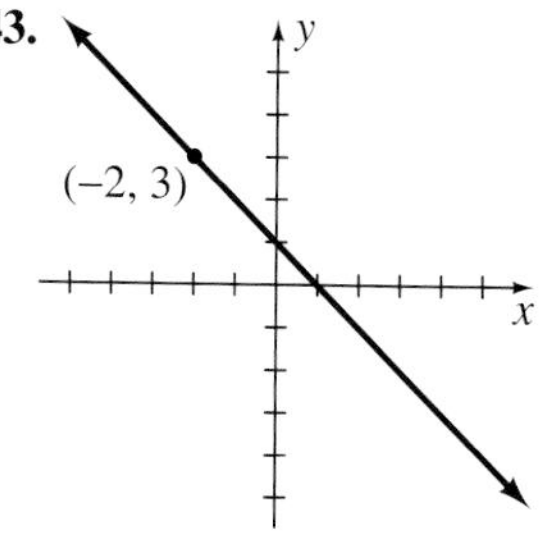

45.

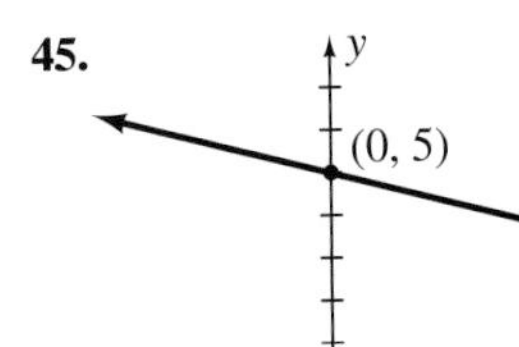

47.

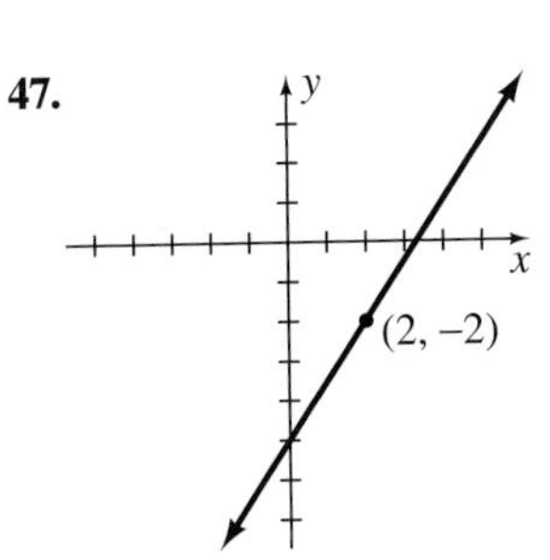

49. $-\frac{3}{2}$ **51.** $\frac{5}{4}$ **53.** $-\frac{1}{5}$ **55.** 2 **57.** 0 **59.** $\frac{2}{5}$ **61.** $\frac{6}{5}$

63. −3 **65.** 4 **67.** $\frac{2}{3}$ **69.** 5.1% **71.** 32 cm

77. 1.0 feet

Problem Set 8.4 (page 356)

1. $2x - 3y = -5$ **3.** $x - 2y = 7$ **5.** $x + 3y = 20$
7. $y = -7$ **9.** $4x + 9y = 0$ **11.** $3x - y = -16$
13. $7x - 5y = -1$ **15.** $3x + 2y = 5$ **17.** $x - y = 1$

19. $5x - 3y = 0$ **21.** $4x + 7y = 28$ **23.** $y = \frac{3}{5}x + 2$

25. $y = 2x - 1$ **27.** $y = -\frac{1}{6}x - 4$ **29.** $y = -x + \frac{5}{2}$

31. $y = -\frac{5}{9}x - \frac{1}{2}$

33. $m = -2, b = -5$

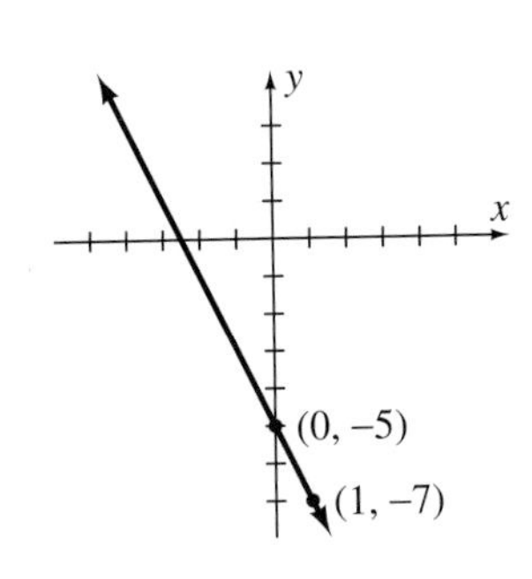

35. $m = \frac{3}{5}, b = -3$

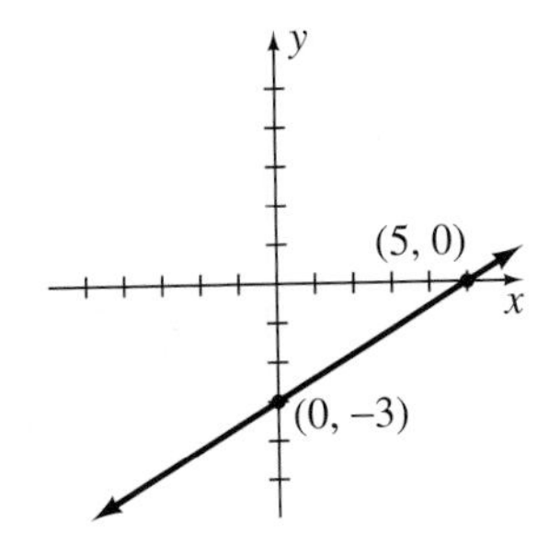

37. $m = \frac{4}{9}, b = 2$

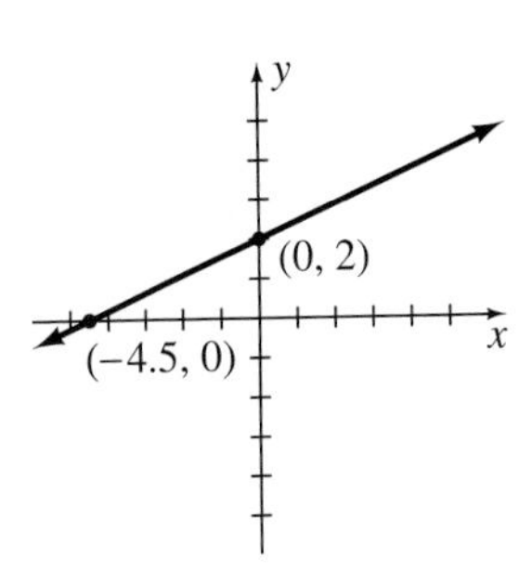

39. $m = \frac{3}{4}, b = -4$

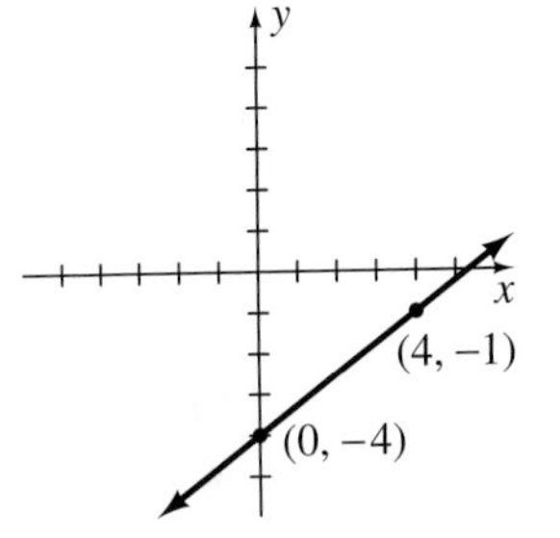

41. $m = -\frac{2}{11}, b = -1$

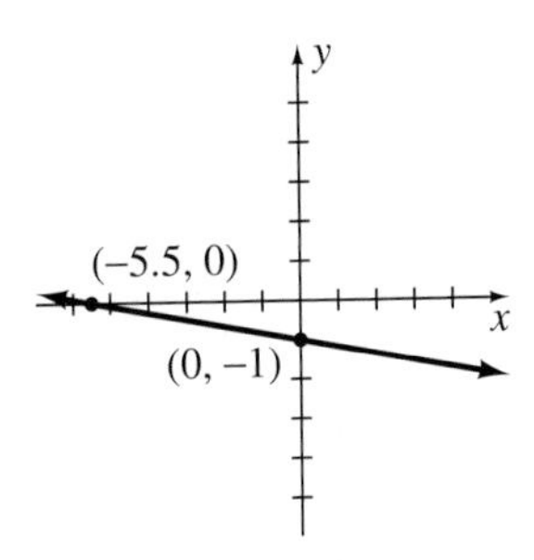

43. $m = -\frac{9}{7}, b = 0$

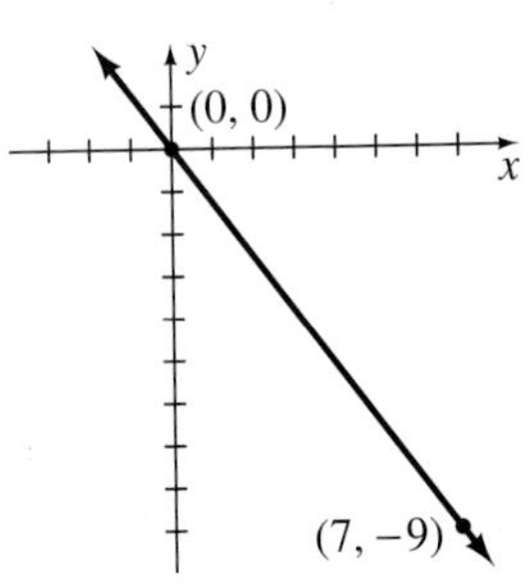

49. Perpendicular
51. Intersecting lines that are not perpendicular
53. Parallel **55.** $2x - 3y = -1$

Problem Set 8.5 (page 362)

1. No **3.** Yes **5.** Yes **7.** Yes **9.** No **11.** $\{(2, -1)\}$
13. $\{(2, 1)\}$ **15.** $\varnothing$ **17.** $\{(0, 0)\}$ **19.** $\{(1, -1)\}$
21. Infinitely many **23.** $\{(1, 3)\}$ **25.** $\{(3, -2)\}$
27. $\{(2, 4)\}$ **29.** $\{(-2, -3)\}$

35.

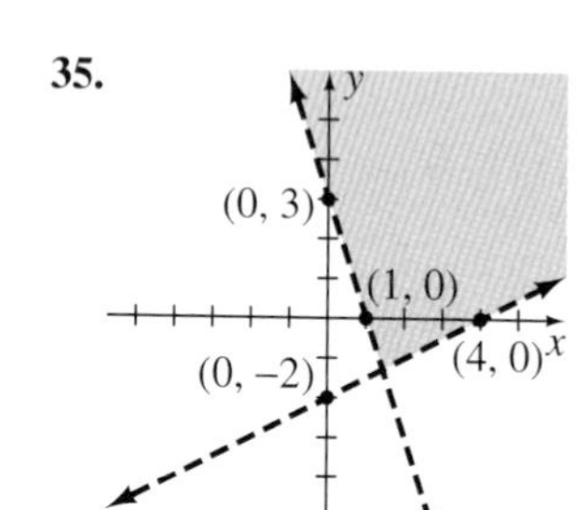

37.

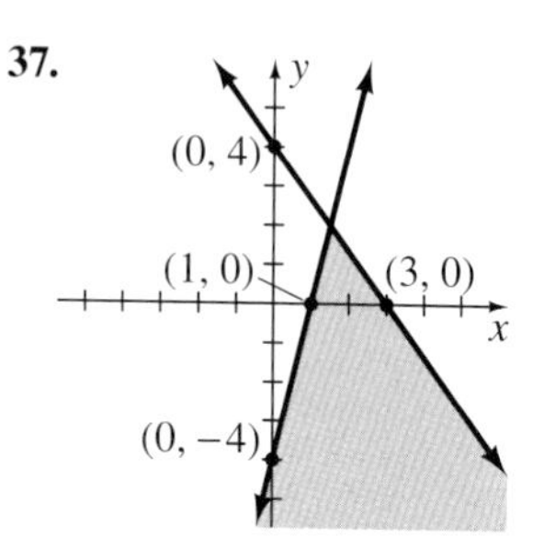

39.

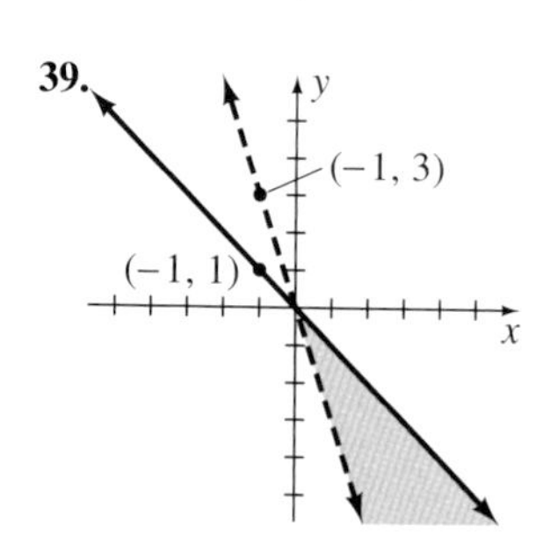

41.

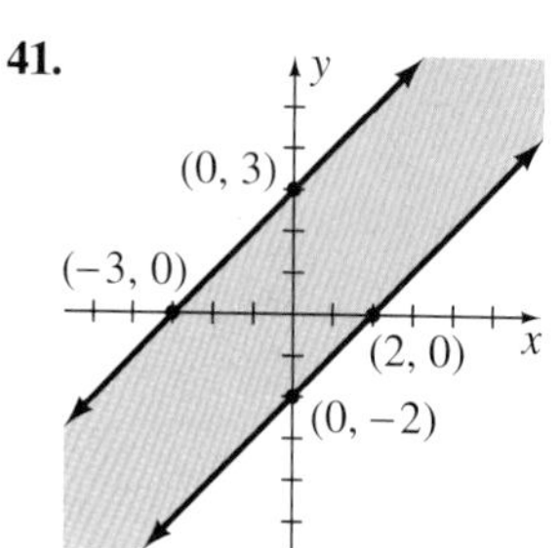

43.

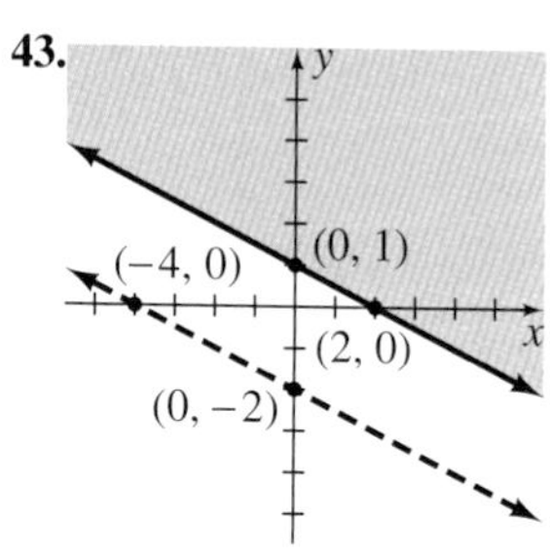

Problem Set 8.6 (page 369)

1. $\{(6, 8)\}$ **3.** $\{(-5, -4)\}$ **5.** $\{(-6, 12)\}$ **7.** $\{(5, -2)\}$
9. $\left\{\left(\frac{11}{4}, \frac{9}{8}\right)\right\}$ **11.** $\left\{\left(-\frac{2}{3}, \frac{2}{3}\right)\right\}$ **13.** $\{(-4, 5)\}$
15. $\{(4, 1)\}$ **17.** $\left\{\left(\frac{3}{2}, -3\right)\right\}$ **19.** $\left\{\left(-\frac{18}{71}, \frac{5}{71}\right)\right\}$
21. $\{(250, 500)\}$ **23.** $\{(100, 200)\}$ **25.** 9 and 21
27. 8 and 15 **29.** 6 and 12
31. \$.45 per lemon and \$.50 per apple
33. 7 dimes and 3 quarters
35. 22 hardcover and 14 softcover
37. 4 gallons of 10% and 6 gallons of 15%
39. \$5000 at 5% and \$8000 at 6% **43.** $\{(12, 24)\}$ **45.** $\varnothing$

Problem Set 8.7 (page 379)

1. $\{(5, 9)\}$ **3.** $\{(-4, 10)\}$ **5.** $\left\{\left(\frac{3}{2}, 4\right)\right\}$
7. $\left\{\left(-\frac{1}{3}, \frac{4}{3}\right)\right\}$ **9.** $\{(18, 24)\}$ **11.** $\{(-10, -15)\}$
13. $\varnothing$ **15.** $\{(-2, 6)\}$ **17.** $\{(-6, -13)\}$ **19.** $\left\{\left(\frac{9}{8}, \frac{7}{12}\right)\right\}$
21. $\left\{\left(\frac{6}{31}, \frac{9}{31}\right)\right\}$ **23.** $\{(100, 400)\}$ **25.** $\{(3, 10)\}$
27. $\{(-2, -6)\}$ **29.** $\left\{\left(-\frac{6}{7}, \frac{6}{7}\right)\right\}$ **31.** $\{(10, 12)\}$
33. $\left\{\left(\frac{3}{5}, \frac{12}{5}\right)\right\}$ **35.** $\left\{\left(-\frac{5}{2}, 6\right)\right\}$ **37.** $\left\{\left(\frac{1}{2}, 4\right)\right\}$
39. Infinitely many solutions **41.** $\left\{\left(5, \frac{5}{2}\right)\right\}$
43. $\{(12, 4)\}$ **45.** $\left\{\left(-\frac{1}{11}, -\frac{10}{11}\right)\right\}$ **47.** 12 and 34
49. 35 double rooms and 15 single rooms
51. 42 women **53.** 20 inches by 27 inches
55. 60 five-dollar bills and 40 ten-dollar bills
57. 90 nickels and 30 pennies
59. 20 liters of 40% alcohol and 10 liters of 70% alcohol
61. The length is 29 meters and the width is 8 meters.
65. $\left\{\left(-1, -\frac{7}{6}\right)\right\}$ **67.** Infinitely many solutions

Chapter 8 Review Problem Set (page 382)

1. 2.

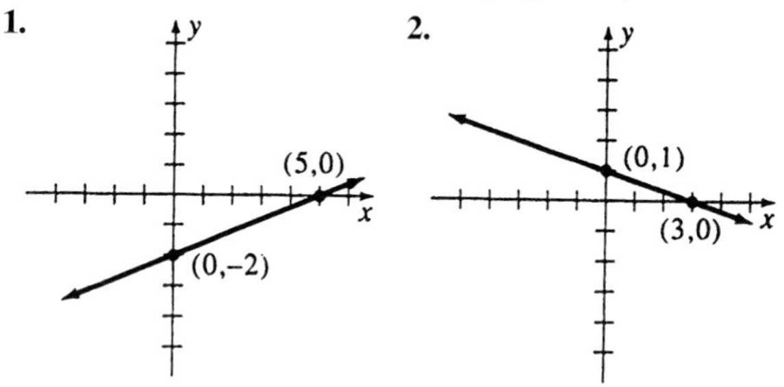

3. 4.

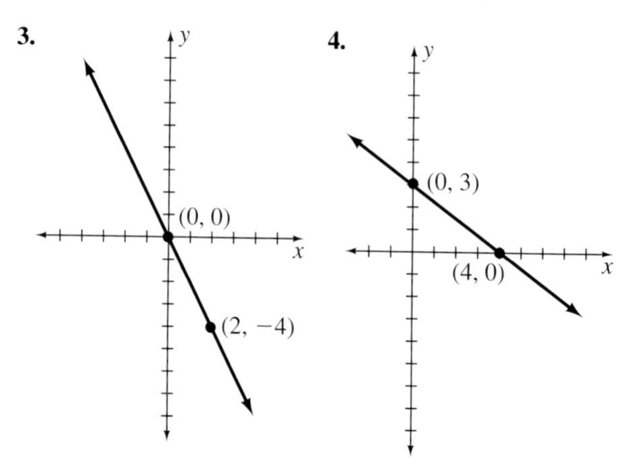

5.

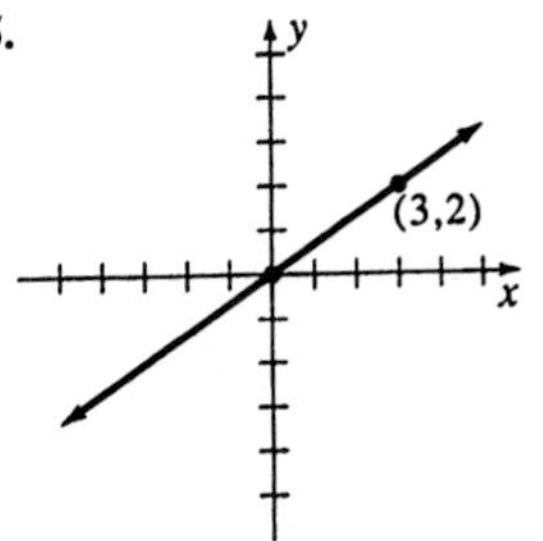

6.

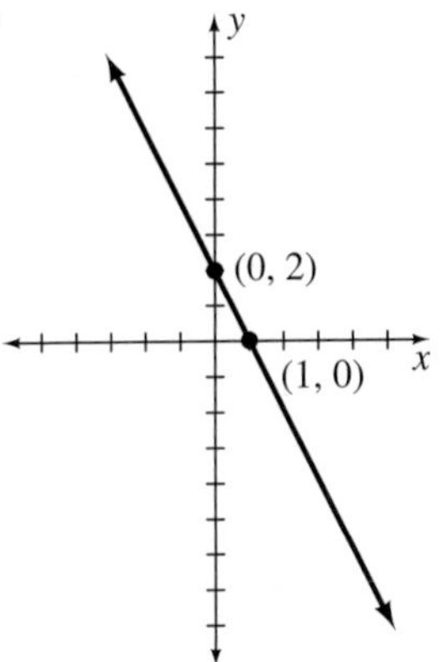

7. **8.**

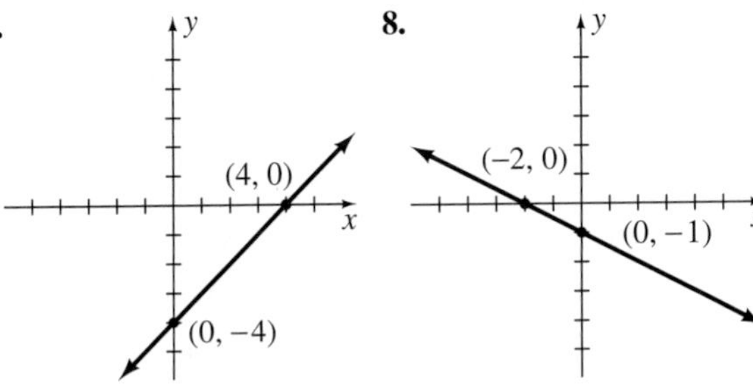

9.

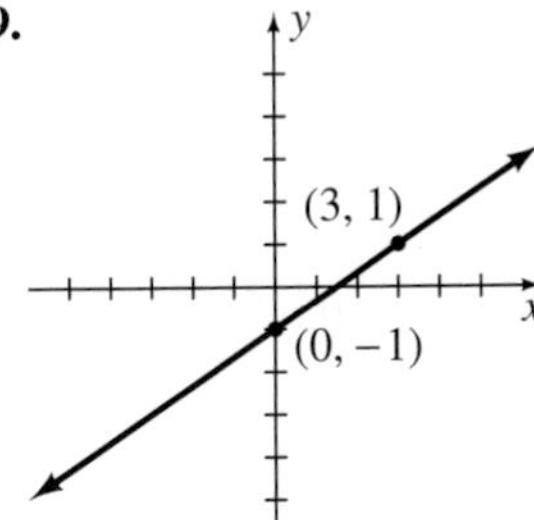

10.

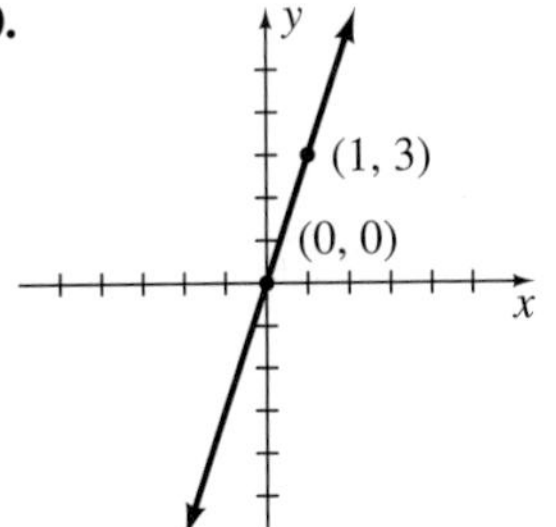

11. $m = \frac{2}{5}, b = -2$

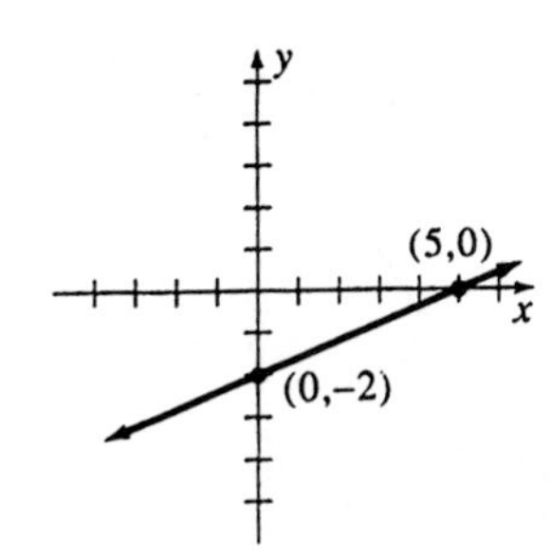

12. $m = -\frac{1}{3}, b = 1$

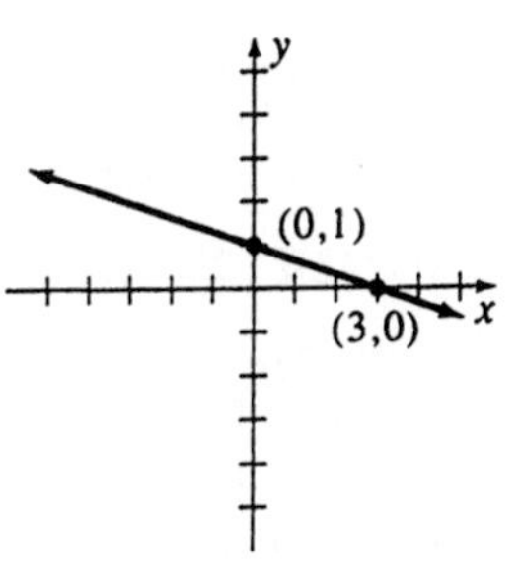

13. $m = -\frac{1}{2}, b = 1$

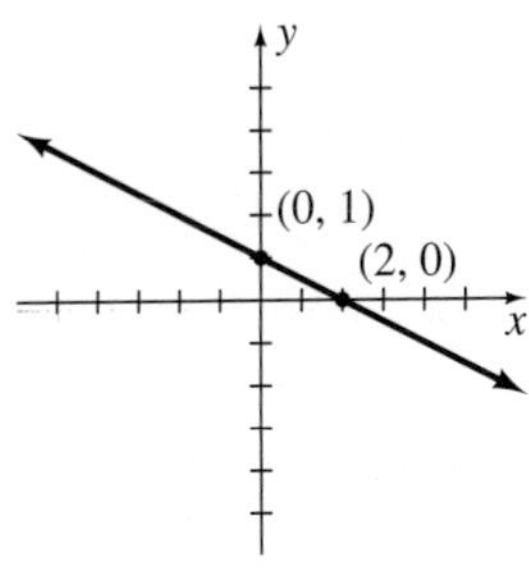

14. $m = -3, b = -2$

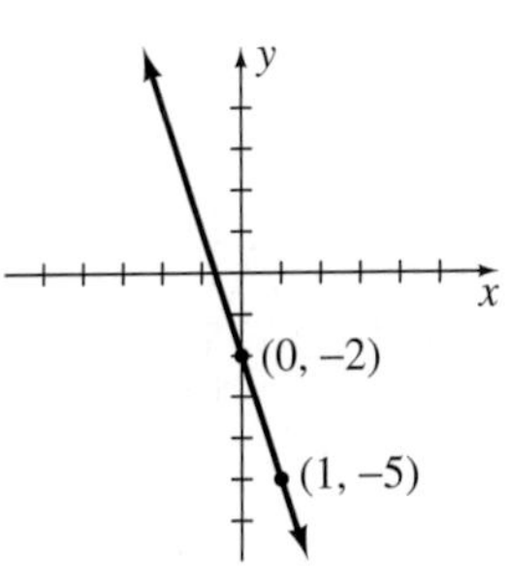

15. $m = 2, b = -4$

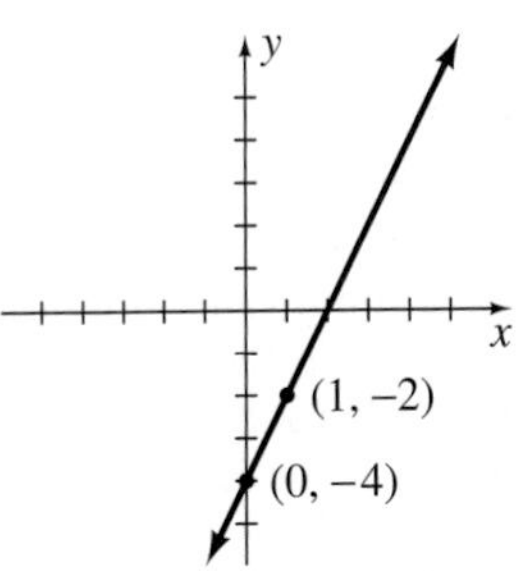

16. $m = \frac{3}{4}, b = -3$

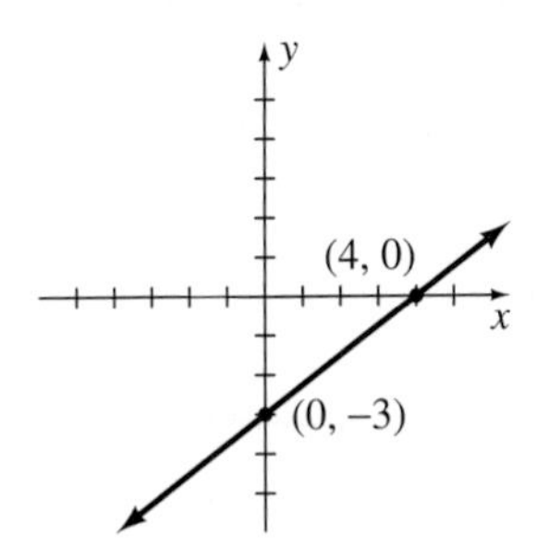

17. $-\frac{9}{5}$ **18.** $\frac{5}{6}$ **19.** $5x + 7y = -11$ **20.** $8x - 3y = 1$
21. $2x - 9y = 9$ **22.** $x = 2$ **23.** $\{(3, -2)\}$
24. $\{(7, 13)\}$ **25.** $\{(16, -5)\}$ **26.** $\left\{\left(\frac{41}{23}, \frac{19}{23}\right)\right\}$
27. $\{(10, 25)\}$ **28.** $\{(-6, -8)\}$ **29.** $\{(400, 600)\}$ **30.** $\varnothing$
31. $\left\{\left(\frac{5}{16}, -\frac{17}{16}\right)\right\}$ **32.** $t = 4$ and $u = 8$
33. $t = 8$ and $u = 4$ **34.** $t = 3$ and $u = 7$ **35.** $\{(-9, 6)\}$

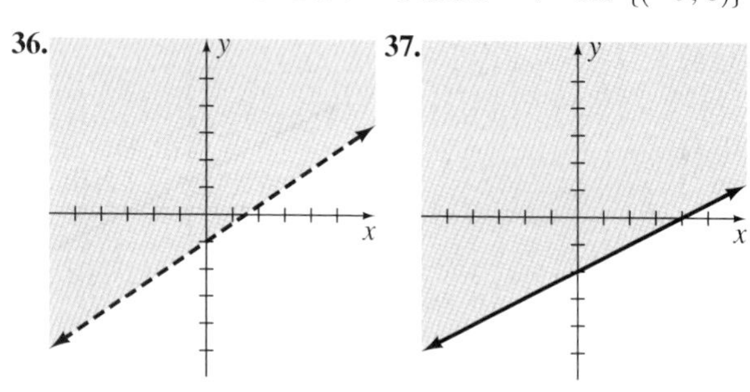

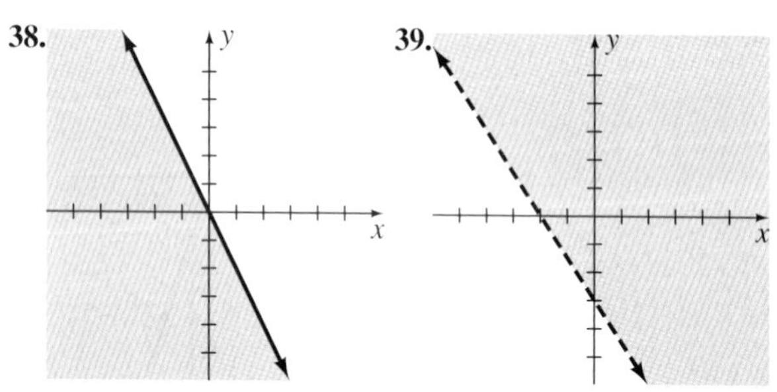

40. 38 and 75 **41.** \$2500 at 6% and \$3000 at 8%
42. 18 nickels and 25 dimes
43. Length of 19 inches and width of 6 inches
44. Length of 12 inches and width of 7 inches
45. 21 dimes and 11 quarters **46.** 32° and 58°
47. 50° and 130°
48. \$3.25 for a cheeseburger and \$2.50 for a milkshake
49. \$1.59 for orange juice and \$.99 for water

Chapter 8 Test (page 384)

1. $m = -\frac{5}{3}, b = 5$

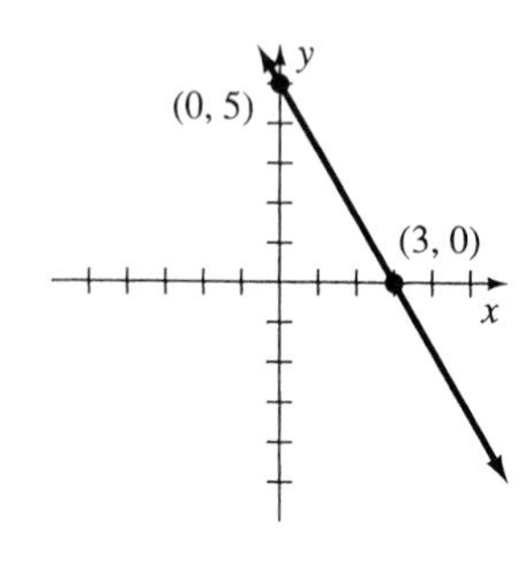

2. $m = 2, b = -4$

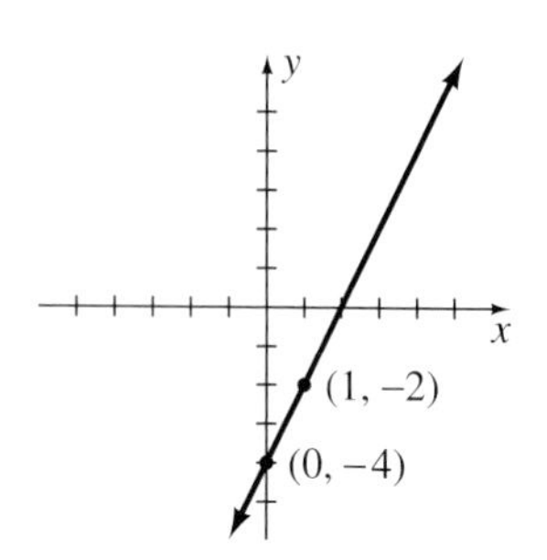

3. $m = -\frac{1}{2}, b = -2$

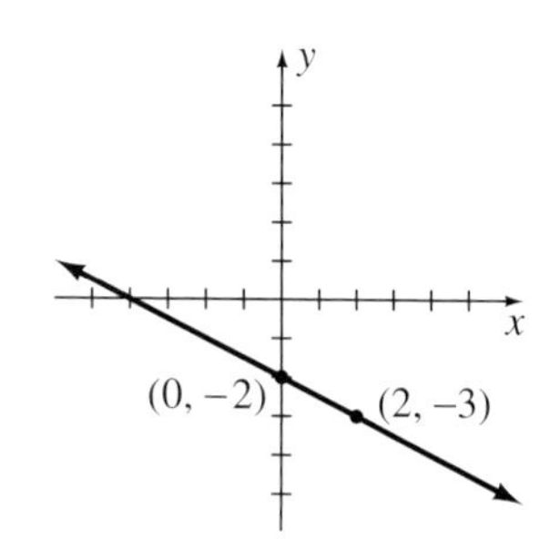

4. $m = -3, b = 0$

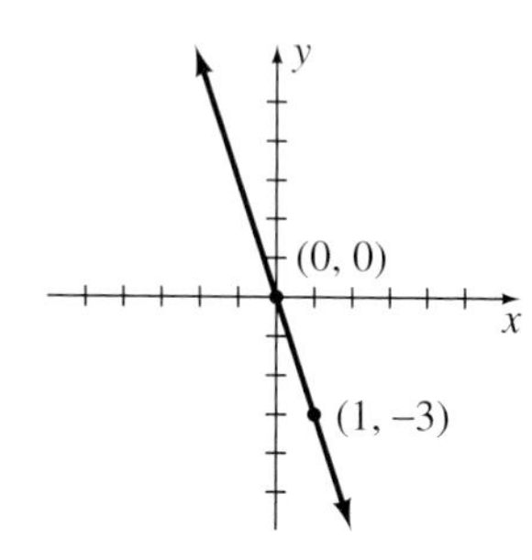

5. 8 **6.** 8 **7.** (7, 6), (11, 7) or others
8. (3, −2), (4, −5) or others **9.** 4.6% **10.** −2
11. $3x + 5y = 20$ **12.** $4x - 9y = 34$ **13.** $3x - 2y = 0$

14.

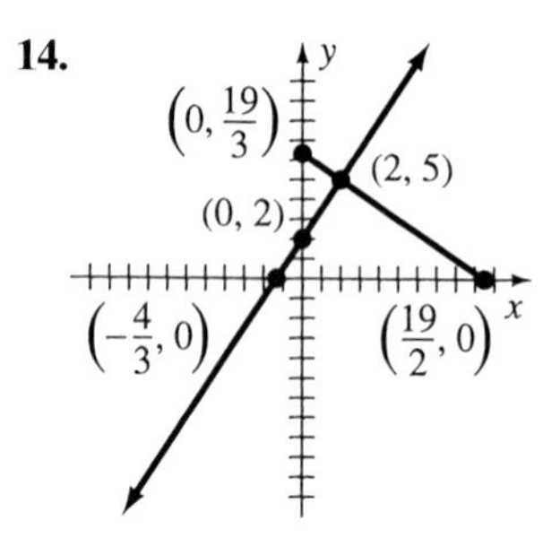

15. $\{(3, 4)\}$ **16.** $\{(-4, 6)\}$ **17.** $\{(-1, -4)\}$
18. $\{(3, -6)\}$

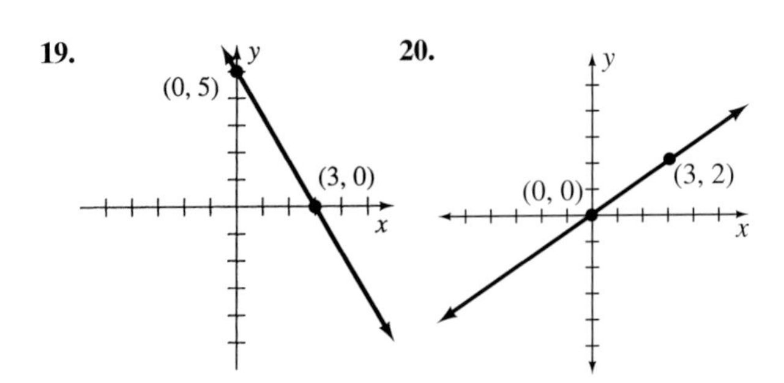

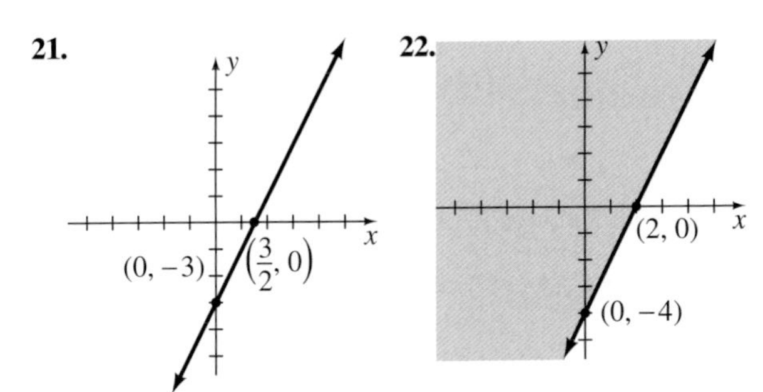

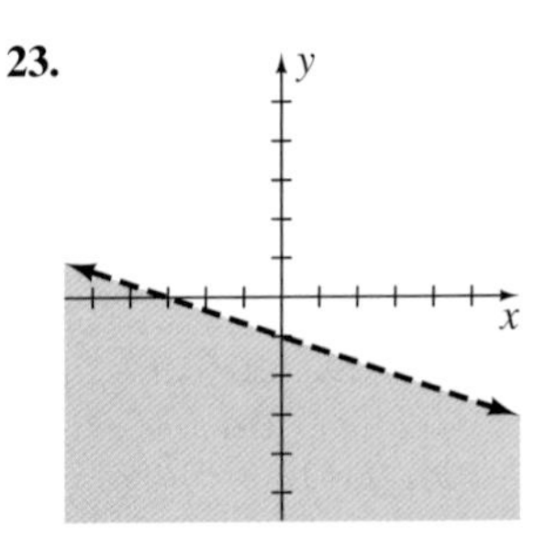

24. Paper \$3.49, notebook \$2.29 **25.** 13 inches

CHAPTER 9

Problem Set 9.1 (page 391)

1. 7 **3.** −8 **5.** 11 **7.** 60 **9.** −40 **11.** 80 **13.** 18 **15.** $\frac{5}{3}$ **17.** 0.4 **19.** 3 **21.** 2 **23.** −9 **25.** −6 **27.** 24 **29.** 48 **31.** 28 **33.** 65 **35.** 58 **37.** 15 **39.** 21 **41.** 4.36 **43.** 7.07 **45.** 8.66 **47.** 9.75 **49.** 66 **51.** 34 **53.** 97 **55.** 20 **57.** 32 **59.** $21\sqrt{2}$ **61.** $8\sqrt{7}$ **63.** $11\sqrt[3]{2}$ **65.** $3\sqrt[3]{7} + 2\sqrt[3]{5}$ **67.** $-\sqrt{2} + 2\sqrt{3}$ **69.** $-9\sqrt{7} - 2\sqrt{10}$ **71.** 17.3 **73.** 13.4 **75.** −1.4 **77.** 26.5 **79.** 4.1 **81.** 2.1 **83.** −29.8 **85.** 1.6 seconds; 2.1 seconds; 2.2 seconds **87.** 2.2 seconds; 2.8 seconds; 18.2 seconds

Problem Set 9.2 (page 397)

1. $2\sqrt{6}$ **3.** $3\sqrt{2}$ **5.** $3\sqrt{3}$ **7.** $2\sqrt{10}$ **9.** $-3\sqrt[3]{2}$ **11.** $4\sqrt{5}$ **13.** $3\sqrt{13}$ **15.** $24\sqrt{2}$ **17.** $6\sqrt[3]{5}$ **19.** $-10\sqrt{5}$ **21.** $-32\sqrt{6}$ **23.** $3\sqrt{2}$ **25.** $\frac{3}{2}\sqrt{3}$ **27.** $-2\sqrt{5}$ **29.** $-\frac{1}{2}\sqrt[3]{4}$ **31.** $xy\sqrt{y}$ **33.** $x\sqrt{2y}$ **35.** $2x\sqrt{2}$ **37.** $3a\sqrt{3ab}$ **39.** $4x\sqrt[3]{xy^2}$ **41.** $3x^2y\sqrt{7}$ **43.** $12x\sqrt{3}$ **45.** $-36x^3\sqrt{2x}$ **47.** $\frac{2}{3}\sqrt{6xy}$ **49.** $\frac{5x}{8}\sqrt[3]{2x}$ **51.** $-\frac{26}{3}a^4$ **53.** $33\sqrt{2}$ **55.** $3\sqrt{5}$ **57.** $18\sqrt[3]{2}$ **59.** $-2\sqrt{7}$ **61.** $9\sqrt{3}$ **63.** $2\sqrt{5}$ **65.** $-15\sqrt{2} - 55\sqrt{5}$ **69.** **(a)** $9\sqrt{2}$ **(c)** $5\sqrt{11}$

Problem Set 9.3 (page 403)

1. $\frac{4}{5}$ **3.** −3 **5.** $\frac{1}{8}$ **7.** $\frac{5}{4}$ **9.** $-\frac{5}{16}$ **11.** $\frac{\sqrt{19}}{5}$ **13.** $\frac{2\sqrt{2}}{7}$ **15.** $\frac{5\sqrt[3]{3}}{6}$ **17.** $\frac{\sqrt{3}}{3}$ **19.** $\frac{\sqrt{6}}{2}$ **21.** $\frac{\sqrt{10}}{4}$ **23.** $\sqrt{7}$ **25.** 3 **27.** $\frac{\sqrt{10}}{6}$ **29.** $\frac{2\sqrt{3}}{9}$ **31.** $\frac{\sqrt{6}}{12}$ **33.** $\frac{2\sqrt{15}}{5}$ **35.** $\frac{4\sqrt{6}}{9}$ **37.** $\frac{\sqrt{21}}{8}$ **39.** $\frac{\sqrt{37}}{3}$ **41.** $\frac{3\sqrt{x}}{x}$ **43.** $\frac{5\sqrt{2x}}{2x}$ **45.** $\frac{\sqrt{3x}}{x}$ **47.** $\frac{2\sqrt{3}}{x}$ **49.** $\frac{\sqrt{10xy}}{5y}$ **51.** $\frac{\sqrt{15xy}}{9y}$ **53.** $\frac{x\sqrt{xy}}{2y}$ **55.** $\frac{3\sqrt{x}}{x^2}$ **57.** $\frac{4\sqrt{x}}{x^4}$ **59.** $\frac{3\sqrt{xy}}{2y^2}$ **61.** $\frac{22\sqrt{3}}{3}$ **63.** $\frac{19\sqrt{10}}{5}$ **65.** $-3\sqrt{5}$ **67.** $-\frac{14\sqrt{6}}{3}$ **69.** $-6\sqrt{3}$ **71.** $-7\sqrt{15}$ **75.** $\frac{\sqrt[3]{63}}{3}$ **77.** $\frac{2\sqrt[3]{5}}{5}$ **79.** $\frac{\sqrt[3]{14}}{4}$

Problem Set 9.4 (page 409)

1. $\sqrt{35}$ **3.** $4\sqrt{3}$ **5.** $5\sqrt{2}$ **7.** $3\sqrt[3]{2}$ **9.** $4\sqrt{6}$ **11.** $15\sqrt{21}$ **13.** $-10\sqrt[3]{3}$ **15.** 72 **17.** $40\sqrt{6}$ **19.** $24\sqrt{5}$ **21.** $\sqrt{6} + \sqrt{10}$ **23.** $2\sqrt{3} - 5\sqrt{6}$ **25.** $2 + \sqrt[3]{20}$ **27.** $6\sqrt{2} - 4\sqrt{6}$ **29.** $4\sqrt{6} - 8\sqrt{15}$ **31.** $56 + 15\sqrt{2}$ **33.** $-9 - 2\sqrt{6}$ **35.** $3\sqrt{2} + 2\sqrt{6} + 4\sqrt{3} + 6$ **37.** 15 **39.** 15 **41.** 51 **43.** $x\sqrt{y}$ **45.** $5x$ **47.** $12a\sqrt{b}$ **49.** $x\sqrt{6} - 2\sqrt{3xy}$ **51.** $x + 2\sqrt{x} - 15$ **53.** $x - 49$ **55.** $\frac{-3\sqrt{2} + 12}{14}$

57. $4\sqrt{6}+8$ **59.** $\sqrt{5}-\sqrt{3}$ **61.** $\frac{-20-30\sqrt{3}}{23}$
63. $\frac{4\sqrt{x}+8}{x-4}$ **65.** $\frac{x-3\sqrt{x}}{x-9}$
67. $\frac{a+7\sqrt{a}+10}{a-25}$ **69.** $\frac{6+2\sqrt{2}+3\sqrt{3}+\sqrt{6}}{7}$

Problem Set 9.5 (page 415)

1. {49} **3.** {18} **5.** ∅ **7.** $\left\{\frac{9}{4}\right\}$ **9.** $\left\{\frac{4}{9}\right\}$ **11.** {14}
13. ∅ **15.** $\left\{\frac{7}{3}\right\}$ **17.** {36} **19.** {6} **21.** {2}
23. $\left\{\frac{21}{4}\right\}$ **25.** {−3, −2} **27.** {6} **29.** {2} **31.** {12}
33. {25} **35.** {3} **37.** $\left\{-\frac{1}{2}\right\}$ **39.** {2}
41. 56 feet; 106 feet; 148 feet
43. 3.2 feet; 5.1 feet; 7.3 feet **47.** {18} **49.** {4, 12}
51. {57} **53.** {110} **55.** {252}

Chapter 9 Review Problem Set (page 417)

1. 8 **2.** −7 **3.** 40 **4.** $\frac{9}{5}$ **5.** $-\frac{2}{3}$ **6.** $\frac{7}{6}$ **7.** $2\sqrt{5}$
8. $4\sqrt{2}$ **9.** $10\sqrt{2}$ **10.** $4\sqrt{5}$ **11.** −10 **12.** $\sqrt[3]{5}$
13. $\frac{6\sqrt{7}}{7}$ **14.** $\frac{\sqrt{14}}{4}$ **15.** $\frac{\sqrt{3}}{3}$ **16.** $\frac{3\sqrt{10}}{5}$ **17.** 2
18. $\frac{5\sqrt{6}}{6}$ **19.** $\frac{-\sqrt{6}}{3}$ **20.** $\frac{2\sqrt{2}}{3}$ **21.** 5.2 **22.** 1.2
23. 17.3 **24.** −6.9 **25.** $2ab\sqrt{3b}$ **26.** $5y^2\sqrt{2x}$
27. $4xy\sqrt{3x}$ **28.** $5\sqrt[3]{a^2b}$ **29.** $4y\sqrt{3x}$ **30.** $\frac{3x\sqrt[3]{3}}{2}$
31. $\frac{\sqrt{10xy}}{5y}$ **32.** $\frac{3\sqrt{2xy}}{2y}$ **33.** $\frac{2\sqrt{x}}{x}$ **34.** $\frac{x\sqrt{2x}}{3}$
35. $\frac{3\sqrt{xy}}{4y^2}$ **36.** $\frac{-2\sqrt{x}}{5}$ **37.** $6\sqrt{2}$ **38.** $18\sqrt{2}$
39. −40 **40.** $10\sqrt[3]{28}$ **41.** $\sqrt[3]{6}+2$
42. $6\sqrt{10}-12\sqrt{15}$ **43.** $3+\sqrt{21}+\sqrt{15}+\sqrt{35}$
44. $-24-7\sqrt{6}$ **45.** $4+5\sqrt{42}$ **46.** $-18-\sqrt{5}$
47. $\frac{5(\sqrt{7}+\sqrt{5})}{2}$ **48.** $3\sqrt{2}+2\sqrt{3}$ **49.** $\frac{3\sqrt{2}+\sqrt{6}}{6}$
50. $\frac{3\sqrt{42}-4\sqrt{15}}{23}$ **51.** $18\sqrt{2}$ **52.** $-7\sqrt{2x}$
53. $-\sqrt[3]{2}$ **54.** $\frac{16\sqrt{10}}{5}$ **55.** $\frac{52\sqrt{5}}{5}$ **56.** $\frac{-17\sqrt{6}}{3}$
57. {6} **58.** {4} **59.** {0, 9} **60.** {−5, −4}
61. {3} **62.** {1} **63.** 46 **64.** 66 **65.** 72 **66.** 26
67. 47 **68.** 74

Chapter 9 Test (page 419)

1. $-\frac{8}{7}$ **2.** 0.05 **3.** 2.8 **4.** −5.6 **5.** 2.1 **6.** $3\sqrt{5}$
7. $-12\sqrt[3]{2}$ **8.** $\frac{\sqrt{2}}{3}$ **9.** $\frac{5\sqrt{2}}{2}$ **10.** $\frac{\sqrt{6}}{3}$ **11.** $\frac{\sqrt{10}}{4}$
12. $-5xy\sqrt[3]{2x}$ **13.** $\frac{\sqrt{15xy}}{5y}$ **14.** $3xy\sqrt{3x}$ **15.** $4\sqrt{6}$
16. $24\sqrt[3]{10}$ **17.** $12\sqrt{2}-12\sqrt{3}$ **18.** $1-5\sqrt{15}$
19. $\frac{3\sqrt{2}-\sqrt{3}}{5}$ **20.** $4\sqrt{6}$ **21.** 22 **22.** {5} **23.** ∅
24. {3} **25.** {−2, 1}

Cumulative Review Problem Set (page 420)

1. −64 **2.** 64 **3.** 144 **4.** −8 **5.** $\frac{2}{3}$ **6.** $\frac{13}{9}$ **7.** −9
8. −49 **9.** −29 **10.** 49 **11.** $-\frac{15}{4x}$ **12.** $\frac{-x+17}{(x-2)(x+3)}$
13. $\frac{5y}{2}$ **14.** $\frac{1}{x-4}$ **15.** $\frac{-8x-41}{(x+6)(x-3)}$ **16.** $60x^4y^4$
17. $\frac{-8}{x^2}$ **18.** $\frac{-3a^3}{b}$ **19.** $\frac{1}{3n^4}$ **20.** $27x^2+30x-8$
21. $-5x^2-12x-7$ **22.** $6x^3-x^2-13x-4$
23. $3x^3y^6-4y^3$ **24.** $2x^2-2x-3$ **25.** $y-x$
26. 4.67 gallons **27.** 25% **28.** 108 **29.** 130 feet
30. 314 square inches **31.** **(a)** $(8.5)(10^4)$ **(b)** $(9)(10^{-4})$
(c) $(1.04)(10^{-6})$ **(d)** $(5.3)(10^7)$ **32.** $2x(2x+5)(3x-4)$
33. $3(2x+3)(2x-3)$ **34.** $(y+3)(x-2)$
35. $(5-x)(6+5x)$ **36.** $4(x+1)(x-1)(x^2+1)$
37. $(7x-2)(3x+4)$ **38.** $8\sqrt{7}$ **39.** $-3\sqrt{5}$ **40.** $\frac{6\sqrt{5}}{5}$
41. $\frac{5\sqrt{6}}{18}$ **42.** $6y^2\sqrt{2xy}$ **43.** $-\frac{2\sqrt{ab}}{5}$ **44.** 48
45. $216-36\sqrt{6}$ **46.** 11 **47.** $4\sqrt{3}-4\sqrt{2}$
48. $-\frac{2(3\sqrt{5}+\sqrt{6})}{13}$ **49.** $\sqrt{2}$ **50.** $\frac{37\sqrt{5}}{12}$

51. **52.**

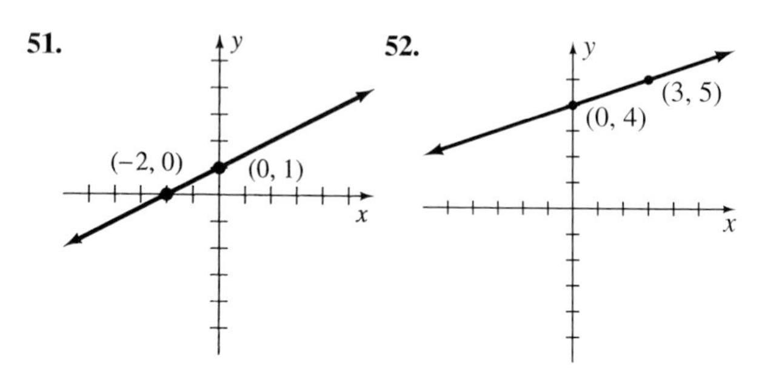

53. **54.**

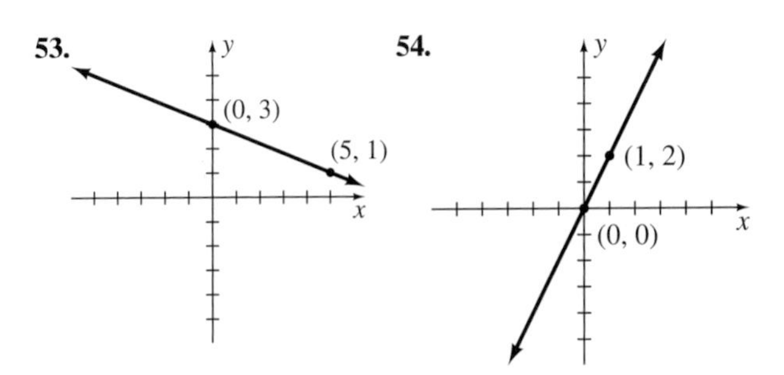

55. **56.**

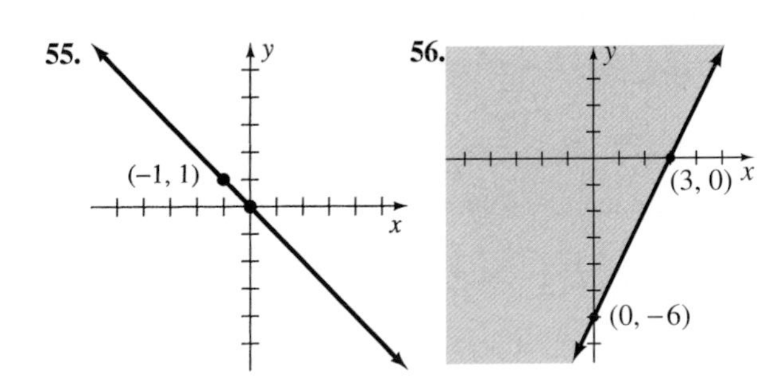

57. **58.**

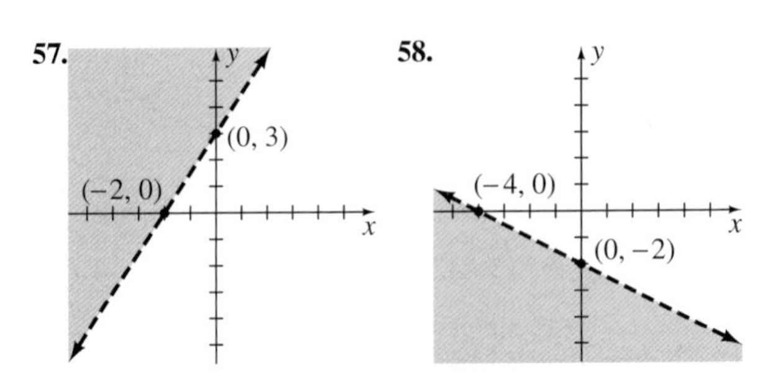

59. -2 **60.** $\frac{4}{7}$ **61.** $2x - 3y = 8$ **62.** $4x + 3y = -13$
63. $x + 4y = -12$ **64.** 3/2 **65.** $\{(1, -2)\}$
66. $\{(-2, 4)\}$ **67.** $\{(-6, 12)\}$ **68.** $\left\{\left(\frac{1}{2}, 3\right)\right\}$ **69.** $\{17\}$
70. $\left\{-\frac{23}{5}\right\}$ **71.** $\{5\}$ **72.** $\left\{-\frac{19}{10}\right\}$ **73.** $\left\{\frac{47}{7}\right\}$
74. $\{500\}$ **75.** $\{53\}$ **76.** $\left\{-\frac{2}{3}\right\}$ **77.** $\{-6, 2\}$
78. $\{-2, 2\}$ **79.** $\{29\}$ **80.** $\{27\}$
81. $\{n|n \geq -5\}$ or $[-5, \infty)$ **82.** $\left\{n|n > \frac{1}{4}\right\}$ or $\left(\frac{1}{4}, \infty\right)$
83. $\left\{x|x > -\frac{2}{5}\right\}$ or $\left(-\frac{2}{5}, \infty\right)$
84. $\{n|n > 6\}$ or $(6, \infty)$ **85.** $\left\{x|x < \frac{5}{16}\right\}$ or $\left(-\infty, \frac{5}{16}\right)$
86. $\left\{x|x \geq -\frac{16}{3}\right\}$ or $\left[-\frac{16}{3}, \infty\right)$ **87.** 65° and 115°
88. 13 and 37 **89.** 7, 9
90. 7 nickels, 15 dimes, and 25 quarters **91.** \$1350
92. \$48 **93.** $4\frac{1}{2}$ hours **94.** 12.5 milliliters
95. 91 or better **96.** More than 11 **97.** 12 minutes

CHAPTER 10

Problem Set 10.1 (page 430)

1. $\{-15, 0\}$ **3.** $\{0, 12\}$ **5.** $\{0, 5\}$ **7.** $\{1, 8\}$ **9.** $\{-2, 7\}$
11. $\{-6, 1\}$ **13.** $\left\{-\frac{5}{3}, \frac{1}{2}\right\}$ **15.** $\left\{\frac{2}{5}, \frac{5}{6}\right\}$ **17.** $\left\{\frac{1}{2}\right\}$
19. $\{-8, 8\}$ **21.** $\left\{-\frac{5}{3}, \frac{5}{3}\right\}$ **23.** $\{-4, 4\}$
25. $\{-\sqrt{14}, \sqrt{14}\}$ **27.** No real number solutions
29. $\{-4\sqrt{2}, 4\sqrt{2}\}$ **31.** $\{-3\sqrt{2}, 3\sqrt{2}\}$
33. $\left\{-\frac{3\sqrt{2}}{2}, \frac{3\sqrt{2}}{2}\right\}$ **35.** $\left\{-\frac{5\sqrt{2}}{4}, \frac{5\sqrt{2}}{4}\right\}$ **37.** $\{-1, 3\}$
39. $\{-8, 2\}$ **41.** $\left\{-\frac{5}{3}, 3\right\}$ **43.** $\{-6 - \sqrt{5}, -6 + \sqrt{5}\}$
45. $\{1 - 2\sqrt{2}, 1 + 2\sqrt{2}\}$
47. $\left\{\frac{-3 - 2\sqrt{5}}{2}, \frac{-3 + 2\sqrt{5}}{2}\right\}$
49. No real number solutions
51. $\left\{\frac{5 - 2\sqrt{10}}{3}, \frac{5 + 2\sqrt{10}}{3}\right\}$ **53.** $\left\{-\frac{3}{7}, \frac{5}{7}\right\}$
55. $\{-8 - 5\sqrt{2}, -8 + 5\sqrt{2}\}$ **57.** $5\sqrt{2}$ inches
59. $2\sqrt{7}$ meters **61.** $2\sqrt{11}$ feet
63. $a = 4$ inches and $b = 4\sqrt{3}$ inches
65. $c = 12$ feet and $b = 6\sqrt{3}$ feet
67. $a = 4\sqrt{3}$ meters and $c = 8\sqrt{3}$ meters
69. $a = 10$ inches and $c = 10\sqrt{2}$ inches
71. $a = b = \frac{9\sqrt{2}}{2}$ meters **73.** 8.2 feet **75.** 30 meters
77. 35 meters **81.** 5.2 centimeters **83.** 10.8 centimeters

Problem Set 10.2 (page 437)

1. $\{-4 - \sqrt{17}, -4 + \sqrt{17}\}$
3. $\{-5 - \sqrt{23}, -5 + \sqrt{23}\}$ **5.** $\{2 - 2\sqrt{2}, 2 + 2\sqrt{2}\}$
7. No real number solutions
9. $\{-1 - 3\sqrt{2}, -1 + 3\sqrt{2}\}$
11. $\left\{\frac{-1 - \sqrt{13}}{2}, \frac{-1 + \sqrt{13}}{2}\right\}$

13. $\left\{\frac{5-\sqrt{33}}{2}, \frac{5+\sqrt{33}}{2}\right\}$
15. $\left\{\frac{-4-\sqrt{22}}{2}, \frac{-4+\sqrt{22}}{2}\right\}$
17. $\left\{\frac{-6-\sqrt{42}}{3}, \frac{-6+\sqrt{42}}{3}\right\}$
19. $\left\{\frac{2-\sqrt{2}}{2}, \frac{2+\sqrt{2}}{2}\right\}$
21. No real number solutions
23. $\left\{\frac{9-\sqrt{65}}{2}, \frac{9+\sqrt{65}}{2}\right\}$
25. $\left\{\frac{-3-\sqrt{17}}{4}, \frac{-3+\sqrt{17}}{4}\right\}$
27. $\left\{\frac{-1-\sqrt{7}}{3}, \frac{-1+\sqrt{7}}{3}\right\}$ **29.** $\{-14, 12\}$
31. $\{-11, 15\}$ **33.** $\{-6, 2\}$ **35.** $\{-9, -3\}$
37. $\{-5, 8\}$ **39.** $\left\{\frac{1}{2}, 4\right\}$ **41.** $\left\{-\frac{5}{2}, \frac{3}{2}\right\}$
45. $\left\{\frac{-b+\sqrt{b^2-4ac}}{2a}, \frac{-b-\sqrt{b^2-4ac}}{2a}\right\}$

Problem Set 10.3 (page 442)

1. $\{-1, 6\}$ **3.** $\{-9, 4\}$ **5.** $\{1-\sqrt{6}, 1+\sqrt{6}\}$
7. $\left\{\frac{5-\sqrt{33}}{2}, \frac{5+\sqrt{33}}{2}\right\}$ **9.** No real number solutions
11. $\{-2-\sqrt{2}, -2+\sqrt{2}\}$ **13.** $\{0, 6\}$ **15.** $\left\{0, \frac{7}{2}\right\}$
17. $\{16, 18\}$ **19.** $\{-10, 8\}$ **21.** $\{-2\}$ **23.** $\left\{-\frac{2}{3}, \frac{1}{2}\right\}$
25. $\left\{-1, \frac{2}{5}\right\}$ **27.** $\left\{-\frac{5}{4}, -\frac{1}{3}\right\}$
29. $\left\{\frac{-5-\sqrt{73}}{4}, \frac{-5+\sqrt{73}}{4}\right\}$
31. $\left\{\frac{-2-\sqrt{7}}{3}, \frac{-2+\sqrt{7}}{3}\right\}$ **33.** $\left\{-\frac{3}{4}\right\}$
35. $\left\{\frac{-2-\sqrt{5}}{2}, \frac{-2+\sqrt{5}}{2}\right\}$
37. $\left\{\frac{-9-\sqrt{57}}{12}, \frac{-9+\sqrt{57}}{12}\right\}$
39. $\left\{\frac{1-\sqrt{33}}{4}, \frac{1+\sqrt{33}}{4}\right\}$
41. No real number solutions
43. $\left\{\frac{-5-\sqrt{137}}{14}, \frac{-5+\sqrt{137}}{14}\right\}$
45. $\left\{\frac{1-\sqrt{85}}{6}, \frac{1+\sqrt{85}}{6}\right\}$ **47.** $\{-14, -9\}$
53. $\{-2.52, 7.52\}$ **55.** $\{-8.10, 2.10\}$ **57.** $\{-3.55, 1.22\}$
59. $\{-1.33, 3.58\}$ **61.** $\{-1.95, 2.15\}$

Problem Set 10.4 (page 448)

1. $\{-9, 5\}$ **3.** $\left\{-\frac{13}{5}, \frac{1}{5}\right\}$ **5.** $\{-1, 2\}$ **7.** $\left\{0, \frac{8}{3}\right\}$
9. $\left\{\frac{1}{3}\right\}$ **11.** $\left\{0, \frac{2\sqrt{2}}{5}\right\}$ **13.** $\{7-2\sqrt{17}, 7+2\sqrt{17}\}$
15. $\left\{-1, \frac{7}{5}\right\}$ **17.** $\left\{-\frac{5}{3}, -\frac{1}{5}\right\}$ **19.** $\{\sqrt{2}-3, \sqrt{2}+3\}$
21. $\{-12, 7\}$ **23.** $\left\{\frac{3-\sqrt{33}}{4}, \frac{3+\sqrt{33}}{4}\right\}$ **25.** $\{-1, 4\}$
27. No real number solutions **29.** $\{16, 30\}$
31. $\left\{\frac{-1-\sqrt{13}}{2}, \frac{-1+\sqrt{13}}{2}\right\}$ **33.** $\left\{\frac{3}{4}, \frac{4}{3}\right\}$
35. $\{-13, 1\}$ **37.** $\{11, 17\}$ **39.** $\left\{\frac{3-\sqrt{3}}{2}, \frac{3+\sqrt{3}}{2}\right\}$
41. $\left\{-1, -\frac{2}{3}\right\}$ **43.** $\left\{\frac{-5-\sqrt{193}}{6}, \frac{-5+\sqrt{193}}{6}\right\}$
45. $\{-5, 3\}$

Problem Set 10.5 (page 454)

1. 17 and 18 **3.** 19 and 25 **5.** $3+\sqrt{5}$ and $3-\sqrt{5}$
7. $\sqrt{2}$ or $\frac{\sqrt{2}}{2}$ **9.** 12, 14, and 16 **11.** -8 or 8
13. 8 meters by 12 meters **15.** 15 centimeters by 25 centimeters **17.** 27 feet by 78 feet **19.** 15 rows and 20 seats per row **21.** 7 feet by 9 feet **23.** 6 inches and 8 inches **25.** $1\frac{1}{2}$ inches **27.** 30 students **29.** 24 inches and 32 inches **31.** 18 miles per hour

Chapter 10 Review Problem Set (page 456)

1. $\{-6, -1\}$ **2.** $\{-4-\sqrt{13}, -4+\sqrt{13}\}$ **3.** $\left\{\frac{2}{7}, \frac{1}{3}\right\}$
4. $\{0, 17\}$ **5.** $\{-4, 1\}$ **6.** $\{11, 15\}$
7. $\left\{\frac{-7-\sqrt{61}}{6}, \frac{-7+\sqrt{61}}{6}\right\}$ **8.** $\left\{\frac{1}{2}\right\}$
9. No real number solutions **10.** $\{-5, -1\}$
11. $\{2-\sqrt{2}, 2+\sqrt{2}\}$
12. $\{-2-3\sqrt{2}, -2+3\sqrt{2}\}$ **13.** $\{-3\sqrt{5}, 3\sqrt{5}\}$
14. $\{-3, 9\}$ **15.** $\{0, 1\}$ **16.** $\{2-\sqrt{13}, 2+\sqrt{13}\}$
17. $\{20, 24\}$ **18.** $\{2-2\sqrt{2}, 2+2\sqrt{2}\}$ **19.** $\left\{-\frac{8}{5}, 1\right\}$

20. $\left\{\frac{-1-\sqrt{73}}{12}, \frac{-1+\sqrt{73}}{12}\right\}$ **21.** $\left\{\frac{1}{2}, 4\right\}$

22. $\left\{-\frac{4}{3}, -1\right\}$ **23.** 9 inches by 12 inches

24. 18 and 19 **25.** 7, 9, and 11

26. $\sqrt{5}$ meters and $3\sqrt{5}$ meters

27. 4 yards and 6 yards **28.** 40 shares at \$20 per share

29. 10 meters

30. Jay's rate was 45 miles per hour and Jean's rate was 48 miles per hour; or Jay's rate was $7\frac{1}{2}$ miles per hour and Jean's rate was $10\frac{1}{2}$ miles per hour.

31. $6\sqrt{2}$ inches **32.** $\frac{16\sqrt{3}}{3}$ centimeters

Chapter 10 Test (page 458)

1. $2\sqrt{13}$ inches **2.** 13 meters **3.** 7 inches

4. $4\sqrt{3}$ centimeters **5.** $\left\{-3, \frac{5}{3}\right\}$ **6.** $\{-4, 4\}$

7. $\left\{\frac{1}{2}, \frac{3}{4}\right\}$ **8.** $\left\{\frac{3-\sqrt{29}}{2}, \frac{3+\sqrt{29}}{2}\right\}$

9. $\{-1-\sqrt{10}, -1+\sqrt{10}\}$

10. No real number solutions **11.** $\{-12, 2\}$

12. $\left\{\frac{3-\sqrt{41}}{4}, \frac{3+\sqrt{41}}{4}\right\}$

13. $\left\{\frac{1-\sqrt{57}}{2}, \frac{1+\sqrt{57}}{2}\right\}$ **14.** $\left\{\frac{1}{5}, 2\right\}$ **15.** $\{13, 15\}$

16. $\left\{\frac{3}{4}, 4\right\}$ **17.** $\left\{0, \frac{1}{6}\right\}$ **18.** $\left\{-1, \frac{3}{7}\right\}$

19. $\left\{\frac{1-3\sqrt{3}}{4}, \frac{1+3\sqrt{3}}{4}\right\}$

20. No real number solutions **21.** 15 seats per row

22. 14 miles per hour **23.** 15 and 17 **24.** 9 feet

25. 20 shares

CHAPTER 11

Problem Set 11.1 (page 463)

1. $\{-4, 4\}$

-4 4

3. $\{x|x > -1 \text{ and } x < 1\}$ or $(-1, 1)$

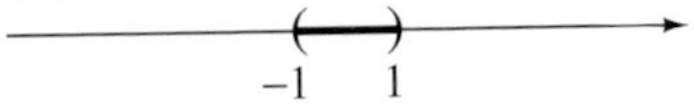

-1 1

5. $\{x|x \le -2 \text{ or } x \ge 2\}$ or $(-\infty, -2] \cup [2, \infty)$

-2 2

7. $\{-3, -1\}$

-3 -1

9. $\{-1, 3\}$

-1 3

11. $\{x|x \ge 0 \text{ and } x \le 4\}$ or $[0, 4]$

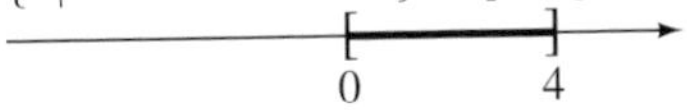

0 4

13. $\{x|x < -4 \text{ or } x > 2\}$ or $(-\infty, -4) \cup (2, \infty)$

-4 2

15. $\{-2, 1\}$

-2 1

17. $\left\{-\frac{2}{5}, \frac{6}{5}\right\}$

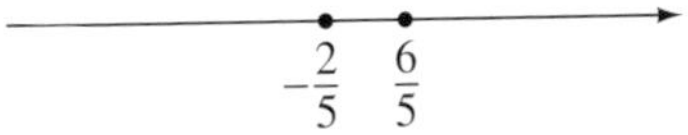

$-\frac{2}{5}$ $\frac{6}{5}$

19. $\{x|x \le 1 \text{ or } x \ge 2\}$ or $(-\infty, 1] \cup [2, \infty)$

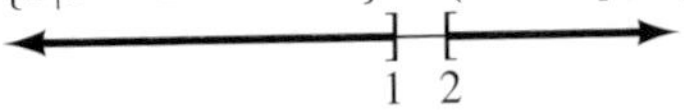

1 2

21. $\left\{x|x > -\frac{5}{4} \text{ and } x < -\frac{1}{4}\right\}$ or $\left(-\frac{5}{4}, -\frac{1}{4}\right)$

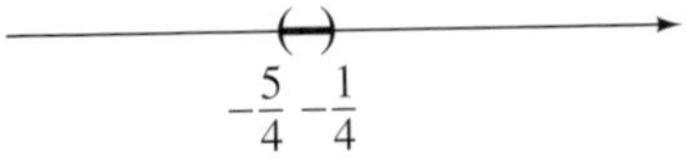

$-\frac{5}{4}$ $-\frac{1}{4}$

23. $\{-2\}$

-2

25. $\{x|x \ne \frac{2}{3}\}$ or $\left(-\infty, \frac{2}{3}\right) \cup \left(\frac{2}{3}, \infty\right)$

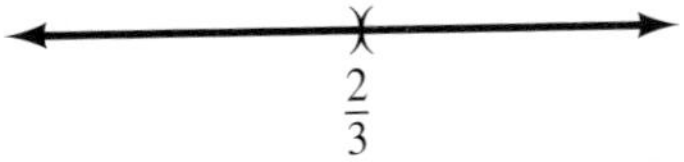

$\frac{2}{3}$

27. $\left\{-\frac{16}{3}, 6\right\}$

29. $\{x|x < -5 \text{ or } x > 4\}$ or $(-\infty, -5) \cup (4, \infty)$

31. $\left\{x|x > -\frac{14}{3} \text{ and } x < 8\right\}$ or $\left(-\frac{14}{3}, 8\right)$

33. $\left\{-6, \frac{16}{3}\right\}$ **35.** $\left\{x|x \ge -6 \text{ and } x \le \frac{19}{2}\right\}$ or $\left[-6, \frac{19}{2}\right]$

37. $\left\{x|x \le -\frac{21}{5} \text{ or } x \ge 3\right\}$ or $\left(-\infty, -\frac{21}{5}\right] \cup [3, \infty)$

39. $\{x|x > -6 \text{ and } x < 2\}$ or $(-6, 2)$

41. $\left\{x|x < -\frac{5}{2} \text{ or } x > \frac{7}{2}\right\}$ or $\left(-\infty, -\frac{5}{2}\right) \cup \left(\frac{7}{2}, \infty\right)$
43. $\{0\}$ **45.** $\{x|x \text{ is any real number}\}$ **47.** $\varnothing$
49. $\{-6\}$ **53.** $\{x|4 < x < 14\}$ or $(4, 14)$
55. $\{x|-1 \le x \le 4\}$ or $[-1, 4]$
57. $\{x|-11 < x < 7\}$ or $(-11, 7)$
59. $\{x|-7 < x < -1\}$ or $(-7, -1)$
61. $\{x|-1 < x < 6\}$ or $(-1, 6)$

Problem Set 11.2 (page 473)

1. $(3, 1, -2)$ **3.** $(-1, 3, 4)$ **5.** $(5, -3, 1)$ **7.** $(-2, 3, 5)$
9. $(3, -2, -4)$ **11.** $(-1, 3, 1)$ **13.** $(3, 0, -2)$
15. $(1, 4, 2)$ **17.** 10 quarters, 8 dimes, 2 nickels
19. $\angle A = 120°$, $\angle B = 24°$, $\angle C = 36°$
21. Plumber = \$50 per hour; Apprentice = \$20 per hour; Laborer = \$10 per hour
23. Peaches = \$1.29 per pound; Cherries = \$.99 per pound; Pears = \$.69 per pound
25. Helmet = \$250, Jacket = \$350, Gloves = \$50
29. Infinitely many solutions **31.** $\varnothing$

Problem Set 11.3 (page 479)

1. 9 **3.** -10 **5.** 5 **7.** -4 **9.** $\frac{4}{7}$ **11.** 3 **13.** -3
15. 8 **17.** 16 **19.** 16 **21.** 32 **23.** $\frac{1}{2}$ **25.** -3 **27.** $\frac{1}{8}$
29. $\frac{27}{8}$ **31.** 2 **33.** 625 **35.** -32 **37.** $\frac{1}{8}$ **39.** 2
41. 27 **43.** 1 **45.** $\frac{1}{3}$ **47.** 4 **49.** 49 **51.** $x^{\frac{3}{4}}$ **53.** $a^{\frac{17}{12}}$
55. $15x^{\frac{7}{12}}$ **57.** $24x^{\frac{11}{12}}$ **59.** $2y^{\frac{5}{12}}$ **61.** $10n^{\frac{1}{4}}$ **63.** $\frac{2}{x^{\frac{1}{6}}}$
65. $25xy^2$ **67.** $64x^{\frac{3}{4}}y^{\frac{3}{2}}$ **69.** $2x^2y$ **71.** $4x^{\frac{4}{15}}$ **73.** $\frac{4}{b^{\frac{5}{12}}}$
75. 3 **77.** $\frac{9}{4x^{\frac{1}{3}}}$ **79.** $\frac{125x^{\frac{3}{2}}}{216y}$
83. **(a)** 28 **(b)** 35 **(c)** 17 **(d)** 42 **85.** **(a)** 11.18 **(b)** 5.28 **(c)** 3.11 **(d)** 1573.56 **(e)** 6.45 **(f)** 5.81

Problem Set 11.4 (page 483)

1. $8i$ **3.** $\frac{5}{3}i$ **5.** $i\sqrt{11}$ **7.** $5i\sqrt{2}$ **9.** $4i\sqrt{3}$ **11.** $3i\sqrt{6}$
13. $8 + 17i$ **15.** $10 - 10i$ **17.** $4 + 2i$ **19.** $-2 - 6i$
21. $-5 + 3i$ **23.** $-12 - 16i$ **25.** $-10 - 4i$
27. $-1 + 12i$ **29.** $-20 - 8i$ **31.** $\frac{5}{6} + \frac{5}{12}i$
33. $-\frac{1}{15} + \frac{7}{12}i$ **35.** $-56 + 0i$ **37.** $-6 + 12i$
39. $-24 + 20i$ **41.** $-2 + 23i$ **43.** $59 - 17i$
45. $-21 - 12i$ **47.** $-26 + 15i$ **49.** $-9 + 40i$
51. $61 + 0i$ **53.** $5 + 0i$

Problem Set 11.5 (page 487)

1. $\{-8i, 8i\}$ **3.** $\{2 - i, 2 + i\}$
5. $\{-5 - i\sqrt{13}, -5 + i\sqrt{13}\}$
7. $\{3 - 3i\sqrt{2}, 3 + 3i\sqrt{2}\}$ **9.** $\left\{-\frac{2}{5}, \frac{4}{5}\right\}$ **11.** $\{-1, 4\}$
13. $\{-3 - i\sqrt{3}, -3 + i\sqrt{3}\}$ **15.** $\{3 - 2i, 3 + 2i\}$
17. $\{2 - 4i, 2 + 4i\}$ **19.** $\left\{\frac{1 - i\sqrt{2}}{3}, \frac{1 + i\sqrt{2}}{3}\right\}$
21. $\left\{-1, \frac{5}{2}\right\}$ **23.** $\{1 - 3i\sqrt{2}, 1 + 3i\sqrt{2}\}$
25. $\{2 - i\sqrt{3}, 2 + i\sqrt{3}\}$ **27.** $\left\{\frac{1 - i\sqrt{31}}{8}, \frac{1 + i\sqrt{31}}{8}\right\}$
29. $\left\{\frac{-1 - i\sqrt{5}}{6}, \frac{-1 + i\sqrt{5}}{6}\right\}$

Problem Set 11.6 (page 494)

1. 14% **3.** 480 **5.** 66% **7.** Physics **9.** 21%
11. 62% **13.** 1000 people **15.** 1500 people
17. Jan. and Feb. **19.** Feb. and March **21.** 0.4%
23. \$390 **25.** Monday or Wednesday **27.** \$580
29. 2% **31.** 3% **33.** 2002 and 2003
35. **(a)** 12% **(b)** 12% **(c)** Neither, they are the same
39.

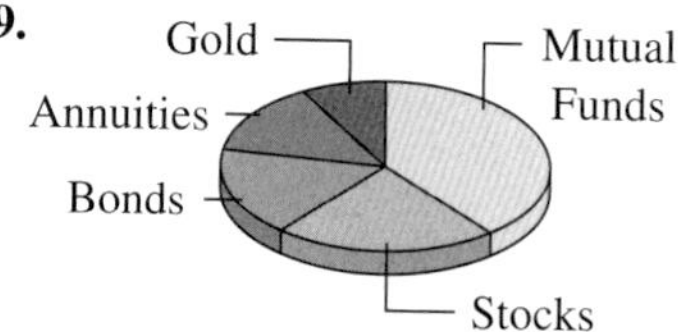

Problem Set 11.7 (page 501)

1. Domain: $\{4, 6, 8, 10\}$
Range: $\{7, 11, 20, 28\}$
It is a function.
3. Domain: $\{-2, -1, 0, 1\}$
Range: $\{1, 2, 3, 4\}$
It is a function.
5. Domain: $\{4, 9\}$
Range: $\{-3, -2, 2, 3\}$
It is not a function.
7. Domain: $\{3, 4, 5, 6\}$
Range: $\{15\}$
It is a function.

9. Domain: {Carol}
Range: {22400, 23700, 25200}
It is not a function.

11. Domain: $\{-6\}$
Range: {1, 2, 3, 4}
It is not a function.

13. Domain: $\{-2, -1, 0, 1, 2\}$
Range: {0, 1, 4}
It is a function.

15. All reals **17.** All reals except -8

19. All reals **21.** All reals except 5

23. All reals except $\frac{8}{5}$ **25.** All reals

27. All reals except 0 **29.** All reals **31.** 4; 7; 1; 22

33. -16; 19; 24; $-5t - 1$ **35.** $\frac{11}{4}; \frac{13}{12}; \frac{19}{36}; -\frac{7}{12}$

37. 0; 0; 45; -4 **39.** 0; -3; -3; -8 **41.** 23; -21; 3; 8

43. 8; 48; 8; -7

Problem Set 11.8 (page 504)

1. 9; 289; 72.25; 430.56; 126.56 **3.** \$464,500

5. \$6.75; \$10.13; \$13.50; \$24.60 **7.** \$30; \$15; \$27

9. \$4.00; \$1.50; \$0; \$6.00 **11.** \$367.50; \$420; \$577.50; \$210

13. \$118.50; \$138.50; \$98.50; \$158.50

15.

x	0	1	2	3	4
$f(x)$	0	1/10	1/5	3/10	2/5

17.

n	10,000	15,000	20,000	25,000	30,000
$f(n)$	3	4.5	6	7.5	9

19. (a) 12 **(b)** 20

(c)

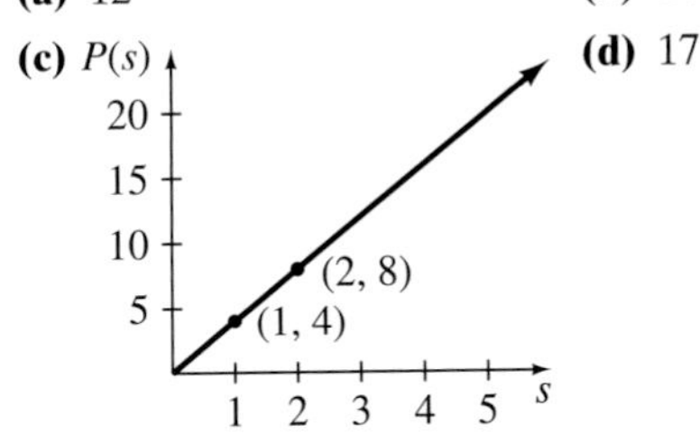

(d) 17

21. (a) \$20,800 **(b)** \$14,950

(c)

(d) \$13,000

(f) 16.7 years

23. (a)

t	50	41	-4	212	95	77	59
$f(t)$	10	5	-20	100	35	25	15

(b)

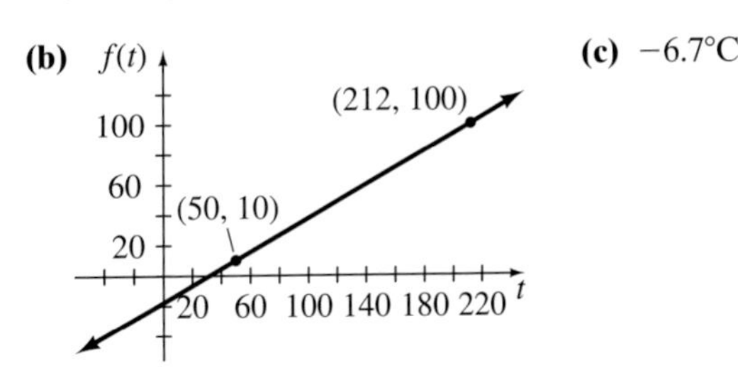

(c) $-6.7°C$

Chapter 11 Review Problem Set (page 508)

1. 55% **2.** 44% **3.** 52% **4.** 25 students

5. Shopping and workout

6. Dinner out, television, movie, shopping, workout, sports, study

7. 2001 **8.** \$1500 **9.** \$5750 **10.** \$5583

11. $\left\{-\frac{2}{3}, 4\right\}$

12. $\{x|x > 3 \text{ and } x < 5\}$ or $(3, 5)$

13. $\{x|x \leq -1 \text{ or } x \geq 2\}$ or $(-\infty, -1] \cup [2, \infty)$

14. $\left\{x|x \geq -\frac{2}{3} \text{ and } x \leq 2\right\}$ or $\left[-\frac{2}{3}, 2\right]$

15. $\{-4, 5\}$

16. $\left\{x|x \leq -\frac{4}{5} \text{ or } x \geq \frac{8}{5}\right\}$ or $\left(-\infty, -\frac{4}{5}\right] \cup \left[\frac{8}{5}, \infty\right)$

17. $\frac{4}{3}$ **18.** -1 **19.** $\frac{3}{4}$ **20.** -5 **21.** $\frac{3}{2}$ **22.** 125 **23.** 32

24. -32 **25.** $\frac{1}{16}$ **26.** $\frac{1}{2}$ **27.** $\frac{1}{4}$ **28.** $\frac{3}{2}$ **29.** 8 **30.** 9

31. $\frac{1}{3}$ **32.** $x^{\frac{5}{3}}$ **33.** $6x^{\frac{17}{20}}$ **34.** $36a^{\frac{1}{6}}$ **35.** $27xy^2$ **36.** $5x^2y^3$

37. $13n^{\frac{7}{20}}$ **38.** $\frac{4}{n^{\frac{1}{4}}}$ **39.** $8x^3$ **40.** $\{(0, 2, 5)\}$

41. $\{(2, 1, 3)\}$ **42.** $1 + 2i$ **43.** $-7 - 5i$ **44.** $2 - 4i$

45. $3 - 4i$ **46.** $-1 + 3i$ **47.** $-6 + 10i$ **48.** $-34 + 31i$

49. $-2 - 11i$ **50.** $-4 - 8i$ **51.** $3 - 45i$ **52.** 85

53. 58 **54.** $55 - 48i$ **55.** $3 - 15i$ **56.** $\{6 - 5i, 6 + 5i\}$

57. $\{-1 - i\sqrt{6}, -1 + i\sqrt{6}\}$ **58.** $\{1 - 4i, 1 + 4i\}$

59. $\left\{\frac{1 - 3i\sqrt{3}}{2}, \frac{1 + 3i\sqrt{3}}{2}\right\}$

60. $\left\{\frac{1 - i\sqrt{23}}{4}, \frac{1 + i\sqrt{23}}{4}\right\}$ **61.** $\left\{\frac{1}{3}, \frac{3}{2}\right\}$

62. $\left\{\frac{5 - i\sqrt{3}}{2}, \frac{5 + i\sqrt{3}}{2}\right\}$

63. $\left\{\frac{-3 - i\sqrt{39}}{4}, \frac{-3 + i\sqrt{39}}{4}\right\}$

64. $\left\{\frac{-1 - i\sqrt{59}}{6}, \frac{-1 + i\sqrt{59}}{6}\right\}$

65. $\left\{\frac{-1 - i\sqrt{47}}{8}, \frac{-1 + i\sqrt{47}}{8}\right\}$

66. Domain: {red, blue, green}
Range: $\left\{\frac{1}{8}, \frac{1}{4}, \frac{5}{8}\right\}$
It is a function.

67. Domain: {3, 4, 5}
Range: {5, 7, 9}
It is a function.

68. Domain: {1, 2}
Range: {−16, −8, 8, 16}
It is not a function.

69. Domain: {2, 3, 4, 5}
Range: {10}
It is a function.

70. Domain: {−2, −1, 0, 1, 2}
Range: {0, 1, 4}
It is a function.

71. Domain: {1, 2, 3}
Range: {4, 8, 10, 15}
It is not a function.

72. All reals except 6 **73.** All reals

74. All reals **75.** All reals except −4

76. All reals except $\frac{1}{2}$ **77.** All reals except $-\frac{1}{3}$

78. $-14; -2; 13; 3a - 2$ **79.** $-\frac{3}{4}; \frac{3}{2}; -2; -3$

80. $\frac{3}{5}; 0; \frac{2}{5}; \frac{3}{7}$ **81.** $-6; -3; 2; 9$

82. **(a)** \$.80 **(b)** \$1.70

(c)

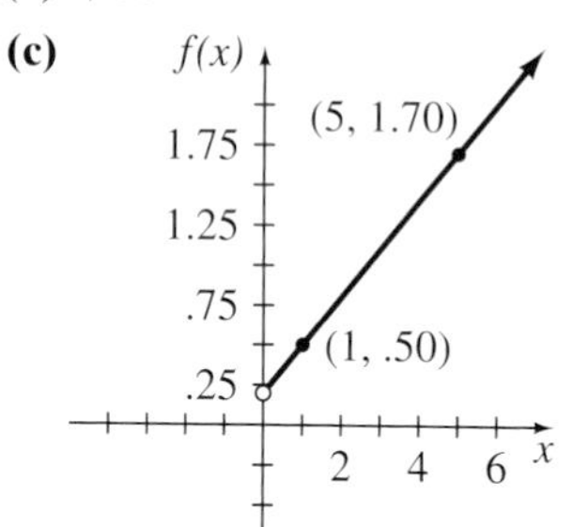

(d) \$1.40

83. **(a)** 22 gallons **(b)** 32 gallons

(c)

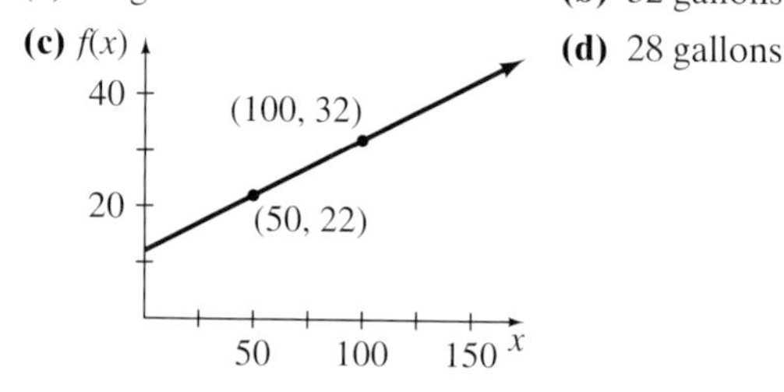

(d) 28 gallons

Chapter 11 Test (page 511)

1. French fries; salad **2.** 30 **3.** 60 **4.** 35

5. $11 - 13i$ **6.** $5i\sqrt{3}$ **7.** 216 **8.** $\frac{27}{8}$ **9.** $10x^{\frac{11}{12}}$

10. $5n^{\frac{1}{10}}$ **11.** $\{2 - 4i, 2 + 4i\}$

12. $\{1 - i\sqrt{2}, 1 + i\sqrt{2}\}$

13. $\{-3 - 2i\sqrt{3}, -3 + 2i\sqrt{3}\}$

14. $\left\{\frac{3 - i\sqrt{11}}{2}, \frac{3 + i\sqrt{11}}{2}\right\}$ **15.** $\{-4, 8\}$

16. $\left\{-\frac{7}{4}, -\frac{3}{4}\right\}$ **17.** $\varnothing$

18. $\left\{\frac{1 - i\sqrt{7}}{4}, \frac{1 + i\sqrt{7}}{4}\right\}$ **19.** $\left\{-4, \frac{7}{3}\right\}$

20. $\{x \mid x \le -5 \text{ or } x \ge -1\}$ or $(-\infty, -5] \cup [-1, \infty)$

21. $\{x \mid x > -3 \text{ and } x < 4\}$ or $(-3, 4)$

22. All real numbers except $\frac{7}{2}$ **23.** 4; −6; −21; 14

24. \$447 **25.** \$11,550; \$16,250

Appendix B Answers (page 516)

1. 16 **2.** 25 **3.** 5 **4.** $\frac{6}{5}$ **5.** 2 **6.** 33 and 34

7. −41 and −42 **8.** 22, 24, and 26 **9.** 47, 49, and 51

10. 3 **11.** −3 **12.** 19 and 20 **13.** 49, 50, and 51

14. 9 **15.** 32 **16.** 48 **17.** 72

18. 4 and 7 or $-\frac{7}{2}$ and -8

19. -2 and -12 or $\frac{12}{5}$ and 10

20. 5 and 6 or -5 and -6 **21.** 3 and 5

22. 18 and 39 **23.** 9 and 56

24. $\frac{5}{6}$ **25.** 4 **26.** $-\frac{1}{5}$ or 5 **27.** $-\frac{4}{3}$ or $\frac{3}{4}$ **28.** $\frac{1}{2}$ or 6

29. $-\frac{3}{2}$ or $\frac{2}{3}$ **30.** 36 and 42 **31.** 14 and 41 **32.** 17 and 85

33. 32 and 67 **34.** 40 and 96 **35.** $5 + \sqrt{3}; 5 - \sqrt{3}$

36. 8 and 9 **37.** 3 and 6 **38.** 20 years **39.** 23 years

40. Kaitlin is 12 years old and Nikki is 4 years old

41. Jesse is 18 years old and Annilee is 12 years old

42. Angie is 22 years old and her mother is 42 years old

43. All numbers greater than $\frac{13}{3}$

44. All numbers less than or equal to 8

45. All numbers less than 3 **46.** 92 or better

47. 100 or better **48.** 77 or less **49.** 166 or better

50. 10 centimeters **51.** 14 inches **52.** 35° and 70°

53. 62° and 118° **54.** 24° and 66° **55.** 60°

56. A = 32°, B = 58°, and C = 90°

57. 55° **58.** 7 inches by 15 inches

59. 14 centimeters by 21 centimeters

60. 7 yard, 12 yards, and 15 yards **61.** 12 inches

62. 5 centimeters **63.** 5 meters by 8 meters

64. Side is 9 inches and altitude is 4 inches

65. 9 inches by 12 inches

66. Width is 8 meters and length is 15 meters

67. 1 meter **68.** 8 inches by 14 inches

69. 9 units by 11 units

70. \$3000 at 5% and \$4500 at 6%

71. \$1000 at 8% and \$3000 at 9%

72. \$1000 **73.** \$800 at 7% and \$1200 at 8%

74. 40 shares at \$20 per share

75. 40 shares at \$15 per share

76. 12 lots at \$10,000 per lot **77.** 14 feet by 17 feet

78. 130 miles **79.** 300 miles **80.** \$4000 **81.** 25 pounds

82. $9\frac{3}{5}$ feet and $14\frac{2}{5}$ feet **83.** $87\frac{1}{2}$ miles

84. \$300 and \$450 **85.** 2750 females and 1650 males

86. 30% **87.** \$115 **88.** 75 **89.** 55 **90.** 90 **91.** \$52

92. \$60 **93.** 40% **94.** 35% **95.** \$8 **96.** \$36.25

97. \$27.50 **98.** \$1.60 per pound

99. 58 mph and 65 mph **100.** $3\frac{1}{2}$ hours **101.** 1 hour

102. $1\frac{3}{4}$ hours

103. Kaitlin at 9 miles per hour and Kent at 12 miles per hour

104. Simon walks 2 hours at 3 miles per hour and Dave walks $3\frac{1}{2}$ hours at 2 miles per hour

105. 12 miles per hour out and 8 miles per hour back or 16 mph out and 12 mph back

106. Walks at $2\frac{1}{2}$ miles per hour and jogs at 5 miles per hour

107. 55 miles per hour **108.** 20 miles per hour

109. Lorraine at 20 mph and Charlotte at 25 mph, or Lorraine at 45 mph and Charlotte at 50 mph

110. 4 liters **111.** $2\frac{2}{9}$ cups **112.** 25 milliliters

113. $6\frac{2}{3}$ quarts **114.** 10 gallons **115.** $43\frac{1}{3}$%

116. 15 quarts of 30% and 5 quarts of 70%

117. $3\frac{1}{5}$ quarts **118.** 12 minutes **119.** $37\frac{1}{2}$ minutes

120. A in 3 hours and B in 2 hours **121.** 60 minutes

122. 6 hours for Tom and 8 hours for Terry

123. 80 minutes

124. 5 pennies, 15 nickels, and 20 dimes

125. 14 nickels and 16 dimes

126. 48 pennies, 24 nickels, 32 dimes

127. 40 quarters, 20 dimes, 10 nickels

128. 30 pennies, 20 nickels, 45 dimes

129. 8 dimes and 15 quarters

130. 12 nickels, 25 dimes, 35 quarters

131. \$9.50 per hour **132.** \$10 **133.** \$30 per hour

134. \$3.69 **135.** \$65 **136.** 730 democrats

137. 360 girls **138.** \$23 per share **139.** \$12 per hour

140. \$10.20 per hour **141.** 5 rows and 13 trees per row

142. 6 rows and 8 desks per row

143. \$1.79 per pound for Gala apples and \$.99 per pound for Fuji apples

144. \$1.69 for tape

145. \$2.29 for Corn flakes and \$2.98 for Wheat flakes

146. 30 hours **147.** 25 students at \$4 each

148. 8 people **149.** 12 mugs at \$7 each **150.** 30 feet